GOVERNMENT SERIES

Energy: Nuclear

Advanced Reactor Concepts and Fuel Cycle Technologies, 2005 Energy Policy Act (P.L. 109-58), Light Water Reactors, Small Modular Reactors, Generation IV Nuclear Energy Systems, Nuclear Power 2010, Nuclear Power Plant Security, Nuclear Regulatory Commission, Radioactive Waste Storage and Disposal, Yucca Mountain

Compiled by TheCapitol.Net

Authors: John Grossenbacher, Carl E. Behrens, Carol Glover, Mark Holt, Marvin S. Fertel, Thomas B. Cochran, Dale E. Klein, Phillip Finck, Anthony Andrews, Fred Sissine, and Todd Garvey

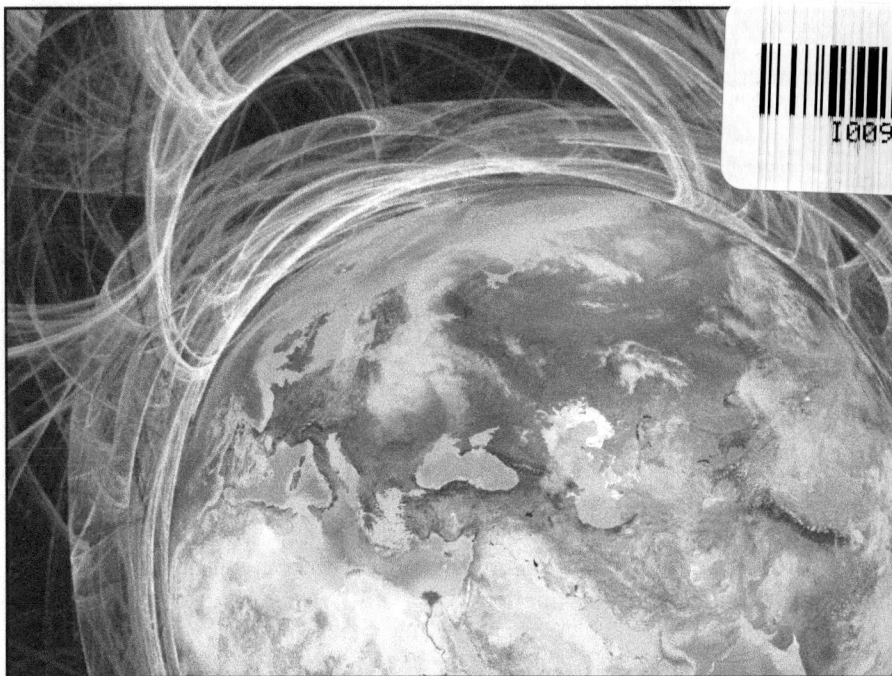

TheCapitol.Net

For over 30 years, TheCapitol.Net and its predecessor, Congressional Quarterly Executive Conferences, have been training professionals from government, military, business, and NGOs on the dynamics and operations of the legislative and executive branches and how to work with them.

Our training and publications include congressional operations, legislative and budget process, communication and advocacy, media and public relations, research, business etiquette, and more.

TheCapitol.Net is a non-partisan firm.

Our publications and courses, written and taught by *current* Washington insiders who are all independent subject matter experts, show how Washington works.™ Our products and services can be found on our web site at *<www.TheCapitol.Net>*.

Additional copies of *Energy: Nuclear* can be ordered online: *<www.GovernmentSeries.com>*.

Design and production by Zaccarine Design, Inc., Evanston, IL; 847-864-3994.

∞ The paper used in this publication exceeds the requirements of the American National Standard for Information Sciences—Permanence of Paper for Printed Library Materials, ANSI Z39.48-1992.

v 1

Energy: Nuclear, softbound:
ISBN: 158733-186-1
ISBN 13: 978-1-58733-186-2

Summary Table of Contents

iii

vi

Table of Contents

Chapter 7:

Most Recent Developments

Nuclear Power Status and Outlook

    Possible New Reactors

    Federal Support

        Nuclear Production Tax Credit

        Standby Support

        Loan Guarantees

Chapter 8:

xii

For Additional Reading

Figure 1. DOE Estimate of Future Liabilities for Nuclear Waste Delays

Table 1. DOE Civilian Spent Fuel Management Funding

Introduction

Energy: Nuclear

Advanced Reactor Concepts and Fuel Cycle Technologies, 2005 Energy Policy Act (P.L. 109-58), Light Water Reactors, Small Modular Reactors, Generation IV Nuclear Energy Systems, Nuclear Power 2010, Nuclear Power Plant Security, Nuclear Regulatory Commission, Radioactive Waste Storage and Disposal, Yucca Mountain

According to the U.S. Department of Energy's (DOE) Office of Nuclear Energy, nuclear energy provides about 20 percent of U.S. electricity through the operation of 104 nuclear reactors. Combined construction and operating license applications have been submitted for 28 new U.S. nuclear power plants, with eight more expected.

Nuclear power started coming online in significant amounts in the late 1960s. By 1975, in the midst of the oil crisis, nuclear power was supplying 9 percent of total electricity generation. Increases in capital costs, construction delays, and public opposition to nuclear power following the Three Mile Island accident in 1979 curtailed expansion of the technology, and many construction projects were canceled. Continuation of some construction increased the nuclear share of generation to 20 percent in 1990, where it remains currently.

Nuclear power is now receiving renewed interest, prompted by volatile fossil fuel prices, possible carbon dioxide controls, and new federal subsidies and incentives. The 2005 Energy Policy Act (P.L. 109-58) authorized streamlined licensing that combines construction and operating permits, and tax credits for production from advanced nuclear power facilities.

All U.S. nuclear plants are light water reactors (LWRs), which are cooled by ordinary water. DOE's nuclear energy research and development program includes advanced reactors, fuel cycle technology and facilities, and infrastructure support. DOE's Generation IV Nuclear Energy Systems Initiative is developing advanced reactor technologies that could be safer than LWRs and produce high-temperature heat to make hydrogen. The Nuclear Power 2010 program is a government-industry, 50-50 cost-shared initiative. It focuses on deploying Generation III+ advanced light-water reactor designs, and is managed by DOE's Office of Nuclear Energy.

Congress designated Yucca Mountain, NV as the nation's sole candidate site for a permanent high-level nuclear waste repository in 1987 amid much controversy. To date no nuclear waste has been transported to Yucca Mountain. In March 2010, the Secretary of Energy filed to withdraw its application for a nuclear-waste repository at Yucca Mountain.

Current law provides no alternative repository site to Yucca Mountain, and it does not authorize the DOE to open temporary storage facilities without a permanent repository in operation. Without congressional action, the default alternative to Yucca Mountain would be indefinite onsite storage of nuclear waste at reactor sites and other nuclear facilities. Private central storage facilities can also be licensed under current law. Such a facility has been licensed in Utah, but its operation has been blocked by the Department of the Interior.

Nuclear energy issues facing Congress include federal incentives for new commercial reactors, radioactive waste management policy, research and development priorities, power plant safety and regulation, nuclear weapons proliferation, and security against terrorist attacks.

Links to Internet resources are available on the book's web site at <www.TCNNuclear.com>.

DOE/NE-0088

THE HISTORY OF
Nuclear Energy

NUCLEAR ENERGY

U.S. Department of Energy
Office of Nuclear Energy - Science,
and Technology

The History of Nuclear Energy

Table of Contents

One the cover:
Albert Einstein (1879-1955)

U.S. Department of Energy
Office of Nuclear Energy, Science and Technology
Washington, D.C. 20585

The History Of Nuclear Energy

Although they are tiny, atoms have a large amount of energy holding their nuclei together. Certain *isotopes* of some elements can be split and will release part of their energy as heat. This splitting is called *fission.* The heat released in fission can be used to help generate electricity in powerplants.

Uranium-235 (U-235) is one of the isotopes that fissions easily. During fission, U-235 atoms absorb loose neutrons. This causes U-235 to become unstable and split into two light atoms called *fission products.*

The combined mass of the fission products is less than that of the original U-235. The reduction occurs because some of the matter changes into energy. The energy is released as heat. Two or three neutrons are released along with the heat. These neutrons may hit other atoms, causing more fission.

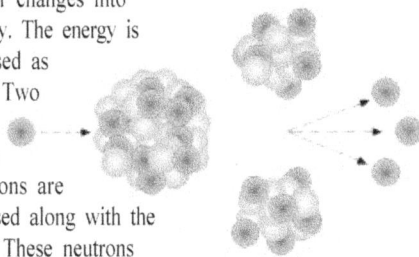

A series of fissions is called a *chain reaction.* If enough uranium is brought together under the right conditions, a continuous chain reaction occurs. This is called a *self-sustaining chain reaction.* A self-sustaining chain reaction creates a great deal of heat, which can be used to help generate electricity.

Nuclear powerplants generate electricity like any other steam-electric powerplant. Water is heated, and steam from the boiling water turns turbines and generates electricity. The main difference in the various types of steam-electric plants is the heat source. Heat from a self-sustaining chain reaction boils the water in a nuclear powerplant. Coal, oil, or gas is burned in other powerplants to heat the water.

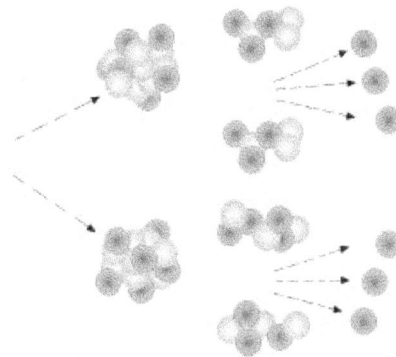

ii

iii

3

Preface

The concept of the atom has existed for many centuries. But we only recently began to understand the enormous power contained in the tiny mass.

In the years just before and during World War II, nuclear research focused mainly on the development of defense weapons. Later, scientists concentrated on peaceful applications of nuclear technology. An important use of nuclear energy is the generation of electricity. After years of research, scientists have successfully applied nuclear technology to many other scientific, medical, and industrial purposes.

This pamphlet traces the history of our discoveries about atoms. We begin with the ideas of the Greek philosophers. Then we follow the path to the early scientists who discovered radioactivity. Finally, we reach modern-day use of atoms as a valuable source of energy.

This pamphlet also includes a detailed chronology of the history of nuclear energy and a glossary. We hope the glossary will explain terms that may be new to some readers and that studying the chronology will encourage readers to explore the resources listed in the bibliography. By doing so, you can discover first-hand our nation's efforts to develop and control this powerful technology.

1

Introduction

It is human nature to test, to observe, and to dream. The history of nuclear energy is the story of a centuries-old dream becoming a reality.

Ancient Greek philosophers first developed the idea that all matter is composed of invisible particles called *atoms*. The word atom comes from the Greek word, *atomos*, meaning indivisible. Scientists in the 18th and 19th centuries revised the concept based on their experiments. By 1900, physicists knew the atom contains large quantities of energy. British physicist Ernest Rutherford was called the father of nuclear science because of his contribution to the theory of atomic structure. In 1904 he wrote:

If it were ever possible to control at will the rate of disintegration of the radio elements, an enormous amount of energy could be obtained from a small amount of matter.

Albert Einstein developed his theory of the relationship between mass and energy one year later. The mathematical formula is $E=mc^2$, or "energy equals mass times the speed of light squared." It took almost 35 years for someone to prove Einstein's theory.

3

The Discovery Of Fission

In 1934, physicist Enrico Fermi conducted experiments in Rome that showed neutrons could split many kinds of atoms. The results surprised even Fermi himself. When he bombarded uranium with neutrons, he did not get the elements he expected. The elements were much lighter than uranium.

Enrico Fermi, an Italian physicist, led the team of scientists who created the first self-sustaining nuclear chain reaction.

In the fall of 1938, German scientists Otto Hahn and Fritz Strassman fired neutrons from a source containing the elements *radium* and *beryllium* into uranium (atomic number 92). They were surprised to find lighter elements, such as *barium* (atomic number 56), in the leftover materials.

These elements had about half the atomic mass of uranium. In previous experiments, the leftover materials were only slightly lighter than uranium.

Hahn and Strassman contacted Lise Meitner in Copenhagen before publicizing their discovery. She was an Austrian colleague who had been forced to flee Nazi Germany. She worked with Niels Bohr and her nephew, Otto R. Frisch. Meitner and Frisch thought the barium and other light elements in the leftover material resulted from the uranium splitting — or fissioning. However, when she added the atomic masses of the fission products, they did not total the uranium's mass. Meitner used Einstein's theory to show the lost mass changed to energy. This proved fission occurred and confirmed Einstein's work.

The First Self-Sustaining Chain Reaction

In 1939, Bohr came to America. He shared with Einstein the Hahn-Strassman-Meitner discoveries. Bohr also met Fermi at a conference on theoretical physics in Washington, D.C. They discussed the exciting possibility of a self-sustaining chain reaction. In such a process, atoms could be split to release large amounts of energy.

Scientists throughout the world began to believe a self-sustaining chain reaction might be possible. It would happen if enough uranium could be brought together under proper conditions. The amount of uranium needed to make a self-sustaining chain reaction is called a *critical mass*.

4

5

Fermi and his associate, Leo Szilard, suggested a possible design for a uranium chain reactor in 1941. Their model consisted of uranium placed in a stack of graphite to make a cube-like frame of fissionable material.

Leo Szilard

Early in 1942, a group of scientists led by Fermi gathered at the University of Chicago to develop their theories. By November 1942, they were ready for construction to begin on the world's first nuclear reactor, which became known as *Chicago Pile-1*. The pile was erected on the floor of a squash court beneath the University of Chicago's athletic stadium. In addition to uranium and graphite, it contained control rods made of *cadmium*. Cadmium is a metallic element that absorbs neutrons. When the rods were in the pile, there were fewer neutrons to fission uranium atoms. This slowed the chain reaction. When the rods were pulled out, more neutrons were available to split atoms. The chain reaction sped up.

On the morning of December 2, 1942, the scientists were ready to begin a demonstration of Chicago Pile-1. Fermi ordered the control rods to be withdrawn a few inches at a time during the next several hours. Finally, at 3:25 p.m., Chicago time, the nuclear reaction became self-sustaining. Fermi and his group had successfully transformed scientific theory into technological reality. The world had entered the nuclear age.

The Development Of Nuclear Energy For Peaceful Applications

The first nuclear reactor was only the beginning. Most early atomic research focused on developing an effective weapon for use in World War II. The work was done under the code name *Manhattan Project*.

Lise Meitner and Otto R. Frisch

6

7

However, some scientists worked on making *breeder reactors*, which would produce fissionable material in the chain reaction. Therefore, they would create more fissionable material than they would use.

Enrico Fermi led a group of scientists in initiating the first self-sustaining nuclear chain reaction. The historic event, which occurred on December 2, 1942, in Chicago, is recreated in this painting.

After the war, the United States government encouraged the development of nuclear energy for peaceful civilian purposes. Congress created the Atomic Energy Commission (AEC) in 1946. The AEC authorized the construction of Experimental Breeder Reactor I at a site in Idaho. The reactor generated the first electricity from nuclear energy on December 20, 1951.

A major goal of nuclear research in the mid-1950s was to show that nuclear energy could produce electricity for commercial use. The first commercial electricity-generating plant powered by nuclear energy was located in Shippingport, Pennsylvania. It reached its full design power in 1957. Light-water reactors like Shippingport use ordinary water to cool the reactor core during the chain reaction. They were the best design then available for nuclear powerplants.

Private industry became more and more involved in developing light-water reactors after Shippingport became operational.

Federal nuclear energy programs shifted their focus to developing other reactor technologies.

The nuclear power industry in the U.S. grew rapidly in the 1960s. Utility companies saw this new form of electricity production as economical, environmentally clean, and safe. In the 1970s and 1980s, however, growth slowed. Demand for electricity decreased and concern grew over nuclear issues, such as reactor safety, waste disposal, and other environmental considerations.

Still, the U.S. had twice as many operating nuclear powerplants as any other country in 1991. This was more than one-fourth of the world's operating plants. Nuclear energy supplied almost 22 percent of the electricity produced in the U.S.

The Experimental Breeder Reactor I generated electricity to light four 200-watt bulbs on December 20, 1951. This milestone symbolized the beginning of the nuclear power industry.

At the end of 1991, 31 other countries also had nuclear powerplants in commercial operation or under construction. That is an impressive world-wide commitment to nuclear power technology.

During the 1990s, the U.S. faces several major energy issues and has developed several major goals for nuclear power, which are:

◆ To maintain exacting safety and design standards;

◆ To reduce economic risk;

◆ To reduce regulatory risk; and

◆ To establish an effective high-level nuclear waste disposal program.

Several of these nuclear power goals were ad-dressed in the Energy Policy Act of 1992, which was signed into law in October of that year.

The U.S. is working to achieve these goals in a number of ways. For instance, the U.S. Department of Energy has undertaken a number of joint efforts with the nuclear industry to develop the next generation of nuclear powerplants. These plants are being designed to be safer and more efficient. There is also an effort under way to make nuclear plants easier to build by standardizing the design and simplifying the licensing requirements, without lessening safety standards.

In the area of waste management, engineers are developing new methods and places to store the radioactive waste produced by nuclear plants and other nuclear processes. Their goal is to keep the waste away from the environment and people for very long periods of time.

Scientists are also studying the power of nuclear fusion. Fusion occurs when atoms join — or fuse — rather than split. Fusion is the energy that powers the sun. On earth, the most promising fusion fuel is deuterium, a form of hydrogen. It comes from water and is plentiful. It is also likely to create less radioactive waste than fission. However, scientists are still unable to produce useful amounts of power from fusion and are continuing their research.

In Oak Ridge, Tennessee, workers package isotopes, which are commonly used in science, industry, and medicine.

10

11

Research in other nuclear areas is also continuing in the 1990s. Nuclear technology plays an important role in medicine, industry, science, and food and agriculture, as well as power generation. For example, doctors use radioisotopes to identify and investigate the causes of disease.
They also use them to enhance traditional medical treatments. In industry, radioisotopes are used for measuring microscopic thicknesses, detecting irregularities in metal casings, and testing welds. Archaeologists use nuclear techniques to date prehistoric objects accurately and to locate structural defects in statues and buildings. Nuclear irradiation is used in preserving food. It causes less vitamin loss than canning, freezing, or drying.

Nuclear research has benefited mankind in many ways. But today, the nuclear industry faces huge, very complex issues. How can we minimize the risk? What do we do with the waste? The future will depend on advanced engineering, scientific research, and the involvement of an enlightened citizenry.

Chronology of Nuclear Research and Development

The '40s

1942 December 2. The first self-sustaining nuclear chain reaction occurs at the University of Chicago.

1945 July 16. The U.S. Army's Manhattan Engineer District (MED) tests the first atomic bomb at Alamogordo, New Mexico, under the code name *Manhattan Project.*

1945 August 6. The atomic bomb nicknamed *Little Boy* is dropped on Hiroshima, Japan. Three days later, another bomb, *Fat Man,* is dropped on Nagasaki, Japan. Japan surrenders on August 15, ending World War II.

1946 August 1. The Atomic Energy Act of 1946 creates the Atomic Energy Commission (AEC) to control nuclear energy development and explore peaceful uses of nuclear energy.

1947 October 6. The AEC first investigates the possibility of peaceful uses of atomic energy, issuing a report the following year.

1949 March 1. The AEC announces the selection of a site in Idaho for the National Reactor Testing Station.

The '50s

1951 December 20. In Arco, Idaho, Experimental Breeder Reactor I produces the first electric power from nuclear energy, lighting four light bulbs.

1952 June 14. Keel for the Navy's first nuclear submarine, *Nautilus,* is laid at Groton, Connecticut.

12

13

1953 March 30. *Nautilus* starts its nuclear power units for the first time.

1953 December 8. President Eisenhower delivers his "Atoms for Peace" speech before the United Nations. He calls for greater international cooper-aton in the development of atomic energy for peaceful purposes.

1954 August 30. President Eisenhower signs The Atomic Energy Act of 1954, the first major amendment of the original Atomic Energy Act, giving the civilian nuclear power program further access to nuclear technology.

1955 January 10. The AEC announces the Power Demonstration Reactor Program. Under the program, AEC and industry will cooperate in constructing and operating experimental nuclear power reactors.

1955 July 17. Arco, Idaho, population 1,000, becomes the first town powered by a nuclear powerplant, the experimental boiling water reactor BORAX III.

1955 August 8-20. Geneva, Switzerland, hosts the first United Nations International Conference on the Peaceful Uses of Atomic Energy.

1957 July 12. The first power from a civilian nuclear unit is generated by the Sodium Reactor Experiment at Santa Susana, California. The unit provided power until 1966.

1957 September 2. The Price-Anderson Act provides financial protection to the public and AEC licensees and contractors if a major accident occurs at a nuclear powerplant.

The Nautilus-the First Atomic-Powered Sub

14

1957 October 1. The United Nations creates the International Atomic Energy Agency (IAEA) in Vienna, Austria, to promote the peaceful use of nuclear energy and prevent the spread of nuclear weapons around the world.

1957 December 2. The world's first large-scale nuclear powerplant begins operation in Shippingport, Pennsylvania. The plant reaches full power three weeks later and supplies electricity to the Pittsburgh area.

1958 May 22. Construction begins on the world's first nuclear-powered merchant ship, the *N.S. Savannah*, in Camden, New Jersey. The ship is launched July 21, 1959.

1959 October 15. Dresden-1 Nuclear Power Station in Illinois, the first U.S. nuclear plant built entirely without government funding, achieves a self-sustaining nuclear reaction.

The '60s

1960 August 19. The third U.S. nuclear powerplant, Yankee Rowe Nuclear Power Station, achieves a self-sustaining nuclear reacton.

Early 1960s. Small nuclear-power generators are first used in remote areas to power weather stations and to light buoys for sea navigation.

NS Savannah

15

1961 November 22. The U.S. Navy commissions the world's largest ship, the *U.S.S. Enterprise*. It is a nuclear-powered aircraft carrier with the ability to operate at speeds up to 30 knots for distances up to 400,000 miles (740,800 kilometers) without refueling.

1964 August 26. President Lyndon B. Johnson signs the Private Ownership of Special Nuclear Materials Act, which allows the nuclear power industry to own the fuel for its units. After June 30, 1973, private ownership of the uranium fuel is mandatory.

1963 December 12. Jersey Central Power and Light Company announces its commitment for the Oyster Creek nuclear powerplant, the first time a nuclear plant is ordered as an economic alternative to a fossil-fuel plant.

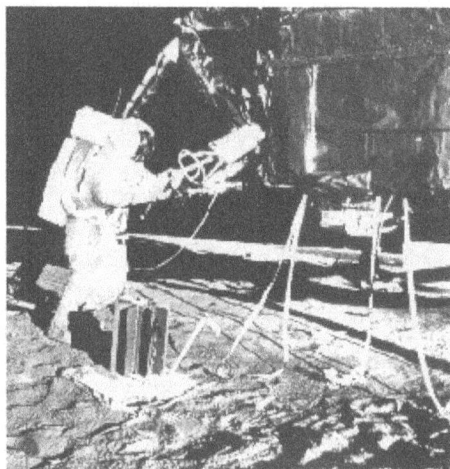

An atomic battery operated on the moon continuously for three years. Nuclear electric power arrived on the moon for the first time on November 19, 1969, when the Apollo 12 astronauts deployed the AEC's SNAP-27 nuclear generator on the lunar surface.

1964 October 3. Three nuclear-powered surface ships, the *Enterprise*, *Long Beach*, and *Bainbridge*, complete "Operation Sea Orbit," an around-the-world cruise.

1965 April 3. The first nuclear reactor in space (SNAP-10A) is launched by the United States. SNAP stands for Systems for Nuclear Auxiliary Power.

The '70s

1970 March 5. The United States, United Kingdom, Soviet Union, and 45 other nations ratify the Treaty for Non-Proliferation of Nuclear Weapons.

1971 Twenty-two commercial nuclear powerplants are in full operation in the United States. They produce 2.4 percent of U.S. electricity at this time.

1973 U.S. utilities order 41 nuclear powerplants, a one-year record.

1974 The first 1,000-megawatt-electric nuclear powerplant goes into service – Commonwealth Edison's Zion 1 Plant.

1974 October 11. The Energy Reorganization Act of 1974 divides AEC functions between two new agencies — the Energy Research and Development Administration (ERDA), to carry out research and development, and the Nuclear Regulatory Commission (NRC), to regulate nuclear power.

1977 April 7. President Jimmy Carter announces the United States will defer indefinitely plans for reprocessing spent nuclear fuel.

16

17

1977 August 4. President Carter signs the Department of Energy Organization Act, which transfers ERDA functions to the new Department of Energy (DOE).

1977 October 1. DOE begins operations.

1979 March 28. The worst accident in U.S. commercial reactor history occurs at the Three Mile Island nuclear power station near Harrisburg, Pennsylvania. The accident is caused by a loss of coolant from the reactor core due to a combination of mechanical malfunction and human error. No one is injured, and no overexposure to radiation results from the accident. Later in the year, the NRC imposes stricter reactor safety regulations and more rigid inspection procedures to improve the safety of reactor operations.

1979 Seventy-two licensed reactors generate 12 percent of the electricity produced commercially in the United States.

The '80s

1980 March 26. DOE initiates the Three Mile Island research and development program to develop technology for disassembling and de-fueling the damaged reactor. The program will continue for 10 years and make significant advances in developing new nuclear safety technology.

1982 October 1. After 25 years of service, the Shippingport Power Station is shut down. Decommissioning would be completed in 1989.

1983 January 7. The Nuclear Waste Policy Act (NWPA) establishes a program to site a repository for the disposal of high-level radioactive waste, including spent fuel from nuclear powerplants. It also establishes fees for owners and generators of radioactive waste and spent fuel, who pay the costs of the program.

1983 Nuclear power generates more electricity than natural gas.

1984 The atom overtakes hydropower to become the second largest source of electricity, after coal. Eighty-three nuclear power reactors provide about 14 percent of the electricity produced in the United States.

1985 The Institute of Nuclear Power Operations forms a national academy to accredit every nuclear powerplant's training program.

1986 The Perry Power Plant in Ohio becomes the 100th U.S. nuclear powerplant in operation.

1986 April 26. Operator error causes two explosions at the Chernobyl No. 4 nuclear powerplant in the former Soviet Union. The reactor has an inadequate containment building, and large amounts of radiation escape. A plant of such design would not be licensed in the United States.

1987 December 22. The Nuclear Waste Policy Act (NWPA) is amended. Congress directs DOE to study only the potential of the Yucca Mountain, Nevada, site for disposal of high-level radioactive waste.

1988 U.S. electricity demand is 50 percent higher than in 1973.

1989 One hundred and nine nuclear powerplants provide 19 percent of the electricity used in the U.S.; 46 units have entered service during the decade.

18

19

1989 April 18. The NRC proposes a plan for reactor design certification, early site permits, and combined construction and operating licenses.

The '90s

1990 March. DOE launches a joint initiative to improve operational safety practices at civilian nuclear powerplants in the former Soviet Union.

1990 America's 110 nuclear powerplants set a record for the amount of electricity generated, surpassing all fuel sources combined in 1956.

The Omaha Public District Fort Calhoun Nuclear Power Station located at Fort Calhoun, Nebraska

1990 April 19. The final shipment of damaged fuel from the Three Mile Island nuclear plant arrives at a DOE facility in Idaho for research and interim storage. This ends DOE's 10-year Three Mile Island research and development program.

1991 One hundred and eleven nuclear powerplants operate in the United States with a combined capacity of 99,673 megawatts. They produce almost 22 percent of the electricity generated commercially in the United States.

1992 One hundred and ten nuclear powerplants account for nearly 22 percent of all electricity used in the U.S.

1992 February 26. DOE signs a cooperative agreement with the nuclear industry to co-fund the development of standard designs for advanced light-water reactors.

1992 October 24. The Energy Policy Act of 1992 is signed into law. The Act makes several important changes in the licensing process for nuclear powerplants.

1992 December 2. The 50th anniversary of the historic Fermi experiment is observed worldwide.

1993 March 30. The U.S. nuclear utility consortium, the Advanced Reactor Corporation (ARC), signs a contract with Westinghouse Electric Corporation to perform engineering work for an advanced, standardized 600-megawatt pressurized-water reactor. Funding for this next-generation plant comes from ARC, Westinghouse, and DOE.

1993 September 6. The U.S. nuclear utility consortium, ARC, signs a contract with General Electric Company for cost-shared, detailed engineering of a standardized design for a large, advanced nuclear powerplant.
The engineering is being funded under a joint program among utilities, General Electric, and DOE.

20

21

Selected References

Cantelon, Philip, and Robert C. Williams. **Crisis Contained: The Department of Energy at Three Mile Island: A History**. Washington, D.C.: U.S. Department of Energy, 1980.

Cohen, Bernard L. **Before It's Too Late, A Scientist's Case for Nuclear Energy**. New York: Plenum Press, 1983. This 1981 recipient of the American Physical Society Bonner Prize for basic research in nuclear physics explains nuclear energy to the layman.

Edelson, Edward The Journalist's Guide to Nuclear Energy. Nuclear Energy Institute, 1994.

Glasstone, Samuel. **Sourcebook on Atomic Energy**. Princeton: D. Van Nostrand Company, 3rd ed., 1979. An encyclopedic compilation of useful atomic energy information.

Groves, Leslie R. **Now It Can Be Told, The Story of the Manhattan Project**. New York: Harper, 1975. The history of the Manhattan Engineering District's wartime project by the man who directed it.

23

Hewlett, Richard, and Oscar Anderson.
The New World, 1939-1946. Pennsylvania:
The Pennsylvania State University Press,
1990. Vol. I of the official history of the AEC
tells the story, from the vantage point of
unrestricted access to the records, of the
early efforts of scientists to understand the
nature of atomic fission, the control of such
fission in the exciting and successful
wartime atomic bomb project, and the
immediate postwar problems with the
control of atomic energy.

Hewlett, Richard, and Francis Duncan.
Atomic Shield, 1947-1952. Pennsylvania:
The Pennsylvania State University Press,
1990. Vol. II of the official history of the AEC
begins with the Commission's assumption
of responsibility for the Nation's atomic
energy program, and follows the course of
developments on both the national and
international scene to the end of the Truman
Administration and the first test of a
thermonuclear device.

Holl, Jack M., Roger M. Anders, Alice L. Buck,
and Prentice D. Dean. **United States
Civilian Nuclear Power Policy, 1954-1984**:
A History. Washington, D.C.: U.S.
Department of Energy, 1985.

Kruschke, Earl Roger and Byron M. Jackson.
**Nuclear Energy Policy: A Reference
Handbook**. Santa Barbara, Calif.: ABC-
CLIO, 1990. Designed to serve as both a
one-stop information source and a guide
to in-depth exploration.

Mazuzan, George, and J. Samuel Walker.
**Controlling the Atom: The Beginnings
of Nuclear Regulation, 1946-1962.**
University of California Press, 1985.
The first comprehensive study of the
early history of nuclear power regulation.

Rhodes, Richard
The Making of the Atomic Bomb.
Touchstone, 1988.

Rhodes, Richard
**Nuclear Renewal: Common Sense about
Energy.** Viking, 1993.

Smyth, Henry D.
Atomic Energy for Military Purposes.
Princeton: Princeton University Press,
1976. The classic account of the atomic
energy program in the United States,
published at the end of World War II.

24

25

Glossary

atom The smallest unit of an element. It is made up of electrons, protons, and neutrons. Protons and neutrons make up the atom's nucleus. Electrons orbit the nucleus.

breeder reactor A nuclear reactor that makes more fuel than it uses. It is designed so that one of the fission products of the U-235 used in fission is plutonium-239 (Pu-239). Pu-239 is also a fissionable isotope.

cadmium A soft, blue-white metal. The control rods in the first nuclear power reactor were made of cadmium because it absorbs neutrons.

chain reaction A continuous fissioning of atoms.

critical mass The amount of uranium needed to cause a self-sustaining chain reaction.

deuterium An isotope of hydrogen used in fusion.

fission The process in which the nucleus of an atom is split to produce heat.

fission products Light atoms that result from fission. The combined mass of fission products is less than that of the original whole atom because energy and neutrons are released.

fusion The process in which atoms are joined to produce energy.

27

isotope A form of an element that contains an unusual number of neutrons in its nucleus.

light-water reactor (LWR) The typical commercial nuclear power reactor. It uses ordinary water (light water) to produce steam. The steam turns turbines and generates electricity.

Manhattan Project The code name for production of the atomic bombs developed during World War II. The name comes from the Manhattan Engineering District, which ran the program.

radioisotope A radioactive isotope of an element.

radium-beryllium source A combination of the elements radium and beryllium. Radium is a rare, brilliant-white, luminescent, highly radioactive metal. Beryllium is a high-melting, lightweight, corrosion-resistant, steel-gray metal.

self-sustaining chain reaction A continuous chain reaction.

uranium A heavy, silver-white, radioactive metal.

uranium-235 (U-235) An isotope of uranium that is used as fuel in nuclear powerplants.

28

The Department of Energy produces
publications to fulfill a statutory mandate
to disseminate information to the public on
all energy sources and energy
conservation technologies. These materials
are for public use and do not purport to
present an exhaustive treatment of the
subject matter.

This is one in a series of publications on
nuclear energy.

U.S. Department of Energy

Printed on recycled paper

DOE/NE-0074

UNDERSTANDING
Radiation

NUCLEAR
ENERGY

U.S. Department of Energy
Office of Nuclear Energy, Science,
and Technology

Understanding
Radiation

Table of Contents

On the cover:
Marie Curie
(1867–1934)

U.S. Department of Energy
Office of Nuclear Energy,
Science, and Technology

Natural and Man-made Sources of Radiation

We are constantly exposed to radiation, whether cosmic radiation from outer space or radiation from radioactive elements in the earth's rocks and soil. These sources contribute to the *natural background radiation* that has always been around us. But there are also man-made sources of radiation, such as medical and dental x-rays, household smoke detectors, and materials released from nuclear and coal-fired power plants.

Cosmic radiation from outer space contributes to the level of natural background radiation.

1

Radiation: The Activity of Atoms

All matter in the universe is made of atoms, and radiation comes from the activity of these tiny particles. Atoms are made up of even smaller particles called *protons*, *neutrons*, and *electrons*. The arrangement of these particles distinguishes one atom from another.

Atoms of different types are known as *elements*. There are over 100 natural and man-made elements. Some of these elements, such as uranium, radium, and thorium, share a very important quality— they are unstable. As they change into more stable forms, they release invisible waves of energy or particles. This is called *ionizing radiation*. Radioactivity is the emitting of this radiation.

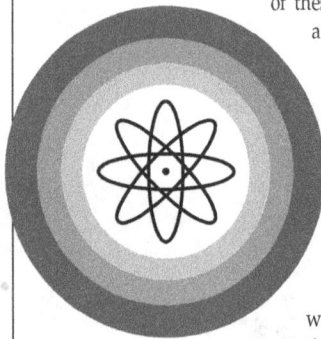

Radioactive atoms release an invisible energy called ionizing radiation.

Ionizing Radiation

Ionizing radiation is energy that can ionize, or electrically charge, an atom by stripping off electrons. Ionizing radiation can change the chemical composition of many things — including living tissue.

The three main types of ionizing radiation are alpha and beta particles and gamma rays:

* *Alpha* particles are the most energetic of the three types of ionizing radiation. But despite their energy, they can travel only a few inches in the air. Alpha particles lose their energy almost as soon as they collide with anything. A sheet of paper or your skin's surface can easily stop them.

* *Beta* particles are much smaller than alpha particles. They can travel in the air as much as 100,000 miles per second for a distance of about 10 feet. Beta particles can pass through a sheet of paper, but may be stopped by a thin sheet of aluminum foil or glass.

* *Gamma* rays, unlike alpha or beta particles, are waves of electromagnetic energy. Gamma rays travel at the speed of light (186,000 miles per second). Gamma radiation is very penetrating and is best shielded by a thick wall of concrete, lead, or steel.

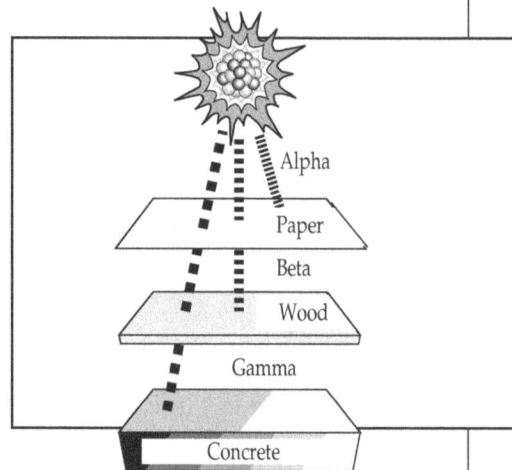

Alpha
Paper
Beta
Wood
Gamma
Concrete

Alpha particles can easily be stopped by a sheet of paper. Beta particles can pass through paper, but not aluminum foil or glass. Gamma rays are more powerful and can be stopped by thick concrete, lead or steel.

2

3

Radioactive "Half-Lives"

The radioactivity of a material decreases with time. The time that it takes a material to lose half of its original radioactivity is referred to as its *half-life*. For example, a quantity of iodine-131, a material that has a half-life of 8 days, will lose half of its radioactivity in that amount of time. In 8 more days, it will lose half of the remaining radioactivity, and so on. Eventually, the radioactivity will essentially disappear. Each radioactive element has a characteristic half-life. The half-lives of various radioactive elements may vary from millionths of a second to millions of years.

As a radioactive element gives up its radioactivity, it often changes to an entirely different element — one that may or may not be radioactive. Eventually, a stable element forms. This transformation may take place in several steps and is known as a *decay chain*. Radium, for example, is a naturally radioactive element with a half-life of 1,622 years. It emits an alpha particle and becomes radon, a radioactive gas with a half-life of only 3.8 days. Radon decays into polonium and, through a series of steps, into bismuth and ultimately to lead. Lead is a stable, non-radioactive element.

Radiation: Units of Measure

Scientists and engineers use a variety of units to measure radiation. These different units can be used to determine the amount, type, and intensity of radiation. Just as heat can be measured in terms of its intensity or its effect using units like degrees and calories, amounts of radiation can be measured in *curies*, *rems*, *millirems*, and *rads*.

The *curie*, named after the scientists Marie and Pierre Curie, describes the intensity of a sample of radioactive material in terms of atoms of the material that decay each second. This rate — 37 billion atoms per second for one gram of radium — is the basis of this measurement.

A *rem* is a measurement of the effects of radiation on the body, much as degrees Celsius are measurements of the effects of sunlight heating sand on a beach.

The unit used most often to measure the radiation exposure for a person is the *millirem* (mrem). It is one-thousandth of a rem. The millirem is used because usually very small amounts of radiation are being measured.

A *rad* is the unit of measure for the physical absorption of radiation. Much like sunlight heats pavement by giving up energy to it, radiation gives up rads of energy to objects in its path. The international units of absorbed dose are sieverts and grays.

Metric Prefixes Used With Radiation Units:

milli- $1/1,000$ (thousandth)

micro- $1/1,000,000$ (millionth)

nano- $1/1,000,000,000$ (billionth)

pico- $1/1,000,000,000,000$ (trillionth)

4

5

Amounts We Receive

Most Americans receive about 360 mrem per year from all sources of radiation, including radon and medical exposure. About 40 mrem per year comes from the natural radioactivity in our own bodies. Most of our natural background radiation comes from cosmic radiation from outer space and from radioactive materials in the earth's rocks and soil.

Radon is a naturally occurring radioactive gas that results from the decay of radium. Radon is the single largest source of radiation exposure. On average, radon accounts for 200 mrem of our exposure each year.

The actual amount of background radiation depends on the location, elevation, rock and soil content, and weather conditions. For example, a person living on the Atlantic coast receives about 65 mrem of natural background radiation per year, while a person in Denver, Colorado, receives about 125 mrem, excluding radon. The difference in the natural background radiation is due in large part to Denver's higher elevation. The higher the elevation, the thinner the atmosphere, meaning that the atmosphere filters out less cosmic radiation. Naturally radioactive minerals in the rocks and soil also add to the higher level of background radiation.

We are also exposed to man-made sources of radiation, principally dental and medical x-rays, medical tests, and radiotherapy used in treating disease. About 15%, or about 50 mrem, of the radiation exposure that the average American receives comes from medical sources.

Tools To Measure Radiation

Ionizing radiation cannot be detected by our senses. However, we can measure and monitor it with simple scientific instruments. Three commonly used devices are dosimeters, film badges, and Geiger counters.

People who routinely work with or around radiation sources wear dosimeters or film badges. These include workers in medicine, research, industry, and nuclear energy. Thermoluminescent dosimeters contain a special semiconductor that becomes energized when it is exposed to radiation. Film badges contain a piece of film that is sensitive to radiation. These badges are read regularly to determine a worker's total exposure to radiation over a period of time.

Geiger counters measure radiation levels at a given time.

6

7

To measure radiation levels at any given time, a Geiger counter is used. A Geiger counter contains a special gas-filled tube that separates two electrodes. When radiation passes through the tube, it interacts with the gas, causing an electrical pulse that can be measured on a meter or by audible clicks. The number of pulses in a given time is a measure of the intensity of radiation.

The Health Effects of Radiation

Radiation is one of the most widely studied of all natural phenomena. Scientists understand the health effects of high levels of radiation. However, the effects of low levels of radiation are more difficult to determine because the major effect is a very slight increase in cancer risk. But because so many other factors also increase the risk of cancer, it is difficult to know which is the cause in many cases.

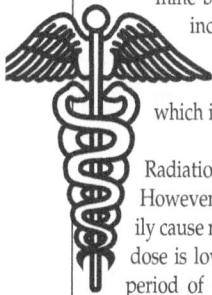

Radiation can chemically change living cells. However, cell transformation does not necessarily cause noticeable health effects. If the radiation dose is low, or a person receives it over a long period of time, the body can usually repair or replace the damaged cells without any detectable health effects.

Exposure to high levels of radiation can cause serious health effects, including burns, cell damage, and death. The degree of the effect depends on the intensity of the dose, the length of the exposure, and the type of body cells exposed. Sudden large doses exceeding 100,000 mrem can cause radiation sickness, with short-term symptoms including nausea, vomiting, extreme tiredness, and hair loss. Long-term effects include increased cancer risks. A dose of over 500,000 mrem at one time is usually fatal unless medical treatment is available.

Radiation Limits

Government regulations limit the amount of man-made radiation that the public may receive. Most people in the United States receive radiation doses of about 25,000 mrem over their entire lifetimes. Most scientists believe that radiation absorbed in small doses over several years is not harmful. However, U.S. Government regulations assume that the effects of all radiation exposures are cumulative and should be limited as much as reasonably possible.

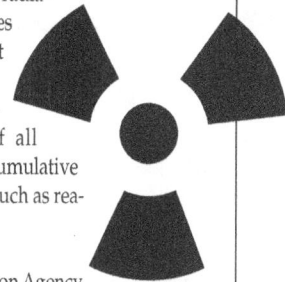

The Environmental Protection Agency limits the amount of radiation the general public may receive from the nuclear fuel cycle. This amount is limited to an annual exposure of 25 mrem in addition to the natural background and medical radiation normally received. For people working in an occupation that involves radiation, regulations forbid exposures above 5,000 mrem in any one year.

Nuclear Power Plants

Nuclear power plants are fueled by uranium. Uranium is used because, under precise conditions, its atoms can be split apart to release large amounts of energy in the form of heat. This process is known as *nuclear fission*. In a nuclear reactor, uranium atoms are split apart in a controlled, continuous process called a *chain reaction*. The heat from the chain reaction inside the reactor is used in a power plant to turn water into steam, which drives a turbine-generator to produce electricity.

8

9

Fort Calhoun Plant (Omaha Public Power District) is shown

The nuclear fuel inside an operating nuclear reactor is highly radioactive. For this reason, these radioactive materials are sealed inside ceramic, encased in metal, and held inside a heavy steel reactor vessel during operation. Nuclear power plants also have multiple protective barriers, including a massive concrete-and-steel containment building. These barriers prevent the radioactive materials inside the reactor from entering the environment in the event of an accident.

Radiation limits for nuclear power plants have been established by the Nuclear Regulatory Commission, an independent Government agency. To be licensed for commercial operation, nuclear power plants must limit the maximum annual radiation exposure at the plant's boundary to no more than 10 mrem above the natural background radiation level. In practice, however, nuclear power plants release only a tiny fraction of the amount permitted by regulations.

Nuclear power plants have generated electricity commercially for over three decades. Nuclear power plant safety compares favorably with other methods of producing electricity.

Nuclear Power Plant Safety Barriers

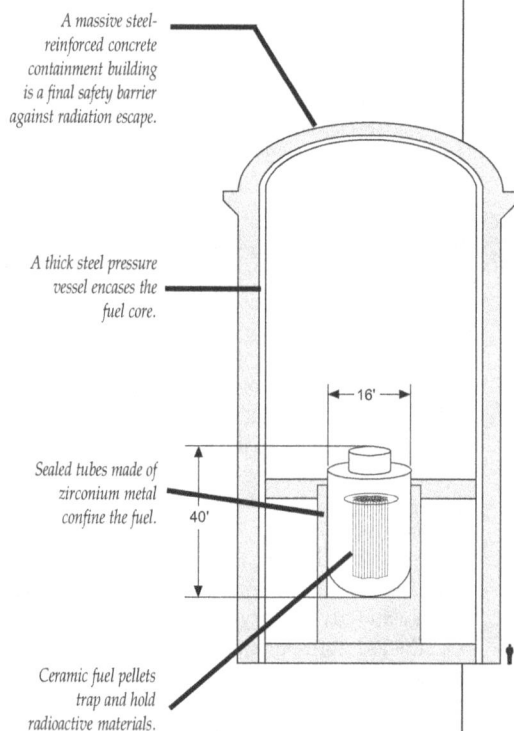

A massive steel-reinforced concrete containment building is a final safety barrier against radiation escape.

A thick steel pressure vessel encases the fuel core.

Sealed tubes made of zirconium metal confine the fuel.

Ceramic fuel pellets trap and hold radioactive materials.

16'

40'

10

11

Using Radiation

Radiation is one of the many natural energy forces that has been used to benefit humanity. With it, major contributions have been made possible in the fields of medicine and industry. However, its use requires special precautions, such as protective barriers and careful transportation regulations.

Medical and dental x-rays have been used for over 50 years to diagnose broken bones and tooth decay. Carefully focused radiation can destroy cancer cells without causing major damage to healthy cells nearby. Radioisotopes and computer imaging devices allow doctors to examine internal organs that are not normally visible by x-rays.

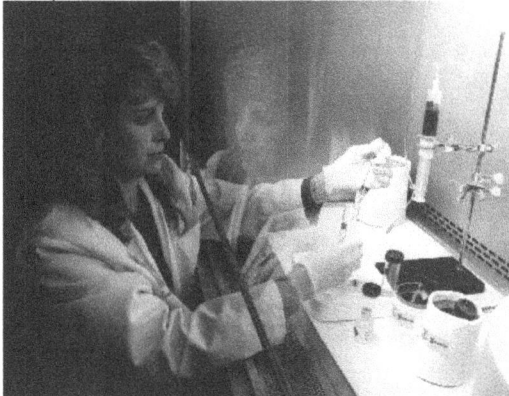

Shannon Seals, a nuclear pharmacist at Syncor International Corporation, prepares a radiopharmaceutical. Seals, who manages a facility, says demand is rapidly growing for radiopharmacists.

It is now standard practice to use radiation to sterilize medical products such as syringes. Radiation may also be preferable to heat for sterilizing bandages and ointments, which can be damaged by high temperatures.

In industry, radiography is used in much the same way as doctors use x-rays. This technique is used to locate defects in metal casings and welds that might not show up otherwise, and to determine microscopic thicknesses of materials such as metal foils. Radiography can also be used to locate structural defects in statues and buildings.

Radiation helps prevent certain foods from spoiling.

Radiation has applications in a variety of other fields. We use it to test the authenticity of art and date prehistoric objects accurately. We also use it to prevent certain foods from spoiling without significantly reducing the nutritional value, or making the food radioactive.

12

13

Courtesy of The Frank H. McClung Museum, The University of Tennessee, Knoxville.

Archaeologists have found successive layers of Indian occupation buried in the river bottoms of Tennessee. Within these layers are abundant fragments of carbonized wood and hickory nut shell. By radiocarbon dating the fragments of charcoal in each layer, historians can establish a chronnogy of settlement. The bottom of this excavation dates to 7500 B.C..

Radiation: A Natural Energy Force

Radiation has been a part of our environment since the earth was formed. Yet, we have only learned of its existence in the last century. Through research and education, we have learned about its effects and potential hazards. We understand the need to monitor and control the amounts of radiation we receive. We also understand that radiation can be beneficial and improve our quality of life.

14

The Department of Energy produces publications to fulfill a statutory mandate to disseminate information to the public on all energy sources and energy conservation technologies. These materials are for public use and do not purport to present an exhaustive treatment of the subject matter.

This is one in a series of publications on nuclear energy.

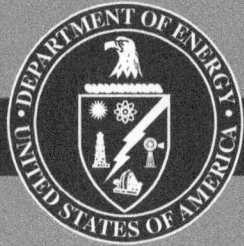

U.S. Department of Energy

Printed on recycled paper

NUCLEAR ENERGY — AN OVERVIEW

The U.S. Department of Energy's Office of Nuclear Energy

T he demand for energy in the United States is rising. By the year 2030, domestic demand for electrical energy is expected to grow to levels of 16 to 36 percent higher than 2007 levels.

A plentiful, reliable, and affordable supply of energy is the cornerstone of sustained economic growth and prosperity.

Nuclear energy today provides about 20 percent of U.S. electricity, and 70 percent of its carbon-free electricity. It does not produce greenhouse gases, and so does not contribute to climate change. Nuclear energy produces large quantities of continuous, affordable electricity.

Today in the United States, 104 nuclear reactors provide carbon-free electricity to help drive the American economy.

Globally, nuclear energy is undergoing renewed growth, with 13 countries constructing 53 new nuclear power units and 27 countries in the planning stages for an additional 142 units. In the United States, a renewed interest in nuclear energy has resulted in blueprints for the first new nuclear power plants in over 30 years. Combined Construction and Operating license applications have been submitted for 28 new U.S. nuclear power plants, with 8 more expected.

Despite the advantages of nuclear energy, the question of how to deal safely and securely with nuclear waste over the long term remains a concern. The Department of Energy's (DOE) Office of Nuclear Energy (NE) is sponsoring the research and technology development that will allow nuclear energy to continue to deliver large quantities of safe, reliable electricity to the marketplace, while developing options to address waste disposal and non-proliferation concerns.

A RECORD OF DISTINCTION

Over the past 15 years, consolidation of plant ownership to a smaller number of excellent operators has made the operation of U.S. plants:

- Safer;
- More cost-effective; and
- More reliable.

Efficiency improvements and power uprates have allowed existing U.S. nuclear plants to produce more energy than in previous decades, adding the equivalent of nearly 5 to 6 new nuclear reactors to the electrical grid. U.S. nuclear plants, which were available to produce energy only 70 percent of the time on average

www.nuclear.energy.gov
February 2010

NUCLEAR ENERGY — AN OVERVIEW

The U.S. Department of Energy's Office of Nuclear Energy

Program Budget

Nuclear Energy
($ in Millions)

	FY 2010 Actual	FY 2011 Request
	$870.0	$912.3

in the early 1990s, are now producing power around 92 percent of the time. Nuclear power plants do not release air pollutants or carbon dioxide in the production of electricity, providing an important option for improving air and environmental quality.

As a result of this success, essentially all U.S. nuclear plants are expected to apply for renewed licenses that will keep most plants in operation into the middle of the century.

DOE'S ROLE

The role of DOE is to work with the private sector, overseas partners, and other agencies to assure that the benefits of nuclear technology continue to contribute to the security and quality of life for Americans — and other citizens of the world — now and into the future. By focusing on the development of advanced nuclear technologies, NE supports the Administration's goals of providing domestic sources of secure energy, reducing greenhouse gases, and enhancing national security.

NE is focused on five goals:

- Extend the life, improve the performance, and maintain the safety of the current fleet of nuclear power plants.
- Enable new plant builds for electricity production and improve the affordability of nuclear energy.
- Enable the transition away from fossil fuels by producing process heat for use in the transportation and industrial sectors. Process heat — now produced by greenhouse gas-emitting fossil fuels — is required to refine oil into gasoline and to produce glass, plastics, steel, and many other materials.
- Enable sustainable fuel cycles. This includes research to make used nuclear fuel less toxic, recycle it, and create widely acceptable solutions to the challenges of nuclear waste.
- Understand and minimize proliferation risk.

NE serves present and future U.S. energy needs by providing the critical nuclear research infrastructure that will help regain U.S. technology leadership and train tomorrow's workforce. These capabilities and technologies will help meet the needs of a growing economy and address climate change by reducing greenhouse gas emissions.

The benefits of nuclear power as a safe, low-carbon, reliable, and secure source of energy make it an essential element in the Nation's energy and environmental future.

www.nuclear.energy.gov
February 2010

NUCLEAR ENERGY ENABLING TECHNOLOGIES

The U.S. Department of Energy's Office of Nuclear Energy

The Nuclear Energy Enabling Technologies (NEET) program will focus on innovative research relevant to multiple reactor and fuel cycle concepts that offer the promise of dramatically improved performance.

The new Nuclear Energy Enabling Technologies (NEET) program proposed in FY 2011 will develop crosscutting technologies that directly support and complement the Office of Nuclear Energy's (NE) development of new and advanced reactor concepts and fuel cycle technologies. It will encourage the development of transformative, "outside-the-box" solutions across the full range of nuclear energy technology issues.

BENEFITS OF THE INITIATIVE

Pursuing crosscutting and transformative nuclear technologies and capabilities for incorporation into advanced reactor and fuel cycle concepts offers the promise of revolutionary improvements in safety, performance, reliability, economics, and proliferation risk reduction. It promotes creative solutions to the broad array of nuclear energy problems related to reactor and fuel cycle development. The activities undertaken in this program complement those within the Reactor Concepts RD&D and Fuel Cycle R&D programs by providing a mechanism for pursuing broadly applicable R&D in areas that may ultimately benefit specific reactor and fuel cycle technology development.

PROGRAM ELEMENTS

The NEET program consists of three elements:

- Crosscutting Technology Development;
- Transformative Nuclear Energy Concepts R&D; and the
- Energy Innovation Hub for Modeling and Simulation

The Crosscutting Technology Development —

activity provides R&D support for the various nuclear energy concepts (existing and future) in areas such as reactor materials, advanced methods for manufacturing and field installation, new sensor technologies for monitoring material and equipment conditions in existing reactors, and creative approaches to further reduce proliferation risks.

The Transformative Nuclear Concepts R&D —

will support, via an open, competitive solicitation process, investigator-initiated projects that relate to any aspect of nuclear energy generation — reactor and power conversion technologies, enrichment, fuels and fuel management, waste disposal, nonproliferation, and so forth — ensuring that good ideas have sufficient outlet for exploration. The research on

www.nuclear.energy.gov
February 2010

NUCLEAR ENERGY ENABLING TECHNOLOGIES

The U.S. Department of Energy's Office of Nuclear Energy

Program Budget

Nuclear Energy Enabling Technologies
($ in Millions)

Crosscutting Technology Development

FY 2010 Actual	FY 2011 Request
$0.0	$43.3

Transformative Nuclear Concepts R&D

FY 2010 Actual	FY 2011 Request
$0.0	$28.9

Energy Innovation Hub for Modeling and Simulation

FY 2010 Actual *	FY 2011 Request
$0.0	$24.3

Small Business Innovation Research (SBIR)/ Small Business Technology Transfer Program (STTR)

FY 2010 Actual	FY 2011 Request
$0.0	$2.8

Total, Nuclear Energy Enabling Technologies

FY 2010 Actual	FY 2011 Request
$0.0	$99.3

* In FY 2010, $21.4 million was included in the Generation IV budget

transformative nuclear concepts will pursue non-traditional nuclear energy ideas that offer the potential for improved system performance and may radically alter nuclear system configuration and development needs. This could include the development of specialized nuclear fuels, revolutionary materials, new enrichment techniques, tailored coolants, new techniques for energy conversion, or other innovations.

The Energy Innovation Hub for Modeling and Simulation —

will apply existing modeling and simulation capabilities to create a "virtual" reactor user environment for engineers to simulate a currently operating reactor. A separate fact sheet is available for the Energy Innovation Hub for Modeling and Simulation.

PLANNED PROGRAM ACCOMPLISHMENTS[a]

FY 2011

- Evaluate and prioritize innovative structural materials for use in radiation environments and high temperature applications.
- Consider approaches such as the use of ion beams to simulate accelerated aging of materials.
- Develop a detailed project plan for quantification of proliferation risk and initiate studies on current risk assessment methodologies.
- Complete the advanced manufacturing and field installation technology study and research roadmap.
- Initiate competitively selected high-potential R&D activities that improve nuclear plant manufacturing and field installation efficiency.
- Perform research to develop advanced sensors to improve physical measurement accuracy and reduce uncertainty.
- Perform research on digital monitoring and control technology, fiber optic and wireless digital instruments and highly integrated control systems to improve performance and reliability.
- Solicit, competitively select and initiate R&D project awards from national laboratories, universities, research institutions, and industry proposals.

[a] See separate Fact sheet for FY 2011 accomplishments associated with the Energy Innovation Hub for Modeling

www.nuclear.energy.gov
February 2010

Nuclear Energy

John Grossenbacher
Director, Idaho National Laboratory

October 1, 2009

INL— The National Nuclear Laboratory

Developing world-class Nuclear Energy capabilities

Fostering education, research, industry, government and international collaborations to produce the needed investment, programs and expertise

Preeminent Internationally-Recognized Nuclear Energy RDD&D Laboratory

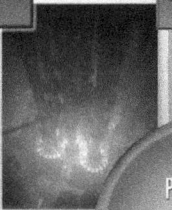

Major center for National and Homeland Security technology RDD&D

Lead clean energy systems RDD&D laboratory and a regional resource

Research - Development - Demonstration - Deployment

2

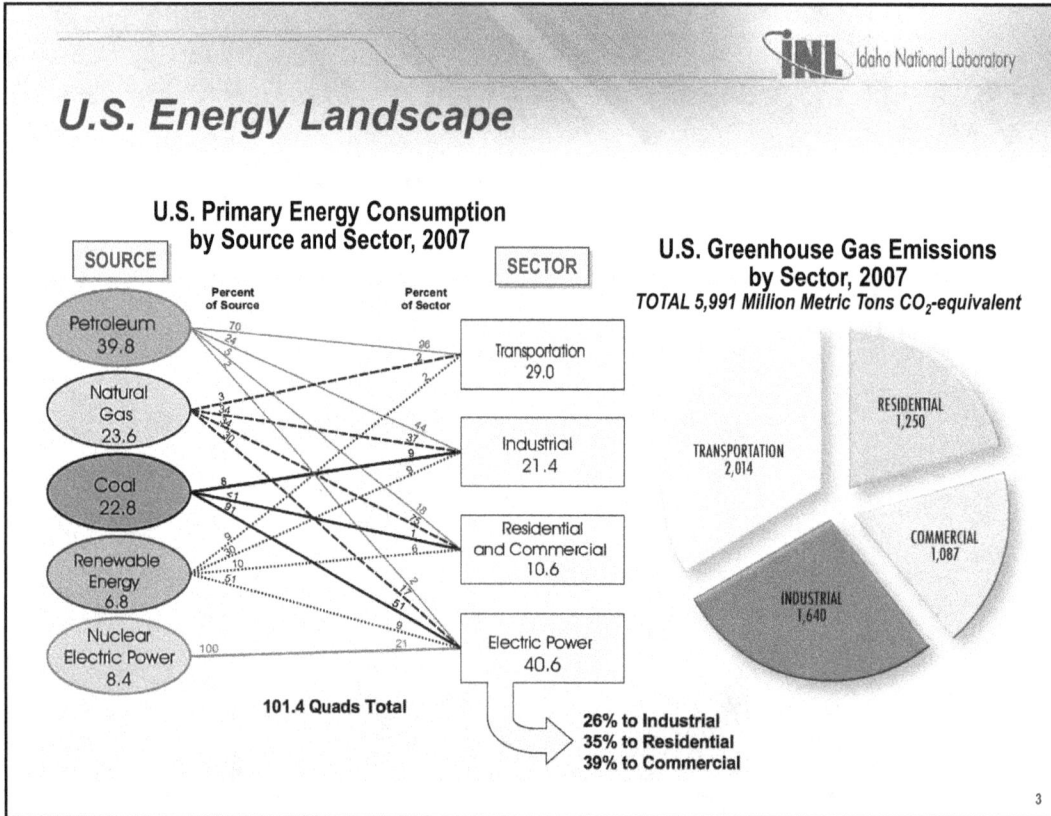

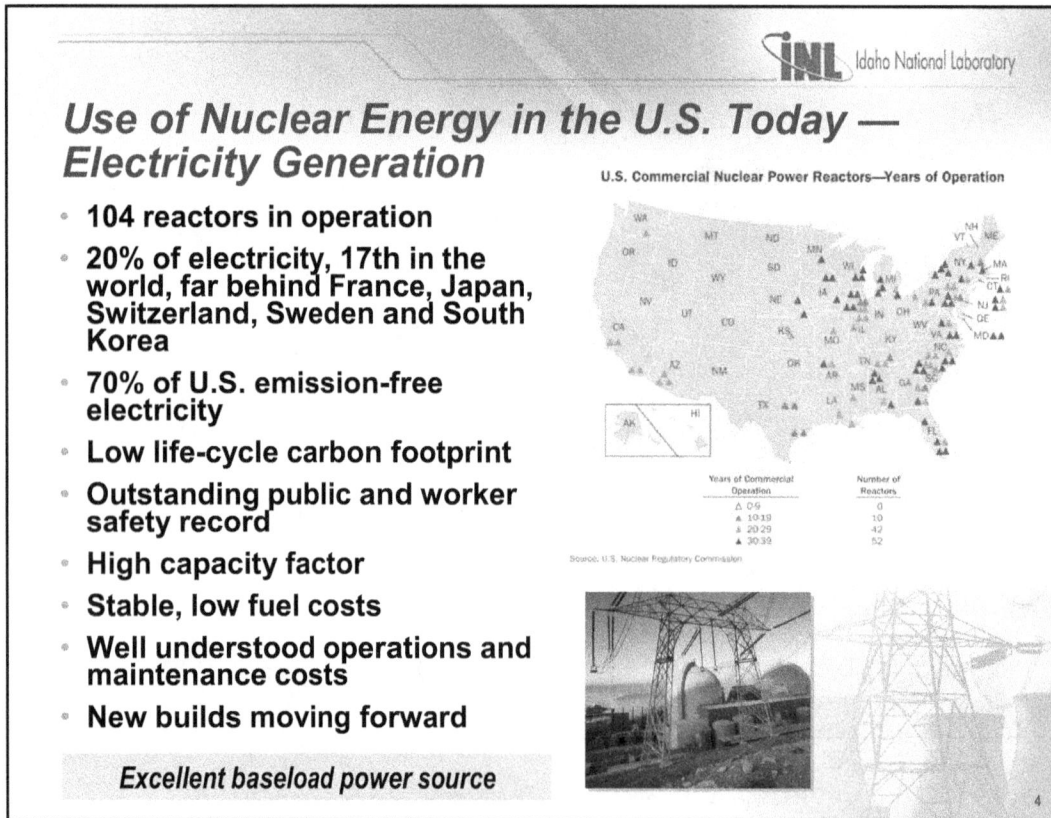

Nuclear Energy Opportunity

- Increased share of carbon free electricity generation
- Extension of the environmental and energy security benefits of nuclear energy to the transportation and industrial sectors

Nuclear Energy Assessment

Cost

- Large capital costs
- Fuel cost low and not volatile
- Operating and management cost understood
- Used fuel disposition uncertain
- Subsidies — Government's role in technology deployment

Construction

- Supply chain choke points — Heavy components and nuclear equipment manufacturing
- Time to license and construct

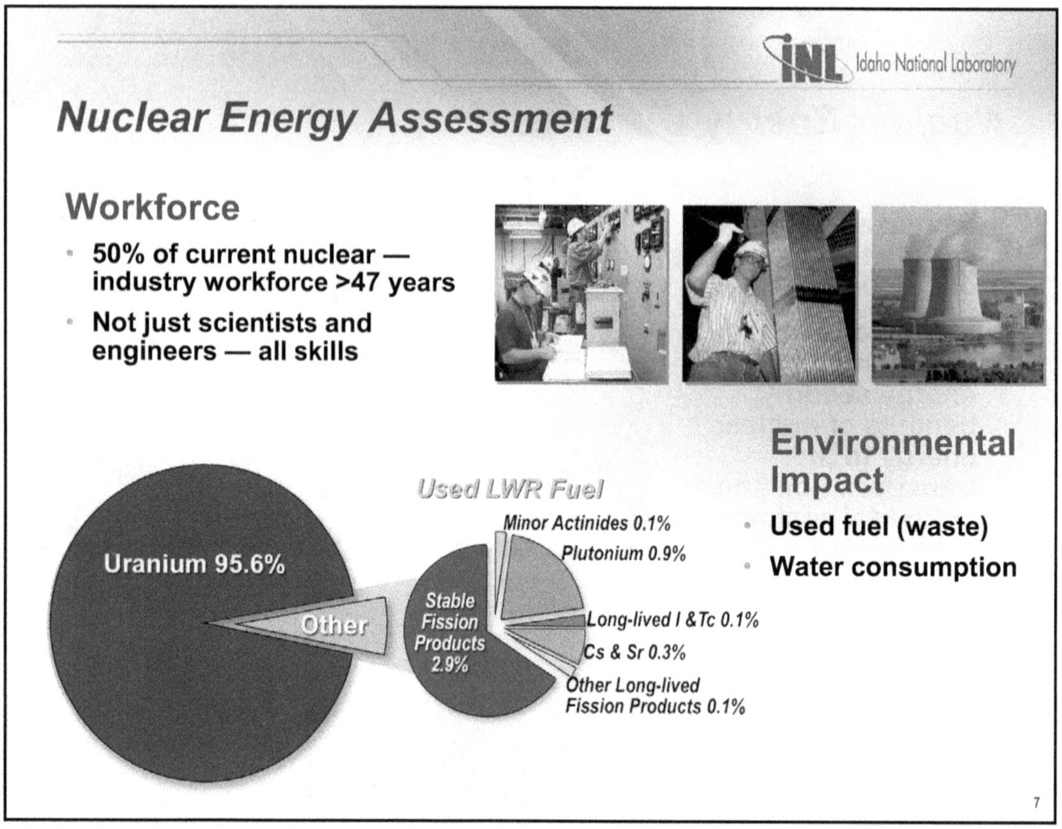
Idaho National Laboratory
Nuclear Energy Assessment
Workforce
50% of current nuclear — industry workforce >47 years
Not just scientists and engineers — all skills
Environmental Impact
Used fuel (waste)
Water consumption
Used LWR Fuel
Uranium 95.6%
Other
Stable Fission Products 2.9%
Minor Actinides 0.1%
Plutonium 0.9%
Long-lived I &Tc 0.1%
Cs & Sr 0.3%
Other Long-lived Fission Products 0.1%
7

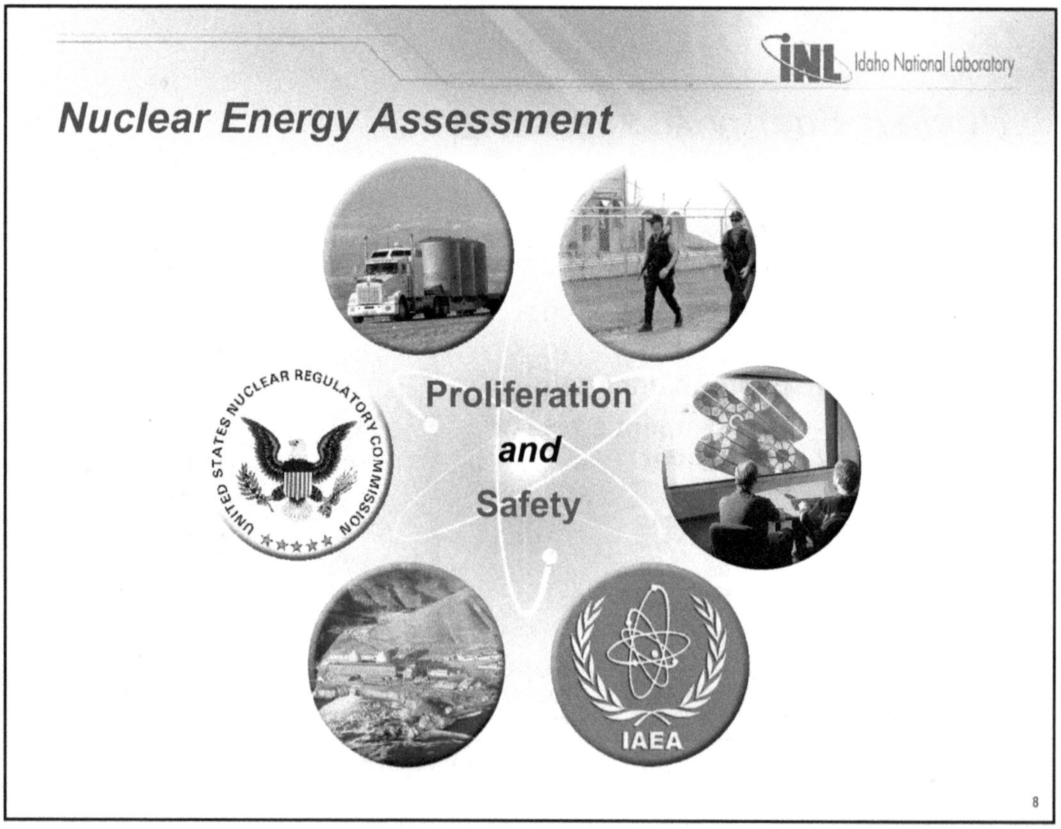
Idaho National Laboratory
Nuclear Energy Assessment
Proliferation
and
Safety
UNITED STATES NUCLEAR REGULATORY COMMISSION
IAEA
8

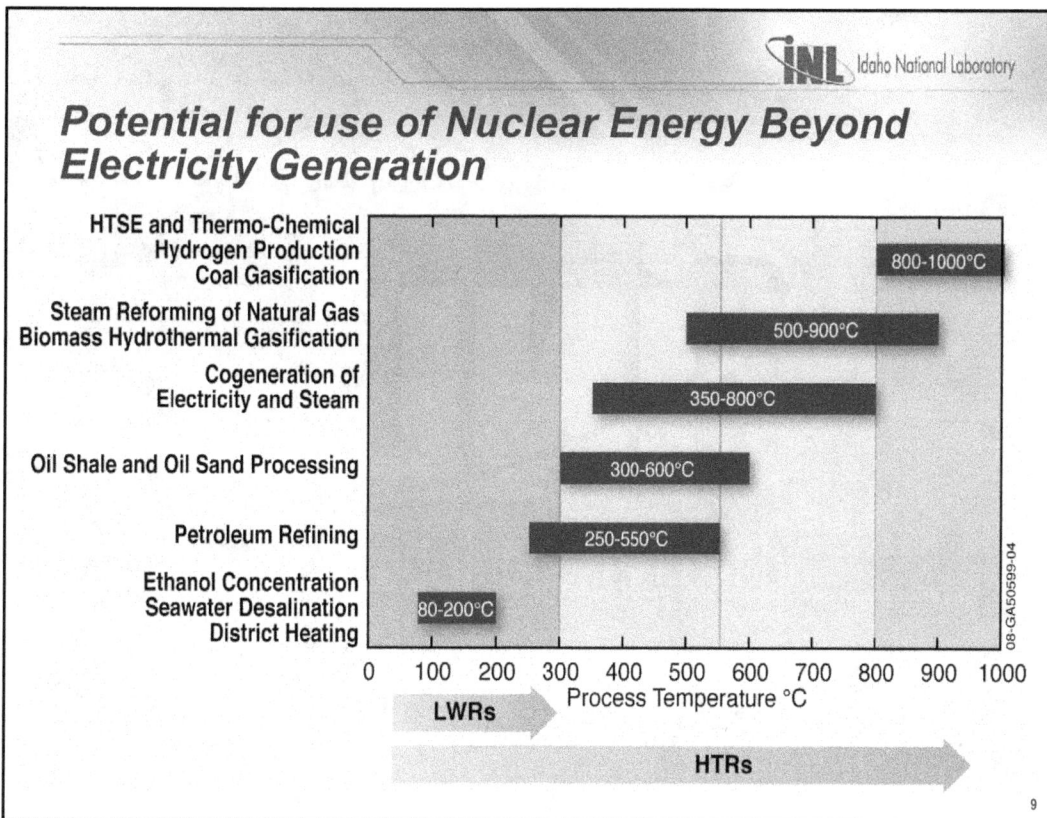

Potential for use of Nuclear Energy Beyond Electricity Generation

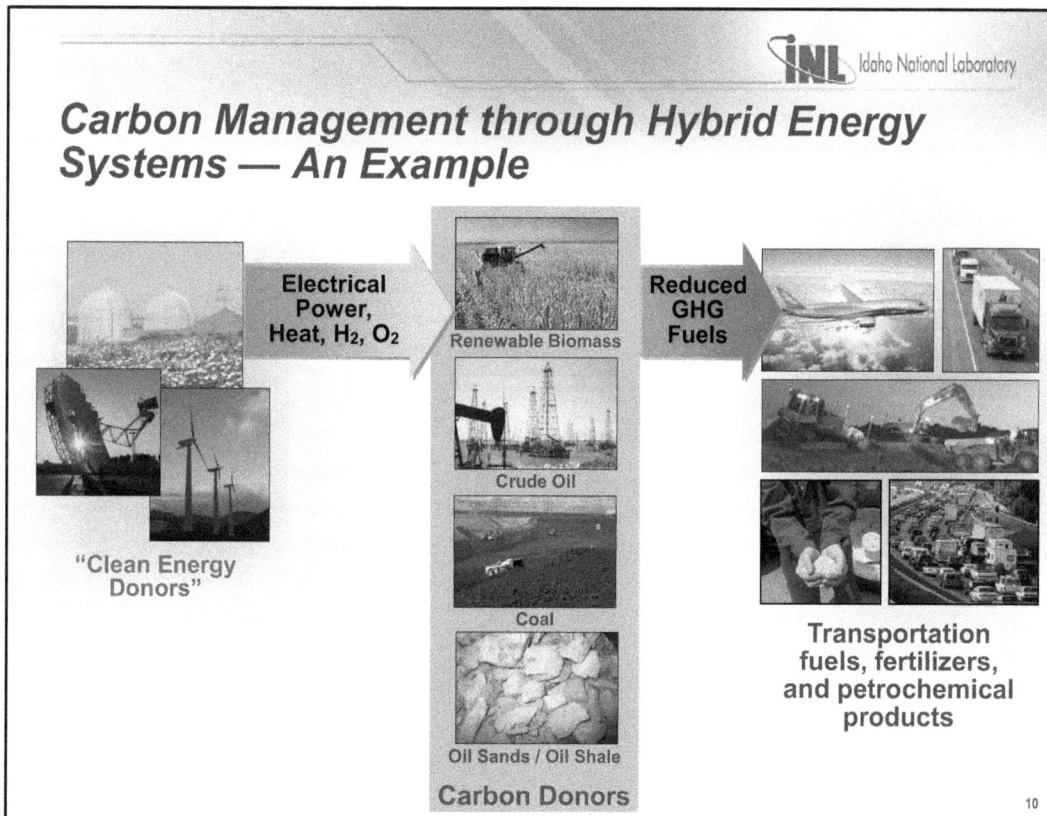

Carbon Management through Hybrid Energy Systems — An Example

U.S. Energy: Overview and Key Statistics

Carl E. Behrens
Specialist in Energy Policy

Carol Glover
Information Research Specialist

October 28, 2009

Congressional Research Service

7-5700

www.crs.gov

R40187

CRS Report for Congress ———
Prepared for Members and Committees of Congress

Summary

Energy supplies and prices are major economic factors in the United States, and energy markets are volatile and unpredictable. Thus, energy policy has been a recurring issue for Congress since the first major crisis in the 1970s. As an aid in policy making, this report presents a current and historical view of the supply and consumption of various forms of energy.

The historical trends show petroleum as the major source of energy, rising from about 38% in 1950 to 45% in 1975, then declining to about 40% in response to the energy crisis of the 1970s. Significantly, the transportation sector has been and continues to be almost completely dependent on petroleum, mostly gasoline. The importance of this dependence on the volatile world oil market was revealed over the past five years as perceptions of impending inability of the industry to meet increasing world demand led to relentless increases in the prices of oil and gasoline. With the downturn in the world economy and a consequent decline in consumption, prices collapsed, but the dependence on imported oil continues as a potential problem.

Natural gas followed a similar pattern at a lower level, increasing its share of total energy from about 17% in 1950 to more than 30% in 1970, then declining to about 20%. Consumption of coal in 1950 was 35% of the total, almost equal to oil, but it declined to about 20% a decade later and has remained at about that proportion since then. Coal currently is used almost exclusively for electric power generation.

Nuclear power started coming online in significant amounts in the late 1960s. By 1975, in the midst of the oil crisis, it was supplying 9% of total electricity generation. However, increases in capital costs, construction delays, and public opposition to nuclear power following the Three Mile Island accident in 1979 curtailed expansion of the technology, and many construction projects were cancelled. Continuation of some construction increased the nuclear share of generation to 20% in 1990, where it remains currently. The first new reactor license applications in nearly 30 years were recently submitted, but no new plants are currently under construction or on order.

Construction of major hydroelectric projects has also essentially ceased, and hydropower's share of electricity generation has gradually declined, from 30% in 1950 to 15% in 1975 and less than 10% in 2000. However, hydropower remains highly important on a regional basis.

Renewable energy sources (except hydropower) continue to offer more potential than actual energy production, although fuel ethanol has become a significant factor in transportation fuel, and wind power has recently grown rapidly. Conservation and energy efficiency have shown significant gains over the past three decades and offer encouraging potential to relieve some of the dependence on imports that has caused economic difficulties in the past, as well as the present.

After an introductory overview of aggregate energy consumption, this report presents detailed analysis of trends and statistics regarding specific energy sources: oil, electricity, natural gas, coal and renewable energy. A section on trends in energy efficiency is also presented.

Congressional Research Service

Contents

Figures

U.S. Energy: Overview and Key Statistics

Tables

Contacts

Introduction

Tracking changes in energy activity is complicated by variations in different energy markets. These markets, for the most part, operate independently, although events in one may influence trends in another. For instance, oil price movement can affect the price of natural gas, which then plays a significant role in the price of electricity. Since aggregate indicators of total energy production and consumption do not adequately reflect these complexities, this compendium focuses on the details of individual energy sectors. Primary among these are oil, particularly gasoline for transportation, and electricity generation and consumption. Natural gas is also an important energy source, for home heating as well as in industry and electricity generation. Coal is used almost entirely for electricity generation, nuclear and hydropower completely so.[1]

Renewable sources (except hydropower) continue to offer more potential than actual energy production, although fuel ethanol has become a significant factor in transportation fuel, and wind power has recently grown rapidly. Conservation and energy efficiency have shown significant gains over the past three decades, and offer encouraging potential to relieve some of the dependence on imports that has caused economic difficulties in the past as well as the present.

To give a general view of energy consumption trends, **Table 1** shows consumption by economic sector—residential, commercial, transportation, and industry—from 1950 to the present. To supplement this overview, some of the trends are highlighted by graphs in **Figure 1** and **Figure 2**.

In viewing these figures, a note on units of energy may be helpful. Each source has its own unit of energy. Oil consumption, for instance, is measured in million barrels per day (mbd),[2] coal in million tons per year, natural gas in trillion cubic feet (tcf) per year. To aggregate various types of energy in a single table, a common measure, British Thermal Unit (Btu), is often used. In **Table 1**, energy consumption by sector is given in units of quadrillion Btus per year, or "quads," while per capita consumption is given in million Btus (MMBtu) per year. One quad corresponds to one tcf of natural gas, or approximately 50 million tons of coal. One million barrels per day of oil is approximately 2 quads per year. One million Btus is equivalent to approximately 293 kilowatt-hours (Kwh) of electricity. Electric power generating capacity is expressed in terms of kilowatts (Kw), megawatts (Mw, equals 1,000 Kw) or gigawatts (Gw, equals 1,000 Mw). Gas-fired plants are typically about 250 Mw, coal-fired plants usually more than 500 Mw, and large nuclear powerplants are typically about 1.2 Gw in capacity.

Table 1 shows that total U.S. energy consumption almost tripled since 1950, with the industrial sector, the heaviest energy user, growing at the slowest rate. The growth in energy consumption per capita (i.e., per person) over the same period was about 50%. As **Figure 1** illustrates, much of the growth in per capita energy consumption took place before 1970.

---

[1] This report focuses on current and historical consumption and production of energy. For a description of the resource base from which energy is supplied, see CRS Report R40872, *U.S. Fossil Fuel Resources: Terminology, Reporting, and Summary*, by Gene Whitney, Carl E. Behrens, and Carol Glover.

[2] Further complications can result from the fact that not all sources use the same abbreviations for the various units. The Energy Information Administration (EIA), for example, abbreviates "million barrels per day" as "MMbbl/d" rather than "mbd." For a list of EIA's abbreviation forms for energy terms, see http://www.eia.doe.gov/neic/a-z/a-z_abbrev/a-z_abbrev.html.

Table 1 does not list the consumption of energy by the electricity sector separately because it is both a producer and a consumer of energy. For the residential, commercial, industrial, and transportation sectors, the consumption figures given are the sum of the resources (such as oil and gas) that are directly consumed plus the total energy used to produce the electricity each sector consumed—that is, both the energy value of the kilowatt-hours consumed and the energy lost in generating that electricity. As **Figure 2** demonstrates, a major trend during the period was the electrification of the residential and commercial sectors and, to a lesser extent, industry. By 2007, electricity (including the energy lost in generating it) represented about 70% of residential energy consumption, about 80% of commercial energy consumption, and about a third of industrial energy consumption.[3]

Table I. U.S. Energy Consumption, 1950-2008

	Energy Consumption by Sector (Quadrillion Btu)					Population (millions)	Consumption Per Capita (Million Btu)		
	Resid.	Comm.	Indust.	Trans.	Total		Total	Resid.	Trans.
1950	6.0	3.9	16.2	8.5	34.6	152.3	227.3	39.4	55.8
1955	7.3	3.9	19.5	9.6	40.2	165.9	242.3	44.0	57.6
1960	9.1	4.6	20.8	10.6	45.1	80.7	249.6	50.2	58.7
1965	10.7	5.8	25.1	12.4	54.0	94.3	278.0	55.0	64.0
1970	13.8	8.3	29.6	16.1	67.8	205.1	330.9	67.3	78.5
1975	14.8	9.5	29.4	18.2	72.0	216.0	333.4	68.7	84.5
1980	15.8	10.6	32.1	19.7	78.1	227.2	343.8	69.5	86.7
1985	16.1	11.4	28.9	20.1	76.5	237.9	321.5	67.6	84.4
1990	17.0	13.3	31.9	22.4	84.7	249.6	339.1	68.2	89.8
1995	18.6	14.7	34.0	23.8	91.2	266.3	342.4	69.8	89.6
2000	20.5	17.2	34.8	26.6	99.0	282.2	350.7	72.6	94.1
2005	21.7	17.9	32.5	28.4	100.5	295.6	340.0	73.4	96.0
2006	20.8	17.7	32.5	28.8	99.9	298.4	334.7	69.6	96.7
2007	21.6	18.3	32.5	29.1	101.6	301.3	337.1	71.8	96.7
2008P	21.6	18.5	31.2	27.9	99.3	304.1	326.6	71.2	91.8

Source: Energy Information Administration (EIA), *Annual Energy Review 2008*, Tables 2.1a and D1. Per capita data calculated by CRS.

Notes: Data for 2008 are preliminary.

---

[3] In calculating these percentages, "electric energy consumption" includes both the energy value of the kilowatt-hours consumed and the energy lost in generating that electricity.

Figure 1. Per Capita Energy Consumption in Transportation and Residential Sectors, 1949-2008

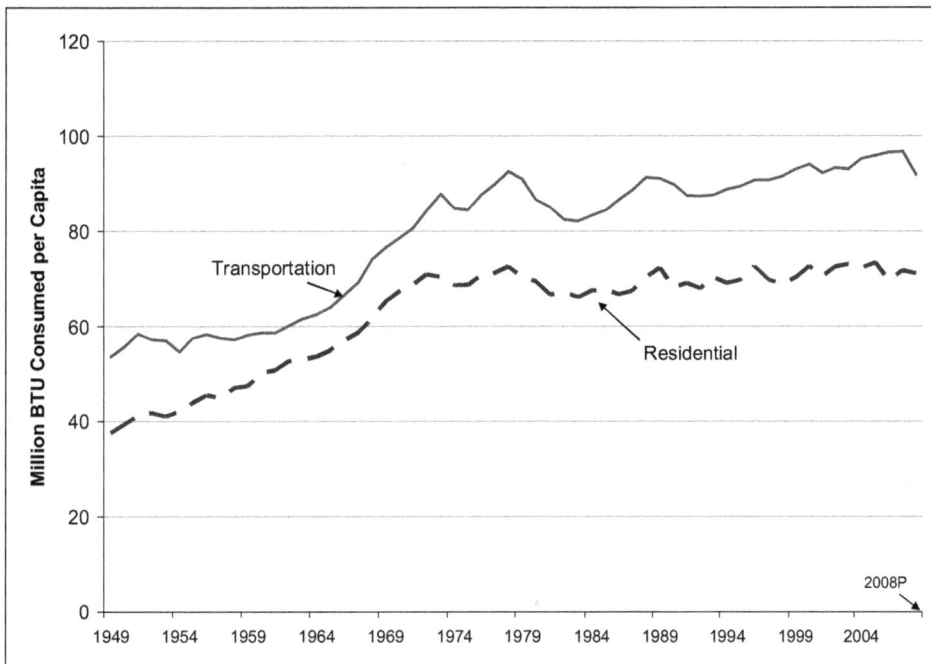

Source: Energy Information Administration (EIA), *Annual Energy Review 2008*, Tables 2.1a and D1. Per capita data calculated by CRS.

Notes: Data for 2008 are preliminary.

Figure 2. Electricity Intensity: Commercial, Residential, and Industrial Sectors, 1949-2008

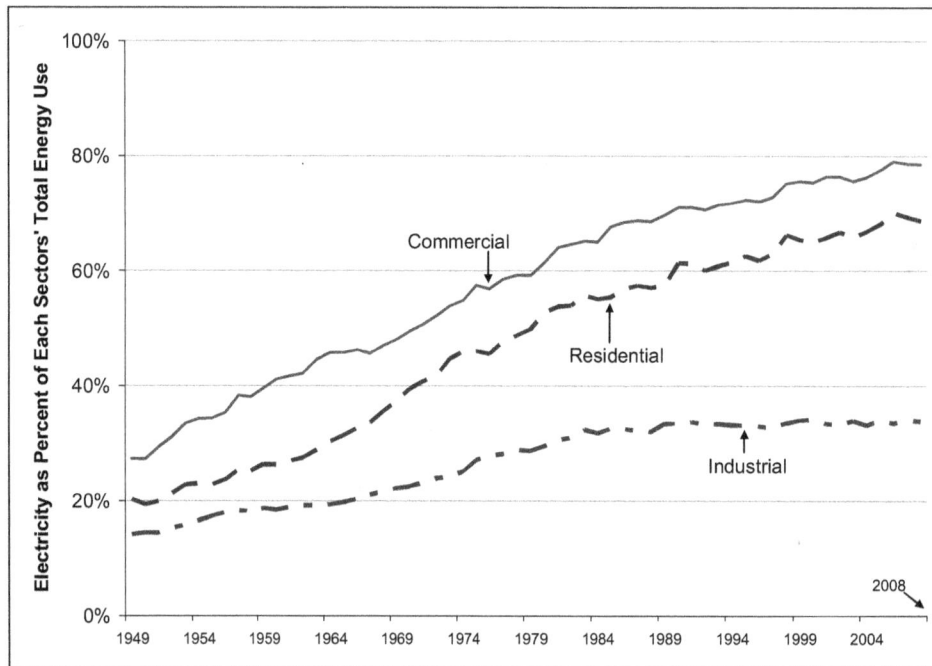

Source: Energy Information Administration (EIA), Annual Energy Review 2008, Tables 2.1a and D1. Per capita data calculated by CRS.

Notes: Data for 2008 are preliminary.

Consumption of major energy resources—petroleum, natural gas, and coal—is presented in **Table 2** and **Figure 3**. The historical trends show that petroleum has been and continues to be the major source of energy, rising from about 38% in 1950 to 45% in 1975, then declining to about 40% in response to the energy crisis of the 1970s. Natural gas followed a similar pattern at a lower level, increasing its share of total energy from about 17% in 1950 to over 30% in 1970, then declining to about 20%. Consumption of coal in 1950 was 35% of the total, almost equal to oil, but it declined to about 20% a decade later and has remained at about that proportion since then.

Table 2. Energy Consumption in British Thermal Units (BTU) and as a Percentage of Total, 1950-2008
(Quadrillion BTU)

	Petroleum		Natural Gas		Coal		Other		
	Quads	% of total	Quads	% of total	Quads	% of total	Quads	% of total	Total
1950	13.3	38.4	6.0	17.3	12.3	35.5	3.0	8.6	34.6
1955	17.3	43.0	9.0	22.4	11.2	27.8	2.8	7.0	40.2
1960	19.9	44.1	12.4	27.5	9.8	21.7	3.0	6.5	45.1
1965	23.2	42.9	15.8	29.2	11.6	21.4	3.4	6.4	54.0
1970	29.5	43.5	21.8	32.1	12.2	18.0	4.3	6.4	67.8
1975	32.7	45.4	19.9	27.6	12.7	17.7	6.6	9.2	72.0
1980	34.2	43.8	20.4	26.1	15.4	19.7	8.3	10.6	78.1
1985	30.9	40.4	17.7	23.1	17.5	22.9	10.4	13.6	76.5
1990	33.6	39.7	19.7	23.3	19.2	22.7	12.3	14.6	84.7
1995	34.6	37.9	22.8	25.0	20.2	22.1	13.9	15.3	91.2
2000	38.4	38.8	23.9	24.1	22.7	22.9	14.2	14.4	99.0
2005	40.4	40.2	22.6	22.5	22.8	22.7	14.7	14.6	100.5
2006	40.0	40.0	22.2	22.3	22.5	22.5	15.2	15.2	99.9
2007	39.8	39.1	23.6	23.3	22.8	22.4	15.4	15.1	101.6
2008P	37.1	37.4	23.8	24.0	22.5	22.6	15.9	16.0	99.3

Source: EIA, *Annual Energy Review 2008*, Table 1.3.

Notes: Percentages calculated by CRS. "Other" includes nuclear and renewable energy. Data for 2008 are preliminary.

Figure 3. U.S. Energy Consumption, 1950-2005 and 2008

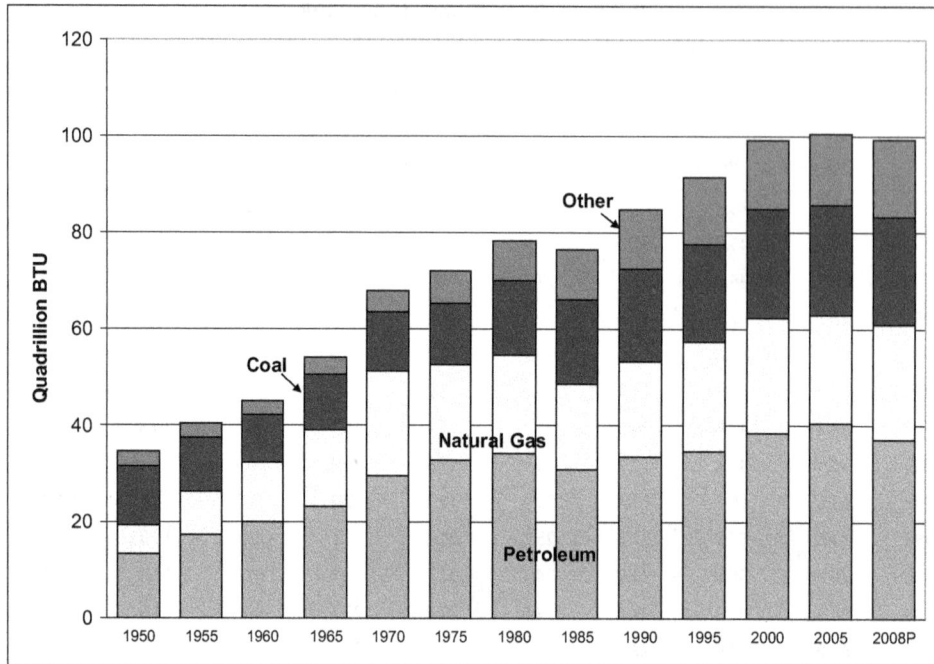

Source: EIA, *Annual Energy Review 2008*, Table 1.3.

Notes: "Other" includes nuclear and renewable energy. Data for 2008 are preliminary.

Oil

About 40% of the energy consumed in the United States is supplied by petroleum, and that proportion has remained approximately the same since 1950, as the data in the previous section show. Also unchanged is the almost total dependence of the transportation sector on petroleum, mostly gasoline.

The perception that the world is on the verge of running out of oil, widespread during the 1970s, has changed, however. The rapid price increases at that time, aided by improved exploration and production technology, stimulated a global search for oil and resulted in the discovery of large amounts of new reserves. Indeed, as concerns about tightening supply and continually increasing prices were at a peak, proven reserves actually increased by about 50% between 1973 and 1990. Some of the increase was in the Western Hemisphere, mostly in Mexico, but most was located in the region that already dominated the world oil market, the Middle East. With prices essentially steady during the 1990s, the search for oil slowed, but additions to reserves during the decade exceeded the amount of oil pumped out of the ground. By 2003, improved technology for retrieving petroleum from oil sands in Canada and, to a lesser extent, from heavy oil in Venezuela led to significant production from these resources, and by 2005, approximately 200 billion barrels of resources from oil sands and heavy oil were added to the total of proven world reserves, 20% of the total 1991 figure. These trends are illustrated in **Figure 4.**

Figure 4. World Crude Oil Reserves, 1973, 1991, and 2008

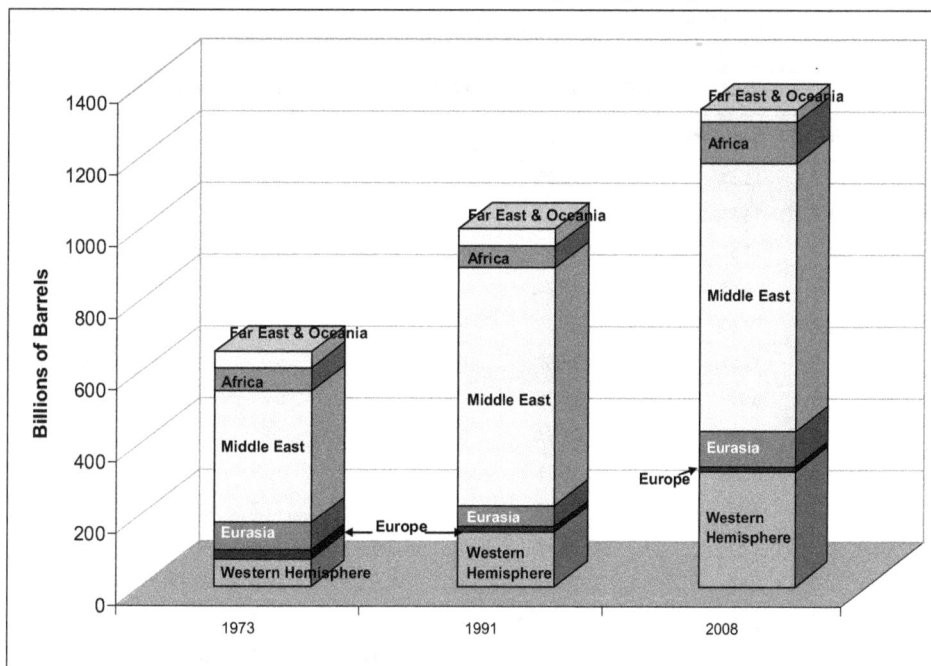

Source: EIA, International Energy Annual (IEA) 1990, Table 32 and IEA 2007 Table 8.1 Table of World Proved Oil and Natural Gas Reserves, Most Recent Estimates. (data is from Oil and Gas Journal and is not certified by EIA, except for the data for the United States in the Western Hemisphere category).

Notes: The categories "Eastern Europe and Former Soviet Union" and "Western Europe," in the data for 1973 and 1991, were changed to "Eurasia" and "Europe" respectively for 2005. Seven countries (Albania, Bulgaria, Czech Republic, Hungary, Poland, Romania, and Slovakia) were moved from the former to the latter.

Petroleum Consumption, Supply, and Imports

Consumption of petroleum by sector reflects a variety of trends (see **Table 3**). In the residential and commercial sectors, petroleum consumption grew steadily from 1950 to 1970, while accounting for about 15% of total petroleum consumption. After the price surge in the 1970s, consumption in those sectors declined, falling to less than 7% of total petroleum consumption by 1995. When oil prices surged again after 2005, consumption declined further, to about 5%. Usage in the electric power sector followed a similar but more abrupt pattern. Until 1965 only about 3% of petroleum went to power generation. In the late 1960s efforts to improve air quality by reducing emissions led utilities to convert a number of coal-fired power plants to burn oil, and many new plants were designed to burn oil or natural gas. Utilities found themselves committed to increasing dependence on oil just at the time of shortages and high prices; in 1975 almost 9% of oil consumption went for power production. Consumption then fell sharply as alternate sources became available, declining to about 2%-3% of total consumption and falling even lower after 2005 as oil prices increased sharply.

Table 3. Petroleum Consumption by Sector, 1950-2008
(Million Barrels per Day (MBD) and Percentage of Total)

	Residential & Commercial		Industrial		Electric		Transportation		Total
	MBD	% of total	MBD	% of total	MBD	% of total	MBD	% of total	MBD
1950	1.1	16.5%	1.8	28.0%	0.2	3.2%	3.4	51.6%	6.5
1955	1.4	16.5%	2.4	28.1%	0.2	2.4%	4.5	52.4%	8.5
1960	1.7	17.5%	2.7	27.6%	0.2	2.5%	5.1	52.4%	9.8
1965	1.9	16.6%	3.2	27.2%	0.3	2.7%	6.0	52.5%	11.5
1970	2.2	14.9%	3.8	25.9%	0.9	6.3%	7.8	52.9%	14.7
1975	1.9	11.9%	4.0	24.8%	1.4	8.5%	9.0	54.9%	16.3
1980	1.5	8.9%	4.8	28.3%	1.2	6.7%	9.5	55.8%	17.1
1985	1.3	8.6%	4.1	25.9%	0.5	3.0%	9.8	62.7%	15.7
1990	1.2	7.2%	4.3	25.3%	0.6	3.3%	10.9	64.0%	17.0
1995	1.1	6.4%	4.6	26.0%	0.3	1.9%	11.7	65.9%	17.7
2000	1.3	6.5%	4.9	24.9%	0.5	2.6%	13.0	66.1%	19.7
2005	1.2	5.8%	5.1	24.5%	0.5	2.6%	14.0	67.1%	20.8
2006	1.0	5.2%	5.2	24.8%	0.3	1.4%	14.2	68.6%	20.7
2007	1.0	5.2%	5.1	24.4%	0.3	1.4%	14.3	68.9%	20.7
2008	1.0	5.0%	4.6	23.6%	0.2	1.1%	13.7	70.3%	19.4

Source: EIA, *Annual Energy Review 2008*, Tables 5.1 and 5.13a-d.

Notes: Percentages calculated by CRS. Data for 2008 are preliminary.

Industrial consumption of petroleum, which includes such large consumers as refineries and petrochemical industries, has remained about 25% of total consumption since 1970. As other sectors' share fell, transportation, which was a little more than half of total consumption prior to 1975, climbed to two-thirds by 2000 and continued to increase its share since then.

While petroleum consumption increased throughout the period from 1950 to the present (except for a temporary decline following the price surge of the 1970s), U.S. domestic production peaked in 1970 (see **Table 4**). The result, as shown in **Figure 5**, was greater dependence on imported petroleum, which rose from less than 20% in 1960 to about 60% in recent years.

Table 4. U.S. Petroleum Production, 1950-2008
(Million Barrels per Day)

	Crude Oil			Gas Liquids	Other	Total
	48 States	**Alaska**	**Total**			
1950	5.4	—	5.4	0.5	—	5.9
1955	6.8	—	6.8	0.8	—	7.6
1960	7	—	7	0.9	0.2	8.1
1965	7.8	—	7.8	1.2	0.2	9.2
1970	9.4	0.2	9.6	1.7	0.4	11.7
1975	8.2	0.2	8.4	1.6	0.5	10.5
1980	7	1.6	8.6	1.6	0.6	10.8
1985	7.2	1.8	9	1.6	0.6	11.1
1990	5.6	1.8	7.4	1.6	0.7	9.6
1995	5.1	1.5	6.6	1.8	0.8	9.1
2000	4.9	1.0	5.8	1.9	1.0	8.7
2005	4.3	0.9	5.2	1.7	1.0	7.9
2006	4.4	0.7	5.1	1.7	1.0	7.8
2007	4.3	0.7	5.1	1.8	1.0	7.9
2008	4.3	0.7	5.0	1.8	1.0	7.8

Source: EIA, *Annual Energy Review 2008*, Table 5.1.

Notes: "Other" includes processing gain. Data for 2008 are preliminary.

**Figure 5. U.S. Consumption of Imported Petroleum,
1960-2008 and Year-to-Date Average for 2009**

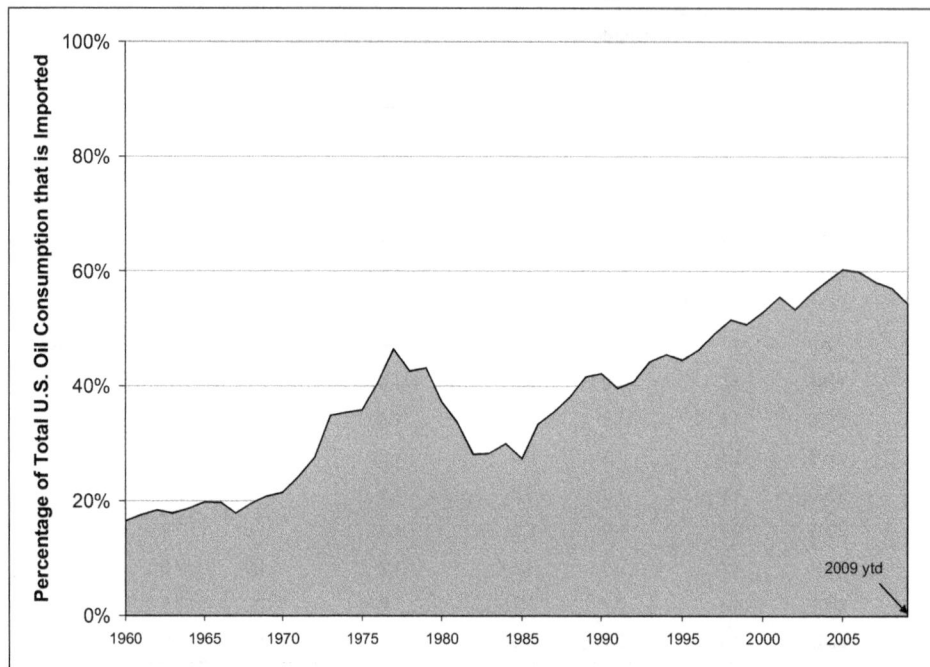

Source: EIA, *Monthly Energy Review*, September 2009, Table 3.3a, and *Annual Energy Review 1986*, Table 51.

Notes: Data for 2009 ytd (year-to-date) is an 8-month average.

Petroleum and Transportation

Since the transportation sector is so heavily dependent on petroleum, and uses so much of it,
Table 5 and **Figure 6** present a more detailed breakdown of the various types of petroleum used.

Table 5. Transportation Use of Petroleum, 1950-2008
(Million barrels per day)

	Aviation	Diesel Fuel	Gasoline	Other	Total
1950	0.1	0.2	2.4	0.6	3.4
1955	0.3	0.4	3.2	0.5	4.5
1960	0.5	0.4	3.7	0.4	5.1
1965	0.7	0.5	4.4	0.4	6.0
1970	1.0	0.7	5.6	0.4	7.8
1975	1.0	1.0	6.5	0.4	9.0
1980	1.1	1.3	6.4	0.7	9.5
1985	1.2	1.5	6.7	0.4	9.8
1990	1.5	1.7	7.1	0.5	10.9
1995	1.5	2.0	7.7	0.5	11.7
2000	1.7	2.4	8.4	0.5	13.0
2005	1.7	2.9	8.9	0.5	14.0
2006	1.7	3.0	9.0	0.5	14.2
2007	1.6	3.0	9.1	0.5	14.3
2008P	1.5	2.9	8.8	0.5	13.7

Source: EIA, *Annual Energy Review 2008*, Table 5.13c.

Notes: Data for 2008 are preliminary.

Aviation fuel includes both aviation gasoline and kerosene jet fuel. In 1950 aviation was almost entirely gasoline powered; by 2000 it was 99% jet fueled. The growth in flying is illustrated by the fact that aviation fuel was only 3% of petroleum consumption for transportation in 1950, but had grown to 12% in 1965 and has maintained that share since then.

Diesel fuel consumption showed a similar dramatic increase. About 6% of total petroleum consumption for transportation in 1950, it rose to 11% by 1975 and to 20% in recent years. Diesel fuel is used by a number of transportation sectors. Part of the increase involved the change of railroads from coal-fired steam to diesel and diesel-electric power. Diesel fuel is used also in the marine transportation sector, and some private automobiles are diesel-powered. The major part of diesel fuel consumption in transportation is by large commercial trucks. Total diesel fuel consumption increased from about 200,000 barrels per day in 1950 to about 3.0 million barrels per day in 2008.

Most of the petroleum consumed in the transportation sector is motor gasoline. In 1950 it was 71% of total sector petroleum consumption, and in recent years, despite the increase in aviation fuel and diesel, it has been about 65%. Since 1950, gasoline consumption has almost quadrupled.

Of the other petroleum products consumed in the transportation sector, the largest is residual fuel oil, most of which is used in large marine transport. Consumption of residual fuel oil in the transportation sector was about 500,000 barrels in 1950, and declined gradually to about 400,000 in 2000.

Figure 6. Transportation Use of Petroleum, 1950-2008

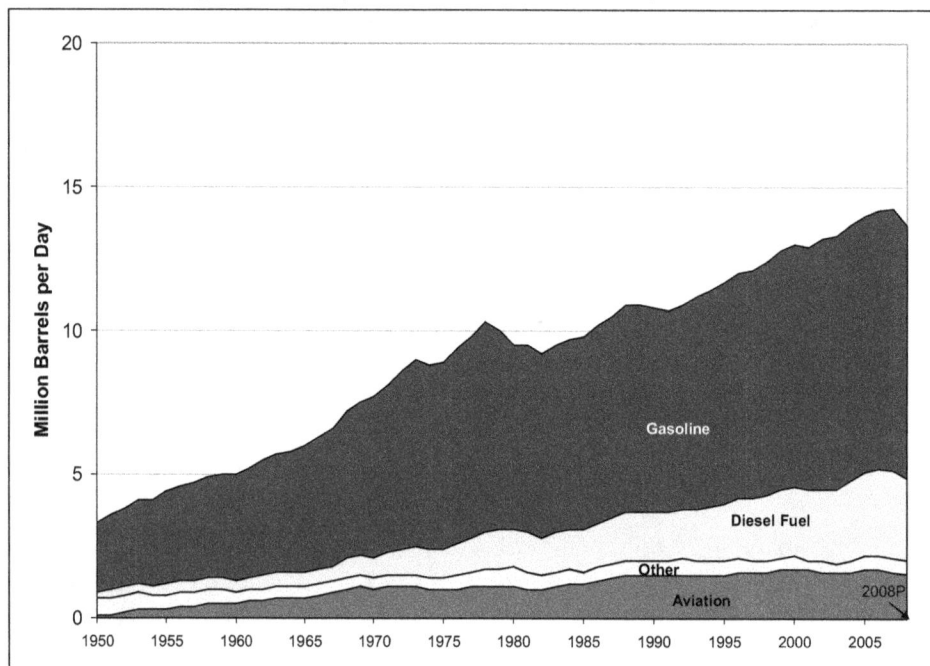

Source: EIA, *Annual Energy Review 2008*, Table 5.13c.

Notes: Data for 2008 are preliminary.

Petroleum Prices: Historical Trends

Most commodity prices are typically volatile. Because oil is widely consumed, and is so important at all levels of the economy, its price is closely watched and analyzed. Especially since the 1970s, when a generally stable market dominated by a few large oil companies was broken by the Organization of Petroleum Exporting Countries (OPEC) cartel and a relatively open world market came into being, the price of crude oil has been particularly volatile. **Figure 7** and **Figure 8** show the long-term trends of crude oil and gasoline prices, in both current dollars and deflated dollars. The data for these charts do not show the collapse in oil prices that began in October 2008. (See "Petroleum Prices: The 2004-2008 Bubble" below.)

Figure 7. Nominal and Real Cost of Crude Oil to Refiners, 1968-2008

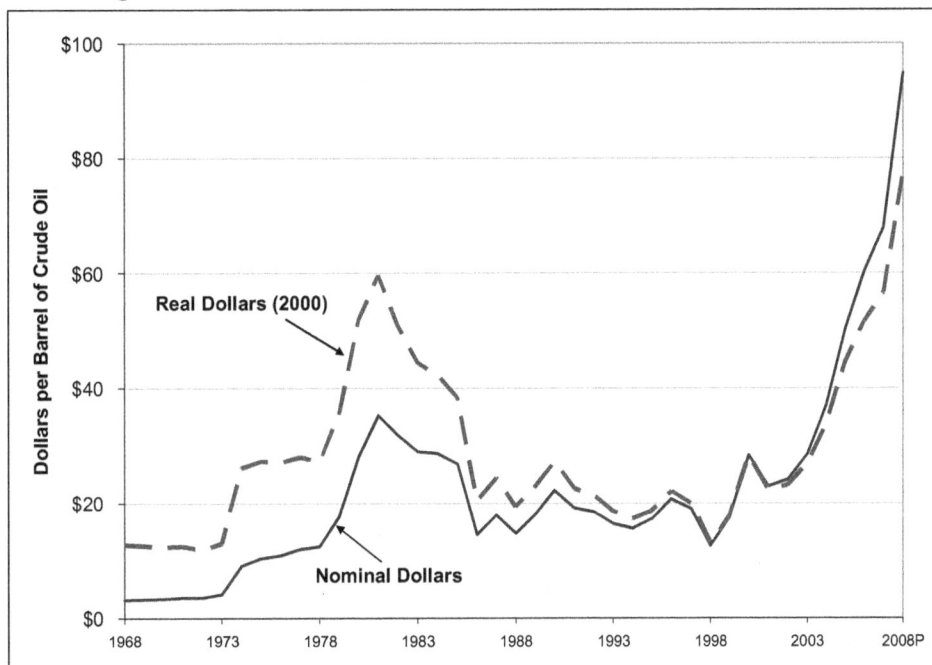

Source: EIA, *Annual Energy Review 2008*, Table 5.21.

Notes: Composite crude oil refiner acquisition cost as reported by EIA. Data for 2008 are preliminary.

At the consumer level, prices of products such as motor gasoline and heating oil have reacted to price and supply disruptions in ways that have been modulated by various government and industry policies and international events. A significant and not often noted fact is that, like many commodities, the long-term trend in gasoline prices, adjusted for inflation and excluding temporary surges, has been down. As shown in **Figure 8**, the real price of gasoline peaked in 1980, then fell precipitously in the mid-1980s. The recent surge in prices brought the price above the peak of 1980 (in real dollars).

Figure 8. Nominal and Real Price of Gasoline, 1950-2008 and August 2009

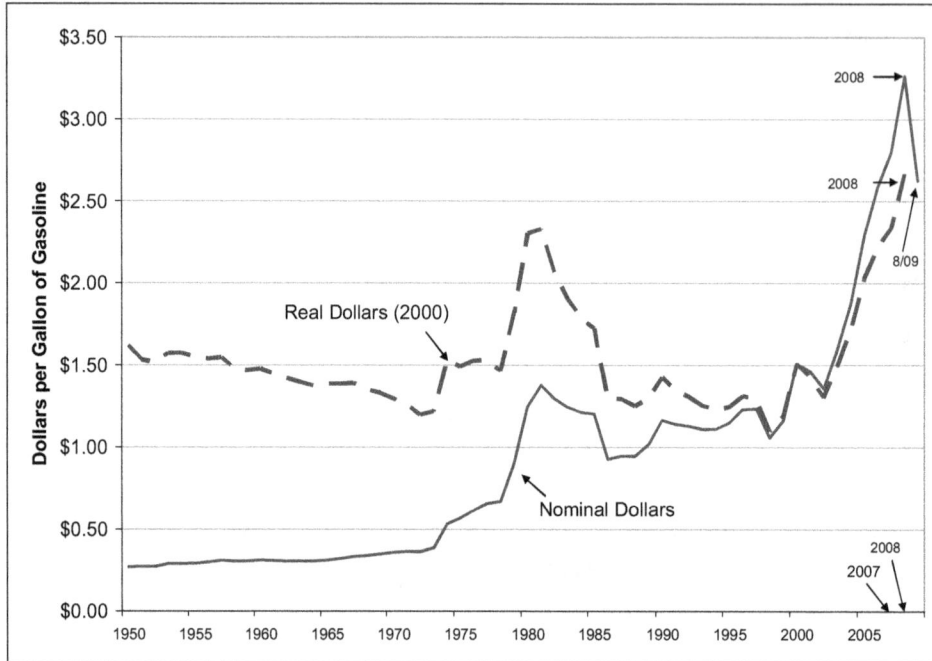

Source: EIA, *Annual Energy Review 2008*, Table 5.24 and *Monthly Energy Review*, September 2009, Table 9.4.

Notes: Average national retail price per gallon of unleaded regular gasoline, including taxes.

Figure 9 illustrates the proportion of the gross domestic product (GDP) dedicated to consumer spending on oil. The price surges in the 1970s pushed this ratio from about 4.5% before the Arab oil embargo to about 8.5% following the crisis in Iran late in the decade. Following that, it declined to less than 4%; during the recent run-up of prices the trend started back up again.

Figure 9. Consumer Spending on Oil as a Percentage of GDP, 1970-2006

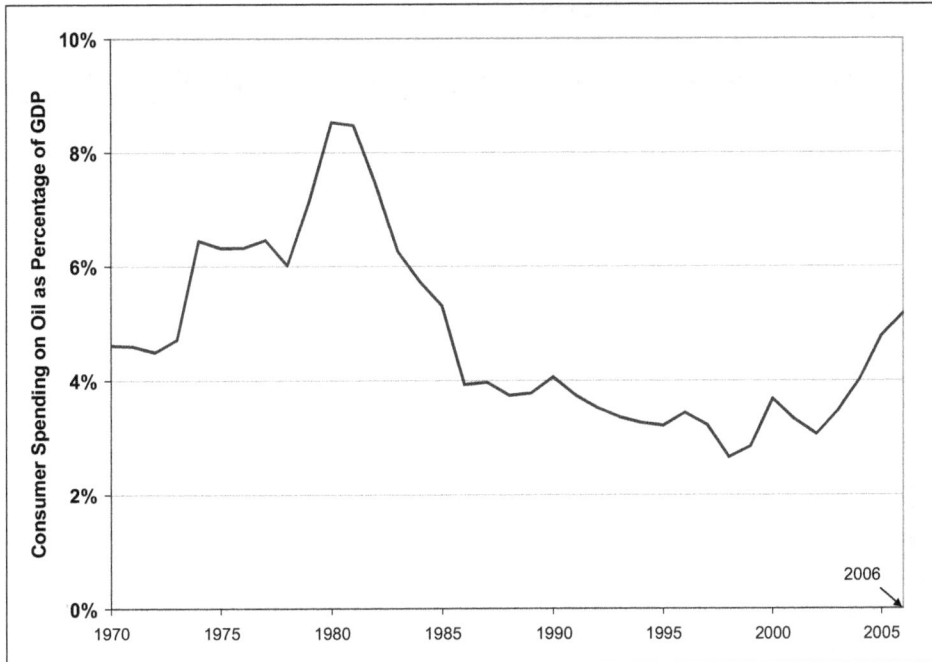

Source: EIA, *Annual Energy Review 2008*, Table 3.5 and Table D1 for GDP in billions of nominal dollars. Percentages calculated by CRS.

Petroleum Prices: The 2004-2008 Bubble

Beginning in 2004 the world price of crude oil, and with it the price of gasoline, began to increase. Unlike the previous increases in the 1970s, there was no interruption or shortage in the supply of either petroleum or its products, except for a few months in the fall of 2005 when Hurricane Katrina shut down a major portion of U.S. refinery capacity. Nevertheless, an unexpected surge in demand for oil imports to China, added to continuing increases in demand from Europe and the United States as economies continued to grow, tightened the production capacity of the major oil producing nations and signaled that demand in the near future might not be met. In addition, turmoil in the Middle East and elsewhere, as well as the possibility of further natural disasters like Katrina, threatened supply interruptions and put further upward pressure on prices. (See **Figure 10** and **Figure 11**.)

As prices continued to climb, it became apparent that demand for gasoline was relatively insensitive to its cost to the consumer. Throughout the period, as illustrated in **Figure 12**, consumption of gasoline varied seasonally but continued an upward trend on an annual basis. In the summer of 2008 crude oil prices soared far beyond the actual cost of production, and the market took on features of a classical commodities bubble, with expectations of indefinitely rising prices and participation in the market by many who would not normally enter it.

The bubble burst in October 2008 with the onset of a financial crisis in the housing and banking sectors and the evidence that consumption of gasoline was finally faltering. As the economic crisis became more acute, crude prices fell in a few months from $135 per barrel to close to $40, where they had been at the start of the run-up five years earlier.

Figure 10. Crude Oil Futures Prices, January 2000 to September 2009

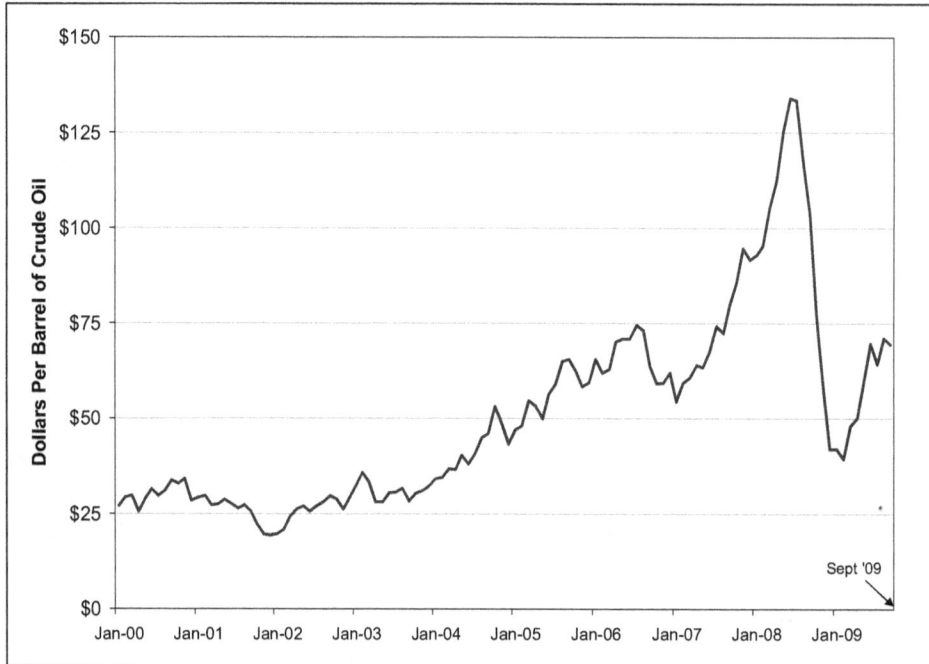

Source: EIA, NYMEX Futures Prices Crude Oil (Light-Sweet, Cushing, Oklahoma) Cushing, OK Crude Oil Future Contract I

Notes: The futures prices shown are the official daily closing prices at 2:30 p.m. from the trading floor of the New York Mercantile Exchange (NYMEX) for a specific delivery month for each product listed.

**Figure 11. Average Daily Nationwide Price of Unleaded Gasoline,
January 2002-October 2009**

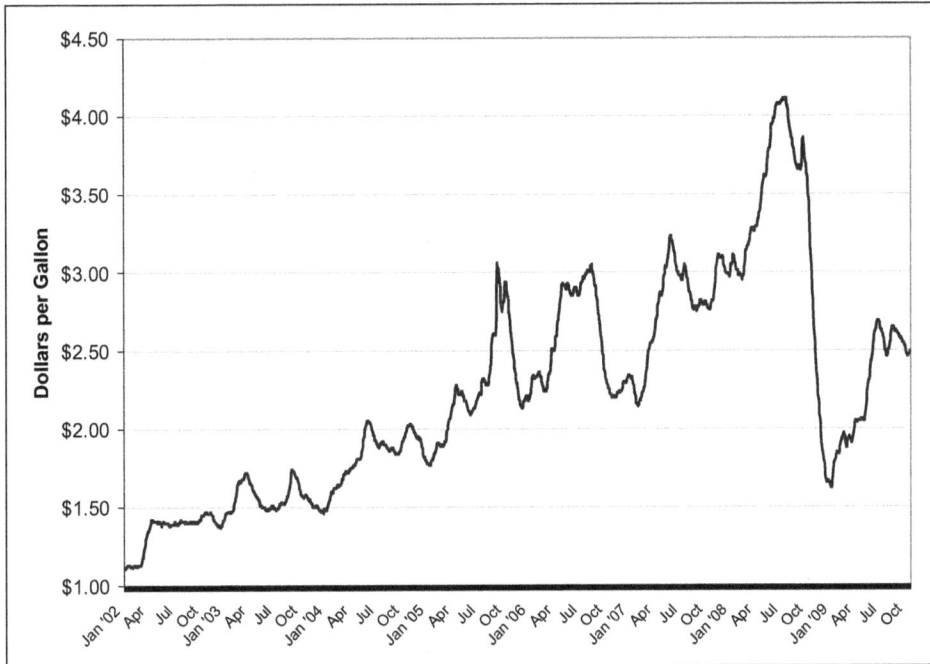

Source: Daily Fuel Gauge Report, American Automobile Association, http://www.fuelgaugereport.com, compiled by CRS.

Notes: Prices include federal, state, and local taxes. Last date above is October 15, 2009; $2.49.

Figure 12. U.S. Gasoline Consumption, January 2000-September 2009

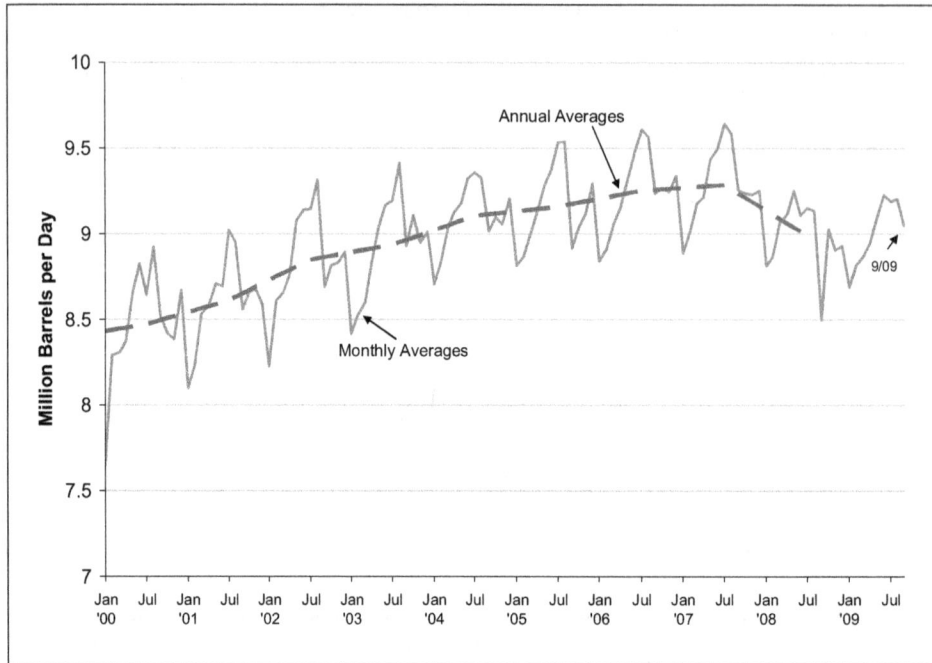

Source: EIA, *Monthly Energy Review*, June 2009, Table 3.5 and EIA, Weekly Petroleum Status Report, September 30, 2009, Table 10.

Gasoline Taxes

The federal tax on gasoline is currently 18.4 cents per gallon. An extensive list of the gasoline and diesel fuel tax rates imposed by each state per gallon of motor fuel is maintained and updated by the American Petroleum Institute (API), "Notes to State Motor Fuel Excise and Other Tax Rates," at http://www.api.org/statistics/fueltaxes/upload/MotorFuelNotesJan20092.pdf.

Electricity

While overall energy consumption in the United States increased nearly three-fold since 1950, electricity consumption increased even more rapidly. Annual power generation is ten times what it was in 1950. **Figure 13** illustrates the trend.

Figure 13. Electricity Generation by Source, Selected Years, 1950-2007

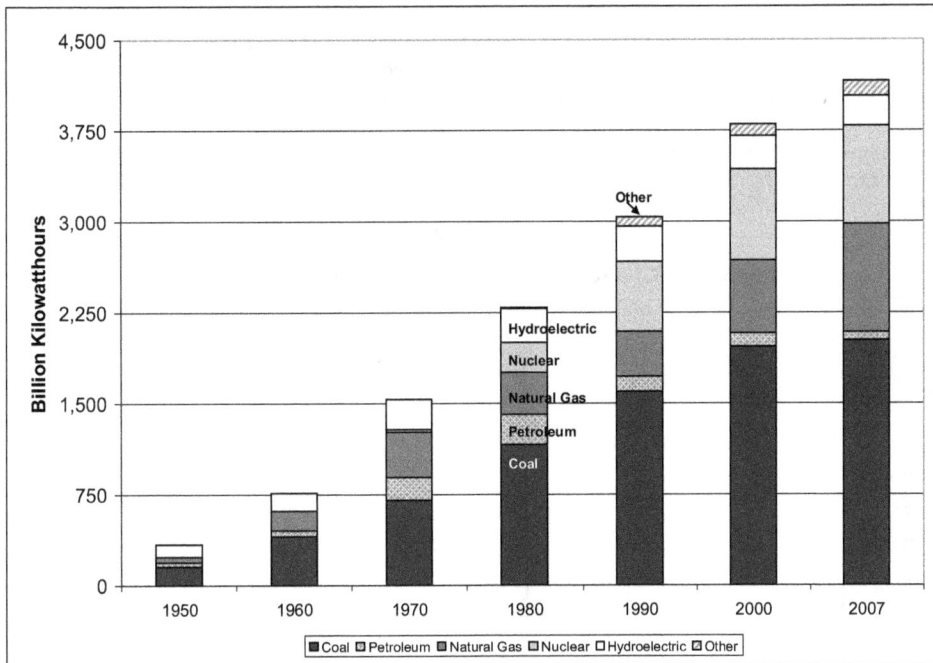

Source: EIA, *Annual Energy Review 2007*, Table 8.2a

Throughout this period, coal was used to generate about half the rapidly increasing amount of electricity consumed. Petroleum became briefly important as a source of power generation in the late 1960s because it resulted in lower emissions of air pollutants, and consumption continued in the 1970s despite the price surge because natural gas was in short supply. By the 1980s, however, oil consumption by utilities dropped sharply, and in 2007 only 1.2% of power generation was oil-fired.

Natural gas generation has a more complicated history. Consumption by the electric power industry increased gradually as access by pipeline became more widespread. With the price increase in oil in the 1970s, demand for gas also increased, but interstate prices were regulated, and gas availability declined. In addition, federal energy policy viewed generation of electricity by gas to be a wasteful use of a diminishing resource. The Fuel Use Act of 1978 prohibited new power generators from using gas and set a timetable for shutting down existing gas-fired plants. Gas prices were later deregulated, resulting in increased production, and the Fuel Use Act was repealed, but in the meantime generation of electricity from gas fell from 24% in 1970 to 12% in 1985. In the 1990s gas became more popular as technology improved, and by 2000 was supplying 16% of total electric generation. Most capacity additions in the last decade have been gas-fired, as illustrated in **Figure 14**. The increased demand contributed to high prices in 2000 that were felt particularly in California.

Nuclear power started coming on line in significant amounts in the late 1960s, and by 1975, in the midst of the oil crisis, was supplying 9% of total generation. However, increases in capital costs, construction delays, and public opposition to nuclear power following the Three Mile Island accident in 1979 curtailed expansion of the technology, and many construction projects were cancelled. Continuation of some construction increased the nuclear share of generation to 20% in 1990, where it remains currently. Recently, plans have been announced for license applications for up to 30 new reactors, and several have been submitted to the Nuclear Regulatory Commission, but no new plants are currently under construction or on order.

Construction of major hydroelectric projects has also essentially ceased, and hydropower's share of electricity generation has gradually declined from 30% in 1950 to 15% in 1975 and less than 10% in 2000. However, hydropower remains highly important on a regional basis.

Figure 14. Changes in Generating Capacity, 1995-2007

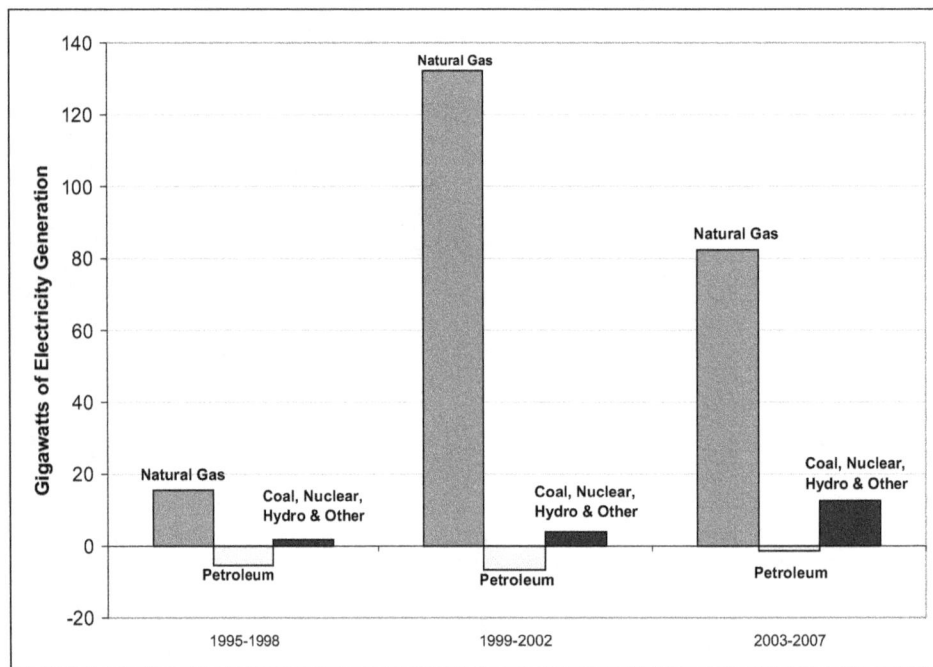

Source: EIA, Annual Energy Review 2007, Table 8.11a.

Sources of power generation vary greatly by region (see **Table 6**). Hydropower in the Pacific Coast states, for instance, supplies over 40% of total generation, and natural gas almost 35%. In 2000, the combination of a drought-caused shortage of hydropower, a tightening of gas supply, and California's new electric regulatory scheme and market manipulation caused very sharp increases in electricity prices in that region. Other regions are heavily dependent on coal generation: The North Central and East South Central states, as well as the Mountain states, generate more than 60% of their electricity from coal, whereas other regions, such as New England and the Pacific Coast, use relatively little coal. The West South Central region (Arkansas, Louisiana, Oklahoma, and Texas) generates 45% of its electricity from gas. New England in the 1970s and 1980s was heavily dependent on oil-generated power; in 2005, despite an increased use of natural gas, oil produced 10% of New England's power, compared with the national average of 2.5%. By 2007, the proportion had dropped to 4.4%, and the national average to 1.2%

Table 6. Electricity Generation by Region and Fuel, 2008

	Total Generation (billion kwh)	Percentage by					
		Coal	Petroleum	Natural Gas	Nuclear	Hydro	Other
New England	59.6	14.9%	2.9%	38.4%	28.9%	7.1%	7.8%
Middle Atlantic	206.0	35.6%	1.0%	17.5%	36.2%	7.4%	2.3%
East North Central	325.0	70.1%	0.2%	3.6%	23.5%	0.6%	2.0%
West North Central	155.4	75.1%	0.1%	3.3%	14.0%	2.7%	4.7%
South Atlantic	398.3	54.3%	1.5%	15.8%	24.3%	1.7%	2.4%
East South Central	189.5	63.5%	0.2%	10.3%	19.7%	4.0%	2.3%
West South Central	305.0	37.2%	0.1%	44.4%	11.1%	1.9%	5.3%
Mountain	180.1	56.6%	0.1%	23.3%	8.0%	9.3%	2.8%
Pacific Contiguous	184.2	3.4%	0.1%	35.0%	11.2%	39.6%	10.8%
Pacific Noncontiguous	8.6	13.5%	51.4%	21.7%	0.0%	7.5%	5.9%
U.S. Total	2,011.7	49.0%	0.8%	20.0%	19.5%	6.8%	3.9%

Source: EIA, Electric Power Monthly, September 2009, Tables 1.6B, 1.7B, 1.8B, 1.10B, 1.12B, and 1.13B.

Note: "Other" includes renewables other than hydro, plus hydro from pumped storage, petroleum coke, gases other than natural gas, and other sources.

The price of electricity varies by region, depending on the fuel mix and the local regulatory system, among other factors. The nationwide average retail price to residential consumers increased during the 1970s energy crises but declined starting in the 1980s, as indicated by **Figure 15**. An increase starting in 2000 resulted from the expiration in numerous regions of price caps that had been previously imposed when utilities were deregulated; the recent runup in oil and natural gas prices, and to a lesser extent in coal prices, has maintained the trend.

Figure 15. Price of Retail Residential Electricity, 1960-2007

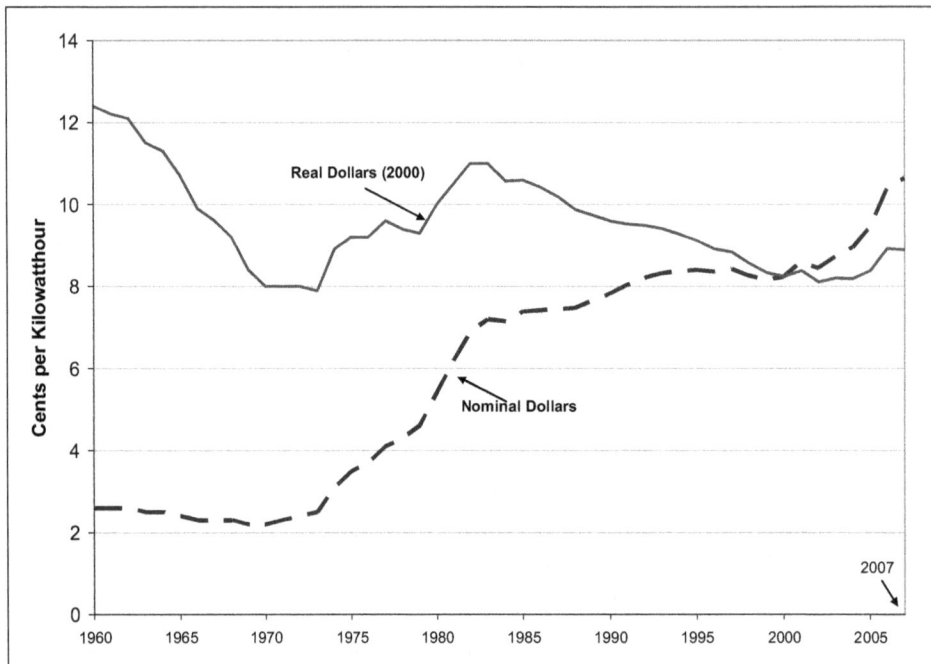

Source: EIA, *Annual Energy Review 2007*, Table 8.10.

Notes: Price includes taxes.

Other Conventional Energy Resources

Natural Gas

Consumption of natural gas was almost four times as great in 2007 as it was in 1950. Throughout the period, consumption in the residential and commercial sector grew at about the same rate as total consumption, in the range of 30% to 40% of the total. As shown in **Table 7**, consumption for electric power generation increased from about 10% in 1950 to more than 20% at the end of the century. The proportion of total gas consumption by the industrial sector declined correspondingly, from more than 50% in 1950 to about 35% in recent years.

Table 7. Natural Gas Consumption by Sector, 1950-2008

	Total Consumption	Percent Consumed by Sector		
	trillion cubic feet (tcf)	Residential - Commercial	Industrial	Electric
1950	5.77	27.5%	59.4%	10.9%
1955	8.69	31.7%	52.2%	13.3%
1960	11.97	34.5%	48.2%	14.4%
1965	15.28	35.0%	46.5%	15.2%
1970	21.14	34.2%	43.8%	18.6%
1975	19.54	38.0%	42.8%	16.2%
1980	19.88	37.0%	41.2%	18.5%
1985	17.28	39.7%	39.7%	17.6%
1990	19.17	36.6%	43.1%	16.9%
1995	22.21	35.5%	42.3%	19.1%
2000	23.33	35.0%	39.8%	22.3%
2001	22.24	35.0%	38.1%	24.0%
2002	23.01	34.9%	37.5%	24.6%
2003	22.38	37.1%	36.9%	22.9%
2004	22.39	35.7%	37.3%	24.4%
2005	22.01	35.6%	35.0%	26.7%
2006	21.69	33.2%	35.3%	28.7%
2007	23.05	33.6%	33.9%	29.7%
2008P	23.05	34.4%	34.1%	28.7%

Source: EIA, *Annual Energy Review 2008*, Table 6.5.

Notes: Data for 2008 are preliminary. Percentages do not add to 100. The remaining amount is used by the transportation sector.

In part because of increased demand by electric utilities, natural gas prices have become extremely volatile in recent years, as illustrated by **Figure 16**, which shows high, low, and yearly average prices for gas delivered to electricity generators.

Figure 16. Natural Gas Prices to Electricity Generators, 1978-2007

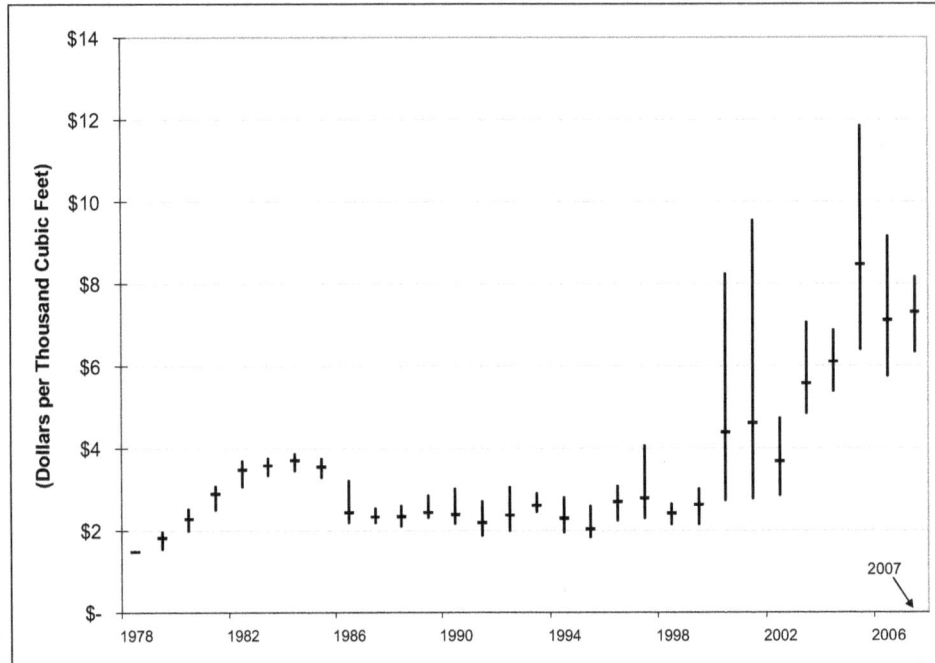

Source: EIA, *Monthly Energy Review*, December 2008, Table 9.11.

Because rates for residential natural gas are regulated, they have been less volatile than those for electric utility consumers, although considerable seasonal fluctuations are common, as shown in **Figure 17**. The long-term trend in residential natural gas prices, both in current dollars and in constant 2008 dollars, is shown in **Figure 18**.

Figure 17. Monthly and Annual Residential Natural Gas Prices, 2000-June 2009

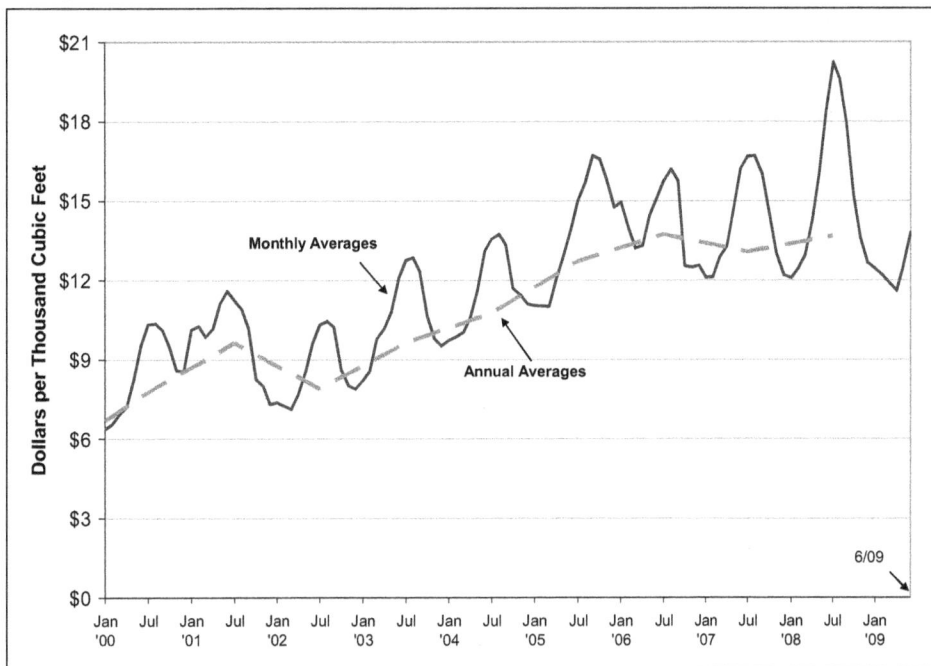

Source: EIA, *Monthly Energy Review*, September 2009, Table 9.11.

Figure 18. Annual Residential Natural Gas Prices, 1973-2008

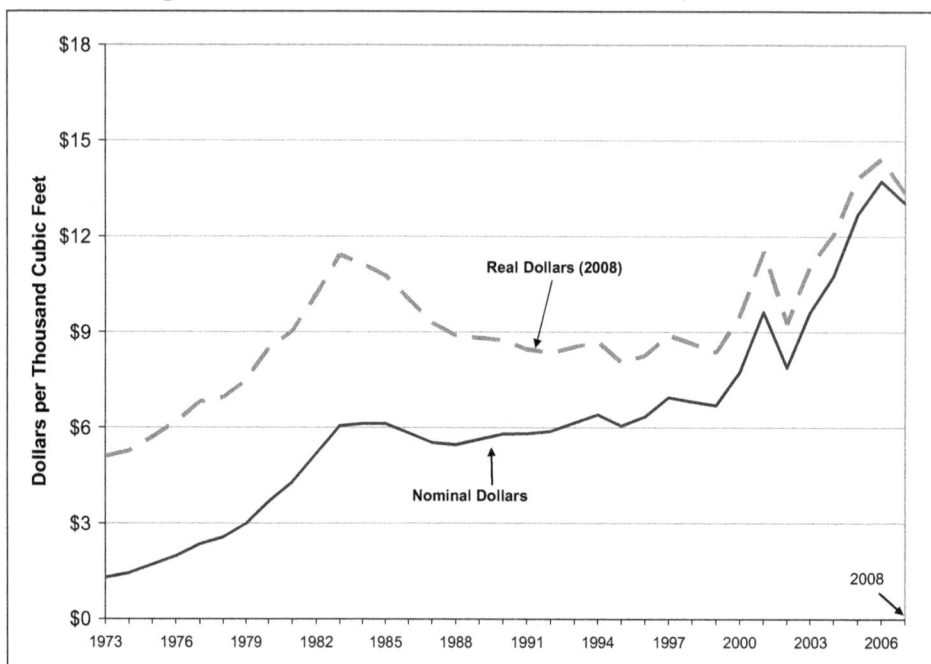

Source: EIA, *Monthly Energy Review*, September 2009, Table 9.11 and FY2010 Budget, *Historical Tables*, Table 10.1 for GDP Chained Price Index.

Coal

Consumption of coal has more than doubled since 1950, but during that period coal as an energy source changed from a widely used resource to a single-use fuel for generating electricity. (See **Table 7**.) In 1950 the residential and commercial sector consumed almost a quarter of total coal consumed; by 1980 less than 1% of coal went to that sector. In transportation, steam locomotives (and some coal-fired marine transportation) consumed 13% of coal; by 1970 they were all replaced with diesel-burning or electric engines. Industry consumed 46% of coal in 1950; by 2000 less than 10% of coal was consumed by that sector. Meanwhile, the electric power sector, which consumed less than 20% of the half-billion tons of coal burned in 1950, used more than 90% of the billion-plus tons consumed in 2008.

Table 8. Coal Consumption by Sector, 1950-2008

	Total Consumption	Percent Consumed by Sector			
	(million tons)	Residential-Commercial	Industrial	Transportation	Electric
1950	494.1	23.2%	45.5%	12.8%	18.6%
1955	447.0	15.3%	48.7%	3.8%	32.2%
1960	398.1	10.3%	44.6%	0.8%	44.4%
1965	472.0	5.4%	42.6%	0.1%	51.9%
1970	523.2	3.1%	35.7%	0.1%	61.2%
1975	562.6	1.7%	26.2%	–	72.2%
1980	702.7	0.9%	18.1%	–	81.0%
1985	818.0	1.0%	14.2%	–	84.8%
1990	904.5	0.7%	12.7%	–	86.5%
1995	962.1	0.6%	11.0%	–	88.4%
2000	1,084.1	0.4%	8.7%	–	90.9%
2005	1,126.0	0.4%	7.4%	–	92.1%
2006	1,112.3	0.3%	7.4%	–	92.3%
2007	1,128.0	0.3%	7.0%	–	92.7%
2008P	1,121.7	0.3%	6.8%	–	92.9%

Source: EIA, *Annual Energy Review 2008*, Table 7.3

Notes: Data for 2008 are preliminary.

Renewables

The major supply of renewable energy in the United States, not counting hydroelectric power generation, is fuel ethanol. Consumption in the United States in 2008 was about 9.0 billion gallons, mainly blended into E10 gasohol (a blend of 10% ethanol and 90% gasoline). This figure represents 6.5% of the approximately 138 billion gallons of gasoline consumption in the same year. As **Figure 19** indicates, fuel ethanol production has increased rapidly in recent years, especially since the phasing out of the fuel additive methyl tertiary butyl ether (MTBE).

Figure 19. U.S. Ethanol Production, 1990-2008

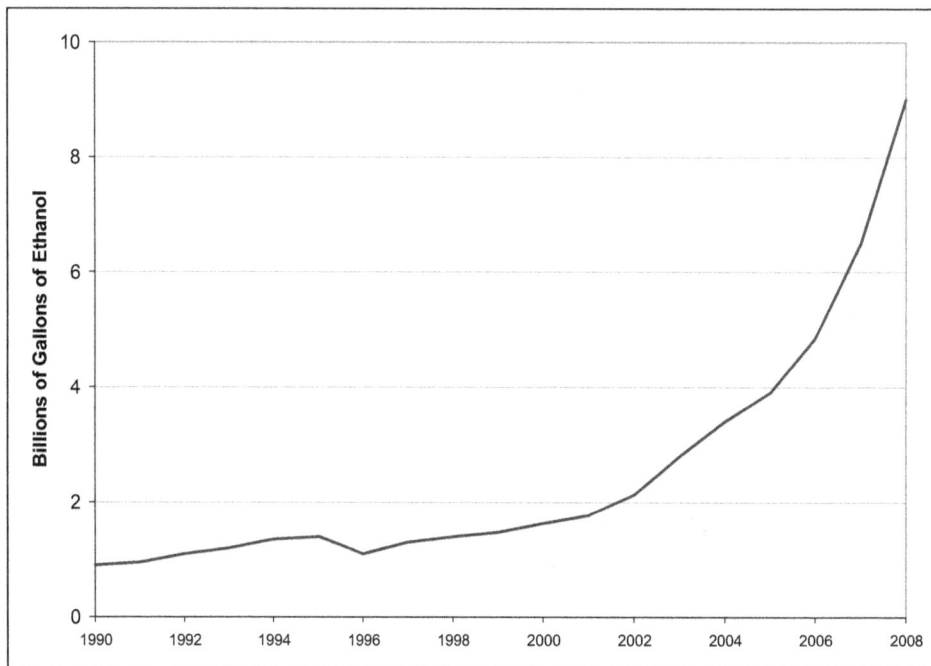

Source: Renewable Fuels Association, September 29, 2009. http://www.ethanolrfa.org/industry/statistics/.

Another rapidly growing renewable resource is wind-generated electric power, as shown in **Figure 20**. The 500 trillion Btu's of wind energy in 2008 is equivalent to approximately 145 billion kilowatt hours, about 3.6% of the 4,000 billion kwh of total electricity generation in that year.

Figure 20. Wind Electricity Net Generation, 1989-2008

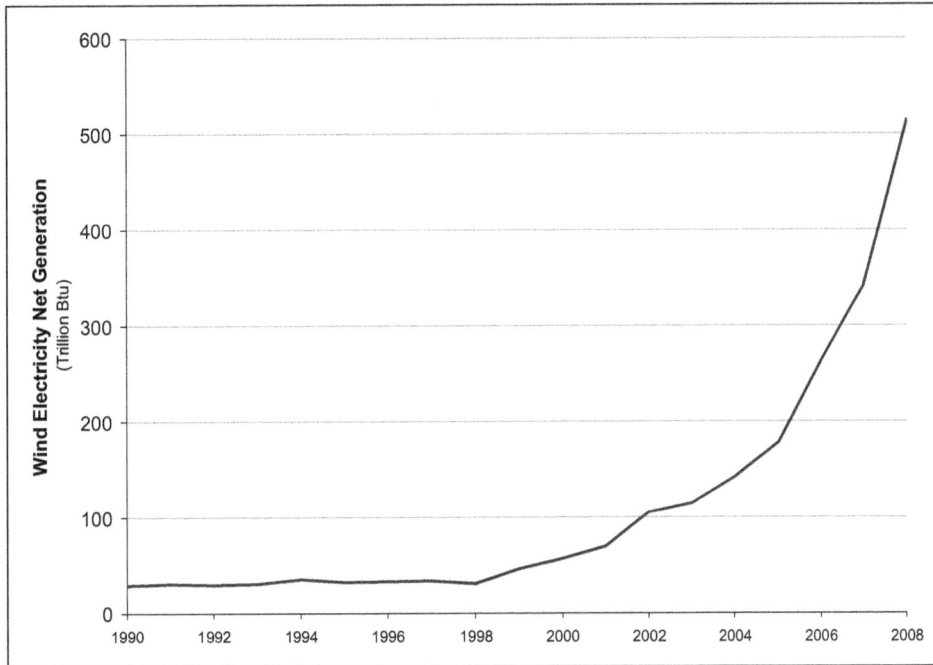

Source: *Monthly Energy Review*, September 2009, Table 10.1.

Notes: Wind electricity net generation converted to Btu using the fossil-fueled plants' heat rate.

Conservation and Energy Efficiency

Vehicle Fuel Economy

Energy efficiency has been a popular goal of policy makers in responding to the repeated energy crises of recent decades, and efforts to reduce the energy intensity of a broad spectrum of economic activities have been made both at the government and private level. Because of the transportation sector's near total dependence on vulnerable oil supplies, improving the efficiency of motor vehicles has been of particular interest. (For an analysis of legislative policies to improve vehicle fuel economy, see CRS Report R40166, *Automobile and Light Truck Fuel Economy: The CAFE Standards*, by Brent D. Yacobucci and Robert Bamberger.) **Figure 21** illustrates the trends in this effort for passenger cars and for light trucks, vans, and sport utility vehicles, as well as the general lack of improvement in heavy trucks.

Figure 21. Motor Vehicle Efficiency Rates, 1973-2007

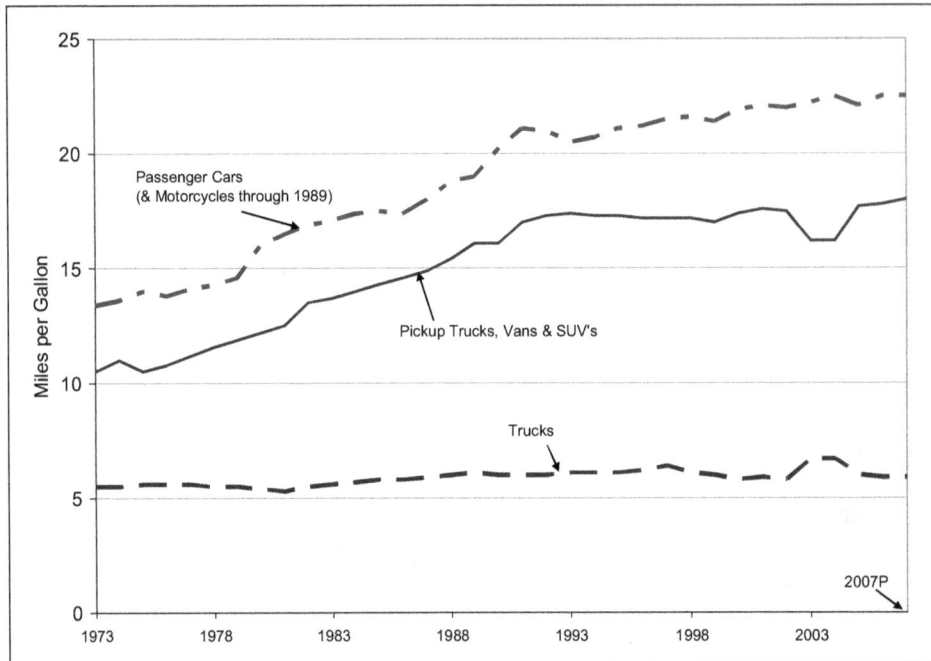

Source: EIA, *Monthly Energy Review*, September 2009, Table 1.8.

Note: Data for 2007 are preliminary.

Further analysis by the Environmental Protection Agency (EPA), involving the composition of the fleet as well as the per-vehicle fuel rates, indicates that light vehicle fuel economy has declined on average between 1988 and 2003. This is largely because of increased weight, higher performance, and a higher proportion of sport utility vehicles and light trucks sold. In 2003, SUVs, pickups, and vans comprised 48% of all sales, more than twice their market share in 1983. (The EPA study is available online at http://www.epa.gov/otaq/fetrends.htm.)

Energy Consumption and GDP

A frequent point of concern in formulating energy policy is the relationship between economic growth and energy use. It seems obvious that greater economic activity would bring with it increased energy consumption, although many other factors affecting consumption make the short-term relationship highly variable. Over a longer period, for some energy-related activities, the relationship with economic growth has been essentially level. For the period from 1973 to 2003, for instance, consumption of electricity remained close to 0.45 kwh per constant dollar of GDP. Similarly, the number of miles driven by all vehicles was close to 3 miles per constant dollar of GDP throughout the same period.

In the case of oil and gas, however, a remarkable drop took place in the ratio of consumption to economic growth following the price spikes and supply disruptions, as illustrated in **Figure 22**. Consumption of oil and gas declined from 14,000 Btus per constant dollar of GDP in 1973 to a little more than 8,000 in 1985, and has continued to decline at a slower rate since then.

Figure 22. Oil and Natural Gas Consumption per Dollar of GDP, 1973-2008

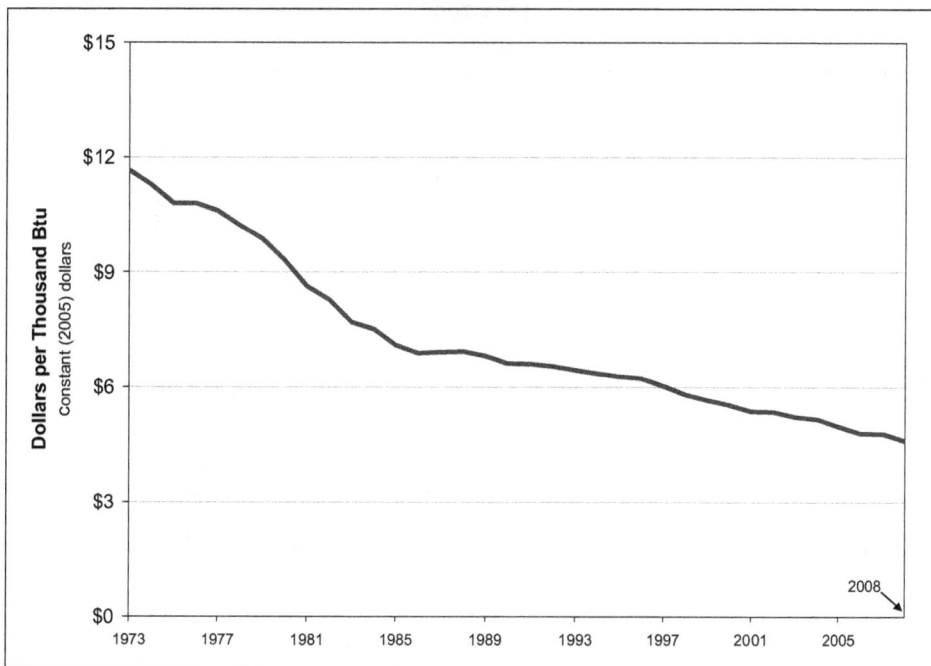

Source: EIA, *Monthly Energy Review*, September 2009, Table 1.7.

During the earlier period, oil and gas consumption actually declined 15% while GDP, despite many economic problems with inflation and slow growth, was increasing by 44% (see **Figure 22**). During the period 1987 to 2007, oil and gas consumption increased by about 25%, while GDP increased 76%.

Figure 23. Change in Oil and Natural Gas Consumption and Growth in GDP, 1973-2008

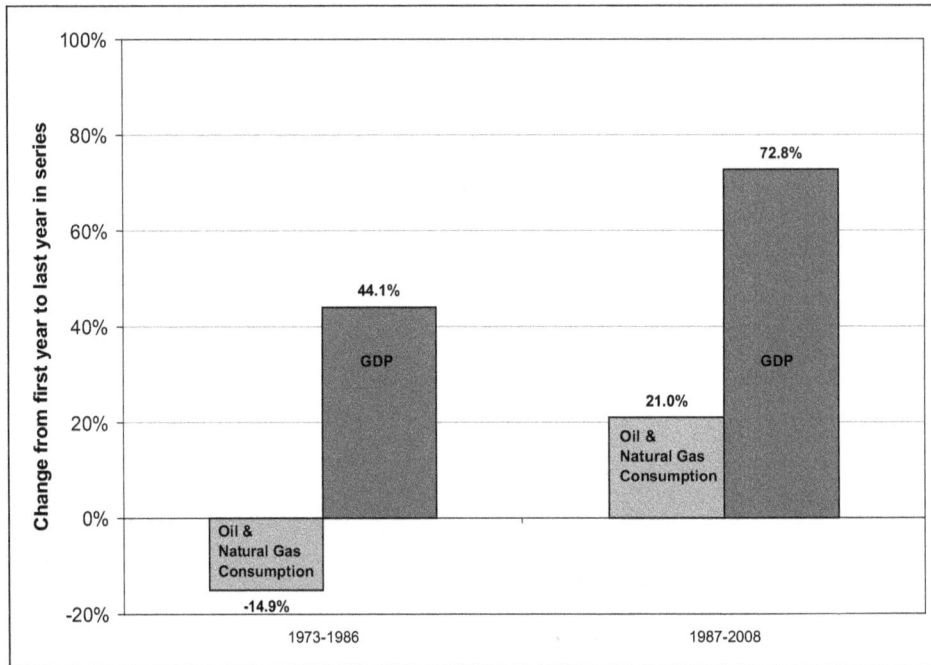

Source: EIA, *Monthly Energy Review*, September 2009, Table 1.7.

Notes: Percentages calculated by CRS. Percent change in oil and natural gas consumption measured in quadrillion Btu. Percent change in GDP based on billion chained (2005) dollars.

Major Statistical Resources

Energy Information Administration (EIA)

EIA home page—http://www.eia.doe.gov

Most of the tables and figures in this report are derived from databases maintained by the Energy Information Administration (EIA), an independent agency of the Department of Energy. EIA's Website presents the complete text of its many statistical reports in PDF's and Excel files.

EIA, Publications and Reports—http://www.eia.doe.gov/bookshelf.html

EIA's most frequently requested reports include the following:

Annual Energy Review: all the historical yearly energy data across fuels

Annual Energy Outlook: energy projections out to 2030

Country Analysis Briefs: country-level energy overviews

Electric Power Monthly: monthly summary of electric power generation and capacity

International Energy Annual: international historical yearly energy data across fuels

International Energy Outlook: worldwide energy projections to 2025

Monthly Energy Review: all the latest monthly energy data across fuels

This Week in Petroleum: weekly prices and analytical summary of the petroleum industry

Weekly Petroleum Status Report: weekly petroleum prices, production and stocks data

Other Sources

Nuclear Regulatory Commission Information Digest: http://www.nrc.gov/reading-rm/doc-collections/nuregs/staff/sr1350/

Updated annually, this official NRC publication (NUREG-1350) includes general statistics on U.S. and worldwide nuclear power production, U.S. nuclear reactors, and radioactive waste

American Petroleum Institute (API): http://api-ec.api.org/newsplashpage/index.cfm

The primary trade association of the oil and natural gas industry representing more than 400 members. Research, programs, and publications on public policy, technical standards, industry statistics, and regulations.

API: State Gasoline Tax Reports: http://www.api.org/statistics/fueltaxes/index.cfm

Bloomberg.Com, Market Data: Commodities, Energy Prices: http://www.bloomberg.com/energy/index.html

Displays four tables:

- Petroleum ($/bbl) for crude oil. The generally accepted price for crude oil is "WTI Cushing $" which is listed third in the table.
- Petroleum (¢/gal) for heating oil and gasoline.
- Natural Gas ($/MMBtu)
- Electricity ($/megawatt hour)

This site is updated two to three times per day.

AAA's Daily Fuel Gauge Report: http://www.fuelgaugereport.com/index.asp

At-the-pump retail fuel prices for gasoline and diesel fuel. Gives average price for today, yesterday, a month ago and a year ago for wholesale and crude oil. Also displays line chart showing the averages for the previous 12 months. National, state, and metropolitan data.

International Energy Agency: http://www.iea.org

The International Energy Agency is an autonomous body within the Organization for Economic Co-operation and Development (OECD). It gathers and analyzes statistics and "disseminates information on the world energy market and seeks to promote stable international trade in energy."

A subscription is required to access most of the information on this Website, although a limited amount of information is available to nonsubscribers. Members of Congress and their staff should contact CRS for a copy of anything that requires a subscription.

Author Contact Information

Carl E. Behrens
Specialist in Energy Policy
cbehrens@crs.loc.gov, 7-8303

Carol Glover
Information Research Specialist
cglover@crs.loc.gov, 7-7353

Key Policy Staff

Area of Expertise	Name	Phone	E-mail
Introduction and General	Carl Behrens	7-8303	cbehrens@crs.loc.gov
Oil	Robert Pirog	7-6847	rpirog@crs.loc.gov
Energy Taxes	Robert Pirog	7-6847	rpirog@crs.loc.gov
Electricity	Stan Kaplan	7-9529	skaplan@crs.loc.gov
Other Conventional Energy Sources			
Natural Gas	Robert Pirog	7-6847	rpirog@crs.loc.gov
Coal	Anthony Andrews	7-6843	aandrews@crs.loc.gov
Nuclear Energy	Mark Holt	7-1704	mholt@crs.loc.gov
Conservation and Energy Efficiency			
Renewable Energy	Fred Sissine	7-7039	fsissine@crs.loc.gov
	Larry Parker	7-7238	lparker@crs.loc.gov
	Brent Yacobucci	7-9662	byacobucci@crs.loc.gov
CAFE Standards (vehicle fuel economy)	Brent Yacobucci	7-9662	byacobucci@crs.loc.gov
	Robert Bamberger	7-7240	rbamberger@crs.loc.gov
Statistics, Tables, Figures	Carol Glover	7-7353	cglover@crs.loc.gov

*Congressional
Research
Service*

Nuclear Energy Policy

Mark Holt
Specialist in Energy Policy

December 10, 2009

Congressional Research Service

7-5700

www.crs.gov

RL33558

CRS Report for Congress

Prepared for Members and Committees of Congress

Summary

Nuclear energy issues facing Congress include federal incentives for new commercial reactors, radioactive waste management policy, research and development priorities, power plant safety and regulation, nuclear weapons proliferation, and security against terrorist attacks.

Significant incentives for new commercial reactors were included in the Energy Policy Act of 2005 (EPACT05, P.L. 109-58). These include production tax credits, loan guarantees, insurance against regulatory delays, and extension of the Price-Anderson Act nuclear liability system. Together with higher fossil fuel prices and the possibility of greenhouse gas controls, the federal incentives for nuclear power have helped spur renewed interest by utilities and other potential reactor developers. Plans for as many as 31 reactor license applications have been announced, although it is unclear how many of those projects will move forward.

The EPACT05 Title XVII loan guarantees, administered by the Department of Energy (DOE), are widely considered crucial by the nuclear industry to obtain financing for new reactors. However, opponents contend that nuclear loan guarantees would provide an unjustifiable subsidy to a mature industry and shift investment away from environmentally preferable energy technologies. The total amount of loan guarantees to be provided to nuclear power projects has been a continuing congressional issue. Nuclear power plants are currently allocated $18.5 billion in loan guarantees, enough for two or three reactors.

DOE's nuclear energy research and development program includes advanced reactors, fuel cycle technology and facilities, and infrastructure support. The FY2010 Energy and Water Development Appropriations Act (P.L. 111-8) provides $786.6 million for those activities, $10 million above the Obama Administration request and about $5 million below the FY2009 level.

Disposal of highly radioactive waste has been one of the most controversial aspects of nuclear power. The Nuclear Waste Policy Act of 1982 (P.L. 97-425), as amended in 1987, requires DOE to conduct a detailed physical characterization of Yucca Mountain in Nevada as a permanent underground repository for high-level waste. DOE submitted a license application for the Yucca Mountain repository to the Nuclear Regulatory Commission (NRC) on June 3, 2008, with the repository to open by 2020 at the earliest.

The Obama Administration has decided to "terminate the Yucca Mountain program while developing nuclear waste disposal alternatives," according to the DOE FY2010 budget justification. Alternatives to Yucca Mountain are to be evaluated by a "blue ribbon" panel of experts convened by the Administration.

The FY2010 budget request of $198.6 million for DOE's Office of Civilian Radioactive Waste Management provides only enough funding to continue the Yucca Mountain licensing process and to evaluate alternative policies, according to DOE. The request is about $90 million below the FY2009 funding level, which was nearly $100 million below the FY2008 level. All work related solely to preparing for construction and operation of the Yucca Mountain repository is being halted, according to the DOE budget justification. The FY2010 Energy and Water Development Appropriations Act includes the requested cuts in the waste program and provides $5 million for the blue ribbon panel. A draft of the DOE FY2011 budget request indicates that Yucca Mountain licensing is to be halted by the end of 2010.

Contents

Congressional Research Service

Tables

Contacts

Most Recent Developments

Funding for Department of Energy (DOE) nuclear energy research and development activities is included in the FY2010 Energy and Water Development Appropriations Act (P.L. 111-85), signed by President Obama on October 28, 2009. DOE's nuclear R&D program includes advanced reactors, fuel cycle technology and facilities, and infrastructure support. P.L. 111-85 provides $786.6 million for those activities, $10 million above the Obama Administration request and about $5 million below the FY2009 level.

The Obama Administration's FY2010 budget request called for termination of DOE's proposed nuclear waste repository at Yucca Mountain, NV, and for a "blue ribbon" panel of experts to develop alternative waste strategies. The FY2010 budget request of $198.6 million for DOE's Office of Civilian Radioactive Waste Management provides only enough funding to continue the Yucca Mountain licensing process before the Nuclear Regulatory Commission (NRC). All work related solely to preparing for construction and operation of the Yucca Mountain repository is being halted, according to the DOE budget justification. The FY2010 Energy and Water Development Appropriations Act includes the requested cuts in the waste program and provides $5 million for the blue ribbon panel. A draft of the DOE FY2011 budget request indicates that Yucca Mountain licensing is to be halted by the end of 2010.

Seventeen applications for combined construction permits and operating licenses (COLs) for 26 new nuclear power units have been submitted to NRC, although two applications were suspended by Entergy on January 9, 2009 (see **Table 1**). NRC is anticipating COL applications for as many as 31 new reactors through 2009. None of the applicants has yet committed to actual plant construction, although some preliminary contracts have been signed.

Nuclear Power Status and Outlook

The outlook for the U.S. nuclear power industry appears to have brightened after decades of uncertainty. No nuclear power plants have been ordered in the United States since 1978, and more than 100 reactors have been canceled, including all ordered after 1973. The most recent U.S. nuclear unit to be completed was TVA's Watts Bar 1 reactor, ordered in 1970 and licensed to operate in 1996. But nuclear power is now receiving renewed interest, prompted by volatile fossil fuel prices, possible carbon dioxide controls, and new federal subsidies and incentives.

The U.S. nuclear power industry currently comprises 104 licensed reactors at 65 plant sites in 31 states and generates about 20% of the nation's electricity.[1] That number includes TVA's Browns Ferry 1, which restarted May 22, 2007, after a 22-year shutdown and $1.8 billion refurbishment. TVA's board of directors voted August 1, 2007, to resume construction on Watts Bar 2, which had been suspended in 1985; the project is to cost about $2.5 billion and be completed in 2013. At TVA's request, NRC in March 2009 reinstated the construction authorization for the two-unit Bellefonte (AL) nuclear plant, which had been deferred in 1988 and canceled in 2006.[2]

---

[1] U.S. Nuclear Regulatory Commission, *Information Digest 2008-2009*, NUREG-1350, Vol. 20, August 2008, p. 32, http://www.nrc.gov/reading-rm/doc-collections/nuregs/staff/sr1350/v20/sr1350v20.pdf.

[2] Nuclear Regulatory Commission, "In the Matter of Tennessee Valley Authority (Bellefonte Nuclear Plant Units 1 and 2)," 74 *Federal Register* 10969, March 13, 2009.

Annual electricity production from U.S. nuclear power plants is greater than that from oil and hydropower, and slightly below natural gas, although it remains well behind coal, which accounts for about half of U.S. electricity generation. Nuclear plants generate more than half the electricity in six states. The near-record 842 billion kilowatt-hours of nuclear electricity generated in the United States during 2008[3] was more than the nation's entire electrical output in the early 1960s, when the oldest of today's operating U.S. commercial reactors were ordered.[4]

Reasons for the 30-year halt in U.S. nuclear plant orders include high capital costs, public concern about nuclear safety and waste disposal, and regulatory compliance costs.

High construction costs may pose the most serious obstacle to nuclear power expansion. Construction costs for reactors completed since the mid-1980s ranged from $2 to $6 billion, averaging more than $3,700 per kilowatt of electric generating capacity (in 2007 dollars). The nuclear industry predicts that new plant designs could be built for less than that if many identical plants were built in a series, but current estimates for new reactors show little if any reduction in cost.[5]

Average U.S. nuclear plant operating costs, however, dropped substantially since 1990, and costly downtime has been steadily reduced. Licensed commercial reactors generated electricity at an average of 90% of their total capacity in 2008, according to industry statistics.[6]

Fifty-seven commercial reactors have received 20-year license extensions from the Nuclear Regulatory Commission (NRC), giving them up to a total of 60 years of operation. License extensions for 20 additional reactors are currently under review, and more are anticipated, according to NRC.[7] The FY2010 Energy and Water Development Appropriations Act provides $10 million for DOE to study further reactor life extension to 80 years.

Existing nuclear power plants appear to hold a strong position in electricity wholesale markets. In most cases, nuclear utilities have received favorable regulatory treatment of past construction costs, and average nuclear operating costs are estimated to be competitive with those of fossil fuel technologies.[8] Although eight U.S. nuclear reactors were permanently shut down during the 1990s, none has been closed since 1998. Despite the shutdowns, annual U.S. nuclear electrical output increased by more than one-third from 1990 to 2006, according to the Energy Information Administration and industry statistics. The increase resulted primarily from reduced downtime at the remaining plants, the startup of five new units (most recently Watts Bar 1 in 1996), and reactor modifications to boost capacity.

---

[3] "World Nuclear Performance in 2008 Close to Output in 2007," *Nucleonics Week*, March 5, 2009, p. 1.

[4] All of today's 104 operating U.S. commercial reactors were ordered from 1963 through 1973; see "Historical Profile of U.S. Nuclear Power Development," U.S. Council for Energy Awareness, 1992.

[5] CRS Report RL34746, *Power Plants: Characteristics and Costs*, by Stan Mark Kaplan

[6] "World Nuclear Performance in 2008 Close to Output in 2007," *Nucleonics Week*, March 5, 2009, p. 1.

[7] http://www.nrc.gov/reactors/operating/licensing/renewal/applications.html

[8] Energy Information Administration, *Nuclear Power: 12 percent of America's Generating Capacity, 20 percent of the Electricity*, July 17, 2003, at http://www.eia.doe.gov/cneaf/nuclear/page/analysis/nuclearpower.html.

Possible New Reactors

The improved performance of existing reactors, the possibility of carbon dioxide controls that could affect coal plants, and volatile prices for natural gas—the favored fuel for new power plants for most of the past 15 years—have prompted renewed electric industry consideration of the feasibility of building new reactors. Electric utilities and other firms have announced plans to apply for combined construction permits and operating licenses (COLs) for about 30 reactors (see **Table 1**).[9]

No firm commitments have been made to build the proposed plants if the COLs are issued, but the sponsors of four nuclear projects have signed preliminary engineering, procurement, and construction (EPC) contracts. However, Entergy suspended further license review of its planned GE ESBWR reactors at River Bend, LA, and Grand Gulf, MS, and Dominion is seeking other potential vendors for its planned ESBWR at North Anna, VA, although it is continuing with the licensing process. AmerenUE suspended review of a COL for its proposed new Callaway unit in Missouri, and Exelon announced June 30, 2009, that it would no longer pursue a COL for a proposed two-unit plant in Victoria County, TX, but would seek an early site permit instead, laying the groundwork for possible future licensing. TVA announced August 7, 2009, that it would consider building one of the two new reactors it had proposed for the Bellefonte site in Alabama, or completing one of two partially built reactors at the site. The Department of Energy (DOE) is assisting Dominion's COL application as part of a program to encourage new commercial reactor orders by 2010, a program discussed in more detail below.

Table I. Announced Nuclear Plant License Applications

Announced Applicant	Site	Planned Application	Reactor Type	Units	Status
Alternate Energy	Hammett (ID)	2009	Areva EPR	I	
AmerenUE	Callaway (MO)	Submitted 7/24/08	Areva EPR	I	Construction plans suspended 4/23/09; NRC license review suspended 6/23/09
Amarillo Power	Near Amarillo (TX)	2009	Areva EPR	2	
Dominion	North Anna (VA)	Submitted 11/27/07	GE ESBWR	I	Other reactor vendors being considered 1/9/09
DTE Energy	Fermi (MI)	Submitted 9/18/08	GE ESBWR	I	
Duke Energy	Cherokee (SC)	Submitted 12/13/07	Westing.house AP1000	2	
Entergy	River Bend (LA)	Submitted 9/25/08	Not specified	I	Licensing suspended 1/9/09
Luminant Power (formerly TXU)	Comanche Peak (TX)	Submitted 9/19/08	Mitsubishi US-APWR	2	
FPL	Turkey Point (FL)	Submitted 6/30/09	Westinghouse AP1000	2	

[9] Nuclear Regulatory Commission, New Reactors, http://www.nrc.gov/reactors/new-reactors.html

Announced Applicant	Site	Planned Application	Reactor Type	Units	Status
NRG Energy	South Texas Project	Submitted 9/20/07	GE ABWR	2	EPC contract signed with Toshiba 2/12/09
NuStart	Grand Gulf (MS), Entergy	Submitted 2/27/08	Not specified	1	Licensing suspended Jan. 9, 2009
	Bellefonte (AL), TVA	Submitted 10/30/07	Westinghouse AP1000	2	NuStart announced shift of lead unit to Vogtle 4/30/09
PPL	Bell Bend (PA)	Submitted 10/10/08	Areva EPR	1	
Progress Energy	Harris (NC)	Submitted 2/19/08	Westinghouse AP1000	2	EPC contract signed 1/5/09
	Levy County (FL)	Submitted 7/30/08	Westinghouse AP1000	2	
SCE&G	Summer (SC)	Submitted 3/31/08	Westinghouse AP1000	2	EPC contract signed 5/27/08
Southern	Vogtle (GA)	Submitted 3/31/08	Westinghouse AP1000	2	EPC contract signed 4/8/08; Vogtle to be NuStart lead unit
UniStar (Constellation Energy and EDF)	Calvert Cliffs (MD)	Submitted 7/13/07 (Part 1), 3/13/08 (Part 2)	Areva EPR	1	
	Nine Mile Point (NY)	Submitted 9/30/08	Areva EPR	1	
Total Units				**29**	

Sources: NRC, *Nucleonics Week*, *Nuclear News*, Nuclear Energy Institute, company news releases.

NRC's current schedules indicate that the first COLs could be issued by 2011 or 2012, depending on the time required for hearings and other factors.[10] Issuance of a COL allows construction to begin and also is a prerequisite for federal loan guarantees and "regulatory risk insurance" as described below. If full-scale construction were to begin soon after receipt of the COLs, the first new reactors could begin operating before 2020. Southern Company is projecting that its planned two new reactors at the Vogtle site, currently scheduled to get the first COLs, will begin commercial operation by 2016 and 2017.[11]

How many of the reactors listed above are likely to move toward construction after receiving COLs remains highly uncertain. Major variables include construction costs, the availability of financing, construction capacity, fossil fuel prices, and federal incentives and carbon control policy. Recent projections of U.S. electric generating capacity show a wide variation in the amount of new nuclear generation that could be built by 2030—from none to 100 gigawatts (approximately double current capacity). (See Table 9 of CRS Report R40809, *Climate Change:*

---

[10] http://www.nrc.gov/reactors/new-reactors/col.html

[11] http://www.southerncompany.com/nuclearenergy/timeline.aspx

Costs and Benefits of the Cap-and-Trade Provisions of H.R. 2454, by Larry Parker and Brent D. Yacobucci.)

Federal Support

The nuclear power industry contends that support from the federal government would be needed for "a major expansion of nuclear energy generation."[12] Significant incentives for building new nuclear power plants were included in the Energy Policy Act of 2005 (EPACT05, P.L. 109-58), signed by President Bush on August 8, 2005. These include production tax credits, loan guarantees, insurance against regulatory delays, and extension of the Price-Anderson Act nuclear liability system (discussed in the "Nuclear Accident Liability" section of this report). Relatively low prices for natural gas—nuclear power's chief competitor—and rising estimated nuclear plant construction costs have decreased the likelihood that new reactors would be built without federal support. As a result, many draft proposals are currently circulating in Congress to strengthen or add to the EPACT incentives, possibly as part of climate change legislation. Nuclear power critics have denounced the federal support programs as a "bailout" of the nuclear industry, contending that federal efforts should focus instead on renewable energy and energy efficiency.[13]

Nuclear Production Tax Credit

EPACT05 provides a 1.8-cents/kilowatt-hour tax credit for up to 6,000 megawatts of new nuclear capacity for the first eight years of operation, up to $125 million annually per 1,000 megawatts.

The Treasury Department published interim guidance for the nuclear production tax credit on May 1, 2006.[14] Under the guidance, the 6,000 megawatts of eligible capacity (enough for about four or five reactors) are to be allocated among reactors that filed license applications by the end of 2008. If more than 6,000 megawatts of nuclear capacity ultimately qualify for the production tax credit, then the credit is to be allocated proportionally among any of the qualifying reactors that begin operating before 2021.

By the end of 2008, license applications had been submitted to NRC for more than 34,000 megawatts of nuclear generating capacity,[15] so if all those reactors were built before 2021 they would receive less than 20% of the maximum tax credit. However, the Energy Information Administration estimates that 8,000 megawatts of new nuclear capacity will ultimately qualify for the credit;[16] in this case the credit amount drops to 1.35 cents per kilowatt-hour once all the qualifying plants are on line. The credit is not adjusted for inflation.

---

[12] Nuclear Energy Institute, "NEI Unveils Package of Policy Initiatives Needed to Achieve Climate Change Goals," press release, October 26, 2009, http://www.nei.org/newsandevents/newsreleases/nei-unveils-package-of-policy-initiatives-needed-to-achieve-climate-change-goals/.

[13] Nuclear Information and Resource Service, "Senate Appropriators Lard President Obama's Stimulus Package with up to $50 Billion in Nuclear Reactor Pork," press release, January 30, 2009, http://www.nirs.org/press/01-30-2009/1.

[14] Department of the Treasury, Internal Revenue Service, *Internal Revenue Bulletin*, No. 2006-18, "Credit for Production From Advanced Nuclear Facilities," Notice 2006-40, May 1, 2006, p. 855.

[15] Energy Information Administration, *Status of Potential New Commercial Nuclear Reactors in the United States*, February 19, 2009.

[16] For a discussion of the operation of the credit, see EIA, *Annual Energy Outlook 2007*, p. 21. For the forecast of 8,000 MW of nuclear capacity on-line before 2021, see the *Annual Energy Outlook 2008*, p. 70.

The Nuclear Energy Institute (NEI) is urging Congress to remove the 6,000 megawatt capacity limit for the production tax credit, index it for inflation, and extend the deadline for plants to begin operation to the start of 2025. NEI is also proposing that a 30% investment tax credit be available for new nuclear construction as an alternative to the production credit.[17]

Standby Support

Because the nuclear industry has often blamed licensing delays for past nuclear reactor construction cost overruns, EPACT05 authorizes the Secretary of Energy to provide "standby support," or regulatory risk insurance, to help pay the cost of regulatory delays at up to six new commercial nuclear reactors, subject to funding availability. For the first two reactors that begin construction, the DOE payments could cover all the eligible delay-related costs, such as additional interest, up to $500 million each. For the next four reactors, half of the eligible costs could be paid by DOE, with a payment cap of $250 million per reactor. Delays caused by the failure of a reactor owner to comply with laws or regulations would not be covered. Project sponsors will be required to pay the "subsidy cost" of the program, consisting of the estimated present value of likely future government payments.

DOE published a final rule for the "standby support" program August 11, 2006.[18] According to a DOE description of the final rule:

> Events that would be covered by the risk insurance include delays associated with the Nuclear Regulatory Commission's reviews of inspections, tests, analyses and acceptance criteria or other licensing schedule delays as well as certain delays associated with litigation in federal, state or tribal courts. Insurance coverage is not available for normal business risks such as employment strikes and weather delays. Covered losses would include principal and interest on debt and losses resulting from the purchase of replacement power to satisfy contractual obligations.[19]

Under the program's regulations, a project sponsor may enter into a conditional agreement for standby support before NRC issues a combined operating license. The first six conditional agreements to meet all the program requirements, including the issuance of a COL and payment of the estimated subsidy costs, can be converted to standby support contracts. No conditional agreements have yet been reached, according to DOE, primarily because the subsidy cost estimates have not been approved by the Office of Management and Budget.[20]

The Nuclear Energy Institute has called for expanding the Standby Support program to $500 million for all six covered plants, rather than just the first two. In addition, NEI proposed that if a plant begins operating without any delay payments, that plant's Standby Support coverage,

[17] Nuclear Energy Institute, *Legislative Proposal to Help Meet Climate Change Goals by Expanding U.S. Nuclear Energy Production*, Washington, DC, October 28, 2009, p. 4, http://www.nei.org/resourcesandstats/documentlibrary/newplants/policybrief/2009-nuclear-policy-initiative.

[18] Department of Energy, "Standby Support for Certain Nuclear Plant Delays," *Federal Register*, August 11, 2006, p. 46306.

[19] DOE press release, August 4, 2006 http://nuclear.gov/home/08-04-06.html.

[20] Meeting with Rebecca F. Smith-Kevern, Director, DOE Office of Light Water Reactor Deployment, October 7, 2009.

instead of expiring unused, be allowed to "roll over" to the next plant with a conditional agreement.[21]

Loan Guarantees

Title XVII of EPACT05 authorizes federal loan guarantees for up to 80% of construction costs for advanced energy projects that reduce greenhouse gas emissions, including new nuclear power plants. The Title XVII loan guarantees are widely considered crucial by the nuclear industry to obtain financing for new reactors. However, opponents contend that nuclear loan guarantees would provide an unjustifiable subsidy to a mature industry and shift investment away from environmentally preferable energy technologies.[22]

The FY2007 continuing resolution (P.L. 110-5) established an initial cap of $4 billion in loan guarantees under the program. DOE issued final rules for the program October 4, 2007,[23] and finalized the first loan guarantee on September 4, 2009, totaling $535 million for a plant to produce photovoltaic panels.[24]

DOE's proposed loan guarantee rules, published May 16, 2007, had been sharply criticized by the nuclear industry for limiting the guarantees to 90% of a project's debt. The industry contended that EPACT05 allows all of a project's debt to be covered, as long as debt does not exceed 80% of total construction costs. In its explanation of the proposed rules, DOE expressed concern that guaranteeing 100% of a project's debt could reduce lenders' incentive to perform adequate due diligence and therefore increase default risks. In the final rule, however, DOE agreed to guarantee up to 100% of debt, but only for loans issued by the Federal Financing Bank.

Title XVII requires that estimated future government costs resulting from defaults on guaranteed loans be covered up-front by appropriations or by payments from project sponsors. These "subsidy costs" are calculated as the present value of probable future net costs to the government for each loan guarantee. If those calculations are accurate, the subsidy cost payments for all the guaranteed projects together should cover the future costs of the program. However, the Congressional Budget Office has predicted that the up-front subsidy cost payments will prove too low by at least 1% and is scoring bills accordingly.[25] For example, appropriations bills that provide loan guarantee authorizations include an adjustment totaling 1% of the loan guarantee ceiling.

DOE loan guarantees for renewable energy and electricity transmission projects under EPACT05 section 1705, added by the American Recovery and Reinvestment Act of 2009 (P.L. 111-5), do not require payments by project sponsors, because potential losses are covered by advance

---

[21] Nuclear Energy Institute, *op. cit.*

[22] Thomas B. Cochran and Christopher E. Paine, *Statement on Nuclear Developments Before the Committee on Energy and Natural Resources, United States Senate*, Natural Resources Defense Council, March 18, 2009, http://energy.senate.gov/public/index.cfm?FuseAction=Hearings.Testimony&Hearing_ID=f25ddd10-c1f5-9e2e-528e-c4321cca4c1b&Witness_ID=9f14a78d-58d0-43fb-bf5b-21426d1d888e.

[23] Published October 23, 2007 (72 *Federal Register* 60116).

[24] Department of Energy, "Vice President Biden Announces Finalized $535 Million Loan Guarantee," press release, September 4, 2009, http://www.lgprogram.energy.gov/press/090409.pdf.

[25] Congressional Budget Office, *S. 1321, Energy Savings Act of 2007*, CBO Cost Estimate, Washington, DC, June 11, 2007, pp. 7-9, http://www.cbo.gov/ftpdocs/82xx/doc8206/s1321.pdf.

appropriations in the act. No such appropriations are currently available for nuclear power projects, so it is anticipated that nuclear loan guarantee subsidy costs would be paid by the project sponsors. As a result, the level of the subsidy costs could have a powerful effect on the viability of nuclear power projects, which are currently expected to cost between $5 billion and $10 billion per reactor. For example, a 10% subsidy cost for a $7 billion loan guarantee would require an up-front payment of $700 million.

The amount of loan guarantees to be available for nuclear power has been the subject of considerable congressional debate. Under the Federal Credit Reform Act (FCRA), federal loan guarantees cannot be provided without an authorized level in an appropriations act. The Senate-passed version of omnibus energy legislation in the 110th Congress (H.R. 6) would have explicitly eliminated FCRA's applicability to DOE's planned loan guarantees under EPACT05 (Section 124(b)). That provision would have given DOE essentially unlimited loan guarantee authority for guarantees whose subsidy costs were paid by project sponsors, but it was dropped from the final legislation (P.L. 110-140). Similar language has been included in subsequent legislative proposals, such as energy legislation reported by the Senate Committee on Energy and Natural Resources July 16, 2009 (S. 1462).

The explanatory statement for the FY2008 omnibus funding act (P.L. 110-161) directed DOE to limit the loan guarantees for nuclear power plants to $18.5 billion through FY2009—enough for about two or three large reactors under current cost estimates. An additional $2 billion in loan guarantee authority was provided for uranium enrichment plants, and $18 billion in authority was provided for non-nuclear energy technologies, such as renewable energy.[26]

The FY2009 omnibus funding act increased DOE's total loan guarantee authority to $47 billion, in addition to the previously authorized $4 billion. Of the $47 billion, $18.5 billion continued to be reserved for nuclear power, $18.5 was for energy efficiency and renewables, $6 billion was for coal, $2 billion was for carbon capture and sequestration, and $2 billion was for uranium enrichment. The time limits on the loan guarantee authority were eliminated. The loan guarantee ceilings remain the same for FY2010.

DOE issued a solicitation for up to $20.5 billion in nuclear power and uranium enrichment plant loan guarantees on June 30, 2008.[27] According to the nuclear industry, 10 nuclear power projects are currently seeking $93.2 billion in loan guarantees, and two uranium enrichment projects are asking for $4.8 billion in guarantees, several times the amount available.[28] Several of the proposed projects listed in **Table 1** have been reported to be finalists for the first conditional nuclear loan guarantee commitments, including the South Texas Project, Calvert Cliffs, Summer, and Vogtle.[29] Under the program's regulations, a conditional loan guarantee commitment cannot become a binding loan guarantee agreement until the project receives a COL and all other regulatory requirements are met; as noted above, the first COLs are not expected until late 2011 at the earliest.

[26] *Congressional Record*, December 17, 2007, p. H15585.

[27] http://www.lgprogram.energy.gov/keydocs.html

[28] Marvin S. Fertel, *Statement for the Record to the Committee on Energy and Natural Resources, U.S. Senate*, Nuclear Energy Institute, March 18, 2009, p. 9, http://energy.senate.gov/public/index.cfm?FuseAction=Hearings.Testimony&Hearing_ID=f25ddd10-c1f5-9e2e-528e-c4321cca4c1b&Witness_ID=4de5e2df-53fe-49ba-906e-9b69d3674e41.

[29] Eileen O'Grady, "DOE Drops Luminant Texas from Nuclear Loan Talks," *Reuters*, May 7, 2009.

Global Climate Change

Global climate change that may be caused by carbon dioxide and other greenhouse gas emissions is cited by nuclear power supporters as an important reason to develop a new generation of reactors. Nuclear power plants emit relatively little carbon dioxide, mostly from nuclear fuel production and auxiliary plant equipment. This "green" nuclear power argument has received growing attention in think tanks and academia. As stated by the Massachusetts Institute of Technology in its major study *The Future of Nuclear Power*: "Our position is that the prospect of global climate change from greenhouse gas emissions and the adverse consequences that flow from these emissions is the principal justification for government support of the nuclear energy option."[30]

However, environmental groups have contended that nuclear power's potential greenhouse gas benefits are modest and must be weighed against the technology's safety risks, its potential for nuclear weapons proliferation, and the hazards of radioactive waste.[31] They also contend that energy efficiency and renewable energy would be far more productive investments for reducing greenhouse gas emissions.[32]

Congressional proposals to reduce carbon dioxide emissions, either through taxation or a cap-and-trade system, could significantly increase the cost of generating electricity with fossil fuels and improve the competitive position of nuclear power. Utilities that have applied for nuclear power plant licenses have often cited the possibility of federal greenhouse gas controls as one of the reasons for pursuing new reactors. (For more on federal incentives and the economics of nuclear power and other electricity generation technologies, see CRS Report RL34746, *Power Plants: Characteristics and Costs*, by Stan Mark Kaplan.)

Nuclear Power Research and Development

DOE's nuclear energy research and development program includes advanced reactors, fuel cycle technology and facilities, and infrastructure support. The Obama Administration's initial FY2010 funding request for nuclear energy R&D activities totaled $761.3 million—about $30 million below the comparable FY2009 level. The FY2010 Energy and Water Development Appropriations Act (P.L. 111-85), signed on October 28, 2009, provides $786.6 million.

According to DOE's FY2010 budget justification, the nuclear energy R&D program includes "generation, safety, waste storage and management, and security technologies, to help meet energy and climate goals." However, opponents have criticized DOE's nuclear research program as providing wasteful subsidies to an industry that they believe should be phased out as unacceptably hazardous and economically uncompetitive.

[30] Interdisciplinary MIT Study, *The Future of Nuclear Power*, Massachusetts Institute of Technology, 2003, p. 79.

[31] Gronlund, Lisbeth, David Lochbaum, and Edwin Lyman, *Nuclear Power in a Warming World*, Union of Concerned Scientists, December 2007.

[32] Travis Madsen, Tony Dutzik, and Bernadette Del Chiaro, et al., *Generating Failure: How Building Nuclear Power Plants Would Set America Back in the Race Against Global Warming*, Environment America Research and Policy Center, November 2009, http://www.environmentamerica.org/uploads/39/62/3962c378b66c4552624d09cbd8ebba02/Generating-Failure—Environment-America—Web.pdf.

Although total funding in the FY2010 nuclear energy request was similar to levels in previous years, the Obama Administration proposed significant priority changes. Funding for the Nuclear Power 2010 Program, which assists the near-term design and licensing of new nuclear power plants, was to be closed out during the fiscal year. The Advanced Fuel Cycle Initiative (AFCI), which had been the primary research component of the Bush Administration's Global Nuclear Energy Partnership (GNEP), has been renamed Fuel Cycle Research and Development and shifted away from the design and construction of nuclear fuel recycling facilities toward an emphasis on longer-term research.

Nuclear Power 2010

Under President Bush, DOE's initial efforts to encourage near-term construction of new commercial reactors—for which there have been no new U.S. orders since 1978—focused on the Nuclear Power 2010 Program. The program provided up to half the costs of licensing lead plant sites and reactors and preparing detailed reactor designs. Nuclear Power 2010 also includes the Standby Support Program, authorized by the Energy Policy Act of 2005 (P.L. 109-58) to pay for regulatory delays that might be experienced by new reactors.

The Obama Administration proposed to cut the Nuclear Power 2010 Program's funding from $177.5 million in FY2009 to $20 million in FY2010 and then terminate the program. Administration of the Standby Support Program was to continue under the Office of Nuclear Energy's program direction account.

DOE's budget justification contended that industry interest in new nuclear power plants has now been demonstrated to the extent that federal funding is no longer needed. The $20 million requested for FY2010 was to provide the final assistance to an industry consortium called NuStart for licensing a new reactor at the Vogtle plant in Georgia. No further funding was to be provided for a second industry consortium led by Dominion Resources, or for the design of General Electric-Hitachi's ESBWR reactor or the Westinghouse AP-1000 reactor. "By FY 2010 sufficient momentum will have been created by the cost-shared programs that the vendors (GEH and Westinghouse) and other partners will have adequate incentive to complete any additional work through private funding," according to the DOE justification.

The House approved a funding level of $71.0 million for the program, to "complete the Department's commitment to this effort." The Senate voted to provide $120 million for the program, with no mention of program termination. The conference agreement provides $105.0 million "as the final installment" for the Nuclear Power 2010 program.

Generation IV

Advanced commercial reactor technologies that are not yet close to deployment are the focus of Generation IV Nuclear Energy Systems, for which $191.0 million was requested for FY2010, $11 million above the FY2009 appropriation. The budget request would have cut $24 million from activities previously conducted by the program, a reduction that "reflects the emphasis shifting from near-term R&D activities to those R&D activities aimed at long-term technology advances," according to the DOE justification. The request included $35 million to establish the Energy Innovation Hub for Modeling and Simulation, which would focus on computer assistance for the development, implementation, and management of nuclear power and radioactive waste. The House provided no funding for the Modeling and Simulation Hub, while boosting total

Generation IV funding to $272.4 million. The Senate approved a funding level of $143 million, including the Modeling and Simulation Hub. The conference agreement provides $220.1 million, including $22.0 million for the Modeling and Simulation Hub.

The focus in the budget request on "long-term technology advances" differed sharply from the program's previous emphasis on developing the Next Generation Nuclear Plant (NGNP). Most of the FY2009 appropriation—$169.0 million—was for NGNP research and development. NGNP is currently planned to use Very High Temperature Reactor (VHTR) technology, which features helium as a coolant and coated-particle fuel that can withstand temperatures up to 1,600 degrees Celsius. Phase I research on the NGNP was to continue until 2011, when a decision was to be made on moving to the Phase II design and construction stage, according to the FY2009 DOE budget justification. In its recommendation on the FY2009 budget, the House Appropriations Committee had provided additional funding "to accelerate work" on NGNP.

DOE's proposed FY2010 nuclear research program did not mention NGNP, although it included several research activities related to the development of VHTR technology, including fuel testing, graphite experiments, and development of VHTR simulation software. Fundamental research on other advanced reactor concepts, such as sodium-cooled fast reactors and molten salt reactors, were also to continue. For FY2010, the House Appropriations Committee report noted that NGNP had been one of its priorities and specified that at least $245.0 million of the Generation IV funding be devoted to the project. The Senate Appropriations Committee's FY2010 report did not specifically mention NGNP, but it called for DOE to select two advanced reactor technologies as the focus of future research and potential deployment.

The conference agreement provides $169.0 million for NGNP and directs DOE within 90 days to prepare a detailed plan for moving forward with the NGNP project. The conference agreement also provides $17.8 million for other Generation IV reactor concepts and $10.0 million for research on extending the lives of existing light water reactors. No funding is provided for gas centrifuge enrichment technology.

The Energy Policy Act of 2005 authorized $1.25 billion through FY2015 for NGNP development and construction (Title VI, Subtitle C). The authorization requires that NGNP be based on research conducted by the Generation IV program and be capable of producing electricity, hydrogen, or both. The act's target date for operation of the demonstration reactor is September 30, 2021. The FY2010 budget request anticipated that Generation IV reactors "could be available in the 2030 timeframe."

Fuel Cycle Research and Development

Formerly called the Advanced Fuel Cycle Initiative, DOE's Fuel Cycle Research and Development program is to be redirected from the development of engineering-scale and prototype reprocessing facilities toward smaller-scale "long-term, science-based research." The FY2010 budget request for the program was $192.0 million, nearly $50 million above the FY2009 level, although $35 million of that amount was to go toward establishing an Energy Innovation Hub for Extreme Materials. The House provided no funding for the Extreme Materials Hub and an overall reduction in the request to $129.2 million, citing "the lack of specificity in terms of the direction of the research in this area." The Senate provided $145.0 million, the same as FY2009, and no funding for the Extreme Materials Hub. The conference agreement provides $136.0 million, with nothing for the Extreme Materials Hub.

According to the DOE budget justification, Fuel Cycle R&D will continue previous research on technology that could reduce the long-term hazard of spent nuclear fuel. Such technologies would involve separation of plutonium, uranium, and other long-lived radioactive materials from spent fuel for reuse in a nuclear reactor or for transmutation in a particle accelerator. DOE plans to broaden the program to include waste storage technologies, security systems, and alternative disposal options such as salt formations and deep boreholes. R&D will also focus on needs identified by a planned DOE nuclear waste strategy panel, according to the justification.

In previous years, AFCI had been the primary technology component of the Bush Administration's GNEP program, including R&D on reprocessing technology and fast reactors that could use reprocessed plutonium. Funding for GNEP was eliminated by Congress in FY2009, and GNEP was not mentioned in the FY2010 budget request, although, as noted above, much of the related R&D work is to continue at a smaller scale.

The Energy Innovation Hub for Extreme Materials was intended to support fundamental research on advanced materials for use in high-radiation and high-temperature environments. Such materials could improve the performance of nuclear waste packages, allow advances in nuclear reactor designs, and improve the safety and operation of existing commercial reactors, according to the budget justification.

(For more information about nuclear reprocessing, see CRS Report RL34579, *Advanced Nuclear Power and Fuel Cycle Technologies: Outlook and Policy Options*, by Mark Holt.)

Small Modular Reactors

Rising cost estimates for large conventional nuclear power plants—widely projected to be $6 billion or more—have contributed to growing interest in proposals for smaller, modular reactors. Ranging from about 40 to 350 megawatts of electrical capacity, such reactors would be only a fraction of the size of current commercial reactors. Several modular reactors would be installed together to make up a power block with a single control room, under most concepts.

Modular reactor concepts would use a variety of technologies, including high-temperature gas technology in the NGNP program and the light water (LWR) technology used by today's commercial reactors. According to media reports, DOE plans to request funding for FY2012 to provide licensing and engineering assistance to small reactor designs, in a program that would be similar to Nuclear Power 2010. Priority would be given to designs closest to commercialization, which DOE anticipates to be the small LWR concepts.[33] Legislation to authorize such a program (S. 2812) was introduced by Senator Bingaman November 20, 2009.

The Senate Appropriations Committee included instructions in its report on the FY2010 Energy and Water Appropriations Act that NRC use carryover funds to "support license application reviews of any new reactor designs, including modular reactors." NRC held a two-day workshop on small modular reactor licensing in early October 2009.

Small modular reactors would go against the overall trend in nuclear power technology toward ever-larger reactors intended to spread construction costs over a greater output of electricity.

[33] Randy Woods and Steven Dolley, "DOE to Seek Funds in FY-11 for Small Modular Reactors," *Nucleonics Week*, October 1, 2009.

Proponents of small reactors contend that they would be economically viable despite their far lower electrical output because modules could be assembled in factories and shipped to plant sites, and because their smaller size would allow for simpler safety systems. In addition, although modular plants might have similar or higher costs per kilowatt-hour than large conventional reactors, their ability to be constructed in smaller increments could reduce the financial commitment and risk to electric utilities.

Nuclear Power Plant Safety and Regulation

Safety

Controversy over safety has dogged nuclear power throughout its development, particularly following the March 1979 Three Mile Island accident in Pennsylvania and the April 1986 Chernobyl disaster in the former Soviet Union. In the United States, safety-related shortcomings have been identified in the construction quality of some plants, plant operation and maintenance, equipment reliability, emergency planning, and other areas. In one serious case, it was discovered in March 2002 that leaking boric acid had eaten a large cavity in the top of the reactor vessel in Ohio's Davis-Besse nuclear plant. The corrosion left only the vessel's quarter-inch-thick stainless steel inner liner to prevent a potentially catastrophic loss of reactor cooling water. Davis-Besse remained closed for repairs and other safety improvements until NRC allowed the reactor to restart in March 2004.

NRC's oversight of the nuclear industry is an ongoing issue; nuclear utilities often complain that they are subject to overly rigorous and inflexible regulation, but nuclear critics charge that NRC frequently relaxes safety standards when compliance may prove difficult or costly to the industry.

Domestic Reactor Safety

In terms of public health consequences, the safety record of the U.S. nuclear power industry in comparison with other major commercial energy technologies has been excellent. During approximately 3,000 reactor-years of operation in the United States,[34] the only incident at a commercial nuclear power plant that might lead to any deaths or injuries to the public has been the Three Mile Island accident, in which more than half the reactor core melted. A study of 32,000 people living within 5 miles of the reactor when the accident occurred found no significant increase in cancer rates through 1998, although the authors noted that some potential health effects "cannot be definitively excluded."[35]

The relatively small amounts of radioactivity released by nuclear plants during normal operation are not generally believed to pose significant hazards, although some groups contend that routine emissions are unacceptably risky. There is substantial scientific uncertainty about the level of risk posed by low levels of radiation exposure; as with many carcinogens and other hazardous substances, health effects can be clearly measured only at relatively high exposure levels. In the

---

[34] *Nuclear Engineering International*, "Country Averages to the End of December 2008," April 2009, p. 38.

[35] Evelyn O. Talbott et al., "Long Term Follow-Up of the Residents of the Three Mile Island Accident Area: 1979-1998," Environmental Health Perspectives, published online October 30, 2002, at http://ehp.niehs.nih.gov/docs/2003/5662/abstract.html.

case of radiation, the assumed risk of low-level exposure has been extrapolated mostly from health effects documented among persons exposed to high levels of radiation, particularly Japanese survivors of nuclear bombing in World War II.

NRC's safety regulations are designed to keep the probability of accidental core damage (fuel melting) below one in 10,000 per year for each reactor. The regulations also are intended to ensure that reactor containments would be successful at least 90% of the time in preventing major radioactive releases during a core-damage accident. Therefore, the probability of a major release at any given reactor is intended to be below one in 100,000 per year.[36] (For the current U.S. fleet of about 100 reactors, that rate would yield an average of one core-damage accident every 100 years and a major release every 1,000 years.) On the other hand, some groups challenge the complex calculations that go into predicting such accident frequencies, contending that accidents with serious public health consequences may be more frequent.[37]

Reactor Safety in the Former Soviet Bloc

The Chernobyl accident was by far the worst nuclear power plant accident to have occurred anywhere in the world. At least 31 persons died quickly from acute radiation exposure or other injuries, and thousands of additional cancer deaths among the tens of millions of people exposed to radiation from the accident may occur during the next several decades.

According to a 2006 report by the Chernobyl Forum organized by the International Atomic Energy Agency, the primary observable health consequence of the accident was a dramatic increase in childhood thyroid cancer. The Chernobyl Forum estimated that about 4,000 cases of thyroid cancer have occurred in children who after the accident drank milk contaminated with high levels of radioactive iodine, which concentrates in the thyroid. Although the Chernobyl Forum found only 15 deaths from those thyroid cancers, it estimated that about 4,000 other cancer deaths may have occurred among the 600,000 people with the highest radiation exposures, plus an estimated 1% increase in cancer deaths among persons with less exposure. The report estimated that about 77,000 square miles were significantly contaminated by radioactive cesium.[38] Greenpeace issued a report in 2006 estimating that 200,000 deaths in Belarus, Russia, and Ukraine resulted from the Chernobyl accident between 1990 and 2004.[39]

Licensing and Regulation

For many years, a top priority of the nuclear industry was to modify the process for licensing new nuclear plants. No electric utility would consider ordering a nuclear power plant, according to the industry, unless licensing became quicker and more predictable, and designs were less subject to mid-construction safety-related changes required by NRC. The Energy Policy Act of 1992 (P.L. 102-486) largely implemented the industry's licensing goals.

---

[36] U.S. NRC, Regulatory Guide 1.174, "An Approach for Using Probabilistic Risk Assessment in Risk-Informed Decisions on Plant-Specific Changes to the Licensing Basis," July 1998.

[37] Public Citizen Energy Program, "The Myth of Nuclear Safety" http://www.citizen.org/cmep/energy_enviro_nuclear/nuclear_power_plants/reactor_safety/articles.cfm?ID=4454

[38] The Chernobyl Forum: 2003-2005, *Chernobyl's Legacy: Health, Environmental and Socio-Economic Impacts*, International Atomic Energy Agency, April 2006.

[39] Greenpeace. *The Chernobyl Catastrophe: Consequences on Human Health*, April 2006, p. 10.

Nuclear plant licensing under the Atomic Energy Act of 1954 (P.L. 83-703; U.S.C. 2011-2282) had historically been a two-stage process. NRC first issued a construction permit to build a plant and then, after construction was finished, an operating license to run it. Each stage of the licensing process involved complicated proceedings. Environmental impact statements also are required under the National Environmental Policy Act.

Over the vehement objections of nuclear opponents, the Energy Policy Act of 1992 provided a clear statutory basis for one-step nuclear licenses, which would combine the construction permits and operating licenses and allow completed plants to operate without delay if they met all construction requirements—called "inspections, tests, analyses, and acceptance criteria," or ITAAC. NRC would hold preoperational hearings on the adequacy of plant construction only in specified circumstances.

DOE's Nuclear Power 2010 initiative (discussed above) has been paying up to half the cost of combined construction and operating licenses for two advanced reactors to demonstrate the process. However, the new licensing process cannot be fully tested until construction of new reactors is completed. At that point, it could be seen whether completed plants will be able to operate without delays or whether adjudicable disputes over construction adequacy may arise. As discussed above, Section 638 of the Energy Policy Act of 2005 authorizes federal payments to the owner of a completed reactor whose operation is delayed by regulatory action. The nuclear industry is asking Congress to require NRC to use informal procedures in determining whether ITAAC have been met, eliminate mandatory hearings for COLs on uncontested issues, and make other changes in the licensing process.[40]

A fundamental concern in the nuclear regulatory debate is the performance of NRC in issuing and enforcing nuclear safety regulations. The nuclear industry and its supporters have regularly complained that unnecessarily stringent and inflexibly enforced nuclear safety regulations have burdened nuclear utilities and their customers with excessive costs. But many environmentalists, nuclear opponents, and other groups charge NRC with being too close to the nuclear industry, a situation that they say has resulted in lax oversight of nuclear power plants and routine exemptions from safety requirements.

Primary responsibility for nuclear safety compliance lies with nuclear plant owners, who are required to find any problems with their plants and report them to NRC. Compliance is also monitored directly by NRC, which maintains at least two resident inspectors at each nuclear power plant. The resident inspectors routinely examine plant systems, observe the performance of reactor personnel, and prepare regular inspection reports. For serious safety violations, NRC often dispatches special inspection teams to plant sites.

In response to congressional criticism, NRC has reorganized and overhauled many of its procedures. The Commission has moved toward "risk-informed regulation," in which safety enforcement is guided by the relative risks identified by detailed individual plant studies. NRC's risk-informed reactor oversight system, inaugurated April 2, 2000, relies on a series of performance indicators to determine the level of scrutiny that each reactor should receive.[41]

---

[40] Nuclear Energy Institute, *Legislative Proposal to Help Meet Climate Change Goals by Expanding U.S. Nuclear Energy Production*, Washington, DC, October 28, 2009, p. 5, http://www.nei.org/resourcesandstats/documentlibrary/newplants/policybrief/2009-nuclear-policy-initiative.

[41] For more information about the NRC reactor oversight process, see http://www.nrc.gov/NRR/OVERSIGHT/ (continued...)

Reactor Security

Nuclear power plants have long been recognized as potential targets of terrorist attacks, and critics have long questioned the adequacy of the measures required of nuclear plant operators to defend against such attacks. All commercial nuclear power plants licensed by NRC have a series of physical barriers against access to vital reactor areas and are required to maintain a trained security force to protect them.

A key element in protecting nuclear plants is the requirement that simulated terrorist attacks, monitored by NRC, be carried out to test the ability of the plant operator to defend against them. The severity of attacks to be prepared for is specified in the form of a "design basis threat" (DBT).

EPACT05 required NRC to revise the DBT based on an assessment of terrorist threats, the potential for multiple coordinated attacks, possible suicide attacks, and other criteria. NRC approved the DBT revision based on those requirements on January 29, 2007. The revised DBT does not require nuclear power plants to defend against deliberate aircraft attacks. NRC contended that nuclear facilities were already required to mitigate the effects of large fires and explosions, no matter what the cause, and that active protection against airborne threats was being addressed by U.S. military and other agencies.[42] After much consideration, NRC voted February 17, 2009, to require all new nuclear power plants to incorporate design features that would ensure that, in the event of a crash by a large commercial aircraft, the reactor core would remain cooled or the reactor containment would remain intact, and radioactive releases would not occur from spent fuel storage pools.[43]

NRC rejected proposals that existing reactors also be required to protect against aircraft crashes, such as by adding large external steel barriers. However, NRC did impose some additional requirements related to aircraft crashes on all reactors, both new and existing, after the 9/11 terrorist attacks of 2001. In 2002, as noted above, NRC ordered all nuclear power plants to develop strategies to mitigate the effects of large fires and explosions that could result from aircraft crashes or other causes. An NRC regulation on fire mitigation strategies, along with requirements that reactors establish procedures for responding to specific aircraft threats, was approved December 17, 2008.[44]

Other ongoing nuclear plant security issues include the vulnerability of spent fuel pools, which hold highly radioactive nuclear fuel after its removal from the reactor, standards for nuclear plant security personnel, and nuclear plant emergency planning. NRC's December 2008 security regulations addressed some of those concerns and included a number of other security enhancements.

(...continued)

ASSESS/index.html.

[42] NRC Office of Public Affairs, *NRC Approves Final Rule Amending Security Requirements*, News Release No. 07-012, January 29, 2007.

[43] Nuclear Regulatory Commission, *Final Rule—Consideration of Aircraft Impacts for New Nuclear Power Reactors, Commission Voting Record*, SECY-08-0152, February 17, 2009.

[44] Nuclear Regulatory Commission, "NRC Approves Final Rule Expanding Security Requirements for Nuclear Power Plants," press release, December 17, 2008, http://www.nrc.gov/reading-rm/doc-collections/news/2008/08-227.html.

EPACT05 required NRC to conduct force-on-force security exercises at nuclear power plants every three years (which was NRC's previous policy), authorized firearms use by nuclear security personnel (preempting some state restrictions), established federal security coordinators, and required fingerprinting of nuclear facility workers.

(For background on security issues, see CRS Report RL34331, *Nuclear Power Plant Security and Vulnerabilities*, by Mark Holt and Anthony Andrews.)

Decommissioning

When nuclear power plants reach the end of their useful lives, they must be safely removed from service, a process called *decommissioning*. NRC requires nuclear utilities to make regular contributions to special trust funds to ensure that money is available to remove radioactive material and contamination from reactor sites after they are closed.

The first full-sized U.S. commercial reactors to be decommissioned were the Trojan plant in Oregon, whose decommissioning completion received NRC approval on May 23, 2005, and the Maine Yankee plant, for which NRC approved most of the site cleanup on October 3, 2005. The Trojan decommissioning cost $429 million, according to reactor owner Portland General Electric, and the Maine Yankee decommissioning cost about $500 million.[45] Decommissioning of the Connecticut Yankee plant cost $790 million and was approved by NRC on November 26, 2007.[46] NRC approved the cleanup of the decommissioned Rancho Seco reactor site in California on October 7, 2009.[47] The decommissioning of Rancho Seco was estimated to cost $500 million, excluding future demolition of the cooling towers and other remaining plant structures.[48] Spent nuclear fuel remains stored in dry casks at the decommissioned plant sites.

The tax treatment of decommissioning funds has been a continuing issue. EPACT05 provided favorable tax treatment to nuclear decommissioning funds, subject to certain restrictions.

Nuclear Accident Liability

Liability for damages to the general public from nuclear incidents is addressed by the Price-Anderson Act (primarily Section 170 of the Atomic Energy Act of 1954, 42 U.S.C. 2210). EPACT05 extended the availability of Price-Anderson coverage for new reactors and new DOE nuclear contracts through the end of 2025. (Existing reactors and contracts were already covered.)

Under Price-Anderson, the owners of commercial reactors must assume all liability for nuclear damages awarded to the public by the court system, and they must waive most of their legal defenses following a severe radioactive release ("extraordinary nuclear occurrence"). To pay any such damages, each licensed reactor with at least 100 megawatts of electric generating capacity must carry the maximum liability insurance reasonably available, currently $300 million. Any

---

[45] Sharp, David, "NRC Signs Off on Maine Yankee's Decommissioning," *Associated Press*, October 3, 2005.

[46] E-mail communication from Bob Capstick, Connecticut Yankee Atomic Power Company, August 28, 2008.

[47] Nuclear Regulatory Commission, "NRC Releases Rancho Seco Nuclear Plant for Unconditional Use," press release, October 7, 2009, http://www.nrc.gov/reading-rm/doc-collections/news/2009/09-165.html.

[48] "20 Years Later, Rancho Seco Ready for Final Shutdown," *Sacramento County Herald*, June 9, 2009, http://m.news10.net/news.jsp?key=190656.

damages exceeding that amount are to be assessed equally against all 100-megawatt-and-above power reactors, up to $111.9 million per reactor. Those assessments—called "retrospective premiums"—would be paid at an annual rate of no more than $17.5 million per reactor, to limit the potential financial burden on reactor owners following a major accident. According to NRC, all 104 commercial reactors are currently covered by the Price-Anderson retrospective premium requirement.[49]

For each nuclear incident, the Price-Anderson liability system currently would provide up to $12.5 billion in public compensation. That total includes the $300 million in insurance coverage carried by the reactor that suffered the incident, plus the $111.9 million in retrospective premiums from each of the 104 currently covered reactors, totaling $11.9 billion. On top of those payments, a 5% surcharge may also be imposed, raising the total per-reactor retrospective premium to $117.5 million and the total available compensation to about $12.5 billion. Under Price-Anderson, the nuclear industry's liability for an incident is capped at that amount, which varies depending on the number of covered reactors, the amount of available insurance, and an inflation adjustment. Payment of any damages above that liability limit would require congressional approval under special procedures in the act.

EPACT05 increased the limit on per-reactor annual payments to $15 million from the previous $10 million, and required the annual limit to be adjusted for inflation every five years. As under previous law, the total retrospective premium limit is adjusted every five years as well. Both the annual and total limits were most recently adjusted October 29, 2008.[50] For the purposes of those payment limits, a nuclear plant consisting of multiple small reactors (100-300 megawatts, up to a total of 1,300 megawatts) would be considered a single reactor. Therefore, a power plant with six 120-megawatt pebble-bed modular reactors would be liable for retrospective premiums of up to $111.9 million, rather than $671.4 million (excluding the 5% surcharge).

The Price-Anderson Act also covers contractors who operate hazardous DOE nuclear facilities. EPACT05 set the liability limit on DOE contractors at $10 billion per accident, to be adjusted for inflation every five years. The first adjustment under EPACT, raising the liability limit to $11.961 billion, took effect October 14, 2009.[51] The liability limit for DOE contractors previously had been the same as for commercial reactors, excluding the 5% surcharge, except when the limit for commercial reactors dropped because of a decline in the number of covered reactors. Price-Anderson authorizes DOE to indemnify its contractors for the entire amount of their liability, so that damage payments for nuclear incidents at DOE facilities would ultimately come from the Treasury. However, the law also allows DOE to fine its contractors for safety violations, and contractor employees and directors can face criminal penalties for "knowingly and willfully" violating nuclear safety rules.

EPACT05 limited the civil penalties against a nonprofit contractor to the amount of management fees paid under that contract. Previously, Atomic Energy Act §234A specifically exempted seven nonprofit DOE contractors and their subcontractors from civil penalties and authorized DOE to

---

[49] Reactors smaller than 100 megawatts must purchase an amount of liability coverage determined by NRC but are not subject to retrospective premiums. Total liability for those reactors is limited to $560 million, with the federal government indemnifying reactor operators for the difference between that amount and their liability coverage (Atomic Energy Act sec. 170 b. and c.).

[50] Nuclear Regulatory Commission, "Inflation Adjustment to the Price-Anderson Act Financial Protection Regulations," 73 *Federal Register* 56451, September 29, 2008.

[51] Department of Energy, "Adjusted Indemnification Amount," 74 *Federal Register* 52793, October 14, 2009.

automatically remit any civil penalties imposed on nonprofit educational institutions serving as DOE contractors. EPACT05 eliminated the civil penalty exemption for future contracts by the seven listed nonprofit contractors and DOE's authority to automatically remit penalties on nonprofit educational institutions.

The Price-Anderson Act's limits on liability were crucial in establishing the commercial nuclear power industry in the 1950s. Supporters of the Price-Anderson system contend that it has worked well since that time in ensuring that nuclear accident victims would have a secure source of compensation, at little cost to the taxpayer. Extension of the act was widely considered a prerequisite for new nuclear reactor construction in the United States. Opponents contend that Price-Anderson inappropriately subsidizes the nuclear power industry by reducing its insurance costs and protecting it from some of the financial consequences of the most severe conceivable accidents.

The United States is supporting the establishment of an international liability system that, among other purposes, would cover U.S. nuclear equipment suppliers conducting foreign business. The Convention on Supplementary Compensation for Nuclear Damage (CSC) will not enter into force until at least five countries with a specified level of installed nuclear capacity have enacted implementing legislation. Such implementing language was included in the Energy Independence and Security Act of 2007 (P.L. 110-140, section 934), signed by President Bush December 19, 2007. Supporters of the Convention hope that more countries will join now that the United States has acted. Aside from the United States, three countries have submitted the necessary instruments of ratification, but the remaining nine countries that so far have signed the convention do not have the required nuclear capacity for it to take effect. Ratification by a large nuclear energy producer such as Japan would allow the treaty to take effect, as would ratification by two significant but smaller producers such as South Korea, Canada, Russia, or Ukraine.

Under the U.S. implementing legislation, the CSC would not change the liability and payment levels already established by the Price-Anderson Act. Each party to the convention would be required to establish a nuclear damage compensation system within its borders analogous to Price-Anderson. For any damages not covered by those national compensation systems, the convention would establish a supplemental tier of damage compensation to be paid by all parties. P.L. 110-140 requires the U.S. contribution to the supplemental tier to be paid by suppliers of nuclear equipment and services, under a formula to be developed by DOE. Supporters of the convention contend that it will help U.S. exporters of nuclear technology by establishing a predictable international liability system. For example, U.S. reactor sales to the growing economies of China and India would be facilitated by those countries' participation in the CSC liability regime.

Nuclear Waste Management

One of the most controversial aspects of nuclear power is the disposal of radioactive waste, which can remain hazardous for thousands of years. Each nuclear reactor produces an annual average of about 20 metric tons of highly radioactive spent nuclear fuel, for a nationwide total of about 2,000 metric tons per year. U.S. reactors also generate about 40,000 cubic meters of low-level radioactive waste per year, including contaminated components and materials resulting from reactor decommissioning.[52]

The federal government is responsible for permanent disposal of commercial spent fuel (paid for with a fee on nuclear power production) and federally generated radioactive waste, whereas states have the authority to develop disposal facilities for most commercial low-level waste. Under the Nuclear Waste Policy Act (42 U.S.C. 10101, et seq.), spent fuel and other highly radioactive waste is to be isolated in a deep underground repository, consisting of a large network of tunnels carved from rock that has remained geologically undisturbed for hundreds of thousands of years. The program is run by DOE's Office of Civilian Radioactive Waste Management (OCRWM). As amended in 1987, NWPA designated Yucca Mountain in Nevada as the only candidate site for the national repository. The act required DOE to begin taking waste from nuclear plant sites by 1998—a deadline that even under the most optimistic scenarios will be missed by more than 20 years.

The Obama Administration has decided to "terminate the Yucca Mountain program while developing nuclear waste disposal alternatives," according to the DOE FY2010 budget justification. Alternatives to Yucca Mountain are to be evaluated by a "blue ribbon" panel of experts convened by the Administration. At the same time, according to the justification, the NRC licensing process for the Yucca Mountain repository is to continue, "consistent with the provisions of the Nuclear Waste Policy Act." However, draft proposals for the FY2011 budget request indicate that DOE will seek only enough funding to terminate all program activities and that repository licensing will end in December 2009.[53]

The FY2010 OCRWM budget request of $198.6 million sought only enough funding to continue the Yucca Mountain licensing process and to evaluate alternative policies, according to DOE. The request was about $90 million below the FY2009 funding level, which was nearly $100 million below the FY2008 level. More than 2,000 waste program contract employees were to be terminated during FY2009, according to the budget justification. Most of the program's remaining work is to be taken over by federal staff.

All work related solely to preparing for construction and operation of the Yucca Mountain repository is being halted, according to the DOE budget justification. Such activities include development of repository infrastructure, waste transportation preparations, and system engineering and analysis.

The House agreed with the Administration's plans to provide funding solely for Yucca Mountain licensing activities and for a blue-ribbon panel to review waste management options. The House approved the Administration budget request, including $5 million for the blue-ribbon review. However, the House-passed bill specified that the review must include Yucca Mountain as one of the alternatives, despite the Administration's contention that the site should no longer be considered. According to the House Appropriations Committee report, "It might well be the case that an alternative to Yucca Mountain better meets the requirements of the future strategy, but the review does not have scientific integrity without considering Yucca Mountain." The House panel

---

[52] DOE, Manifest Information Management System http://mims.apps.em.doe.gov. Average annual utility disposal from 2002 through 2007.

[53] Letter from Joe Barton, Ranking Member, House Committee on Energy and Commerce, and Greg Walden, Ranking Member, Subcommittee on Oversight and Investigations, to Steven Chu, Secretary of Energy, November 18, 2009, http://republicans.energycommerce.house.gov/Media/file/News/111809_Letter_to_Chu_Yucca.pdf.

also recommended that at least $70 million of the program's funding be devoted to maintaining expertise by the Yucca Mountain Project management contractor to support the licensing effort, rather than relying entirely on federal staff. The Senate also recommended approval of the Administration request, but without any restrictions on the blue-ribbon panel.

Funding for the nuclear waste program is provided under two appropriations accounts. The Administration's FY2010 request is divided evenly between an appropriation from the Nuclear Waste Fund, which holds fees paid by nuclear utilities, and the Defense Nuclear Waste Disposal account, which pays for disposal of high-level waste from the nuclear weapons program. The Senate Appropriations Committee report called for the Secretary of Energy to suspend fee collections, "given the Administration's decision to terminate the Yucca Mountain repository program while developing disposal alternatives."

The conference agreement provides the reduced funding requested by the Administration and includes bill language that states, "$5,000,000 shall be provided to create a Blue Ribbon Commission to consider all alternatives for nuclear waste disposal." That is the same language that appeared in the House-passed bill, along with House Appropriations Committee instructions that the Blue Ribbon panel include Yucca Mountain as a disposal option. However, the Conference Committee Joint Explanatory Statement states that "all guidance provided by the House and Senate reports is superseded by the conference agreement."

Additional funding from the Nuclear Waste Fund for the Yucca Mountain licensing process was included in the NRC budget request. The House provided the full $56 million requested, while the Senate voted to cut the request to $29 million. The conference agreement includes the Senate reduction.

The Yucca Mountain project faces regulatory uncertainty, in addition to the Obama Administration's policy review. A ruling on July 9, 2004, by the U.S. Court of Appeals for the District of Columbia Circuit overturned a key aspect of the Environmental Protection Agency's (EPA's) regulations for the planned repository.[54] The three-judge panel ruled that EPA's 10,000-year compliance period was too short, but it rejected several other challenges to the rules. EPA published new standards on October 15, 2008, that would allow radiation exposure from the repository to increase after 10,000 years.[55] The State of Nevada has filed a federal Appeals Court challenge to the EPA standards. (For more information on the EPA standards, see CRS Report RL34698, *EPA's Final Health and Safety Standard for Yucca Mountain*, by Bonnie C. Gitlin.)

NWPA required DOE to begin taking waste from nuclear plant sites by January 31, 1998. Nuclear utilities, upset over DOE's failure to meet that deadline, have won two federal court decisions upholding the department's obligation to meet the deadline and to compensate utilities for any resulting damages. Utilities have also won several cases in the U.S. Court of Federal Claims. DOE estimates that liability payments would eventually total $11 billion if DOE were to begin removing waste from reactor sites by 2020, the previous target for opening Yucca Mountain.[56] (For more information, see CRS Report R40202, *Nuclear Waste Disposal: Alternatives to Yucca*

[54] *Nuclear Energy Institute v. Environmental Protection Agency*, U.S. Court of Appeals for the District of Columbia Circuit, no. 01-1258, July 9, 2004.

[55] Environmental Protection Agency, "Public Health and Environmental Radiation Protection Standards for Yucca Mountain, Nevada," 73 *Federal Register* 61256, October 15, 2008.

[56] Statement of Edward F. Sproat III, Director of the Office of Civilian Radioactive Waste Management, Before the House Budget Committee, October 4, 2007.

Mountain, by Mark Holt, and CRS Report RL33461, *Civilian Nuclear Waste Disposal*, by Mark Holt.)

Nuclear Weapons Proliferation

Renewed interest in nuclear power throughout the world has led to increased concern about nuclear weapons proliferation, because technology for making nuclear fuel can also be used to produce nuclear weapons material. Of particular concern are uranium enrichment, a process to separate and concentrate the fissile isotope uranium-235, and nuclear spent fuel reprocessing, which can produce weapons-useable plutonium.

The International Atomic Energy Agency (IAEA) conducts a safeguards program that is intended to prevent civilian nuclear fuel facilities from being used for weapons purposes, but not all potential weapons proliferators belong to the system, and there are ongoing questions about its effectiveness. Several proposals have been developed to guarantee nations without fuel cycle facilities a supply of nuclear fuel in exchange for commitments to forgo enrichment and reprocessing, which was one of the original goals of the Bush Administration's GNEP program (discussed above under "Nuclear Power Research and Development").

Several situations have arisen throughout the world in which ostensibly commercial uranium enrichment and reprocessing technologies have been subverted for military purposes. In 2003 and 2004, it became evident that Pakistani nuclear scientist A.Q. Khan had sold sensitive technology and equipment related to uranium enrichment to states such as Libya, Iran, and North Korea. Although Pakistan's leaders maintain they did not acquiesce in or abet Khan's activities, Pakistan remains outside the Nuclear Nonproliferation Treaty (NPT) and the Nuclear Suppliers Group (NSG). Iran has been a direct recipient of Pakistani enrichment technology.

IAEA's Board of Governors found in 2005 that Iran's breach of its safeguards obligations constituted noncompliance with its safeguards agreement, and referred the case to the U.N. Security Council in February 2006. Despite repeated calls by the U.N. Security Council for Iran to halt enrichment and reprocessing-related activities, and imposition of sanctions, Iran continues to develop enrichment capability at Natanz and at a site near Qom disclosed in September 2009. Iran insists on its inalienable right to develop the peaceful uses of nuclear energy, pursuant to Article IV of the NPT. Interpretations of this right have varied over time. Former IAEA Director General Mohamed ElBaradei did not dispute this inalienable right and, by and large, neither have U.S. government officials. However, the case of Iran raises perhaps the most critical question in this decade for strengthening the nuclear nonproliferation regime: How can access to sensitive fuel cycle activities (which could be used to produce fissile material for weapons) be circumscribed without further alienating non-nuclear weapon states in the NPT?

Leaders of the international nuclear nonproliferation regime have suggested ways of reining in the diffusion of such inherently dual-use technology, primarily through the creation of incentives not to enrich uranium or reprocess spent fuel. The international community is in the process of evaluating those proposals and may decide upon a mix of approaches. At the same time, there is debate on how to improve the IAEA safeguards system and its means of detecting diversion of nuclear material to a weapons program in the face of expanded nuclear power facilities worldwide.

(For more information, see CRS Report RL34234, *Managing the Nuclear Fuel Cycle: Policy Implications of Expanding Global Access to Nuclear Power*, coordinated by Mary Beth Nikitin.)

Federal Funding for Nuclear Energy Programs

The following tables summarize current funding for DOE nuclear energy programs and NRC. The sources for the funding figures are Administration budget requests and committee reports on the Energy and Water Development Appropriations Acts, which fund DOE and NRC. FY2009 funding for energy and water programs was included in the Omnibus Appropriations Act for FY2009 (P.L. 111-8), signed March 11, 2009. Detailed funding tables for the act are provided by the Committee Print of the House Committee on Appropriations on H.R. 1105. FY2010 funding is included in the Energy and Water Development and Related Agencies Appropriations Act, 2010 (P.L. 111-85, H.Rept. 111-278), signed October 28, 2009.

Table 2. Funding for the Nuclear Regulatory Commission

(budget authority in millions of current dollars)

	FY2009 Approp.	FY2010 Request	FY2010 House	FY2020 Senate	FY2010 Approp.
Nuclear Regulatory Commission					
Reactor Safety	788.3	799.8	—a	—a	—a
Nuclear Materials and Waste	197.3	205.2	—a	—a	—a
Yucca Mountain Licensing	49.0	56.0	56.0	29.0	29.0
Inspector General	10.9	10.1	10.1	10.9	10.9
Total NRC budget authority	1,045.5	1,071.1	1,071.1	1,071.9	1,066.9
—Offsetting fees	-870.6	887.2	-887.2	-912.2	-912.2
Net appropriation	**174.9**	**183.9**	**183.9**	**159.7**	**154.7**

a. Subcategories not specified.

Table 3. DOE Funding for Nuclear Activities

(budget authority in millions of current dollars)

	FY2009 Approp.	FY2010 Request	FY2010 House	FY2010 Senate	FY2010 Approp.
Nuclear Energy (selected programs)					
Integrated University Program	5.0	0	0	5.0	5.0
Nuclear Power 2010	177.5	20.0	71.0	120.0	105.0
Generation IV Nuclear Systems	180.0	206.0	272.4	143.0	220.1
Nuclear Hydrogen Initiative	7.5	0	0	0	0
Fuel Cycle R&D	145.0	192.0	129.2	145.0	136.0
Radiological Facilities Management	66.1	77.0	67.0	62.0	72.0

	FY2009 Approp.	FY2010 Request	FY2010 House	FY2010 Senate	FY2010 Approp.
Idaho National Laboratory Infrastructure	218.8	286.8	277.4	356.7	173.0
Program Direction	73.0	77.9	77.9	73.0	73.0
Total, Nuclear Energy[a]	792.0	776.6	812.0	761.3	786.6
Civilian Nuclear Waste Disposal[b]	288.4	196.8	196.8	196.8	196.8

a.   Excludes funding provided under other accounts.

b.   Funded by a 1-mill-per-kilowatt-hour fee on nuclear power, plus appropriations for defense waste disposal and homeland security.

Legislation in the 111ᵗʰ Congress

H.R. 513 (Forbes)

New Manhattan Project for Energy Independence. Establishes program to develop new energy-related technologies, including treatment of nuclear waste. Introduced January 14, 2009; referred to Committee on Science and Technology.

H.R. 1698 (Van Hollen)

Establishes a Green Bank to finance qualified clean energy projects. Nuclear power projects could receive financing only after exhausting all other existing federal financial support. Introduced March 24, 2009; referred to Committees on Ways and Means and Energy and Commerce.

H.R. 1812 (Bachmann)

Promoting New American Energy Act of 2009. Provides tax benefits for investments in nuclear power plants and other energy investments. Introduced March 31, 2009; referred to Committee on Ways and Means.

H.R. 1936 (Lowey)

Nuclear Power Licensing Reform Act of 2009. Expands requirements for nuclear plant evacuation plans from a 10-mile radius to a 50-mile radius and makes reactor license renewals subject to the same criteria as a new plant. Introduced April 2, 2009; referred to Committee on Energy and Commerce.

H.R. 1937 (Lowey)

Requires NRC to distribute safety-related fines imposed on a nuclear plant to surrounding counties to help pay for emergency planning. Introduced April 2, 2009; referred to Committee on Energy and Commerce.

H.R. 2454 (Waxman)

American Clean Energy and Security Act. Modifies DOE loan guarantee program and establishes Clean Energy Deployment Administration to administer DOE assistance, including loan guarantees, for nuclear energy and other energy technologies. Establishes cap-and-trade program for carbon dioxide emissions. Introduced May 15, 2009, referred to multiple committees. Reported by Committee on Energy and Commerce June 5, 2009 (H.Rept. 111-137, part I). Passed by House June 26, 2009, by vote of 219-212.

H.R. 2768 (Wamp)

Declares that any reference to clean energy in federal law shall be considered to include nuclear energy. Introduced June 9, 2009; referred to Committee on Energy and Commerce.

H.R. 2828 (Bishop)

American Energy Innovation Act. Amends EPACT Title XVII loan guarantee provisions, modifies DOE standby support program for new reactors, reauthorizes the Nuclear Power 2010 program, establishes a tax credit for investments in manufacturing capacity for nuclear plant components, allows the Nuclear Waste Fund to be used for spent fuel reprocessing, modifies reactor licensing requirements, establishes an investment tax credit for nuclear power plants, authorizes temporary spent fuel storage agreements, requires DOE to offer to settle lawsuits for nuclear waste disposal delays, prohibits NRC from considering nuclear waste storage when licensing new nuclear facilities, and prohibits new waste facilities authorized under the act from being located in Nevada. Introduced June 11, 2009; referred to multiple committees.

H.R. 2846 (Boehner)

American Energy Act. Requires expedited procedures for nuclear plant licensing, establishes goal of licensing 100 new reactors by 2030, establishes uranium reserve, requires continued development of the Yucca Mountain repository unless it is found scientifically unsuitable, removes the statutory limit on Yucca Mountain disposal capacity, allows the Nuclear Waste Fund to be used for reprocessing, requires NRC to determine that sufficient waste disposal capacity will be available for proposed new reactors, establishes a National Nuclear Energy Council to advise the Secretary of Energy, and provides investment tax credit for nuclear power plants. Introduced June 12, 2009; referred to multiple committees.

H.R. 3009 (Ross)

American-Made Energy Act of 2009. Establishes American-Made Energy Trust Fund and includes nuclear power among technologies eligible for expenditures from the fund. Introduced June 23, 2009; referred to Committee on Energy and Commerce.

H.R. 3183 (Pastor)

Energy and Water Development Appropriations Act for FY2010. Includes funding for DOE nuclear energy programs. Introduced July 13, 2009; signed into law October 28, 2009 (see *CRS FY2010 Status Table of Appropriations*, http://www.crs.gov/Pages/appover.aspx).

H.R. 3385 (Barton)

Authorizes DOE to use the Nuclear Waste Fund to pay for grants or long-term contracts for spent nuclear fuel recycling or reprocessing and places the Waste Fund off-budget. Introduced July 29, 2009; referred to committees on Energy and Commerce and the Budget.

H.R. 3448 (Pitts)

Streamline America's Future Energy Nuclear Act. Requires NRC to establish expedited nuclear plant licensing procedures, requires NRC to reduce the time required to certify new reactor designs by half, requires NRC to develop technology-neutral guidelines for nuclear plant licensing, establishes a National Nuclear Energy Council to advise the Secretary of Energy, authorizes a final year of appropriations for the Nuclear Power 2010 program, requires DOE to prepare a schedule for accelerating completion of the Next Generation Nuclear Plant from 2021 to 2015, and limits fees and procedural restrictions on uranium mining on federal lands. Introduced July 31, 2009; referred to Committees on Energy and Commerce and Natural Resources.

H.R. 3505 (Gary Miller)

American Energy Production and Price Reduction Act. Prohibits NRC from considering nuclear waste storage when licensing new nuclear facilities and establishes investment tax credit for the costs of obtaining a nuclear manufacturing certification from the American Society of Mechanical Engineers. Introduced July 31, 2009; referred to multiple committees.

S. 591 (Reid)

National Commission on High-Level Radioactive Waste and Spent Nuclear Fuel Establishment Act of 2009. Establishes a commission to recommend alternative nuclear waste management options in the event that the proposed Yucca Mountain, NV, repository does not become operational. Introduced March 12, 2009; referred to Committee on Environment and Public Works.

S. 807 (Nelson)

SMART Energy Act. Authorizes funds for NRC to expedite nuclear plant license applications, authorizes nuclear workforce training program, establishes interagency working group to increase U.S. nuclear plant component manufacturing base, authorizes construction of a spent nuclear fuel recycling development facility, modifies the Standby Support program for new reactors, modifies the EPACT loan guarantee program, expands the nuclear power production tax credit, and provides accelerated depreciation for new reactors. Introduced April 2, 2009; referred to Committee on Finance.

S. 861 (Graham)

Rebating America's Deposits Act. Requires the President to certify that the Yucca Mountain site continues to be the designated location for a nuclear waste repository under the Nuclear Waste Policy Act. If such a certification is not made within 30 days after enactment or is subsequently revoked, the Treasury is to refund all payments, plus interest, made by nuclear reactor owners to the Nuclear Waste Fund. DOE is to begin shipping defense-related high-level radioactive waste to Yucca Mountain by 2017 or pay $1 million per day to each state in which such waste is located. Introduced April 22, 2009; referred to Committee on Energy and Natural Resources.

S. 1333 (Barrasso)

Clean, Affordable, and Reliable Energy Act of 2009. Includes provisions to take the Nuclear Waste Fund off-budget, authorize DOE to use the Nuclear Waste Fund to pay for grants or long-term contracts for spent nuclear fuel recycling or reprocessing, and prohibit NRC from denying licenses for new nuclear facilities because of a lack of waste disposal capacity. Introduced June 24, 2009; referred to Committee on Finance.

S. 1462 (Bingaman)

American Clean Energy Leadership Act of 2009. Establishes Clean Energy Deployment Administration to administer DOE assistance, including loan guarantees, for nuclear energy and other energy technologies. Bill would also establish a national commission to study nuclear waste management alternatives and requirements for nuclear fuel cycle research. Introduced and reported as an original measure from the Committee on Energy and Natural Resources July 16, 2009 (S.Rept. 111-48).

S. 1733 (Kerry)

Clean Energy Jobs and American Power Act. Authorizes programs for nuclear worker training, nuclear safety, and nuclear waste research. Establishes a carbon dioxide cap-and-trade program. Introduced September 30, 2009; referred to Committee on Environment and Public Works. Ordered reported November 5, 2009.

S. 2052 (Mark Udall)

Nuclear Energy Research Initiative Improvement Act of 2009. Authorizes DOE research to reduce nuclear reactor manufacturing and construction costs. Introduced October 29, 2009; referred to Committee on Energy and Natural Resources.

S. 2776 (Alexander)

Clean Energy Act of 2009. Revises DOE loan guarantee program, authorizes DOE assistance for small modular reactors, requires NRC to consider waste disposal to be adequate for potential new reactors, and authorizes funding for nuclear workforce development and research. Introduced November 16, 2009; referred to Committee on Energy and Natural Resources.

S. 2812 (Bingaman)

Nuclear Power 2021 Act. Establishes a cost-shared program between DOE and the nuclear industry to develop and license standard designs for two reactors below 300 megawatts of electric generating capacity. Introduced November 20, 2009; referred to Committee on Energy and Natural Resources.

Author Contact Information

Mark Holt
Specialist in Energy Policy
mholt@crs.loc.gov, 7-1704

STATEMENT FOR THE RECORD
by
Marvin S. Fertel
President and Chief Executive Officer
Nuclear Energy Institute
to the
Committee on Energy and Natural Resources
U.S. Senate

March 18, 2009

Chairman Bingaman, Ranking Member Murkowski, and members of the committee, thank you for your interest in nuclear energy and in addressing the policies that can facilitate deployment of new nuclear plants to meet national energy needs and reduce carbon emissions.

My name is Marvin Fertel. I am the President and Chief Executive Officer of the Nuclear Energy Institute (NEI). NEI is responsible for establishing unified nuclear industry policy on regulatory, financial, technical and legislative issues affecting the industry. NEI members include all companies licensed to operate commercial nuclear power plants in the United States, nuclear plant designers, major architect/engineering firms, fuel fabrication facilities, materials licensees, and other organizations and individuals involved in the nuclear energy industry.

My testimony will cover five major areas:

1. Current status of the U.S. nuclear energy industry
2. The need for new nuclear generating capacity
3. Progress toward new nuclear power plant construction
4. Financial challenges facing the electric power sector
5. Policy actions necessary to address the challenges facing new nuclear plant development

I.   Current Status of the U.S. Nuclear Power Industry

The U.S. nuclear energy industry's top priority is, and always will be, the safe and reliable operation of our existing plants. Safe, reliable operation drives public and political confidence in the industry, and America's nuclear plants continue to sustain high levels of performance.

Just last week, the Nuclear Regulatory Commission published a Fact Sheet highlighting the dramatic improvements in every aspect of nuclear plant performance over the last two decades: "The average number of significant reactor events over the past 20 years has dropped to nearly zero. Today there are far fewer, much less frequent and lower risk events that could lead to reactor core damage. The average number of times safety systems have had to be activated is about one-tenth of what it was 22 years ago. Radiation exposure levels to plant workers has steadily decreased to about one-sixth of the 1985 exposure levels and are well below federal limits. The average number of unplanned reactor shutdowns has decreased by nearly ten-fold. In 2007, there were two shutdowns compared to about 530 shutdowns in 1985."

This high level of performance continued last year. In 2008, the average capacity factor for our 104 operating nuclear plants was over 90 percent, and output of over 800 billion kilowatt hours represented nearly 75 percent of U.S. carbon-free electricity. According to the quantitative performance indicators monitored by the Nuclear Regulatory Commission, last year's performance was the best ever. This performance represents a solid platform for license renewal of the existing fleet and new nuclear plant construction.

II.   The Need for New Nuclear Generating Capacity

Construction of new nuclear plants will address two of our nation's top priorities: Additional supplies of clean energy and creation of jobs.

Nuclear energy is one of the few bright spots in the U.S. economy – expanding rather than contracting, creating thousands of jobs over the past few years. Over the last several years, the nuclear industry has invested over $4 billion in new nuclear plant development, and plans to invest approximately $8 billion more to be in a position to start construction in 2011-2012.

The investment to date has already created 15,000 jobs over the last two to three years, as reactor designers, equipment manufacturers and fuel suppliers expand engineering centers and build new facilities in New Mexico, North Carolina, Tennessee, Pennsylvania, Virginia and Louisiana. These jobs represent a range of opportunities – from skilled craft employment in component manufacturing and plant construction, to engineering and operation of new facilities. The number of new jobs will expand dramatically early in the next decade when the first wave of new nuclear power projects starts construction. If all 26 reactors currently in licensing by the NRC were built, this would result in over 100,000 new jobs to support plant construction and operations, and does not include additional jobs created downstream in the supply chain. This would be in addition to the 30,000 new hires in the next 10 years to support operation of the existing fleet of plants through the extended license period of 60 years.

New nuclear plants will also help the United States meet its climate change objectives. Predominantly independent assessments of how to reduce U.S. electric sector CO_2 emissions – by the International Energy Agency, McKinsey and Company, Cambridge Energy Research Associates, Pacific Northwest National Laboratory, the Energy Information Administration, the Environmental Protection Agency, the Electric Power Research Institute and others – show that there is no single technology that can slow and reverse increases in CO_2 emissions. A portfolio of technologies and approaches will be required, and that portfolio must include more nuclear power as well as aggressive pursuit of energy efficiency and equally aggressive expansion of renewable energy, advanced coal-based technologies, plug-in hybrid electric vehicles and distributed resources.

NEI is not aware of any credible analysis of the climate challenge that does not include substantial nuclear energy expansion as part of the technology portfolio. In fact, removing any technology from the portfolio places unsustainable pressure on those options that remain.

Analysis last year by the Energy Information Administration of the Lieberman-Warner climate change legislation (S. 2191) demonstrates the value of nuclear energy in a carbon-constrained world. In EIA's "Core" scenario, which included new nuclear plant construction, carbon prices in 2030 were 33 percent lower, residential electricity prices were 20 percent lower and residential natural gas prices were 19 percent lower than in the "Limited Alternatives" scenario, which severely limited new nuclear construction.

It is also clear that the United States will need new baseload electric generating capacity even with major improvements in energy efficiency. Recent analysis by The Brattle Group, an independent consulting firm, showed that the United States will need between 133,000 megawatts of new generating capacity (absent controls on carbon) and 216,000 megawatts (in a carbon-constrained world) by 2030. These numbers assume 0.7 percent per year growth in peak load – a significant reduction from historical

2

performance. Annual growth in peak load between 1996 and 2006 was 2.1 percent, and the Energy Information Administration's Annual Energy Outlook assumes a 1.5-percent annual increase in peak load.

NEI estimates that if the 26 reactors being licensed today (approximately 34,000 MW) were built by 2030, this would simply maintain nuclear at 20 percent of U.S. electricity supply. To increase nuclear energy's contribution to 2050 climate goals, build rates of 4-6 plants per year must be achieved. This was possible in the 1970s and 1980s even with the old licensing process and lack of standardization. With standardized designs and improved construction techniques, this accelerated deployment is feasible after the first wave of plants are constructed.

III. Progress Toward New Nuclear Power Plant Construction

The Nuclear Regulatory Commission is reviewing construction and operating license applications from 17 companies or groups of companies for 26 new reactors totaling 34,200 MW. These new plants will be built at a measured pace over the next 10-15 years. Safety-related construction of the first new nuclear plants will start in 2012, and NEI expects four to eight new nuclear plants in commercial operation in 2016 or so. The exact number will, of course, depend on many factors – U.S. economic growth, forward prices in electricity markets, capital costs of all baseload electric technologies, commodity costs, environmental compliance costs for fossil-fueled generating capacity, natural gas prices, growth in electricity demand, availability of federal and state support for financing and investment recovery, and more. We expect construction of those first plants will proceed on schedule, within budget estimates, and without licensing difficulties, and a second wave will be under construction as the first wave reaches commercial operation.

Supported in part by government-industry cost-shared programs like the Department of Energy's Nuclear Power 2010 program, detailed design and engineering work on advanced reactor designs is nearing completion. This detailed design information will allow companies to develop firm cost estimates. Based on what is known today, however, there is a solid business case for new nuclear generating capacity.

Nuclear energy is a capital-intensive technology. NEI estimates a new nuclear power plant could cost $6 billion to $8 billion, including financing costs. This large capital investment does not mean that new nuclear plants will not be competitive. Capital cost is certainly an important factor in financing, but it is not the sole determinant of a plant's competitive position. The key factor is the cost of electricity from the plant at the time it starts commercial operation relative to the other alternatives available at that time. Based on NEI's own modeling, on the financial analysis performed by companies developing new nuclear projects, and on independent analysis by others, new nuclear capacity will be competitive. (NEI's white paper, "The Cost of New Generating Capacity in Perspective", is attached for further information on this topic.)

Florida Power and Light and Florida Progress demonstrated this in the financial modeling that supported their requests last year to the Florida Public Service Commission for "determinations of need" for new reactors at Turkey Point and Levy County. In FP&L's modeling, the only scenario in which nuclear was not preferred was a world in which natural gas prices were unrealistically low and there was no price on carbon. The Florida PSC has approved both projects. Independent analyses reach the same conclusion. In an integrated resource plan developed for Connecticut last year, The Brattle Group concluded that new nuclear plants are a lower-cost source of electricity in a carbon-constrained world than supercritical pulverized coal with carbon capture and storage (CCS), integrated gasification combined cycle with CCS and gas-fired combined cycle with CCS.

Understanding the Past. Many of the nuclear power plants commissioned in the 1960s and early 1970s completed construction in four to five years with construction costs around $500 million. By the late

3

1970s and early 1980s, however, construction was averaging 10 to 12 years, and construction costs ranged as high as $5 billion. The nuclear industry has conducted detailed and extensive analysis of this experience, which demonstrates that the nuclear plants built after the early 1970s were built under extremely unfavorable conditions – caused by several major factors converging at roughly the same time.

Nuclear energy technology in the United States scaled up quickly. The industry scaled from the first 200-megawatt-scale plants to 1,000-megawatt-plus plants in just a few years. This rapid increase in reactor size occurred at a time when electricity demand was growing at seven percent a year on average, which required a doubling of electric generating capacity every 10 years. In that business environment, bigger was better for new power plants. Larger plants meant greater economies of scale. Larger was also more complex, however, and that complexity coupled with other factors discussed subsequently created project management challenges. Construction times stretched out and economies of scale vanished with schedule delays and rising costs.

Changing regulatory requirements and licensing difficulties added to the challenge of managing these large construction projects to schedule and budget, but licensing and regulatory requirements were not the sole cause of cost increases and schedule delays. Construction started before design work was complete. Some projects were managed by companies with no prior nuclear construction experience. Project planning and management tools equal to the complexity of the task did not exist at the time.

Finally and of significant importance to the increasing cost, the first generation of nuclear power plants were built under difficult business and economic conditions. Growth in electricity demand slowed from six to seven percent a year to one to two percent in the mid-1970s. Many utilities intentionally slowed construction. The prime rate reached 20 percent in the early 1980s. As project schedules stretched out, costs increased and companies were forced to borrow more at double-digit interest rates.

Lessons Learned: Roadmap for a Successful Future. The root causes of past construction delays are well understood and both industry and government have taken steps to ensure that past experience is not repeated.

The licensing process has been restructured to increase efficiency and effectiveness and reduce uncertainty and financial risk. Today's plants were licensed under a two-step process: Electric utilities had to secure two permits—a construction permit to build the plant and a second operating license to operate it. Under the new process, all major safety and regulatory issues – reactor design, site suitability – will be resolved before construction begins, and a company receives a single license to build and operate the plant. The use of certified standardized designs will also reduce licensing and construction times through repetition. Once a design has been certified, the NRC reviews will focus only on site suitability and plant operations. The industry is working together to ensure that the standardization carries over into their license applications, construction practices and operating procedures to fully enjoy the benefits of a standard fleet of plants.

As construction proceeds, inspections and tests are performed to ensure the plant has been built in accordance with the approved design. These inspections, tests, analyses and acceptance criteria – or ITAAC – are included in the plant's construction and operating license. ITAAC are a key risk-management tool. When the ITAAC are met, the NRC and the public know that the plant has been built according to its design and will operate safely.

In addition to an improved licensing process, the next generation of nuclear plants built in the United States will benefit from an industry-wide inventory of lessons-learned. The roadmap for future success includes:

4

Detailed design essentially complete before construction. Companies planning to build new nuclear plants intend to have virtually all detailed design complete before construction is started.

Standardized, design-specific pre-build preparation. Starting in 2006, the nuclear industry formed design-centered working groups (DCWG) with each reactor vendor. These groups are charged with maintaining standardization within each reactor design, which will enhance licensing, preparation for construction and construction.

Focus on quality assurance. While quality assurance is a core competency at existing plants, in 2005, the U.S. nuclear industry formed a New Plant Quality Assurance Task Force. In conjunction with the Institute of Nuclear Power Operations (INPO), this task force is conducting a systematic lessons-learned review of past and present nuclear construction projects in the United States and around the world.

Corrective action programs. The industry is adapting the corrective action program (CAP), which is standard at operating plants, for use in new plant construction. A CAP includes a structured database to capture and categorize potentially safety-significant items, enabling constructors to identify and trend quality deficiencies, record that corrective action was taken, and report to the appropriate levels of management.

Focus on safety culture as part of construction. Safety culture, corrective action programs and programs that encourage employees to raise safety concerns are now an essential part of the operating philosophy at the 104 operating plants. The work force building new plants will have the same safety focus.

Preparation for construction inspection. In 2001, the U.S. nuclear industry formed a New Plant Construction Inspection Program Task Force comprised of utilities, reactor vendors and major construction companies. The task force is formulating guidance and developing programs and processes to implement the inspections, tests, analyses and acceptance criteria that the NRC will use to determine whether the plant is built according to the approved design and is ready to operate safely.

Improved planning and construction management tools. Project and construction management at new nuclear plants will benefit from a suite of sophisticated construction planning and management tools equal to the complexity of the task, none of which were developed when the last nuclear plants were built. Companies did not have computer-aided design (CAD) to enable design changes. Databases for tracking components and resources were not yet mature. Computerized tools that linked resources with design and construction schedules were in their infancy.

Improved construction techniques. Construction of new nuclear plants in the U.S. will also benefit from improved construction techniques (such as modular construction), many of which were developed overseas, for the U.S. nuclear navy or for other industries.

Successful Track Record. Recent construction and operational experience demonstrates that an experienced project management team – with effective quality assurance and corrective action programs, with detailed design completed before the start of major construction, with an integrated engineering and construction schedule – can complete projects on budget and on schedule. The global nuclear industry, including the U.S. nuclear industry, has performed projects ranging from major upgrades to plant restarts to refueling outages efficiently, without delay. As recently as 1990, maintenance and refueling outages at U.S. reactors lasted more than 100 days; today's average is 37 days.

5

There are other examples that provide confidence that new nuclear plant development in the United States will proceed smoothly:

- The Tennessee Valley Authority returned Unit 1 of its Browns Ferry nuclear plant to commercial operation in May 2007. The five-year, $1.8-billion project was completed on schedule and only five percent over the original budget estimate, a significant achievement during a period of rapidly escalating commodity costs. The Browns Ferry 1 restart project was comparable in complexity to the construction of a new nuclear power plant. Most systems, components, and structures were replaced, refurbished, or upgraded, and all had to be inspected and tested.
- At the Fort Calhoun plant in Nebraska, Omaha Public Power District replaced the major primary system components – steam generators, reactor vessel head and rapid refueling package and pressurizer – as well as the low pressure turbines, the main transformer and hydrogen coolers, among other equipment. The outage began in September 2006 and ended in December of that year, lasting 85 days. The $417-million project was completed approximately $40 million under budget and five days ahead of schedule.
- Nuclear construction experience in South Korea over the last 15 years demonstrates the "learning curve" that can be achieved. The "first of a kind" nuclear power plants – Yonggwang Units 3 and 4 – were built in the mid-1990s in 64 months. The next two units – Ulchin 3 and 4 – were built in 60 months at 94 percent of the "first of a kind" cost. The next plants – Yonggwang 5 and 6 – were built in 58 months for 82 percent of the "first of a kind" cost. By 2004, Ulchin 5 and 6 were built in 56 months for 80 percent of the "first of a kind" cost. The next two plants – Shin-Kori 1 and 2 – will be in service next year. Construction duration: 53 months and 63 percent of what it cost to build Yonggwang 3 and 4. South Korea's goal is a 39-month construction schedule.
- Nuclear power plants in Japan achieve construction schedules similar to those in South Korea. The first two Advanced Boiling Water Reactors built were constructed in times that beat the previous world record and both were built on budget. Kashiwazaki-Kariwa Unit 6 began commercial operation in 1996, and Unit 7 began commercial operation in 1997. From first concrete to fuel load, it took 36.5 months to construct Unit 6 and 38.3 months for Unit 7. Unit 6 was built 10 months quicker than the best time achieved for any of the previous boiling water reactors constructed in Japan.
- The Qinshan nuclear power plant in China consists of two 728-megawatt pressurized heavy-water reactors. First concrete was placed on June 8, 1998. Unit 1 began commercial operation on December 31, 2002, 43 days ahead of schedule. The construction period was 54 months from first concrete to full-power operation. Unit 2 began commercial operation on July 24, 2003, 112 days ahead of schedule.

U.S. projects will also benefit from this learning curve in other countries, since most of the reactors being licensed in the United States will be built overseas prior to U.S. construction. South Texas Project Units 3 and 4, for example, are Advanced Boiling Water Reactors of the type already built in Japan. There are 44 nuclear plants under construction worldwide, and 108 more ordered or planned.

IV.   Financial Challenges Facing the Electric Power Sector

The U.S. electric industry faces a formidable investment challenge. Consensus estimates show that the electric sector must invest between $1.5 trillion and $2 trillion in new power plants, transmission and distribution systems, and environmental controls to meet expected increases in electricity demand by 2030. To put these numbers in perspective: the book value of America's entire electric power supply and delivery system today is only $750 billion, which reflects investments made over the last 60 years.

Addressing the financing challenge will require innovative approaches. Meeting these investment needs will require a partnership between the private sector and the public sector, combining all the financing

6

capabilities and tools available to the private sector, the federal government and state governments – particularly at a time when the electric sector is already showing some signs of stress.

The financial crisis has forced investor-owned utilities to reduce capital spending for 2009 by approximately 10 percent, on average. The industry is experiencing downward pressure on equity returns, largely because rate increases have not kept pace with rising costs. Bond spreads are also wider (in some cases, significantly wider) and, although all-in debt costs are not dramatically higher because yields on Treasuries are so low, the cost of debt will be significantly higher than historical norms when Treasury yields recover if bond spreads remain at current levels. Industry leverage is beginning to rise – not to the levels seen in 2003, when debt represented about 61 percent of the investor-owned utilities' capital structure – but it has increased somewhat over the last three years and debt now represents about 56 percent of industry capital structure.

In summary, the electric power sector is in the early stages of a major, 20-year capital investment program, and is not as well positioned for these capital expenditures as it was in the 1970s and 1980s when it last undertook a major capital expansion program.

For new nuclear power plants, the financing challenge is structural. Unlike the many consolidated government owned foreign utilities and the large oil and gas companies, U.S. electric power sector consists of many relatively small companies, which do not have the size, financing capability or financial strength to finance power projects of this scale on their own, in the numbers required. Loan guarantees offset the disparity in scale between project size and company size. Loan guarantees allow the companies to use project-finance-type structures and to employ higher leverage in the project's capital structure. These benefits flow to the economy by allowing the rapid deployment of clean generating technologies at a lower cost to consumers. The recent stimulus bill recognized the need to provide access to low-cost capital to encourage rapid deployment of renewable energy projects. Similar support is required for nuclear energy since, in many cases, new nuclear plants and renewable energy projects are built by the same utilities.

Loan guarantees are a powerful tool and an efficient way to mobilize private capital. The federal government manages a loan guarantee portfolio of approximately $1.1 trillion to ensure necessary investment in critical national needs, including shipbuilding, transportation infrastructure, exports of U.S. goods and services, affordable housing, and many other purposes. Supporting investment in new nuclear power plants and other critical energy infrastructure is a national imperative.

The loan guarantee program created by Title XVII of the Energy Policy Act is an essential and appropriate mechanism to enable financing of clean energy technologies. In fact, an effective and workable loan guarantee program is significantly more important today than it was when the Energy Policy Act was enacted in 2005.

The Title XVII program currently includes 10 technologies that are eligible for loan guarantees. They include renewable energy systems, advanced fossil energy technology (including coal gasification), hydrogen fuel cell technology for residential, industrial, or transportation applications, advanced nuclear energy facilities, efficient electrical generation, transmission, and distribution technologies, efficient end-use energy technologies, production facilities for fuel efficient vehicles, including hybrid and advanced diesel vehicles, and pollution control equipment. Each of these technologies presents different financing challenges.

The financing challenges are, of course, somewhat different for the regulated integrated utilities than for the merchant generating companies in those states that have restructured. But these challenges can be

7

managed, with appropriate rate treatment from state regulators or credit support from the federal government's loan guarantee program, or a combination of both.

Supportive state policies include recovery of nuclear plant development costs as they are incurred, and Construction Work in Progress or CWIP, which allows recovery of financing costs during construction. Many of the states where new nuclear plants are planned – including Florida, Virginia, Texas, Louisiana, Mississippi, North Carolina and South Carolina – have passed legislation or implemented new regulations to encourage construction of new nuclear power plants by providing financing support and assurance of investment recovery. By itself, however, this state support may not be sufficient. The federal government must also provide financing support for deployment of clean energy technologies in the numbers necessary to address growing U.S. electricity needs and reduce carbon emissions.

The Title XVII program also represents an innovative departure from other federal loan guarantee programs. It is structured to be self-financing, so that companies receiving loan guarantees pay the cost to the government of providing the guarantee, and all administrative costs. For this reason, a Title XVII loan guarantee program is not a subsidy. In a well-managed program, in which projects are selected based on creditworthiness, extensive due diligence and strong credit metrics, there is minimal risk of default, and minimal risk to the taxpayer. In fact, the federal government will receive substantial payments from project sponsors.

V.   Policy Actions Necessary for New Nuclear Plant Development

Financing

Since enactment of the Energy Policy Act in August 2005, achieving workable implementation of the Title XVII loan guarantee program has been a challenge. The implementation difficulties predate formation of the Loan Guarantee Program Office. In fact, NEI is impressed with what a relatively small staff in the Loan Guarantee Program Office, operating under chronic budgetary constraints, have been able to accomplish in the time – slightly more than a year – that they have been at work.

Despite this significant progress, implementation of the program by the Executive Branch continues to be difficult, for reasons outside the control of the Loan Guarantee Program Office. The staff is working to address problems with the regulations governing this program that were promulgated by the Department of Energy in 2007, but one of the major difficulties stems from an unnecessarily narrow and restrictive reading of the original statutory language by the DOE Office of General Counsel. Section 1702(g)(2)(B) of Title XVII asserts that "[t]he rights of the Secretary, with respect to any property acquired pursuant to a guarantee or related agreements, shall be superior to the rights of any other person with respect to the property." This language can be misinterpreted as a prohibition on *pari passu* financing structures, and a requirement that the Secretary must have a first lien position on the entire project. Counsel for NEI and many of the project sponsors, with substantial experience in project finance, believe that Section 1702(g)(2)(B) gives the Secretary a "superior right" to the property he guarantees, not to the entire project.

The current interpretation of this language is thus a major obstacle to co-financing of nuclear projects. Projects financed as undivided interests cannot proceed if this interpretation stands. Financing from export credit agencies in other countries like France and Japan, would be equally difficult. This result makes little sense since such co-financing will leverage the existing loan volume of $18.5 billion, and reduce the risk to which the Department of Energy is exposed.

NEI is encouraged by Energy Secretary Steven Chu's intent, expressed before this committee during his confirmation hearing and at other times, to address the difficulties that have arisen during implementation

8

of the Title XVII loan guarantee program. Many of these problems can be corrected through rulemaking, and NEI understands that DOE is developing revised rules to address defects in the current rule and to implement the new loan guarantee program authorized in the economic stimulus legislation. The Energy and Natural Resources Committee can play a key oversight role in ensuring that the necessary revisions to the existing rule are promulgated quickly, and do not become entangled in internal Executive Branch procedural difficulties, as has happened so often in the past. If the necessary changes cannot be implemented through rulemaking, it will, of course, be necessary to seek statutory changes to accomplish the same purpose.

Insufficient Loan Volume. The Title XVII loan guarantee program was an important step in the right direction. That program was designed to jump-start construction of the first few innovative clean energy projects that use "technologies that are new or significantly improved ... as compared to commercial technologies in service in the United States at the time the guarantee is issued." [1]

That goal remains as valid now as it was in 2005, but today the United States faces a larger, additional challenge – financing large-scale deployment of clean energy technologies, modernizing the U.S. electric power supply and delivery system, and reducing carbon emissions. As noted earlier, this is estimated to require investment of $1.5-2.0 trillion between 2010 and 2030.

The omnibus appropriations legislation for FY 2008 and FY2009 authorizes $38.5 billion in loan volume for the loan guarantee program – $18.5 billion for nuclear power projects, $2 billion for uranium enrichment projects, and the balance for advanced coal, renewable energy and energy efficiency projects.

DOE has issued solicitations inviting loan guarantee applications for all these technologies and, in all cases the available loan volume is significantly oversubscribed. For example, the initial nuclear power solicitation resulted in requests from 14 projects seeking $122 billion in loan guarantees, with only $18.5 billion available. NEI understands that 10 nuclear power projects submitted Part II loan guarantee applications, which represented $93.2 billion in loan volume. Two enrichment projects submitted Part II applications, seeking $4.8 billion in loan guarantees, with only $2 billion available. NEI also understands that the solicitation for innovative coal projects resulted in requests for $17.4 billion in loan volume, more than twice the $8 billion available.

It is, therefore, essential that limitations on loan volume – if necessary at all in a program where project sponsors pay the credit subsidy cost – should be commensurate with the size, number and financing needs of the projects. In the case of nuclear power, with projects costs between $6 billion and $8 billion, $18.5 billion is not sufficient.

The scale of the challenge requires a broader financing platform than the program envisioned by Title XVII. An effective, long-term financing platform is necessary to ensure deployment of clean energy technologies in the numbers required, and to accelerate the flow of private capital to clean technology deployment.

During the 110th Congress, Senator Bingaman introduced legislation to create a 21st Century Energy Deployment Corporation. Senator Domenici, ranking member of this committee during the last Congress, introduced legislation to create a Clean Energy Bank. Both proposals address aspects of the financing challenge facing the United States and its electric power industry.

NEI believes that the existing Title XVII program and the DOE Loan Guarantee Program Office, operating under workable rules, could serve as a foundation on which to build a larger, independent

---

[1] Energy Policy Act of 2005, Section 1703(a)(2)

9

financing institution within the Department of Energy. There is precedent for such independent entities, equipped with all the resources necessary to accomplish their missions, in the Federal Energy Regulatory Commission and the Energy Information Administration. This approach could have significant advantages:

1. An independent clean energy financing authority within DOE could take advantage of technical resources available within the Department, to supplement its due diligence on prospective projects and to identify promising technologies emerging from the research, development and demonstration pipeline that might be candidates for loan guarantee support to enable and speed deployment.
2. An independent entity within DOE would have the resources necessary to implement its mission effectively, including its own legal and financial advisers with the training and experience necessary for a financing organization. Providing the independent entity with its own resources would eliminate the difficulties encountered during implementation of the Title XVII program.
3. Programmatic oversight in Congress would remain with the Energy Committees, which have significantly more experience with energy policy challenges, and in structuring the institutions necessary to address those challenges.

Development of a National Used Fuel Strategy

Used nuclear fuel is managed safely and securely at nuclear plant sites today, and can be managed safely and securely for an extended period of time. For this reason, used nuclear fuel does not represent an impediment to new nuclear plant development in the near term. It is, however, an issue that must be addressed for the long-term.

The administration has made it clear that Yucca Mountain "is not an option."

The nuclear industry's position on used fuel management is clear:

- The Nuclear Waste Policy Act establishes an unequivocal federal legal obligation to manage used nuclear fuel, and remains the law of the land. Until that law is changed, the nuclear industry believes the NRC's review of the Yucca Mountain license application should continue.
- If the administration unilaterally decides to abandon the Yucca Mountain project without enacting new legislation to modify or replace existing law, it should expect a new wave of lawsuits seeking further damage payments and refunds of at least $22 billion in the Nuclear Waste Fund already collected from consumers that has not been spent on the program.
- Given the uncertainties associated with the Yucca Mountain project, DOE should reduce the fee paid by consumers to cover only costs incurred by DOE, NRC and local Nevada government units that provide oversight of the program.
- A credible and effective program to manage used nuclear fuel must include three integrated components: interim storage of used nuclear fuel at centralized locations, technology development necessary to demonstrate the technical and business case for recycling used nuclear fuel and, ultimately, the licensing of a permanent disposal facility.

The nuclear energy industry supports creation by the Executive Branch of a bipartisan blue-ribbon commission of credible experts to undertake a reassessment of the federal government's program to manage used nuclear fuel, and produce a roadmap for a sustainable long-term program.

10

Regulatory Effectiveness and Predictability

An objective, effective Nuclear Regulatory Commission is a key factor in ensuring safe and secure operation of the 104 operating nuclear generating plants. An objective regulatory process – i.e., a process that is safety-focused and performance-based – will ensure that nuclear plant operators remain focused on safety-significant issues and that management attention is not diverted by matters of low safety or security significance. For new nuclear plants, a central element of the regulatory process is a predictable licensing process for the review and inspection of new reactor designs and new construction. The industry and the financial community must have confidence that the licensing process provides the level of predictability necessary to support large capital investments.

Research and Development

NEI appreciates this committee's recognition – in the draft research and development legislation published recently – of the strategic importance of increased funding for research and development. Substantial increases in energy R&D investment will be necessary in the years ahead to create a sustainable electric supply infrastructure. Unfortunately, recent trends are in the opposite direction. In a 2007 analysis, the Government Accountability Office found that DOE's budget authority for renewable, fossil and nuclear energy R&D declined by over 85 percent (in inflation-adjusted terms) from 1978 through 2005. The need for new technologies to address critical energy needs has not diminished over the same time period, however, nor have the energy and environmental imperatives facing the United States become any less urgent.

The Electric Power Research Institute (EPRI) has estimated that the United States must increase investment in energy R&D by $1.4 billion annually between now and 2030 to develop and demonstrate the technology portfolio necessary to bring electric sector carbon emissions back to 1990 levels by 2030. That additional cumulative investment of approximately $32 billion in R&D would reduce by $1 trillion the cost to the U.S. economy of bringing electric sector emissions back to 1990 levels, according to EPRI's analysis.

A robust research and development program is necessary if nuclear energy is to realize its full potential in the nation's energy portfolio. In 2008, the directors of the 10 DOE national laboratories, including now Secretary of Energy Chu, published a report recognizing that "nuclear energy must play a significant and growing role in our nation's ... energy portfolio ... in the context of broader global energy, environmental, and security issues." The report also expressed support for the required R&D effort: "The national laboratories, working in collaboration with industry, academia, and the international community, are committed to leading and providing the research and technologies required to support the global expansion of nuclear energy."

The report from the national laboratory directors identified areas of research that were incorporated, earlier this year, into a comprehensive strategy for nuclear R&D developed by EPRI and the Idaho National Laboratory. NEI supports the R&D priorities identified:

- Maintaining the high performance of today's light water reactors and extend their operating life beyond 60 years, to 80 years. R&D will be required, among other items, to develop advanced diagnostic and maintenance techniques, to extend component life and introduce new technologies, and to enhance fuel reliability and performance.
- Completing the cost-shared government-industry Nuclear Power 2010 Program, to complete the design and engineering work that will support the nuclear plants on track to start construction over the next several years.

11

- Developing proliferation-resistant recycling technologies that will capture the vast amount of energy that remains in used nuclear fuel and reduce the volume and toxicity of the waste by-product that requires permanent disposal.
- Developing high-temperature gas-cooled reactors to produce electricity and for non-electric applications. High-temperature reactors can reduce greenhouse gas emissions from large-scale process heat operations in the petroleum and chemical industries currently fired by liquid fuels and natural gas. This technology will also be capable of producing hydrogen economically for fuel-cell vehicles and industrial applications, as well as desalinating water cost-effectively.

The national laboratory directors, EPRI and INL point out that the leadership position of the U.S. in the global nuclear enterprise is at stake. Participation in the development of advanced nuclear energy technologies will allow the U.S. to influence energy technology choices around the world, and to ensure that non-proliferation regimes are in place as other countries develop commercial nuclear capabilities. Therefore, technical leadership is in the interest of the administration, the congress, and the industry.

Supply Chain

During the 1970s, the United States had the manufacturing capability to produce the large vessels, steam generators and other components necessary for nuclear power plant construction. Much of that capability – and the associated jobs – moved offshore over the last 30 years.

In the nuclear sector, there are signs that U.S. manufacturing capability is being rebuilt. In North Carolina, Indiana, Pennsylvania, Virginia, Tennessee, Louisiana, Ohio and New Mexico, among other states, U.S. companies are adding to design and engineering staff, expanding their capability to manufacture nuclear-grade components, or building new manufacturing facilities and fuel facilities – partly in preparation for new reactor construction in the United States, partly to serve the growing world market.

Last year, for example, AREVA and Northrop Grumman Shipbuilding formed a joint venture to build a new manufacturing and engineering facility in Newport News, Va. This $360-million facility will manufacture heavy components, such as reactor vessels, steam generators and pressurizers. Global Modular Solutions, a joint venture of Shaw Group and Westinghouse, is building a fabrication facility at the Port of Lake Charles to produce structural, piping and equipment modules for new nuclear plants using the Westinghouse AP1000 technology. In New Mexico, LES is well along with construction of a $3-billion uranium enrichment facility, scheduled to begin production this year. Even for ultra-heavy forgings, Japan Steel Works is expanding capacity, and companies in South Korea, France and Great Britain are planning new facilities.

Although progress in this area is encouraging, federal government policy could accelerate the process of creating new jobs and generating economic growth. Specifically, the expansion and extension of investment tax credits for investments in manufacturing provided in the stimulus would ensure continued expansion of the U.S. nuclear supply chain and help restore U.S. leadership in this sector.

Work Force

The U.S. nuclear industry recognizes the critical importance of a skilled, well-trained and dedicated work force to operate and maintain the 104 nuclear plants that supply 20 percent of America's electricity, and to build and operate new nuclear plants in the years ahead.

12

The nuclear industry is working with the federal government, state governments, universities and community colleges, high schools, labor unions, utilities, other trade associations and professional organizations to address the work force challenge.

Electric utilities have created 42 partnerships with community colleges to train the next generation of nuclear workers. The industry is developing standardized, uniform curricula to ensure that graduates will be eligible to work at any nuclear plant. Sixteen states have developed programs to promote skilled craft development. Enrollment in nuclear engineering programs has increased over 500 percent since 1999. Grant programs from the NRC, the Department of Energy, the Department of Labor and the Department of Defense for education and training are having a major impact on increasing our trained workforce.

NEI commends Senators Bingaman and Murkowski for the attention to workforce development in the draft legislation published recently on research and development. As with the nuclear supply chain, targeted tax credits to encourage companies to invest in apprenticeship programs and other work force development would accelerate job creation and training in the nuclear energy sector.

VI.   Conclusion

In conclusion, the need for advanced nuclear plants is well established. Nuclear energy clearly can and must play a strategic role in meeting national environmental, energy security and economic development goals. The nuclear energy industry has a limited and well-defined public policy agenda to ensure our nation continues to derive the benefits that nuclear power provides. Those policy conditions include:

1. near-term actions to ensure that the Title XVII loan guarantee program is working as intended, and creation of a broader, permanent financing platform to ensure access to capital for the large-scale deployment of advanced technologies including nuclear facilities that will reduce carbon emissions,
2. a sustainable strategy for the management and ultimate disposal of used nuclear fuel,
3. an effective and predictable licensing process, and
4. a research and development program that will allow the nation to meet environmental goals and provide leadership on issues related to expansion of nuclear technology and non-proliferation.

Mr. Chairman, thank you for the opportunity to testify, and this completes my testimony.

∎

13

Statement
of
Thomas B. Cochran, Ph.D.
Senior Scientist, Nuclear Program,
and
Christopher E. Paine
Director, Nuclear Program
Natural Resources Defense Council, Inc.

on

Nuclear Energy Developments

Before the
Committee on Energy and Natural Resources
United States Senate
Washington, D.C.

March 18, 2009

NRDC
THE EARTH'S BEST DEFENSE

Natural Resources Defense Council, Inc.
1200 New York Avenue, N.W., Suite 400
Washington, D.C. 20005
Tele: 202-289-6868
tcochran@nrdc.org
cpaine@nrdc.org

1

I. Introduction

Mr. Chairman and members of the Committee, thank you for providing the Natural Resources Defense Council (NRDC) the opportunity to present its views on several current issues related to nuclear energy. NRDC is a national, non-profit organization of scientists, lawyers, and environmental specialists, dedicated to protecting public health and the environment. Founded in 1970, NRDC serves more than 1.2 million members and supporters with offices in New York, Washington, D.C., Los Angeles, San Francisco, Chicago and Beijing.

Our testimony focuses on three issues: a) whether additional federal loan guarantees should be provided to construct new nuclear power plants; b) whether the United States should engage in reprocessing of spent nuclear fuel; and c) whether Congress should intervene in the Nuclear Regulatory Commission's proposed rulemakings on temporary storage of spent fuel and so-called "waste confidence," that is, whether sufficient confidence exists today in our long-term ability to isolate spent fuel from the biosphere that we can responsibly license new reactors that will add to the nuclear waste burden.[1]

II. Summary of Recommendations

A. Loan Guarantees. Congress should not provide additional loan guarantees to construct new nuclear plants. Sufficient nuclear loan guarantee authority already exists to accomplish the legitimate public purpose that is involved here. Let us define here what we believe to be the legitimate purpose of loan guarantees—they are intended to shift much of the downside financial risk involved in the initial commercial deployment of new or significantly improved low-carbon energy technologies from private interests to federal taxpayers.

Since the underlying light-water reactor technology to be supported by these guarantees has been around for 45 years, has been the prior recipient of many tens of billions of dollars in government support, and already accounts for 20% of U.S. grid-

---

[1] NRC, *Consideration of Environmental Impacts of Temporary Storage of Spent Fuel After Cessation of Reactor Operation* (hereinafter "Proposed Temporary Storage Rule") 73 Fed. Reg. 59547 (October 9, 2008), and *Waste Confidence Decision Update*, (hereinafter "Proposed Waste Confidence Rule") NRC, 73 Fed. Reg. 59551 (October 9, 2008).

2

connected power generation, the technology innovation case for nuclear loan guarantee support is weak, and at best, a very narrow one. To avoid serious and lasting distortion of the U.S. energy marketplace and an economically inefficient decarbonization effort, nuclear loan guarantees should be limited to *the lead units of new nuclear plant designs*, not previously deployed in the United States or in similar markets abroad with comparable regulatory requirements. These designs must *incorporate substantial design innovations promising improved safety, increased operating efficiencies, significantly reduced capital costs, and lower environmental impacts*.

In our view, few if any of the Gen III + reactors being proposed today plausibly meet this description, but if any of them do, it could only be *the lead units of new passive safety, smaller footprint, less capital intensive designs that have not yet been deployed elsewhere*. Fitting that description currently are the AP-1000 and the Economic Simplified Boiling Water Reactor (ESBWR), and possibly later the Very High-Temperature Gas-Cooled Reactor (VHTGR), now in the early stages of development by the Department of Energy (DOE).

But even here, we find that there are currently three regulated utilities, each proposing to add two AP1000 units to their respective rate bases, which do not appear to *require* loan guarantees for financing, or at least full loan guarantee coverage at 80% of total project cost. We believe that the $18.5 billion is already sufficient to support construction of more than just the lead units of the innovative standardized reactor designs currently available to the U.S. market, and therefore no additional loan guarantee authority is needed.

More loan guarantee support to underwrite the U.S. market penetration of additional designs, already deployed or under construction in foreign markets, would only further distort the energy marketplace and undermine the goal of design standardization, which is a widely shared objective of the DOE, Nuclear Regulatory Commission (NRC), nuclear industry and others concerned about the future effectiveness of the NRC's safety regulation.

Federal loan guarantees should not be abused to insulate an entire industry from competition with a host of new energy technologies that promise comprehensive environmental and social benefits. Unlike improvements in efficiency and renewable

3

technologies, nuclear power is a decarbonization solution packaged with a host of non-carbon environmental, security, and waste problems. For these reasons, nuclear power should not be considered for inclusion in any "Renewable Electricity Standard" Congress may legislate.

B. Spent Fuel Reprocessing. The federal government should not encourage or support commercial spent fuel reprocessing. Putting aside for the moment the serious proliferation and security concerns involved in any future global shift toward reprocessing, it's clear that combating climate change is an urgent task that requires near term investments yielding huge decarbonization dividends on a 5 to 20 year timescale. For thermal reactors, the closed fuel cycle (spent fuel reprocessing and recycling plutonium) is unlikely ever to be less costly than the once-through fuel cycle, even assuming significant carbon controls. But setting aside such near-term cost barriers, commercial viability for a closed fuel cycle employing fast reactors is an even longer-term proposition. So even fervent advocates of nuclear power need to put the reprocessing agenda aside for a few decades, and focus on swiftly deploying and improving the low-carbon energy solutions.

Think about it. In pursuit of closing the fuel cycle, the U.S. government could easily spend on the order of $150 billion over 15 years just to get to the starting line of large-scale commercialization. But all that spending will not yield one additional megawatt of low-carbon electricity beyond what could be obtained by sticking with the current once-through cycle, much less by investing that $150 billion in renewable and efficient energy technologies. Spent-fuel reprocessing, plutonium recycle, and fast reactor waste transmutation are currently uneconomical, higher-risk, 100-year answers to an urgent climate question that now requires low-risk 5 to 20 year solutions. For now, Congress and the new Administration should terminate funding for the Global Nuclear Energy Partnership (GNEP) and its associated efforts to close the nuclear fuel cycle and introduce fast burner reactors in the United States.

At any point along the way, Mr. Chairman, we can revisit this issue to assess whether there may be truly disruptive innovations in nuclear technology that would alter this negative assessment, and induce us to view closing the fuel cycle as a more cost-

130

4

effective pathway to decarbonization than the host of cheaper alternatives we have available to us today.

C. Nuclear Waste Disposal. As the political sun sets on the proposed Yucca Mountain project, the federal government needs to begin identifying alternative geological disposal sites for the country's nuclear waste. Congress should initiate a search for a new geologic repository site for disposal of spent fuel, and insure that adequate federal funding is available to retain the technical community associated with the Yucca Mountain project, so that this expertise will be available to assess and develop new proposed geological waste disposal sites. The Congress should not interfere in the NRC's ongoing Waste Confidence and Temporary Storage rulemakings, and let this regulatory body attempt to fulfill its independent regulatory mandate.

III. Detailed Observations

A. Loan Guarantees—Congress should not further subsidize the construction of new nuclear power plants and not provide additional loan guarantees for this purpose.

In the United States existing nuclear power plants operate efficiently and are profitable either because ratepayers long ago paid the piper for their stranded capital costs, or these assets were heavily discounted when corporate ownership changed in the 1990's and now are carried on the books of the new owners at a small fraction of their original asset value. The domestic nuclear power industry, however, is confronting two big economic dilemmas with respect to new nuclear plants. New plants remain uneconomical when compared to other electricity generating technologies and improvements in end-use efficiency; and the unit costs of new nuclear plants are so high that they are difficult to finance in the private capital markets, especially today.

As a purely commercial proposition, when stripped of all the various forms of federal and state subsidies, new nuclear plants are likely to remain non-competitive with other forms of baseload generation in most areas of the country until the price of carbon emissions exceeds $50 per ton of carbon dioxide. We note, however, that efficiency and many renewable sources are competitive with nuclear now and will only become more so. To bridge this gap, the nuclear industry, through its congressional boosters, has

5

already received production tax credits for the first 6,000 megawatts of new capacity, licensing cost sharing with DOE, "regulatory risk" insurance against delays in construction, and to date some $18.6 billion in federal loan authority to support the construction of new plants. In addition, most new reactor projects are benefitting from additional subsidies and incentives, such as tax abatements and worker training programs, offered by state and county governments.

Now the industry is returning to Congress for yet more support, essentially stipulating that nuclear power "must be part of the energy mix" needed to mitigate climate change and to provide for jobs under the economic stimulus plan. We should reject this categorical imperative, command economy type approach. It reminds us of the mindset we used to encounter in Minatom, the old Soviet Ministry of Atomic Energy. The economically efficient way to mitigate climate change is to internalize the cost of carbon emissions through a declining cap-and-trade program, which NRDC strongly supports.

This Committee should reject any broader attempt to use loan guarantees to recapitalize a technically mature industry, or to shift the overall terms of trade in the electricity marketplace in favor of nuclear power. This runs a serious risk of misdirecting investment capital away from commercialization of low-carbon energy technologies that are cheaper, cleaner, and more versatile than currently available nuclear power plants.

Shifting the overall terms of energy commerce in favor of low-carbon solutions, nuclear power included, is the task of a climate bill, not the federal loan guarantee program. At best, federal loan guarantees should be construed as bridging the gap between successful prototype development and a foothold in the commercial marketplace, by spreading the risk of the initial capital investments required to bring a new technology to commercial scale.

But federal loan guarantees should not be abused to insulate an entire industry from competition with a host of new energy technologies that promise comprehensive environmental and social benefits. Unlike improvements in efficiency and renewable technologies, nuclear power is a decarbonization solution packaged with a host of non-carbon environmental, security, and waste problems. For these reasons, nuclear power

6

should not be considered for inclusion in any "Renewable Electricity Standard" Congress may legislate.

In sum, the economically *inefficient* way to mitigate climate change is to broadly subsidize deployment of currently available nuclear power plant technologies. This will crowd out or slow investment in improved energy efficiency, utility-scale renewable electricity supply, and decentralized smart-grid technologies that can mitigate climate change in less time, with less cost and risk. If Congress is unwilling or unable politically to let a climate bill do the work of sorting out the most cost-effective low-carbon energy technologies, one possible way to mitigate economic inefficiency would be to closely couple any additional federal loan guarantees for nuclear with utility commitments to phase out existing coal capacity, such that future electricity demand growth in the affected service area or regional grid must be met in the first instance by large improvements in less costly energy efficiency, and by the development of renewable sources having environmental impacts and a marginal cost of generation less than nuclear power.

The idea that the nuclear and coal dependent Southeastern region of the United States is without renewable resources worthy of development is a gross distortion that needs to be dispelled. The region has vast distributed potential for photo-voltaic solar development, waste-heat cogeneration, bio-gasification, small hydro, and offshore wind. Above all, with the highest rates in the nation of energy consumption per unit of economic output, the region has a huge energy efficiency resource that can be tapped at far less cost than nuclear. The fact that the dominant utilities and electricity grid in that region are not currently structured to take advantage of these resources does not mean that they do not exist.

We should not use loan guarantees, or any other federal subsidies, to promote the economically inefficient use of nuclear power ahead of low-carbon energy alternatives that will be available sooner, at lesser cost, and with fewer environmental impacts. Under a well designed cap and trade system with competitive open access to the transmission and distribution grid, if nuclear power is needed for decarbonization, the marketplace for low-carbon energy will get around to demanding more of it, but not before it has exhausted the potential of other available energy resources (including all cost-effective

7

avenues for extracting energy savings from improvements in efficiency) that can displace CO_2 at a lower cost per ton than nuclear power.

An appropriate role for direct federal support of low-carbon energy is to underwrite research, development, and demonstration of meritorious new technologies that are unlikely to be developed by private industry acting alone, either because the return on the investment is too distant or because the investment risks are too high. Alternatively, society may reap benefits by using production or investment tax credits to more rapidly expand the market for beneficial emerging technologies, thereby helping to driving down unit costs of production to a level that allows the technology to become self-sustaining in the marketplace.

Further subsidization of new nuclear power plants does not meet either of these criteria. The first 6,000 megawatts of nuclear new-build capacity are already covered by a production tax credit comparable to wind, and sufficient loan guarantee authority ($18.5 billion) has already been made available to support construction of the first 'new' Gen III+ reactor designs proposed for the U.S. market—the Toshiba-Westinghouse AP1000 and the GE-Hitachi ESBWR. All other reactor designs proposed for construction in the United States either don't qualify as innovative, have already been constructed elsewhere, or both.

Furthermore, loan guarantees are not essential for nuclear plants currently being developed by regulated utilities as evidenced by Progress Energy's efforts to build two new units in Levy County, Florida, Georgia Power's efforts to build two units (Alvin W. Vogle Units 3 and 4), and South Carolina Electric & Gas's efforts to build two units (Virgil C. Summer Units 2 and 3). All six of these proposed units are AP1000 designs.

Finally, as NRC Chairman Dale E. Klein noted last week, the "excessive exuberance" for nuclear power has declined because of the global credit and economic crisis. The current economic recession has reduced the projected demand for electricity and there is a reduced need to build new base-load electricity generating capacity.

8

B. Reprocessing—The Federal Government should not encourage or support commercial spent fuel reprocessing.

Reprocessing of commercial spent fuel, as it is practiced today in France, Russia and Japan offers no advantages and numerous disadvantages over continuing to rely on the once-through nuclear fuel cycle as practiced in the United States and most countries with nuclear power plants. The trend in recent years has been for more countries to abandon reprocessing than to initiate reprocessing.

Relative to the existing open fuel cycle, the use of a closed or partially closed mixed-uranium and plutonium oxide (MOX) fuel cycle in thermal reactors has proven to be more costly and less safe. It leads to greater routine releases of radioactivity into the environment, greater worker exposures to radiation, larger inventories of nuclear waste that must be managed, and it doesn't appreciably reduce the geologic repository requirements for spent fuel or high-level nuclear waste.

Because reprocessing as it is practiced today does not appreciably reduce repository requirements it is not an alternative to Yucca Mountain. Should GNEP's advanced reprocessing technologies—essential to the success of the GNEP vision—prove technically feasible, they are unlikely to significantly impact repository requirements, because the fast reactors required for efficient waste transmutation are likely to remain more costly and less reliable than conventional thermal reactors, and hence will not be commercially deployed in sufficient numbers to effect the desired reductions.

The GNEP vision of burning the long-lived actinides, requires that some 30 to 40 percent of all reactor capacity be supplied by fast reactors. In other words, for every 100 thermal reactors of the type used throughout the United States today, some 40 to 75 new fast reactors of similar capacity would have to be built. The commercial use of large numbers of fast reactors for actinide burning is unlikely to occur because—to borrow observations made by U.S. Navy Admiral Hyman Rickover more than 50 years ago that remain true today—fast reactors have proven to be "expensive to build, complex to operate, susceptible to prolonged shutdown as a result of even minor malfunctions, and difficult and time-consuming to repair."

The development of fast reactors to breed plutonium failed in the United States, the United Kingdom, France, Germany, Italy, and Japan. We would argue it failed in the

9

Soviet Union despite the fact that the Soviets operated two commercial-size fast breeder plants, BN-350 (now shut down in Kazakhstan) and BN-600 (still operational in Russia), because the Soviet Union and Russia never successfully closed the fuel cycle and thus never operated these plants using MOX fuel.

Moreover, the advanced reprocessing technologies are even more costly than the conventional PUREX method and produce even larger inventories of intermediate and low-level nuclear wastes.

The closed fuel cycle technologies required by GNEP pose greater proliferation risks than the once-through fuel cycle. Even though GNEP's ambitious vision of deploying new reprocessing plants and fast reactors in large numbers will surely fail to materialize, the partnership's research program will encourage the development in non-weapon states of research facilities well suited for plutonium recovery, i.e., small hot cells and even larger reprocessing centers, as well as the training of experts in plutonium chemistry and metallurgy, all of which pose grave proliferation risks. It is for this reason that we advocate terminating the GNEP research on advanced reprocessing technologies.

For now, Congress and the new Administration should terminate funding for the GNEP and its associated efforts to close the nuclear fuel cycle and introduce fast burner reactors in the United States. This leaves the question of what level of long-term DOE research funding is appropriate to explore advanced nuclear fuel recycling technologies.

We hold the view that even substantial research spending in this area is unlikely to lead to disruptive nuclear technology breakthroughs that actually meet the stated goals of the research—cost-effective and non-proliferative techniques for reprocessing, recycling and transmuting plutonium-based fuels. And while the proliferation risks of this cooperative international research would be ongoing and tangible, we and many others in the nonproliferation community believe that shutting down the current U.S. plutonium recycle research effort, and any support it extends to foreign efforts, is the wisest course, at least until such time as the latent nuclear proliferation risk in the world is much better controlled than it is today.

Others, including Energy Secretary Steven Chu, appear to believe that some level of ongoing advanced fuel cycle research is appropriate and has some chance of yielding the desired disruptive nuclear technology breakthrough, if pursued for perhaps a decade

10

or more. History has not been very kind to this view, but the plutonium fuel cycle community is a lot like the fusion energy community in this respect—hope springs eternal as long as federal research dollars are within reach.

So weighing these contrasting glass-half-full and glass half-empty perspectives, Mr. Chairman, you might conclude that some modest long-term research program, geared to narrowing the technical and cost uncertainties surrounding the toughest unresolved technical, economic, safeguards, and proliferation issues, would be an appropriate and prudent middle path to pursue with respect to closing the fuel cycle. We would emphasize that even more important than the particular choice of technology is a better understanding of the requirements for the international institutional setting in which a large-scale fast reactor roll-out would be attempted. This, more than the technology, is the long pole in the closed fuel cycle tent. If one is serious about wanting to minimize the risks of proliferation, one is more or less driven to consider some form of international ownership and control over nuclear fuel cycle facilities, and this is likely to prove just as demanding a task as the development of more "proliferation resistant" strains of reprocessing. We also note that absent such an international structure for closely regulating the closed fuel cycle, we are unlikely ever to transition to a world free of nuclear weapons.

C. Congress should not interfere in the NRC's ongoing Waste Confidence and Temporary Storage rulemakings.

The issue of whether and how the availability of permanent geologic disposal should factor into the NRC licensing of commercial nuclear power plants has been with us for decades. A compromise on how the issue would be addressed in a scientific and publicly acceptable manner was reached nearly twenty five years ago and the basic framework of that compromise has not changed substantially over the years.

To make a long story short, in June of 1977, the NRC denied a NRDC petition that forced the question of whether there should be a rulemaking proceeding to determine whether high-level radioactive wastes generated in nuclear power reactors can be permanently disposed of without undue risk to public health and safety. NRDC then petitioned the United States Court of Appeals for the Second Circuit to review the NRC

11

decision. The D.C. Circuit remanded the matter to the NRC for further proceedings to determine whether there was reasonable assurance that a permanent disposal facility will be found. This and a related case gave rise to the NRC's "waste confidence" rulemaking. The NRC issued a set of findings in 1984 and subsequently revised them in 1990, and reaffirmed them in 1999. The NRC is now revisiting the issue.

The resolution of this issue properly remains with the NRC which was established to address health and safety issues associated with civil use of atomic energy. We would caution against intervention into this ongoing NRC decision-making process. It may be instructive to remind ourselves that the current failure to develop a geologic disposal facility for high level radioactive waste and spent fuel is due in large part to interventions by Congress subsequent to the passage of the Nuclear Waste Policy Act of 1982.

138 703-739-3790 TCNNaturalGas.com

**WRITTEN TESTIMONY
OF DALE E. KLEIN, CHAIRMAN
UNITED STATES NUCLEAR REGULATORY COMMISSION
TO THE
SENATE COMMITTEE ON ENERGY AND NATURAL RESOURCES**

MARCH 18, 2009

Mr. Chairman, Senator Murkowski, and Members of the Committee, I am pleased to appear before you today to discuss the Nuclear Regulatory Commission's new reactor licensing processes.

Let me begin by noting that just last week the NRC hosted our annual Regulatory Information Conference, which was attended by nearly three thousand individuals, including regulators, members of industry, stakeholders, and representatives from 31 other nations. Our annual conference is part of the NRC's ongoing efforts to share information, best practices and lessons learned to enhance nuclear safety and security both domestically and abroad.

Mr. Chairman, my testimony will explain the current licensing process for new reactor applications; contrast this with the agency's older, less efficient, two-step process; and discuss the current status of new reactor applications.

Congress has provided the NRC with the resources needed to meet the growing renewed interest in additional commercial nuclear power in the United States. These resources have enabled the NRC to successfully complete, on schedule, significant new reactor licensing activities. Over a number of years, NRC has taken steps to improve the licensing process. These actions have served to increase the effectiveness, efficiency and predictability of licensing a new reactor while maintaining our focus on safety and security. All currently operating commercial nuclear power plants in the United States were licensed under a two-step process for approval of construction and later for operation. But, all of the new reactor license applications have been submitted under a new combined license application approach (also known as "COL"), which essentially takes the previous two-step review process down to one

1

step. To date, the NRC has received 17 COL applications for 26 new nuclear reactors. A map depicting the locations and types of proposed reactors is attached. Based on industry information submitted to the NRC, we could see up to five more COL applications for seven more reactors by the end of 2010.

In the simplest terms, under the original two-step licensing approach the NRC would first issue a construction permit, based on evaluation of preliminary safety and design information, to allow construction of a nuclear power plant, and then later issue an operating license upon completion of construction. The applicant was not required to submit a complete design at the construction permit phase. Before the scheduled completion of construction, (typically when the plant was 50% completed), the applicant filed an application for an operating license. At this point, the applicant had to provide the complete design bases and other information related to the safe operation of the plant, technical specifications for operation of the plant, and description of operational programs.

Criticism of the two-step process centered on a design-as-you-go approach to constructing the plant, which deferred resolution of important safety issues until plant construction was well underway. The deferral of design details until after construction was authorized allowed commercial reactors to be built with an unusual degree of variability and diversity – in effect, a set of custom-designed and custom-built plants. Other criticisms included regulatory requirements that kept changing, and a seemingly inefficient and duplicative review and hearing process.

To address these problems, the process set forth in Part 52 of the NRC's regulations allows an applicant to seek a combined license, which authorizes construction based on a complete design and provides conditional authority to operate the plant, subject to verification that the plant has been constructed in accordance with the license, design, and the Commission's regulations. Part 52 maintains significant public participation throughout the licensing application process. A graphic depiction of the licensing process is attached.

2

Part 52 provides two other significant procedures: (1) review and approval of standardized designs through a Design Certification rulemaking, and (2) review and approval of a site's suitability, prior to a decision whether to build a particular plant, through an Early Site Permit (ESP). The applicant may also request a Limited Work Authorization (LWA), which allows applicants to perform limited work activities to prepare the site before approval of the COL.

So far, only one of the five designs currently being referenced in COL applications -- the Advanced Boiling Water Reactor -- has completed the certification process and is only referenced in one COL application. It should be noted that although the Westinghouse AP1000 is also a certified reactor design, the design that was approved in 2006 has two revisions under review by the NRC. A final decision on the design changes is expected in 2010.

In addition, the design certification applications and some COL applications received to date initially lacked information that the staff needs to complete its review. Staff reviews have been further complicated because some applicants are revising submission dates and submitting modifications to their applications, often with late notice to the staff, which is disruptive to the work planning process. The result is that the early COL applications are unlikely to achieve the full benefits of the Part 52 process. The NRC is working with stakeholders to overcome these challenges and is confident that the agency will be prepared to make timely regulatory decisions. As this process matures, the Commission will seek the continued support of Congress to sustain these efforts.

I would like to focus my comments briefly on improvements we have made to date, and what we expect down the road in new reactor licensing.

The NRC has sought to position itself strategically to be ready to respond to the new reactor licensing workload. The Commission created the Office of New Reactors, or NRO, to lead the agency effort to establish the regulatory and organizational foundation necessary to

3

address the new reactor licensing demand. Staffing the new office was given high priority, and today NRO has over 475 highly competent and qualified employees.

The NRC has made great strides in addressing the new reactor licensing challenge:

- The NRC published a revised 10 CFR Part 52 (titled, "Licenses, Certifications, and Approvals for Nuclear Power Plants") in August 2007 to clarify the applicability of various requirements and to enhance regulatory effectiveness and efficiency in implementing the licensing and approval processes. The rule also incorporated lessons learned from the reviews of the first design certification and early site permit applications.

- Similarly, the NRC published a final rule on Limited Work Authorizations, or LWAs, which supplements the final rule on 10 CFR Part 52. This rule allows certain early construction activities to commence before a construction permit or combined license is issued. The rule specifies the scope of construction activities that may be performed under an LWA, and specifies activities that no longer require NRC approval. Like the Part 52 revision, these changes were adopted to enhance the efficiency of the licensing and approval process and to reflect more clearly NRC's authority.

- In March 2007, the NRC completed the first comprehensive update to the NRC's Standard Review Plan (SRP), which provides guidance to the staff on how to perform technical reviews. The update brought the SRP into conformance with the Part 52 revision, and extends the applicability of the SRP to the Part 52 licensing process.

- The NRC issued a new regulatory guide, RG 1.206 (titled, "Combined License Applications for Nuclear Power Plants"), which provides guidance to potential applicants on standard format and content of new reactor combined license

4

applications, and also recently issued guidance for applicants on complying with the LWA rule.

- The NRC has implemented a computer-based project management system that significantly enhances the staff's ability to plan and schedule work.

- In 2004, the NRC promulgated substantially revised rules of practice intended to streamline and make the hearing process more effective.

- The NRC promulgated an electronic filing rule that is further increasing the efficiency of the hearing process.

- The NRC created a new reactor construction inspection organization in the Region II Office in Atlanta, Georgia. To prepare for the commencement of construction activities, the staff has observed ongoing new construction activities in China, Finland, France, Japan, Korea, and inspected the refurbishment and startup of the Tennessee Valley Authority (TVA) Browns Ferry Unit 1, which has been idle since 1975, and is currently inspecting the completion of TVA's Watts Bar Unit 2, which had been in a suspended state since 1985.

- Finally, the NRC conducted an efficient review of project management using the Six Sigma problem-solving methodology to streamline the design certification rulemaking process.

With these activities, I believe that the NRC has established a strong regulatory foundation for the review of new reactor license applications.

I should also mention that the agency has made a consistent effort to improve our coordination with other Federal agencies involved in new reactor licensing. For example, consistent with its lead responsibility for off-site nuclear emergency planning and response, the Federal Emergency Management Agency (FEMA) supports the NRC's COL application reviews by providing input to ensure that the off-site emergency plans are adequate.

5

143

In addition to COLs, the NRC staff has completed the review of three early site permit applications and is proceeding with the review of the fourth application. With respect to design certifications, the staff is continuing its review of General Electric's Economic Simplified Boiling Water Reactor, commonly referred to as the ESBWR; Areva Nuclear Power's U.S. Evolutionary Power Reactor, or U.S. EPR; Mitsubishi's U.S. Advanced Pressurized Water Reactor, or US-APWR; and amendments to Westinghouse's AP1000 design certification.

The NRC has completed preliminary work for the licensing of the Next Generation Nuclear Plant, or NGNP. In August 2008, the NRC and DOE delivered a licensing strategy to the Congress, as required by the Energy Policy Act of 2005.

I would like to touch briefly on the GAO's 2007 audit of the NRC's readiness to conduct reviews of COL applications. In general, the GAO's findings were positive assessments, acknowledging the NRC's extensive preparations and the quality of plans. The NRC continues to believe that the GAO assessments provide useful insights to the agency's management. The GAO identified four recommendations:

- Fully develop and implement criteria for setting priorities to allocate resources across applications by January 2008.

- Provide the resources for implementing reviewer and management tools needed to ensure that the most important tools will be available as soon as is practicable, but no later than March 2008.

- Clarify the responsibilities of Office of New Reactor's Resource Management Board in facilitating the coordination and communication of resource allocation decisions.

- Enhance the process for requesting additional information by (1) providing more specific guidance to staff on the development and resolution of requests for additional information within and across design centers and (2) explaining

6

forthcoming workflow and electronic process revisions to combined license

applicants in a timely manner.

I am pleased to report to you that the NRC has completed its work in response to these

recommendations.

The NRC is also working with its international partners on many areas of common

interest. One program that we have initiated is the Multi National Design Evaluation Program

(MDEP) in order to take advantage of international experience in licensing and constructing two

EPR plants in Europe to assist the NRC in its review of the US EPR application. The NRC also

has recently established interactions with regulatory counterparts in China, Canada and the

United Kingdom to exchange information on the licensing review of proposed AP1000 reactors

in the United States.

In addition to focusing on completing licensing reviews, the NRC is working on the

development and implementation of a new Construction and Vendor Inspection Program. The

program is building upon prior experience, including lessons learned during the construction of

the 104 currently operating reactors. Numerous historical lessons provide insights related to

quality and oversight problems during the previous period of construction in the United States,

and abroad. The most important of these lessons is that a commitment to quality, instilled early

in a nuclear construction project, is vital to ensuring that the facility is constructed and will

operate in conformance with its license and the regulations.

The NRC staff is working with the industry to ensure that a strong commitment to quality

is part of the foundation of every new reactor project in the United States. Many of the

components that will be used in the construction of possible new reactors in the U.S. will be

manufactured abroad, so NRC inspectors are also visiting manufacturing facilities and working

with our regulatory counterparts in other countries to ensure the quality of the manufactured

components. Quality assurance (QA) inspections of engineering and site activities are

contributing to the conduct of effective and efficient reviews of design certifications, COLs, and

7

early site permit applications. The agency has also sought stakeholder involvement in an effort to make construction and vendor inspection a timely, accurate and transparent process.

While the Commission is satisfied that we have in place an effective regulatory process, we are always looking for ways to improve. Just as industry can become more efficient, the NRC is constantly working to improve its efficiency with no compromise in safety.

Mr. Chairman and Members of the Committee, this concludes my overview of the NRC's licensing process for new reactor applications, and the current status of license applications. I would be pleased to respond to any questions you may have.

8

New Licensing Process – 10 CFR Part 52

Pre-Construction — Construction Verification

Optional Pre-Application Review

Early Site Permit OR Equivalent Site Information

Standard Design Certification OR Equivalent Design Information

Combined License Review, Hearing, and Decision

Verification of Regulations with ITAAC

Reactor Operation Decision

- Licensing decisions finalized before major construction begins
- Inspections to verify construction
- Limited work may be authorized before license is issued

9

U.S.NRC New Reactors

Location of Projected New Nuclear Power Reactors

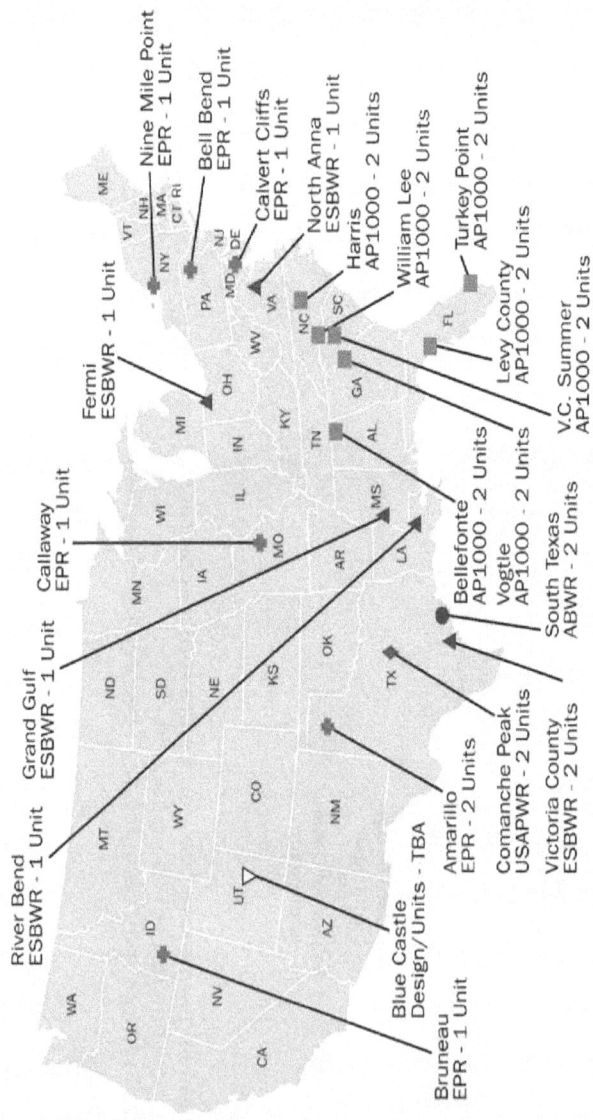

ADVANCED REACTOR CONCEPTS

The U.S. Department of Energy's Office of Nuclear Energy

Advanced reactor concepts will provide substantially enhanced operational performance, safety, security, economics, and proliferation resistance over currently deployed light water reactor technologies. These enhancements will be realized through national and international research and development collaborations.

The Advanced Reactor Concepts program, an expanded version of the Generation IV research and development (R&D) program, sponsors research and development leading to further safety, technical, economical, and environmental advancements of innovative nuclear energy technologies. The Office of Nuclear Energy (NE) will pursue these advancements through R&D activities at the nation's national laboratories, universities, and industrial labs, as well as through partnerships with international organizations and governments. These activities will focus on advancing scientific understanding of these technologies; pioneering the use of advanced modeling and simulation; establishing an international network of user facilities for nuclear R&D; improving economic competitiveness; and reducing the technical and regulatory uncertainties for deploying new nuclear reactor technologies.

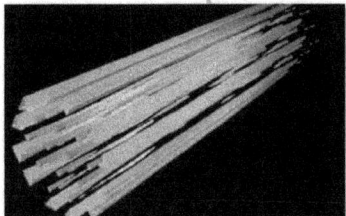

BENEFITS OF R&D FOR ADVANCED REACTORS

As a result of Advanced Reactor Concepts research, nuclear energy will be able to increase its contribution to the reduction of greenhouse gas emissions by introducing advanced designs into new energy and industrial markets. Both advanced thermal and fast neutron spectrum systems will be investigated.

High-temperature reactors will provide significantly improved thermal efficiency to generate higher temperature process heat for power generation and a range of industrial applications that require higher heat. This is achieved through a higher reactor outlet temperature and the use of gas turbine technologies that will utilize helium-based Brayton power conversion cycles.

Fast neutron spectrum systems provide systems that could be employed to consume long-lived, high-activity elements found in used light-water-reactor fuels. These systems could provide used fuel management alternatives to direct geological disposal by "burning" undesirable constituents of used fuel. Some modified open fuel cycle concepts are enabled by fast neutron spectrum systems. Higher outlet temperatures allowed by the liquid-metal coolants, coupled with supercritical CO_2 Brayton cycle energy conversion, will also increase thermal efficiency.

INTERNATIONAL COOPERATION

Key to achieving these advancements is the multiplication effect on investment from international collaboration. By coordinating U.S. efforts with partner nations, our funding is leveraged by multiple factors. The United

www.nuclear.energy.gov
February 2010

ADVANCED REACTOR CONCEPTS

The U.S. Department of Energy's Office of Nuclear Energ

Program Budget

Advanced Reactor Concepts
($ in Millions)

	FY 2010 Actual	FY 2011 Request
	$17.6*	$21.9

appropriated under GEN IV program

States collaborates with the international community through the Generation IV International Forum (GIF), the International Atomic Energy Agency (IAEA), and through a number of bilateral agreements.

PLANNED PROGRAM ACCOMPLISHMENTS

In FY 2010, many of the R&D activities that supported advanced reactor concepts were established under the GEN IV budget.

FY 2010

- Perform post-irradiation examination of highly irradiated metallic fuel samples from the Fast Flux Test Facility and the Phoenix Fast Reactor in France.
- Develop advanced materials such as ceramics, composite materials, and nano-structured ferritic materials for use in structural systems, fuel claddings, and other high-temperature applications.
- Complete the advanced gas reactor fuel irradiation experiment and commence irradiation for the first fuel produced in large-scale production equipment.
- Continue the development of advanced gas reactor system-simulation software and initiate bilateral cooperation with Japan on the use of its gas reactor.
- Continue study of liquid salt as a circulating fluid in reactor cooling loops.
- Demonstrate advanced Brayton-cycle energy conversion systems using supercritical carbon dioxide.
- Continue international R&D through GIF, IAEA, and bilateral agreements.

FY 2011

- Evaluate innovative reactor systems to identify promising candidates for further R&D as part of an integrated system.
- Evaluate and test heat transfer properties for a wide range of operating fluids and conditions.
- Complete post-irradiation examinations on unique material samples obtained from formerly operating fast neutron reactors to obtain mechanical and physical properties of these materials for use in materials model development.
- Continue advanced modeling techniques utilizing the Department's high-speed, parallel computers for the development of close-coupled neutronic and thermofluid codes.
- Demonstrate the technical and economic viability of an advanced Brayton-cycle energy-conversion system using supercritical carbon dioxide as the working fluid.
- Conduct nuclear data measurements and validation, specifically cross section and other nuclear data measurements needed for advanced fast reactor designs and safety validation.
- Conduct research on components or systems applicable to multiple reactor concepts, such as fuel handling, in-service inspection and repair, and energy conversion.
- Continue international collaboration on nuclear safety and cost reduction.
- Evaluate molten salt and other advanced reactor conceptual ideas.

www.nuclear.energy.gov
February 2010

FUEL CYCLE RESEARCH AND DEVELOPMENT

The U.S. Department of Energy's Office of Nuclear Energy

The U.S. Department of Energy (DOE) supports long-term, results-oriented, science-based research and development (R&D) through its Fuel Cycle R&D Program.

The mission of the Fuel Cycle Research and Development program is to perform results-oriented, science-based R&D to provide options for decision-makers for future commercial fuel cycle management strategies. This will enable the safe, secure, economic, and sustainable expansion of nuclear energy while minimizing proliferation risks.

FUEL CYCLE STRATEGIES

The program will examine three fuel cycle strategies: **once-through fuel cycle**, **modified open fuel cycle**, and **full recycle fuel cycle**. Examination of this full range of strategies is critical to provide future decision-makers with adequate information to make decisions on how best to manage used fuel. For each fuel cycle strategy, the Fuel Cycle R&D program has objectives to pursue in fulfilling its mission.

Once-Through Fuel Cycle Strategy —

Nuclear fuel makes a single pass through a reactor, after which the used fuel is removed, stored for some period of time, and then directly disposed in a geologic repository for long-term isolation from the environment. The objectives in this strategy are to develop fuels for use in reactors that would increase the efficient use of uranium resources and reduce the amount of used fuel for direct disposal for each megawatt-hour of electricity produced.

Modified Open Fuel Cycle Strategy —

Limited separations and fuel processing technologies are applied to the used light water reactor fuel to create fuels that enable the extraction of much more energy from the same mass of material and accomplish waste management goals. The objectives in this strategy are to investigate fuel forms, reactors, and fuel/waste management approaches that would dramatically increase utilization of fuel resources and reduce the quantity of long-lived radiotoxic elements in the used fuel to be disposed. Technologies will be considered that require at most limited separation steps and minimize proliferation risks.

Full Recycle Fuel Cycle Strategy —

All of the long-lived, radiotoxic chemical elements important for waste management are recycled in thermal- or fast-spectrum systems to reduce the radiotoxicity of the waste placed in a geologic repository while more fully utilizing uranium resources. In a full recycle system, only those elements that are considered to be waste (primarily the fission

www.nuclear.energy.gov
February 2010

FUEL CYCLE RESEARCH AND DEVELOPMENT

Program Budget

Fuel Cycle R&D Program
($ in Millions)

	FY 2010 Actual	FY 2011 Request
	$136.0	$201.0

products) would be disposed, not used fuel. The objectives in this strategy are to develop techniques that will enable long-lived, radiotoxic chemical elements to be repeatedly recycled. The ultimate goal is to develop a cost-effective and low-proliferation-risk approach that would dramatically decrease the long-term challenges posed by the waste and reduce uncertainties associated with its disposal.

NEW AREAS OF FOCUS

The modified open fuel cycle constitutes a range of technology options in between once-through fuel cycle and full recycle fuel cycle and could be an important part of achieving a sustainable fuel cycle. The full recycle fuel cycle has been the focus of the Fuel Cycle R&D program to date and the once-through fuel cycle is the current practice in the United States. The modified open fuel cycle has not been studied in any depth and that is why it is being introduced as a new focus area for FY 2011.

The Used Nuclear Fuel Disposition technical area is being expanded in order to conduct more R&D for storage, transportation, and disposal options for used nuclear fuel and high-level waste. These activities are consistent with NE responsibilities under the Nuclear Waste Policy Act associated with nuclear waste management.

CONTRIBUTING TO THE SECRETARY'S ENERGY GOAL

The Fuel Cycle R&D program contributes to the Secretary of Energy's goal of *Energy: Build a Competitive, Low-Carbon Economy and Secure America's Energy Future*. The program is designed to develop advanced fuel cycle technologies that will help enable the deployment of nuclear power. The program will help create a safe and sustainable path forward for the expansion of nuclear power and help to optimize the nuclear fuel cycle. Developing these advanced technologies decreases the fuel cycle risks associated with constructing and operating nuclear power plants, increasing the likelihood that new nuclear power plants will be deployed, thus contributing to greenhouse gas abatement efforts.

A SCIENCE-BASED R&D PROGRAM

The Fuel Cycle R&D program is an integrated program to research, develop, and improve fuel cycle waste management options and transformational technologies. It involves small-scale experiments coupled with theory development and advanced modeling and simulation with validation experiments. This science-based R&D program will provide a more complete understanding of the underlying science supporting the development of advanced fuel cycle technologies and waste management options and, therefore, help provide a sound basis for future decision-making. The program will also identify alternatives and conduct scientific research and technology development to enable storage, transportation, and disposal of used nuclear fuel and all radioactive wastes generated by existing and future nuclear fuel cycles.

www.nuclear.energy.gov
February 2010

152

The U.S. Department of Energy's Office of Nuclear Energy

A WELL-MANAGED, SCIENTIFIC COLLABORATION

DOE's NE programs allocate R&D funding to those entities (*e.g.*, national laboratories, universities, and industry) that are best qualified to carry out the work in support of NE's mission. NE R&D activities typically require highly specialized R&D facilities and capabilities that are primarily available only at DOE national laboratories; therefore, a majority of NE activities are carried out here.

Consistent with NE's commitment to supporting R&D activities at university and educational research institutions, NE designates up to 20 percent of funds appropriated to its R&D programs for work to be performed at university and research institutions through open, competitive solicitations for investigator-led projects. Universities often collaborate with the DOE national laboratories, thereby expanding the base of highly qualified engineers and scientists in the future.

The Fuel Cycle R&D program also collaborates with nuclear industry and international partners with advanced fuel cycles to leverage U.S. research investments and pursue common goals towards advanced fuel cycles that are economic, minimize waste, and reduce proliferation risk.

PLANNED PROGRAM ACCOMPLISHMENTS

FY 2010

- Initiate a series of fundamental measurements to serve as a basis for expanding the understanding of actinide separations science.

- Begin to develop advanced safeguards instrumentation for materials accountability measurements with increased accuracy and reliability for future separations facilities.

- Initiate the use of advanced experimental techniques and modeling tools to design novel fuel forms with significantly improved performance.

- Continue R&D on advanced alloy and composite cladding materials to support the improved performance goals of advanced fuels.

- Continue R&D activities on high precision measurements of nuclear data, sensitivity analyses to reduce uncertainty, and development of advanced measurement techniques.

- Initiate development of a system capable of capturing and managing all types of fuel cycle R&D knowledge.

- Conduct system studies for a range of possible fuel cycles and geologic repository environments in order to specify technical requirements for each key step of the fuel cycle to achieve a "near-zero" loss cycle.

At Argonne National Laboratory, research in electrorefining of nuclear fuel is being tested to reduce the amount of high-level waste requiring long-term storage.

www.nuclear.energy.gov
February 2010

FUEL CYCLE RESEARCH AND DEVELOPMENT

The U.S. Department of Energy's Office of Nuclear Energy

FY 2011

- Continue to develop advanced concepts for electrochemical processing to recycle salt for waste minimization, advanced methods for transuranic recovery, and novel product consolidation methods.

- Continue to develop alternative waste forms that are tailored to specific radionuclides and potential geologic media.

- Initiate development of innovative fuel systems that possibly support alternative fuel cycles to the current once-through fuel cycle with the potential for dramatic performance and waste minimization potential.

- Develop an initial modeling and simulation integration framework that facilitates capability transfer by allowing interoperability of existing codes and newly developed capabilities.

- Using a systems engineering approach, conduct systems analyses to define and analyze a broad variety of innovative fuel cycle options including analyzing the effects of a variety of alternative disposal geologies to inform R&D prioritization and program planning.

- Based on a roadmap created in FY 2010, continue development of improved proliferation risk assessment tools to evaluate fuel cycle options.

- Provide technical expertise to inform policy decision-making regarding the storage, transportation, and disposal of used nuclear fuel and radioactive waste that would be generated under existing and potential future nuclear fuel cycles, including the long-term durability of the materials.

- Identify novel approaches to improve resource utilization including new fuel forms, ultra-high burnup fuels, thorium-based fuels, deep burn of transuranic-bearing tristructural-isotropic (TRISO) fuels, new advanced reactors designed for transuranic burn-up such as molten salt reactors and travelling wave reactors, options to declad and reclad used fuel to allow volatile and gaseous fission products to be removed and captured before recycling.

www.nuclear.energy.gov
February 2010

Advanced Nuclear Power and Fuel Cycle Technologies: Outlook and Policy Options

Mark Holt
Specialist in Energy Policy

July 11, 2008

Congressional Research Service

7-5700

www.crs.gov

RL34579

CRS Report for Congress
Prepared for Members and Committees of Congress

Summary

Current U.S. nuclear energy policy focuses on the near-term construction of improved versions of existing nuclear power plants. All of today's U.S. nuclear plants are light water reactors (LWRs), which are cooled by ordinary water. Under current policy, the highly radioactive spent nuclear fuel from LWRs is to be permanently disposed of in a deep underground repository.

The Bush Administration is also promoting an aggressive U.S. effort to move beyond LWR technology into advanced reactors and fuel cycles. Specifically, the Global Nuclear Energy Partnership (GNEP), under the Department of Energy (DOE) is developing advanced reprocessing (or recycling) technologies to extract plutonium and uranium from spent nuclear fuel, as well as an advanced reactor that could fully destroy long-lived radioactive isotopes. DOE's Generation IV Nuclear Energy Systems Initiative is developing other advanced reactor technologies that could be safer than LWRs and produce high-temperature heat to make hydrogen.

DOE's advanced nuclear technology programs date back to the early years of the Atomic Energy Commission in the 1940s and 1950s. In particular, it was widely believed that breeder reactors—designed to produce maximum amounts of plutonium from natural uranium—would be necessary for providing sufficient fuel for a large commercial nuclear power industry. Early research was also conducted on a wide variety of other power reactor concepts, some of which are still under active consideration.

Although long a goal of nuclear power proponents, the reprocessing of spent nuclear fuel is also seen as a weapons proliferation risk, because plutonium extracted for new reactor fuel can also be used for nuclear weapons. Therefore, a primary goal of U.S. advanced fuel cycle programs, including GNEP, has been to develop recycling technologies that would not produce pure plutonium that could easily be diverted for weapons use. The "proliferation resistance" of these technologies is subject to considerable debate.

Much of the current policy debate over advanced nuclear technologies is being conducted in the appropriations process. For FY2009, the House Appropriations Committee recommended no further funding for GNEP, although it increased funding for the Generation IV program. Typically, the Senate is more supportive of GNEP and reprocessing technologies.

Recent industry studies conducted for the GNEP program conclude that advanced nuclear technologies will require many decades of government-supported development before they reach the current stage of LWRs. Key questions before Congress are whether the time has come to move beyond laboratory research on advanced nuclear technologies to the next, more expensive, development stages and what role, if any, the federal government should play.

Congressional Research Service

Advanced Nuclear Power and Fuel Cycle Technologies: Outlook and Policy Options

Contents

Contacts

Congressional Research Service

All commercial nuclear power plants in the United States, as well as nearly all nuclear plants worldwide, use light water reactor (LWR) technology that was initially developed for naval propulsion. Cooled by ordinary water, LWRs in the early years were widely considered to be an interim technology that would pave the way for advanced nuclear concepts. After the early 1960s, the federal government focused most of its nuclear power research and development efforts on breeder reactors and high temperature reactors that could use uranium resources far more efficiently and potentially operate more safely than LWRs.

However, four decades later, LWRs continue to dominate the nuclear power industry, and are the only technology currently being considered for a new generation of U.S. commercial reactors. Federal license applications for as many as 30 new LWRs have been recently announced. The proposed new nuclear power plants would begin coming on line around 2016 and operate for 60 years or longer. Under that scenario, LWRs appear likely to dominate the nuclear power industry for decades to come.

If the next generation of nuclear power plants consists of LWRs, what is the potential role of advanced nuclear reactor and fuel cycle technologies? Do current plans for a new generation of LWRs raise potential problems that advanced nuclear technologies could or should address? Can new fuel cycle technologies reduce the risk of nuclear weapons proliferation? What is the appropriate time frame for the commercial deployment of new nuclear technology? This report provides background and analysis to help Congress address those questions.

Prominent among the policy issues currently before Congress is the direction of the existing nuclear energy programs in the U.S. Department of Energy (DOE). DOE administers programs to encourage near-term construction of new LWRs, such as the Nuclear Power 2010 program, which is paying half the cost of licensing and first-of-a-kind engineering for new U.S. LWR designs, and loan guarantees for new reactors now under consideration by U.S. utilities. DOE's Global Nuclear Energy Partnership (GNEP) is developing advanced fuel cycle technologies that are intended to allow greater worldwide use of nuclear power without increased weapons proliferation risks. Advanced nuclear reactors that could increase efficiency and safety are being developed by DOE's Generation IV program, which is looking beyond today's "Generation III" light water reactors.

The priority given to these options depends not only on the characteristics of existing and advanced nuclear technologies, but on the role that nuclear power is expected to play in addressing national energy and environmental goals. For example, if nuclear energy is seen as a key element in global climate change policy, because of its low carbon dioxide emissions, the deployment of advanced reactor and fuel cycle technologies could be considered to be more urgent than if nuclear power is expected to have a limited long-term role because of economic, non-proliferation, and safety concerns.

Nuclear Technology Overview

As their name implies, light water reactors use ordinary water for cooling the reactor core and "moderating," or slowing, the neutrons in a nuclear chain reaction. The slower neutrons, called thermal neutrons, are highly efficient in causing fission (splitting of nuclei) in certain isotopes of heavy elements, such as uranium 235 and plutonium 239 (Pu-239). Therefore, a smaller percentage of those isotopes is needed in nuclear fuel to sustain a nuclear chain reaction (in which neutrons released by fissioned nuclei then induce fission in other nuclei, and so forth). The downside is that thermal neutrons cannot efficiently induce fission in more than a few specific isotopes.

Natural uranium has too low a concentration of U-235 (0.7%) to fuel an LWR (the remainder is U-238), so the U-235 concentration must be increased ("enriched") to between 3% and 5%. In the reactor, the U-235 fissions, releasing energy, neutrons, and fission products (highly radioactive fragments of U-235 nuclei). Some neutrons are also absorbed by U-238 nuclei to create Pu-239, which itself may then fission.

After several years in an LWR, fuel assemblies will build up too many neutron-absorbing fission products and become too depleted in fissile U-235 to efficiently sustain a nuclear chain reaction. At that point, the assemblies are considered spent nuclear fuel and removed from the reactor. LWR spent fuel typically contains about 1% U-235, 1% plutonium, 4% fission products, and the remainder U-238. Under current policy, the spent fuel is to be disposed of as waste, although only a tiny fraction of the original natural uranium has been used. Long-lived plutonium and other actinides[1] in the spent fuel pose a long-term hazard that greatly increases the complexity of finding a suitable disposal site.

Reprocessing, or recycling, of spent nuclear fuel for use in "fast" reactors—in which the neutrons are not slowed—is intended to address some of the shortcomings of the LWR once-through fuel cycle. Fast neutrons are less effective in inducing fission than thermal neutrons but can induce fission in all actinides, including all plutonium isotopes. Therefore, nuclear fuel for a fast reactor must have a higher proportion of fissionable isotopes than a thermal reactor to sustain a chain reaction, but a larger number of different isotopes can constitute that fissionable proportion.

A fast reactor's ability to fission all actinides makes it theoretically possible to repeatedly separate those materials from spent fuel and feed them back into the reactor until they are entirely fissioned. Fast reactors are also ideal for "breeding" the maximum amount of Pu-239 from U-238, eventually converting virtually all of natural uranium to useable nuclear fuel.

Current reprocessing programs are generally viewed by their proponents as interim steps toward a commercial nuclear fuel cycle based on fast reactors, because the benefits of limited recycling with LWRs are modest. Commercial-scale spent fuel reprocessing is currently conducted in France, Britain, and Russia. The Pu-239 they produce is blended with uranium to make mixed-oxide (MOX) fuel, in which the Pu-239 largely substitutes for U-235. Two French reprocessing plants at La Hague can each reprocess up to 800 metric tons of spent fuel per year, while Britain's THORP facility at Sellafield has a capacity of 900 metric tons per year. Russia has a 400-ton plant at Ozersk, and Japan is building an 800-ton plant at Rokkasho to succeed a 90-ton demonstration facility at Tokai Mura. Britain and France also have older plants to reprocess gas-cooled reactor fuel, and India has a 275-ton plant.[2] About 200 metric tons of MOX fuel is used annually, about 2% of new nuclear fuel,[3] equivalent to about 2,000 metric tons of mined uranium.[4]

---

[1] Actinides consist of actinium and heavier elements in the periodic table.

[2] World Nuclear Association, *Processing of Used Nuclear Fuel for Recycle*, March 2007, at http://www.world-nuclear.org/info/inf69.html.

[3] World Nuclear Association, *Mixed Oxide Fuel (MOX)*, November 2006, at http://www.world-nuclear.org/info/inf29.html.

[4] World Nuclear Association, *Uranium Markets*, March 2007.

While long a goal of nuclear power proponents, the reprocessing or recycling of spent nuclear fuel is also seen as a weapons proliferation risk, because plutonium extracted for new reactor fuel can also be used for nuclear weapons. Therefore, a primary goal of U.S. advanced fuel cycle programs, including GNEP, has been to develop recycling technologies that would not produce pure plutonium that could easily be diverted for weapons use. The "proliferation resistance" of these technologies is subject to considerable debate.

Removing uranium from spent nuclear fuel through reprocessing would eliminate most of the volume of radioactive material requiring disposal in a deep geologic repository. In addition, the removal of plutonium and conversion to shorter-lived fission products would eliminate most of the long-term (post-1,000 years) radioactivity in nuclear waste. But the waste resulting from reprocessing would have nearly the same short-term radioactivity and heat as the original spent fuel, because the reprocessing waste consists primarily of fission products, which generate most of the radioactivity and heat in spent fuel. Because heat is the main limiting factor on repository capacity, conventional reprocessing would not provide major disposal benefits in the near term.

DOE is addressing that problem with a proposal to further separate the primary heat-generating fission products—cesium 137 and strontium 90—from high level waste for separate storage and decay over several hundred years. That proposal would greatly increase repository capacity, although it would require an alternative secure storage system for the cesium and strontium that has yet to be designed.

Safety and efficiency are other areas in which improvements have long been envisioned over LWR technology. The primary safety vulnerability of LWRs is a loss-of-coolant accident, in which the water level in the reactor falls below the nuclear fuel. When the water is lost, the chain reaction stops, because the neutrons are no longer moderated. But the heat of radioactive decay continues and will quickly melt the nuclear fuel, as occurred during the 1979 Three Mile Island accident. DOE's Generation IV program is focusing on high temperature, gas-cooled reactors that would use fuel whose melting point would be higher than the maximum reactor temperature. The high operating temperature of such reactors would also result in greater fuel efficiency and the potential for cost-effective production of hydrogen, which could be used as a non-polluting transportation fuel. However, the commercial viability of Generation IV reactors remains uncertain.

DOE Advanced Nuclear Programs

DOE's advanced nuclear technology programs date back to the early years of the Atomic Energy Commission in the 1940s and 1950s. In particular, it was widely believed that breeder reactors would be necessary for providing sufficient fuel for a commercial nuclear power industry. Early research was also conducted on a wide variety of other power reactor concepts, some of which are still under active consideration. The U.S. research effort on various advanced nuclear concepts has waxed and waned during subsequent decades, sometimes resulting from changes in Administrations. Technical and engineering advances have appeared to move some of the technologies closer to commercial viability, but significantly greater federal support would be necessary to move them beyond the indefinite research and development stage.

Advanced Nuclear Power and Fuel Cycle Technologies: Outlook and Policy Options

Global Nuclear Energy Partnership

GNEP is the Bush Administration's program for commercial deployment of reprocessing or recycling of spent nuclear fuel. The program's goal is to develop "proliferation resistant" fuel cycle technologies—not producing pure plutonium—that could be used around the world. Previous U.S. commercial reprocessing programs have been blocked at least partly over concerns that they would encourage other countries to begin separating weapons-useable plutonium.

History

The fundamental technology for spent fuel reprocessing is the PUREX process (plutonium-uranium extraction) developed to provide pure plutonium for nuclear weapons. A commercial PUREX plant operated from 1966 through 1972 in West Valley, New York, and two other commercial U.S. plants were built but never operated.

Meanwhile, DOE and its predecessor agencies worked to develop fast breeder reactors that could run on the reprocessed plutonium fuel. Major facilities included Experimental Breeder Reactors I and II, which began operating in Idaho in 1951 and 1964, and the Fast Flux Test Facility (FFTF), a larger fast reactor that began full operation in Hanford, Washington, in 1982. FFTF was designed to pave the way for the first U.S. commercial-scale breeder reactor, planned to begin construction near Clinch River, Tennessee, in 1977. However, the Clinch River Breeder Reactor (CRBR) and the federal government's support for commercial reprocessing were halted by President Carter in 1977 because of the nuclear proliferation issues noted above.

Upon taking office in 1981, President Reagan reversed the Carter policy and restarted preparations for CRBR, but Congress eliminated further funding for the project in 1983. DOE then turned to an alternative technology based on work carried out at Experimental Breeder Reactor II (EBR-II), which used metal fuel that could be recycled through pyroprocessing (melting and electrochemical separation) rather than with the aqueous (water-based) PUREX process. Supporters of this program, called the Integral Faster Reactor (IFR) and the Advanced Liquid Metal Reactor (ALMR), contended that pyroprocessing would not produce a pure plutonium product and could be carried out at a small scale at reactor sites, reducing weapons proliferation risks.

The Clinton Administration, however, moved in 1993 to terminate DOE's advanced reactor programs, including shutdown of EBR-II. Congress agreed to the proposed phaseout but continued funding for pyroprocessing technology as a way to treat EBR-II spent fuel for eventual disposal.

Current Program

The George W. Bush Administration made energy policy a high priority and placed particular emphasis on nuclear energy. The National Energy Policy Development (NEPD) Group, headed by Vice President Cheney, recommended in May 2001 that nuclear power be expanded in the United States and that reprocessing once again become integral to the U.S. nuclear program:

- The NEPD Group recommends that, in the context of developing advanced nuclear fuel cycles and next generation technologies for nuclear energy, the United States should reexamine its policies to allow for research, development and deployment of fuel conditioning methods (such as pyroprocessing) that reduce waste streams and enhance proliferation resistance. In doing so, the United States will continue to discourage the accumulation of separated plutonium, worldwide.

- The United States should also consider technologies (in collaboration with international partners with highly developed fuel cycles and a record of close cooperation) to develop reprocessing and fuel treatment technologies that are cleaner, more efficient, less waste-intensive, and more proliferation-resistant.[5]

The Bush Administration's first major step toward implementing those recommendations was to announce the Advanced Fuel Cycle Initiative in 2003 (AFCI), a DOE program to develop proliferation-resistant reprocessing technologies. The program built on the ongoing pyroprocessing technology development effort and reprocessing research conducted under other DOE nuclear programs. Much of the program's research has focused on an aqueous separations technology called UREX+, in which uranium and other elements are chemically removed from dissolved spent fuel, leaving a mixture of plutonium and other highly radioactive elements.

Congress provided $5 million above the Administration's $63 million initial request in FY2004 for AFCI, and the program received statutory authorization in the Energy Policy Act of 2005 (P.L. 109-58, Sec. 953), including support for international cooperation.

The announcement of the GNEP initiative in February 2006 (as part of the Administration's FY2007 budget request) appeared to further address the 2001 reprocessing goals of the National Energy Policy Development Group. Using reprocessing technologies to be developed by AFCI, GNEP envisioned a consortium of nations with advanced nuclear technology that would guarantee to provide fuel services and reactors to countries that would agree not to conduct fuel cycle activities, such as enrichment and reprocessing.

GNEP has attracted significant international attention, but no country has yet indicated interest in becoming solely a fuel recipient rather than a supplier. The Nuclear Nonproliferation Treaty guarantees the right of all participants to develop fuel cycle facilities, and a GNEP Statement of Principles signed by the United States and 15 other countries on September 16, 2007, preserves that right, while encouraging the establishment of a "viable alternative to acquisition of sensitive fuel cycle technologies."[6] According to DOE, GNEP currently has 21 member countries and 17 candidates and observers.[7]

Although GNEP is largely conceptual at this point, DOE issued a Spent Nuclear Fuel Recycling Program Plan in May 2006 that provided a general schedule for a GNEP Technology Demonstration Program (TDP),[8] which would develop the necessary technologies to achieve GNEP's goals. According to the Program Plan, the first phase of the TDP, running through FY2006, consisted of "program definition and development" and acceleration of AFCI. Phase 2,

[5] National Energy Policy Development Group, *National Energy Policy*, May 16, 2001, p. 5-22.

[6] See GNEP website at http://www.gnep.energy.gov

[7] Members: Australia, Bulgaria, Canada, China, France, Ghana, Hungary, Italy, Japan, Jordan, Kazakhstan, Lithuania, Poland, Republic of Korea, Romania, Russia, Senegal, Slovenia, Ukraine, United Kingdom, and United States. Candidates and Observers: Argentina, Belgium, Brazil, Czech Republic, Egypt, Finland, Germany, Libya, Mexico, Morocco, Netherlands, Slovak Republic, South Africa, Spain, Sweden, Switzerland, and Turkey. http://www.gneppartnership.org

[8] DOE, *Spent Nuclear Fuel Recycling Plan*, Report to Congress, May 2006.

running through FY2008, was to focus on the design of technology demonstration facilities, which then were to begin operating during Phase 3, from FY2008 to FY2020. The National Academy of Sciences in October 2007 strongly criticized DOE's "aggressive" deployment schedule for GNEP and recommended that the program instead focus on research and development.[9] Similar criticism was raised in April 2008 by the Government Accountability Office.[10]

As part of GNEP, AFCI is conducting R&D on an Advanced Burner Reactor (ABR) that could destroy recycled plutonium and other long-lived radioactive elements. The ABR is similar to a breeder reactor, except that its core would be configured to produce less plutonium (from U-238) than it consumes, reducing potential plutonium stockpiles.

Funding

AFCI, the primary funding component of GNEP, has received steadily increased funding from Congress, but far less than requested during the past two budget cycles. For FY2007, DOE sought $243.0 million and received $166.1 million, and for FY2008 the request of $395.0 million was cut to $179.4 million. Typically, the Senate recommends more for the program than the House does, and that pattern appears to be continuing for FY2009.

The FY2009 Advanced Fuel Cycle Initiative funding request is $301.5 million, nearly 70% above the FY2008 appropriation of $179.4 million but below the FY2008 request of $395.0 million. The House Appropriations Committee recommended cutting AFCI to $90.0 million in FY2009, eliminating all funding for GNEP.[11] The remaining funds would be used for research on advanced fuel cycle technology, but none could be used for design or construction of new facilities. The Committee urged DOE to continue coordinating its fuel cycle research with other countries that already have spent fuel recycling capability, but not with "countries aspiring to have nuclear capabilities."

FY2009 funding of $10.4 million was requested for conceptual design work on an Advanced Fuel Cycle Facility (AFCF) to provide an engineering-scale demonstration of AFCI technologies, according to the budget justification. The FY2008 Consolidated Appropriations act rejected funding for development of AFCF, as did the House Appropriations Committee for FY2009. DOE requested $18.0 million for the ABR program for FY2009, up from $11.7 million in FY2008. The program is expected to focus on developing a sodium-cooled fast reactor (SFR). The House Appropriations Committee recommended no FY2009 funding for the ABR.

Generation IV

DOE describes "Generation IV" as advanced reactor technologies that could be available for commercial deployment after 2030. These technologies are intended to offer significant

[9] National Academy of Sciences, *Review of DOE's Nuclear Energy Research and Development Program*, prepublication draft, October 2007.

[10] Government Accountability Office, *Global Nuclear Energy Partnership: DOE Should Reassess Its Approach to Designing and Building Spent Nuclear Fuel Recycling Facilities*, GAO-08-483, April 2008.

[11] The Committee voted on the FY2009 Energy and Water Development Appropriations Bill on June 25, 2008, but has not filed a report. The draft report was accessed on cq.com.

advantages over existing "Generation III" reactors (LWRs in the United States) in the areas of cost, safety, waste, and proliferation. DOE is conducting some Generation IV research in cooperation with other countries through the Generation IV International Forum (GIF), established in 2001.[12]

A technology roadmap issued by GIF and DOE in 2002 identified six Generation IV nuclear technologies to pursue: fast neutron gas-cooled, lead-cooled, sodium-cooled, molten salt, supercritical water-cooled, and very high temperature reactors.[13] These reactor concepts are not new, and some have been demonstrated at the commercial scale, but none has been sufficiently developed for successful commercialization.

The DOE Generation IV Nuclear Energy Systems Initiative (Gen IV) is focusing on a helium-cooled Very High Temperature Gas Reactor (VHTR) and conducting cross-cutting research on materials and other areas that could apply to all reactor technologies, including LWRs. The VHTR technology is being developed for the Next Generation Nuclear Plant (NGNP) authorized by the Energy Policy Act of 2005. Development of sodium-cooled fast reactors is being conducted by the AFCI program as part of the ABR effort described above.

DOE requested $70.0 million for Gen IV for FY2009—$44.9 million below the FY2008 funding level of $114.9 million, which was nearly triple the Administration's FY2008 budget request of $36.1 million. The House Appropriations Committee recommended an increase to $200.0 million.

Most of the FY2009 request—$59.5 million—is for the NGNP program. The VHTR technology being developed by DOE uses helium as a coolant and coated-particle fuel that can withstand temperatures up to 1,600 degrees celsius. Phase I research on the NGNP is to continue until 2011, when a decision will be made on moving to the Phase II design and construction stage, according to the FY2009 DOE budget justification. The House Appropriations Committee provided $196.0 million "to accelerate work" on NGNP—all but $4.0 million of the Committee's total funding level for the Generation IV program. The Energy Policy Act of 2005 authorizes $1.25 billion through FY2015 for NGNP development and construction (Title VI, Subtitle C). The authorization requires that NGNP be based on research conducted by the Generation IV program and be capable of producing electricity, hydrogen, or both.

Time Lines and Options

DOE's plans for commercial nuclear fuel recycling facilities are still being formulated. The Department is currently preparing a draft Programmatic Environmental Impact Statement (PEIS) for GNEP that will lead to decisions about development of an advanced fuel cycle research facility. The PEIS will not consider the next stages of the program, which would include commercial-scale reprocessing/recycling facilities and an advanced fast reactor, according to DOE.[14] A schedule for completing this process has not been announced.

---

[12] GIF active members are Canada, China, Euratom, France, Japan, Republic of Korea, Russia, Switzerland, and the United States. http://www.gen-4.org

[13] DOE Nuclear Energy Research Advisory Committee and Generation IV International Forum, *A Technology Roadmap for Generation IV Nuclear Energy Systems*, GIF-002-00, December 2002.

[14] http://www.gnep.energy.gov/PEIS/gnepPEIS.html accessed July 9, 2008.

Advanced Nuclear Power and Fuel Cycle Technologies: Outlook and Policy Options

Industry Studies

To help determine the future direction of the GNEP program, DOE solicited studies from four industry consortia. The four studies, released by DOE on May 28, 2008, describe concepts for advanced fuel recycling/reprocessing facilities, along with general cost estimates and schedules. The four teams have signed cooperative agreements with DOE to continue developing "conceptual designs, technology development roadmaps, and business plans for potential deployment and commercialization of recycling and reactor technologies" at least through FY2008 and possibly through FY2009. According to DOE, these additional studies will "help inform a decision on the potential path forward for technologies and facilities associated with domestic implementation of GNEP."[15]

EnergySolutions, Shaw, and Westinghouse

EnergySolutions, a waste treatment and disposal firm, Shaw Group, an engineering and construction firm, and Westinghouse Electric Company, a reactor design firm, led an industry team that proposed that aqueous reprocessing facilities to handle 1,500 metric tons per year of LWR spent fuel begin operating by 2023. A fuel fabrication plant would be built to supply MOX fuel to existing LWRs. Recycling facilities during this initial phase would be funded and built by DOE.

The next phase of the EnergySolutions proposal would run from 2030 to 2049. A 410 megawatt (electric) fast reactor would begin operating in 2033, with four additional units starting up by 2045. Aqueous reprocessing capacity would be expanded by 3,000 metric tons per year, and non-aqueous reprocessing facilities would be added. In the final phase, 2050 through 2100, the fast reactor recycling fleet would expand to 96 gigawatts (about the capacity of today's U.S. LWR fleet), and less aqueous reprocessing capacity would be needed.

A federal corporation would be established to sign long-term contracts with industry for spent fuel recycling and fuel fabrication, build and operate a waste repository, and transport spent fuel. The federal corporation's funding would come from nearly doubling the nuclear waste fee currently imposed on nuclear power generation, from 1 mill per kilowatt-hour to 1.95 mills/kwh, assuming the previously collected balance in the Nuclear Waste Fund (the Treasury account that holds the waste fees) is not used. At the current rate of nuclear power generation, the proposed fee would produce revenues of about $1.5 billion per year.

GE-Hitachi

A team led by General Electric Hitachi Nuclear Energy prepared a proposal based on the IFR/ALMR program that was halted in 1993. The pyroprocessing facility that is proposed would use the electrometallurgical separations process developed by the IFR program, with improvements that have been made during the subsequent 15 years. The fast reactor is the Power Reactor Inherently Safe Module (PRISM) that GE developed for the ALMR program, also with subsequent refinements. According to the report, a power plant consisting of six PRISM modules

---

[15] DOE Office of Public Affairs, "DOE Releases Domestic Global Nuclear Energy Partnership (GNEP) Industry Reports and Presentations," May 28, 2008.

(totaling 1,866 megawatts electric, mwe), along with the necessary reprocessing capacity, would consume 5,800 metric tons of LWR spent fuel over its planned 60-year operating life.

The first phase of the GE-Hitachi proposal, taking about 20 years, would consist of construction and operation of one or two PRISM modules. The second phase, lasting about 10 years, would feature commercial deployment of at least one Advanced Recycling Center (ARC), consisting of six PRISM modules and a reprocessing and fuel fabrication facility. Multiple ARCs would be constructed in the third phase, after 30 years.

General Atomics

General Atomics, long associated with gas-cooled reactor technology, led a team that proposed a two-tier spent fuel recycling system. In the first tier, LWR spent fuel would be sent to aqueous reprocessing plants to extract nuclear fuel material to be used in high-temperature gas reactors, such as the type being developed by the DOE Gen IV program. Because of their high fuel burnup, the gas reactors would eliminate most plutonium and minor actinides. In the second tier, spent fuel from the gas reactors would be pyroprocessed so that the remaining plutonium and minor actinides could be fissioned in a fast reactor.

Under the team's preferred scenario, LWRs would continue to be constructed through 2050 (136 in all) and be phased out by 2110. The first gas-cooled reactor module (385 mwe) would start up by 2025, and the first aqueous reprocessing center would begin operation by about 2030. The aqueous reprocessing centers would have a capacity of about 1,500 tons of LWR spent fuel per year and cost about $8.3 billion to construct (in 2006 dollars). The first pyroprocessing facility would open in 2040, and the first fast reactor would open by 2075. The team recommended that initial facilities for the program be developed by a government corporation, which would be privatized by 2035.

Areva

The French nuclear firm Areva, which has long experience with commercial PUREX reprocessing plants in France, led a team that proposed continued reliance on LWRs with a gradual buildup of fast reactors. Through 2019, the team recommended that MOX fuel be tested in existing U.S. reactors, from plutonium extracted from U.S.-origin spent fuel reprocessed overseas. The first 800-ton per year aqueous recycling plant would open in 2023, with additional 800-ton modules starting up in 2045 and 2070. A 500 mwe fast reactor would begin operating in 2025, a 1,000 mwe reactor would open in 2035, and a 1,500 mwe reactor would begin operating in 2050, with additional 1,500 mwe units starting up about every two years thereafter. A government corporation would be established to run the recycling program. Costs are estimated to be 10%-70% higher than the existing 1 mill/kwh nuclear waste fee.

Policy Implications

For Congress and other federal policymakers, issues posed by current GNEP and Gen IV proposals are similar to those of the past several decades. The fundamental policy question is whether the government should encourage the expansion of nuclear power. The industry has long contended that new commercial reactors will not be constructed without increased government incentives or subsidies. After the initial federal push to commercialize nuclear power in the 1950s and 1960s, government support waned to the point where a nuclear phaseout seemed possible.

But nuclear power proponents now contend that dramatic growth will be needed (with federal support) to meet future energy demand in a carbon-constrained environment.

Such high-growth scenarios must overcome many of the same perceived challenges that faced the optimistic initial expectations for nuclear power. If dramatic growth were to finally occur, could light water reactors meet the challenge, or is a transition to advanced nuclear technologies necessary? And if new technologies will be needed, how urgently must the federal government move forward?

As in the early years of the nuclear power program, a primary concern with renewed nuclear power growth is long-term fuel supply, since LWRs can extract energy from only a fraction of natural uranium. During the past two decades of slowed U.S. and world nuclear power expansion, the only problem with uranium was oversupply and chronically low prices. Supply has since tightened, but uranium production capacity is expanding rapidly in response. Whether increased exploration activity will result in higher worldwide resource estimates will have important implications for this issue.

The long-proposed solution to the fuel problem—replacing LWRs with fast breeder reactors—raises the nuclear weapons proliferation issue. LWR spent fuel is highly resistant to proliferation at least for the first 100 years, although the technology requires uranium enrichment facilities that may pose their own risks. GNEP's goal of expanding nuclear power while limiting the proliferation of fuel cycle facilities is widely shared, but the success of the program's current approach remains uncertain.

Nuclear waste management has also been a longstanding problem in the United States and the world. The once-through LWR fuel cycle requires extremely long-term isolation of plutonium and other long-lived radionuclides. Reprocessing could potentially shorten the disposal horizon and make siting easier for waste repositories. But if long-term isolation is determined to be feasible, the waste disposal benefits of reprocessing may become less significant.

Other anticipated benefits of advanced reactor technologies over LWRs include improved safety, lower costs, and high-temperature heat production for hydrogen and other industrial purposes. LWR technology has improved steadily in safety, particularly in its vulnerability to loss-of-coolant accidents, and the projected risks of the latest designs have been reduced one to two orders of magnitude below that of existing reactors. Proposed Generation IV designs are intended to virtually eliminate the major risk factors inherent in LWRs, although they may have other safety risks that have yet to be as fully quantified.

New LWR designs are also intended to reduce costs from those incurred by existing reactors, but cost estimates have recently escalated (along with those of all competing power systems). Generation IV reactors are projected by their designers to reduce both construction and operating costs, but these projections have yet to be demonstrated.

LWRs are limited to relatively low-temperature operation, so high-temperature gas reactors could be the most practical technology for nuclear generation of hydrogen as a transportation fuel. If hydrogen were to become a major transportation fuel—which remains far from certain—nuclear power could begin to play a significant role in replacing petroleum. However, more commercial attention has recently been focused on battery-based electric vehicle systems, which could be recharged by LWRs.

Recent U.S. nuclear energy policy has focused primarily on large government incentives for private-sector construction of new LWRs, such as loan guarantees, tax credits, and regulatory risk insurance. Imposition of federal controls on carbon dioxide emissions would provide additional powerful incentives for LWR construction. As shown by the industry studies described above, the advanced nuclear technologies under development by GNEP and Gen IV will require many years of government-supported development before they reach the current stage of LWRs. The Bush Administration has renewed the federal research effort on these technologies, so now the question before Congress is whether the time has come to move to the next, more expensive, development stages.

Author Contact Information

Mark Holt
Specialist in Energy Policy
mholt@crs.loc.gov, 7-1704

Nuclear Fuels & Materials SPOTLIGHT

703-739-3790 TCNNuclear.com **169**

Nuclear Fuels & Materials SPOTLIGHT

March 2009

Idaho National Laboratory
P.O. Box 1625
Mailstop 3878
Idaho Falls, ID   83415-3878

INLNFMSpotlight@inl.gov

Tel: 208-526-9549

Fax: 208-526-2930

www.inl.gov/nuclearfuels

Interest in nuclear energy is growing in the United States and around the world as nations recognize that it is the most environmentally friendly, large-scale energy source. It is anticipated that we are at the threshold of a nuclear renaissance. To support this renaissance, the U.S. Department of Energy (DOE) identified the Idaho National Laboratory (INL) as its lead laboratory for nuclear energy research, development, and demonstration.

ADVANCED TEST REACTOR

NUCLEAR FUELS & MATERIALS SPOTLIGHT

ABOUT THE NUCLEAR FUELS AND MATERIALS SPOTLIGHT

Kemal O. Pasamehmetoglu

As the use of nuclear energy grows within the United States and internationally, nuclear fuels and materials research is key to the effort to make nuclear power more affordable and to enhance its safety and reliability. Especially important are developments that can increase burnup and reduce the aging of nuclear materials. The next generation of reactors, depending upon the targeted applications (e.g., process heat and hydrogen production), may use totally different fuel types and materials. Over the longer term, closing the fuel cycle and using recycled materials for nuclear fuel will become a necessity as a means of resource conservation and nuclear waste management.

The INL Nuclear Fuels and Materials Division, established in 2007, will play an important role in meeting these emerging research demands.

The Division will publish an annual *Nuclear Fuels and Materials Spotlight* that will include multiple articles that highlight research areas, capabilities, and expertise of the division. While it is impossible to cover all areas in a single publication in a given year, the *Spotlight* will report on the full spectrum of capabilities, expertise, and accomplishments over multiple years. The objective is to provide a forum for stakeholders and the technical community to get to know the INL Nuclear Fuels and Materials Division. Hopefully, the articles included in each publication will trigger ideas for collaborative research with other national laboratories, universities, industry (both the utilities and nuclear system vendors), and international institutions.

Without feedback from the technical community, sustaining and improving such a publication will be impossible. On behalf of the Nuclear Fuels and Materials Division at the INL, I hope you will find this publication valuable and will provide recommendations for improving it.

Kemal O. Pasamehmetoglu, PhD

Director, INL Nuclear Fuels and Materials Division

KEMAL O. PASAMEHMETOGLU received his PhD in Mechanical Engineering from the University of Central Florida in 1986, specializing in nuclear reactor technology with the sponsorship of the U.S. Nuclear Regulatory Commission. He then joined Los Alamos National Laboratory where he worked and served in leadership positions for multiple projects, including the Transient Reactor Analysis Code development, restart analysis for Savannah River K-Reactors, flammable gas mitigation in high-level nuclear waste storage tanks at Hanford, Accelerator Production of Tritium, Accelerator Transmutation of Waste, and Advanced Accelerator Application. While serving as national technical director for advanced fuel development in the Advanced Fuel Cycle Initiative (AFCI), he joined the INL in October 2004. He continues to serve as the national director of advanced fuel development activities, which involves participation by multiple national laboratories and universities. Kemal became the first director of the Nuclear Fuels and Materials Division at the INL in April 2007.

i

---※---

DISCLAIMER

This information was prepared as an account of work sponsored by an agency of the U.S. Government. Neither the U.S. Government nor any agency thereof, nor any of their employees, makes any warranty, express or implied, or assumes any legal liability or responsibility for the accuracy, completeness, or usefulness of any information, apparatus, product, or process disclosed, or represents that its use would not infringe privately owned rights. References herein to any specific commercial product, process, or service by trade name, trademark, manufacturer, or otherwise, does not necessarily constitute or imply its endorsement, recommendation, or favoring by the U.S. Government or any agency thereof. The views and opinions of authors expressed herein do not necessarily state or reflect those of the U.S. Government or any agency thereof.

INL/MIS-08-14888
Volume I

ii

NUCLEAR FUELS & MATERIALS SPOTLIGHT

TABLE OF CONTENTS

iii

The Advanced Test Reactor (ATR), at the INL, is one of the world's premier test reactors for studying the effects of intense neutron and gamma radiation on reactor materials and fuels. The physical configuration of the ATR, a four-leaf-clover shape, allows the reactor to be operated at different power levels in the corner lobes to allow for different testing conditions for multiple simultaneous experiments. The combination of high flux (maximum thermal neutron fluxes of 1×10^{15} neutrons per square centimeter per second and maximum fast [E>1.0 MeV] neutron fluxes of 5×10^{14} neutrons per square centimeter per second) and large test volumes (up to 48 in. long and 5.0 in. diameter) provides unique testing opportunities.

The facility offers three basic experimental configurations. The simplest is the static capsule, wherein the target material is placed in a capsule, or plate form, and the capsule is in direct contact with the primary coolant. The next level of complexity is the instrumented lead experiment, which allows for active monitoring and control of experiment conditions during the irradiation. The highest level of complexity is the pressurized water loop experiment, in which the test sample is subjected to the exact environment of a pressurized water reactor.

Various customers are currently sponsoring experiments in the ATR, including the U.S. government, foreign governments, private researchers, and commercial companies needing neutron irradiation services. The ATR was designated a National Scientific User Facility in 2007, which is enabling greater user access to its capabilities for fuels and materials research. This section provides more details on some of the ATR capabilities, key design features, experiments, and future plans.

ADVANCED TEST REACTOR
EXPERIMENTS AND CAPABILITIES

Frances M. Marshall and S. Blaine Grover

THE ATR IS one of the most versatile operating research reactors in the United States. The ATR is located at the INL, which is owned by the DOE and currently operated by Battelle Energy Alliance (BEA). The ATR has a long history of supporting reactor fuel and material research for the U.S. Government and other test sponsors. The mission of the ATR is to study the effects of intense neutron and gamma radiation on reactor materials and fuels.

The ATR first achieved criticality in 1967 and is expected to continue operation for several more decades. Current experiments are being conducted in the ATR for a variety of customers, including the DOE, foreign governments, private researchers, and commercial companies that need neutrons. The ATR has several unique features that enable it to perform diverse simultaneous tests for multiple sponsors.

The remainder of this section discusses the ATR design features, testing options, previous experiment programs, and future plans for ATR capabilities and experiments. It also provides some information about the INL and DOE's expectations for nuclear research in the future.

ATR DESCRIPTION

The ATR is a pressurized, light-water-moderated, beryllium-reflected reactor that operates at nominally 2.5 Mpa (360 psig) and 71°C (160°F) and has a maximum operating power of 250 MW. The current operating power is typically closer to 110 MW because of test sponsor requirements, but it is still capable of full power operations. The ATR operates an average of 240 days per year, currently with operating cycles of 6 to 8 weeks, followed by a 1- or 2-week outage for refueling and experiment changes. The core comprises 40 curved-plate fuel elements, each containing 19 curved, aluminum-clad uranium plates arranged in a serpentine configuration around a three-by-three array of primary testing locations called flux traps. These locations have the highest flux locations in the core.

The ATR reactor vessel is constructed of solid stainless steel and is located far enough away from the active core that neutron embrittlement of the vessel is not a concern. In addition, the ATR core internals are completely replaced every 7 to 10 years, with the last change having been completed in January 2005. These two major factors – the stainless-steel vessel and regular change-out of core internals – combined with a proactive maintenance and plant equipment replacement program, have resulted in the ATR operational life being essentially unlimited.

The physical configuration of the ATR (Figure 1) (four-leaf-clover shape) allows the reactor to be operated at different power levels in the four corner lobes to allow for different testing conditions for multiple simultaneous experiments. The horizontal rotating control drum system provides stable axial/vertical flux profiles for experiments throughout each reactor operating cycle unperturbed by the typical vertically positioned control components (see Figure 2 for typical flux distributions). This stable axial flux profile, with the peak flux rate at the center of the core, allows experimenters to have specimens positioned in the core at different known flux rates to receive a range of neutron fluences during the same irradiation periods over the duration of the test program. This control system also allows the reactor to operate different sections of the core at different power levels.

1

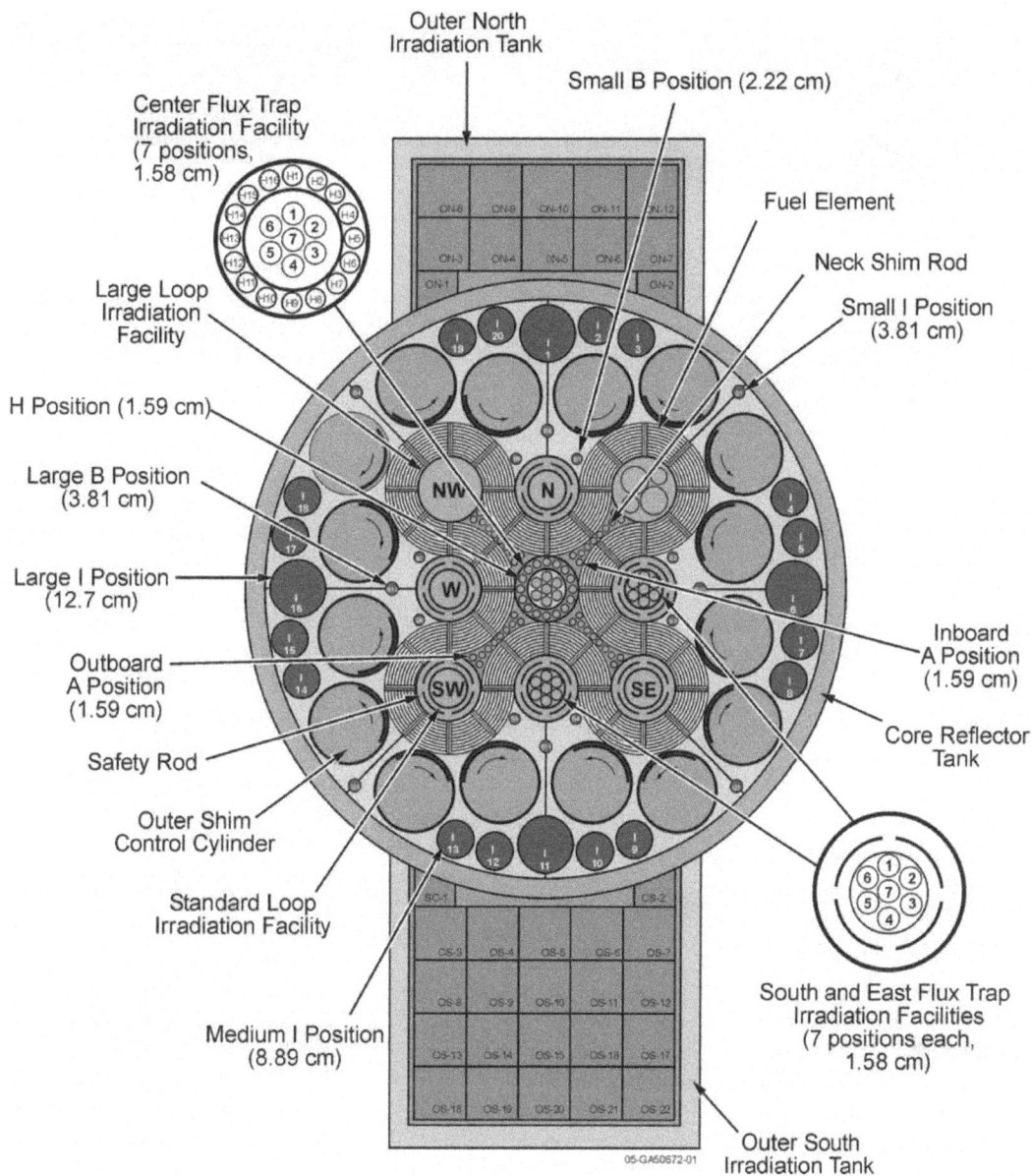

ADVANCED TEST REACTOR EXPERIMENTS & CAPABILITIES

Outer North Irradiation Tank

Small B Position (2.22 cm)

Center Flux Trap Irradiation Facility (7 positions, 1.58 cm)

Fuel Element

Neck Shim Rod

Large Loop Irradiation Facility

Small I Position (3.81 cm)

H Position (1.59 cm)

Large B Position (3.81 cm)

Large I Position (12.7 cm)

Inboard A Position (1.59 cm)

Outboard A Position (1.59 cm)

Core Reflector Tank

Safety Rod

Outer Shim Control Cylinder

Standard Loop Irradiation Facility

South and East Flux Trap Irradiation Facilities (7 positions each, 1.58 cm)

Medium I Position (8.89 cm)

Outer South Irradiation Tank

Figure 1.

Map of the Advanced Test Reactor core.

2

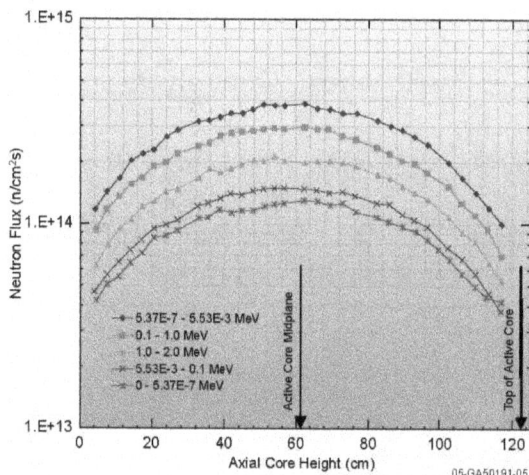

Figure 2.

Unperturbed five-energy-group neutron flux intensity profiles over the active core length of the ATR center flux trap for a total reactor power of 125 MW_{th}.

The ATR design can accommodate a wide variety of testing requirements. The key design features are as follows:

- Large test volumes – 48 in. long (at all testing locations) and up to 5 in. in diameter

- A total of 77 testing positions

- High neutron flux – up to 1×10^{15} n/cm²-s thermal and 5×10^{14} n/cm²-s fast

- Frequent experiment changes

- Variety of fast/thermal flux ratios (0.1 – 1.0)

- Constant and symmetrical axial power profile

- Power tilt capability between lobes – a ratio of 3:1 between different lobes of the reactor in the same operating cycle

- Individual experiment control

- Simultaneous experiments in different test conditions

- Core internals replacement every 10 years

- Accelerated testing for fuel, up to 20-times-actual burnup rate.

As testing has progressed at the ATR since initial operations, several changes to the reactor and plant have been needed. Some of the changes were implemented to offer more testing capabilities to researchers; other changes have upgraded the plant operating characteristics and increased operational reliability. Changes to the reactor to expand the testing capabilities include addition of the Powered Axial Locator Mechanism (PALM), which allows experiments to be moved axially in and out of the reactor core flux region to simulate reactor startups and other transient conditions. Changes to increase operational reliability

include upgrading the instrument and control reactor protection systems (RPS) to more reliable digital systems, resulting in fewer unintentional RPS shutdowns.

ATR TEST CAPABILITIES

The ATR uses three basic experimental configurations: 1) the static capsule, 2) the instrumented lead, and 3) the pressurized water loop. The following sections describe each experiment in more detail and include examples of the experiments performed using each type of configuration. An additional experiment configuration, a hydraulic shuttle irradiation system (also discussed below), was being added in 2008.

Static Capsule Experiments

The simplest experiment performed in the ATR is a static capsule experiment. If the specimens are to be irradiated at a specific elevated temperature, then they are sealed in aluminum, zircaloy, or stainless steel tubing with a cover gas and a specially designed insulating gas gap between the specimen and the capsule tube. The sealed tube is then placed in a holder that sits in the ATR test position. A single capsule can be the full 48–in. core height (or shorter) such that a series of stacked capsules may comprise a single test. Capsules are usually placed in an irradiation basket to facilitate the handling of the experiment in the reactor. Figure 3 shows a simplified drawing of the mixed oxide (MOX) irradiation test capsule and basket assembly. Some capsule experiments

3

ADVANCED TEST REACTOR EXPERIMENTS & CAPABILITIES

contain material that can be in contact with the ATR primary coolant and therefore need the cooling function; these capsules will not be sealed but will remain in an open configuration such that the capsule internals are exposed to and cooled by the ATR primary coolant system. Examples of this are fuel plate testing, in which the fuel to be tested is in a cladding material compatible with the ATR primary coolant chemistry requirements.

Static capsules typically have passive instrumentation, including flux-monitor wires and temperature melt wires for examination following the irradiation. As briefly indicated earlier, limited temperature control can be designed into the capsule through the use of an insulating gas gap between the test specimen and the outside capsule wall. The size of the gap is determined through analysis based upon the experiment temperature requirements, and an appropriate inert gas is sealed into the capsule.

Static capsule experiments are easier to insert, remove, and reposition than more complex experimental configurations. For fuel experiments, it is sometimes necessary to relocate an experiment to a different irradiation location within the ATR to compensate for fuel burnup over the duration of the experiment. A static capsule experiment is also typically less costly than an instrumented one and requires less time for design and analysis prior to insertion into the ATR.

Instrumented Lead Experiments

The next level in complexity is an instrumented lead experiment, which provides active monitoring and control of experiment parameters during irradiation. The primary difference between a static capsule and an instrumented lead experiment is an umbilical tube that connects the experiment in the reactor to a monitoring/control system elsewhere in the reactor building.

In a temperature-controlled experiment, thermocouples continuously monitor the temperature and provide feedback to a gas control system that supplies the necessary insulating gas mixture to the experiment to achieve the desired experiment conditions. The thermocouple leads and the gas tubing are located in the umbilical tube. A conducting (helium) gas and an insulating (typically neon or possibly argon) gas are mixed to control the thermal conductance across a predetermined gas gap. The computer-controlled gas blending system allows for the gas mixture to be up to 98% of one gas and as low as 2% of the other gas to allow for a wide range of experiment

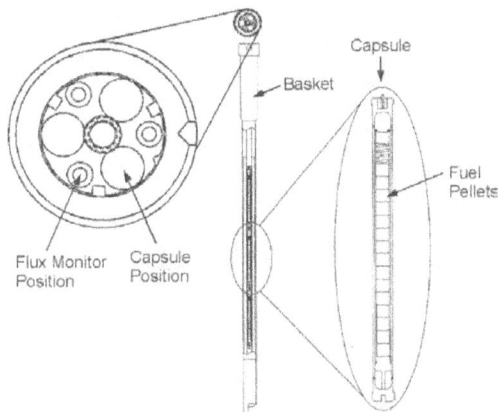

Figure 3.

Static capsule assembly for the MOX experiment.

Capsule
Basket
Fuel Pellets

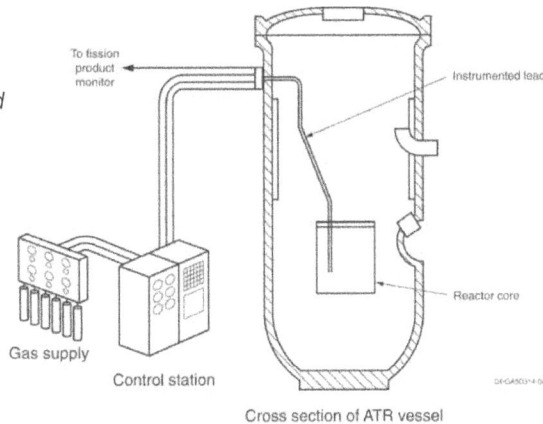

Figure 4.

Example of an instrumented lead experiment configuration.

To fission product monitor
Instrumented lead
Reactor core
Gas supply
Control station
Cross section of ATR vessel

Flux Monitor Position
Capsule Position

4

temperature ranges. Figure 4 shows a typical instrumented lead experiment.

Another feature of the instrumented lead experiment is the ability to monitor the effluent gas from around the test specimen to determine if changes to the experimental conditions are needed. In a fueled experiment, for example, there is sometimes a desire to monitor for fission gases in order to detect specimen failure. Gas chromatography can also be used to monitor oxidation of an experiment specimen. The instrument leads allow for a real-time display of parameters on an operator control panel. The instrumented leads can also provide an alarm to the operators and experimenters if any parameters exceed test limits. A data acquisition and archive capability can be provided for any monitored experiment parameter. Typically, data are saved for 6 months on a circular first-in, first-out format.

The primary advantage to the instrumented lead experiment is the active monitoring and control of parameters, which is not possible in a static capsule experiment. Additionally, the experiment sponsor does not have to wait until the full irradiation has been completed for all experiment results; the instrumentation provides preliminary results of the experiment and specimen condition.

Pressurized Water Loop Experiments

The pressurized water loop experiment is the most complex and comprehensive type of testing performed in the ATR. Five of the ATR flux traps contain in-pile tubes (IPTs) that provide a barrier between the experiment and the reactor primary coolant system, and are connected to a secondary pressurized water loop coolant system. The IPTs extend through the entire reactor vessel and contain closure plugs at the top of the vessel to allow the experiments to be independently inserted and removed.

The secondary cooling system includes pumps, coolers, ion exchangers, heaters (to control experiment temperature), and chemistry control systems. Loop tests can precisely represent conditions in a commercial pressurized water reactor. As in the instrumented lead experiments, all of the secondary loop parameters are continuously monitored and computer controlled to ensure precise testing conditions. Operator control display stations for each loop continuously display information that is monitored by the reactor operations staff. Test sponsors receive preliminary irradiation data before the irradiations are completed, so there are opportunities to modify testing conditions if needed. The data from the experiment instruments are collected and archived similar to the data in the instrumented lead experiments.

There are two PALM drive units that can be connected to specially configured tests in the pressurized water loop facilities so that complex transient testing can be performed. The PALM drive units move a small test section from above the reactor core region into the core region and back out again – either quickly (in approximately 2 seconds) or slowly, depending on test requirements. This process simulates multiple startup and shutdown cycles of test fuels and materials. Thousands of cycles can be simulated during a normal ATR operating cycle. The PALM drive units are also used to position a test precisely within the neutron flux of the reactor and change this position slightly as the reactor fuel burns.

Shuttle Irradiation System

In 2008, a new facility was added to the ATR – a hydraulic shuttle irradiation system. This system will enable experiments to be inserted into, and removed from, the ATR during the irradiation cycle. Currently, experiments can only be inserted and removed during the outage time, so materials that do not need, or cannot withstand, longer irradiation times are not irradiated in the ATR.

5

ADVANCED TEST REACTOR EXPERIMENTS & CAPABILITIES

PREVIOUS AND CURRENT TESTS IN THE ATR

The tests performed in the ATR have been diverse in their designs, objectives, and sponsors. The ATR has supported major nuclear reactor research initiatives for the United States and international collaborations. Some of the more notable experiments are discussed below.

Mixed Oxide Fuel

As part of the nuclear nonproliferation initiatives, it was proposed that weapons grade plutonium be mixed with commercial UO_2 and burned in current light water reactors (LWRs). Some testing was needed on the MOX fuel, however. A simple capsule was prepared to contain nine fuel samples, which were exposed to a variety of burnups to simulate LWR burnup profiles. These test capsules were moved from one experiment position to another in the ATR during the irradiation duration to enable a more stable fuel burnup rate during the course of the experiment. Preliminary fuel analysis results were sufficient to enable commercial nuclear power companies to pursue fabrication and use of MOX fuel in commercial LWR power plants.

MHTGR Fuel

In the late 1980s, the United States was interested in designing and deploying smaller high temperature reactors – specifically, the Modular High Temperature Gas Reactor (MHTGR). Several tests for particle fuels were planned in the ATR. One experiment was performed; however, there was evidence of fuel failure so the test was terminated early. Subsequently, the project was cancelled so these tests were also discontinued. The data obtained from this experiment, however, have been valuable in establishing fuel fabrication techniques and a fuel testing program for the Advanced Gas Reactor (AGR) project as part of the Next Generation Nuclear Plant Fuel Research and Development Program.

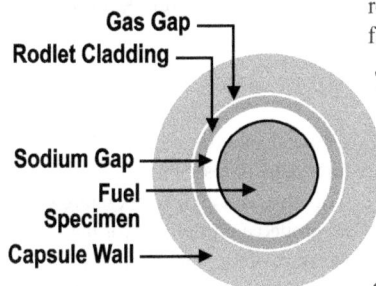

Figure 5.

AFCI capsule cross section.

Advanced Fuel Cycle Initiative Fuel

As part of work on the AFCI, different fuel types are undergoing irradiation in the East Flux Trap of the ATR. Plans are to continue to irradiate fuel specimens for several more years. The objective of the tests is to support development of fuels to minimize the spent fuel volume needed to be stored in a long-term repository. Because the experiments are taking place in one of the high thermal-flux positions of the ATR and have a maximum linear heat generation rate similar to existing power reactors, a cadmium-lined basket is utilized to reduce the thermal flux and, therefore, reduce the fission rate in the fuel. This approach also increases the fast-to-thermal flux ratio to be more representative of the value of future fast reactors.

These tests are static capsule type experiments consisting of short, internal capsules called rodlets (Figure 5), which contain the fuel specimens. The rodlets are filled with sodium to provide good heat transfer and temperature equalization within the capsule and fuel. An inert cover gas plenum is also included in the top of the rodlet to provide room for swelling and collection of any fission gas releases. Several rodlets are loaded into an outer capsule with a precisely designed gas gap between the rodlets and the capsule wall. The gas gap is filled with a suitable gas to control the heat transfer from the rodlets to the capsule wall and into the ATR

6

primary coolant, which determines and controls the temperature in the fuel rodlets. The capsules are loaded into an open top basket that positions the capsules in the proper vertical location within the selected position within the East Flux Trap of the ATR.

Reactor Pressure Vessel Steel

Several stainless steel samples were irradiated in multiple ATR positions to simulate commercial power plant neutron damage. Some samples were welded before being irradiated, while others were welded after the irradiations. The experiment objectives were to determine the effects of irradiation on welds and weld repairs. Some of the very large experiment specimens required a flux rate that was between the rates of the two positions large enough to accommodate the large specimens. Additional fuel was included as part of the experiment in one of the outer "I" test positions to ensure that the flux received by the experiment was appropriate for the test data needs. This fuel booster provided approximately three times the flux in the test position than would have been provided by the ATR driver fuel alone.

RERTR Plate Testing

As part of the Global Threat Reduction Initiative (GTRI), high-enriched uranium fuel is discouraged in all research and test reactors. The Reduced Enrichment for Research and Test Reactors (RERTR) program was initiated to develop and qualify new low-enriched fuels. The ATR has been used as the primary testing location for the new fuel types and will continue to be used until all reactor fuel development is completed and new fuels are fabricated, in approximately 2014. These tests will be static capsule configurations; however, these fuel specimens are in a plate geometry rather than cylindrical pellet, and the fuel plate cladding is in contact with the ATR primary coolant system. These tests are being performed in reflector and flux trap positions.

Magnox Graphite

Graphite samples were irradiated to high-density losses from radiolytic oxidation in a high-temperature gas-controlled environment for the Magnox power stations in the United Kingdom in support of life extension studies. Some samples were irradiated in an inert environment, and others in a CO_2 environment, to assess the effect of the density loss. The experiment successfully achieved the results the customer wanted. One set of samples was extremely degraded due to oxidation, and the other set of samples in the inert environment was relatively intact.

Advanced Gas Reactor Fuel

The AGR Fuel Development and Qualification Program was initiated in 2003, and there will be a total of eight different fuel irradiations throughout the program.

The test train for AGR-1 (Figure 6) consists of six separate capsules vertically centered in the ATR core, each with its own custom-blended gas supply and exhaust for independent temperature control. Each of the six capsules has a diameter of approximately 1.3 in., is 5.2 in. long, and will contain 12 prototypical fuel compacts approximately 0.5 in. in diameter and 1.0 in. long. The fuel compacts are made up of 780-μm-diameter TRISO-coated fuel particles in a graphite matrix compact. The compacts are arranged in four layers in each capsule with three compacts per layer nested in a triad configuration. A graphite spacer surrounds and separates the three fuel compact stacks in each capsule and also provides the inner boundary for the insulating gas jacket. The graphite spacer also contains boron carbide as a consumable neutron poison to limit the initial fission rate in the fuel, providing a more consistent fission rate during the planned 2-year irradiation. In addition to the boron carbide, a thin hafnium shroud is located around the outside portion of the capsule, located toward the center of the ATR core to provide additional neutron absorption and more control of the experiment fission rate.

7

ADVANCED TEST REACTOR EXPERIMENTS & CAPABILITIES

There are three thermocouples in four capsules (top capsule has five and the bottom capsule has only two due to space limitations) located in the top, middle, and bottom of the graphite holder to measure temperatures during irradiation. For the initial test, high-temperature thermocouples were developed that enabled precise temperature control of the experiment (up to 1,200°C).

Gaseous fission products are monitored by routing the outlet gas from each capsule to an individual fission product monitor system, which includes a high purity germanium (HPGe) spectrometer for identifying specific fission gases and a gross gamma (sodium iodide crystal scintillation) detector to provide indication when a small cloud of fission gases passes through the monitor. This small cloud, or wisp, of fission gases typically indicates when a TRISO fuel coating failure may have occurred.

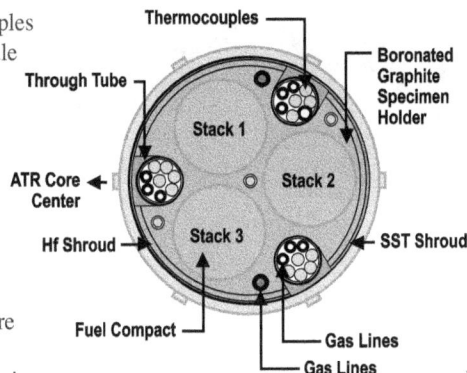

Thermocouples

Through Tube

Boronated
Graphite
Specimen
Holder

Stack 1

ATR Core
Center

Stack 2

Hf Shroud

Stack 3

SST Shroud

Fuel Compact

Gas Lines

Gas Lines

Figure 6.

AGR capsule

cross section.

FUTURE PLANS FOR THE ATR

Planned enhancements will increase ATR testing capabilities in the future. One additional pressurized water loop will be reactivated, and there are several organizations interested in performing boiling water reactor (BWR) simulations in the ATR. These tests will require modification of the loop to simulate the BWR conditions (i.e., voids in the core region of the coolant), as well as modifications to the current safety basis and operating processes of the ATR, but that have the potential to yield valuable information about BWR aging issues and design constraints on new BWRs. Preliminary analysis indicates that this testing configuration can be within the ATR safety basis and achieve the necessary conditions for various tests.

The new hydraulic shuttle irradiation system that was added to the ATR in 2008 will enable irradiation of specimens for a short period of time to perform initial feasibility or scoping studies on small amounts of material. This will decrease the cost and time necessary to obtain preliminary data that could be used to develop more complex test programs. Additionally, the shuttle system can be used to irradiate isotope targets that need high fluxes but not high fluences.

In addition to these enhancements, the INL is working to upgrade the irradiation testing infrastructure to enable the ATR to be a center for nuclear fuels and materials research within the DOE complex. This infrastructure includes the experiment assembly capabilities, development of in-reactor instrumentation (e.g., higher temperature thermocouples, creep testing rigs, and additional temperature instrumentation), procurement of a multi-use shipping container for ATR experiments and other irradiation material testing, and upgraded equipment for post irradiation examinations.

The designation of the ATR as a National Scientific User Facility (NSUF) in April 2007 is opening ATR and other INL facilities to a broader range of researchers for work on fuels and materials. The number of experiments sponsored by the DOE through the NSUF is expected to average 15 per year once the ATR NSUF is fully established.

8

NUCLEAR   FUELS   &   MATERIALS   SPOTLIGHT

CONCLUSION

The ATR is a unique and versatile reactor, and the research taking place there today continues the historic role that the INL has played in nuclear reactor development. The ATR will continue operating well into the 21st century as an important contributor to DOE's nuclear research objectives. Additionally, collaborative and complementary capabilities of other research facilities will be vital to achieve the objectives of several national and international initiatives. DOE's commitment to the INL and BEA's commitment to invest in ATR upgrades will ensure that the ATR is ready and available to meet nuclear research needs for diverse experiment sponsors for many years to come.

ACKNOWLEDGEMENT

This work was supported by the DOE under DOE Idaho Field Office Contract Number DE-AC07-05ID14517.

REFERENCES

INL, 2007, "FY 2008, Advanced Test Reactor National Scientific User Facility Users' Guide," INL/EXT-07-13577, Idaho National Laboratory, Idaho Falls, ID.

Grover, S.B., 2008 "The Advanced Test Reactor Irradiation Capabilities Available as a National Scientific User Facility," International Conference on the Physics of Reactors, Interlaken, Switzerland, September 2008.

FRANCES M. MARSHALL (BS, 1982, Nuclear Engineering, University of Virginia; ME, 1999, Chemical Engineering, University of Idaho; Registered PE) is currently the manager of Irradiation Testing for the ATR at the INL, with responsibility for developing the irradiation experiments performed in the ATR. Most recently, Ms Marshall co-led the team to establish the ATR as a NSUF, and she continues to support the transition to full NSUF operation. She held a reactor operator license and worked in the commercial nuclear power industry as a startup and plant system engineer. Ms. Marshall has worked at the INL since 1991 supporting and leading projects in the areas of irradiation experiments, nuclear power plant engineering, regulatory support, power plant performance assessment, and probabilistic risk assessment for the DOE and the Nuclear Regulatory Commission.

S. BLAINE GROVER (BS, 1978, Nuclear Engineering, Idaho State University; Registered PE) has been involved in designing irradiation experiments and associated support systems for the ATR during almost all of his 30-year tenure at the INL. He is currently the technical lead for the ATR irradiation experiments in support of the DOE Next Generation Nuclear Plant (NGNP) program, which includes both fuel and material irradiation testing. Mr. Grover also supports developing new irradiation experiment programs and external customers for the ATR, as well assisting the ATR NSUF.

9

184

N early all fuels and materials irradiation performance and degradation behavior can be linked to structural changes that occur on the micron and submicron scale. Therefore, it is critically important to have a variety of tools to characterize materials at this scale to improve the fundamental understanding of changes that occur under irradiation. The Electron Microscopy Laboratory (EML), located within the Materials and Fuels Complex at the INL, houses a suite of electron beam instruments primarily dedicated to the structural and chemical characterization (on a micron and nanometer scale) of radioactive nuclear fuels, materials, and waste-forms. Co-located within the EML is a complete set of sample preparation equipment for cutting, grinding, polishing, disc punching, core drilling, dimpling, electro jet-polishing, and precision ion-polishing. Many unique and novel techniques have been developed to prepare challenging activated samples for analysis. Specific examples of analytical scanning and transmission electron microscopy analysis are briefly described in this section.

ELECTRON MICROSCOPY OF FUELS AND MATERIALS

James I. Cole, J. Rory Kennedy, Jian Gan, and Dennis D. Keiser, Jr.

A ZEISS 960A (LaB$_6$ filament) thermal emission scanning electron microscope (SEM) has been a workhorse for the INL since its installation in 1996. The instrument provides a magnification range up to 200,000x at an accelerating voltage of 30kV, employing analytical detectors that permit, in addition to the standard secondary electron (SE) imaging capabilities, backscatter electron (BSE) imaging and both energy-dispersive spectroscopy (EDS) and wavelength-dispersive spectroscopy (WDS).

Figure 1 shows a result obtained from the Zeiss 960 SEM on an as-cast fuel sample of nominal composition 40U-34Pu-4Am-2Np-20Zr (wt%) prepared in the context of the AFCI. The microstructure is homogeneous throughout the fuel sample and composed of a matrix phase that appears to contain both light and dark contrast sub-phases. Observed within the matrix phase are very dark contrast globular precipitates. Higher magnification images of the sample are shown in Figure 2, where grain boundaries can be identified. The globular precipitates seem to be mostly associated with the dark contrast phase, and the light contrast areas favor the grain boundaries (Figure 2). EDS analysis showed that the light-contrast matrix regions are enriched in plutonium (with possible enrichment in uranium and neptunium), and the darker contrast matrix areas are enriched in zirconium. It appears, therefore, that the grain boundaries are enriched in actinides (particularly plutonium) and the center of the grains are enriched in zirconium.

Higher magnification microstructure images of an as-cast 40Pu-60Zr sample, prepared in the context of the AFCI accelerator-driven system transmutation effort, are shown in Figure 3. The microstructure was shown to be uniform from the center of the sample out to the edge. As in the low-fertile, uranium-bearing fuel sample discussed above, grain boundaries can be observed; however, three distinct phases can be observed here: 1) a relatively large globular phase, 2) an acicular phase, and 3) a matrix phase. The globular

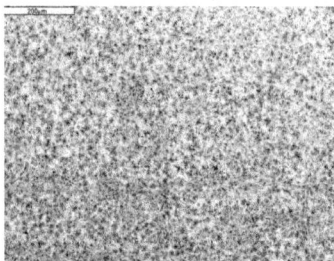

Figure 1.

A BSE image showing the microstructure observed in an as-cast U-34Pu-4Am-2Np-20Zr sample.

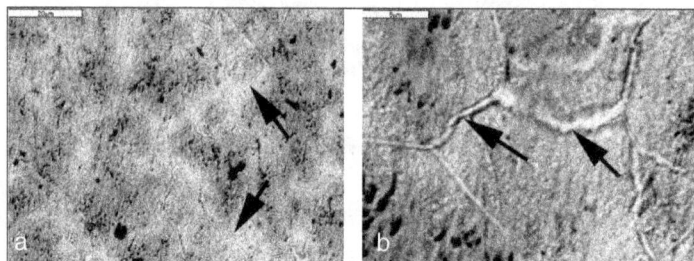

Figure 2.

BSE images showing the grain boundaries (black arrows) in the microstructure observed in an as-cast U-34Pu-4Am-2Np-20Zr sample at two different magnifications.

11

E L E C T R O N   M I C R O S C O P Y   O N   F U E L S   &   M A T E R I A L S

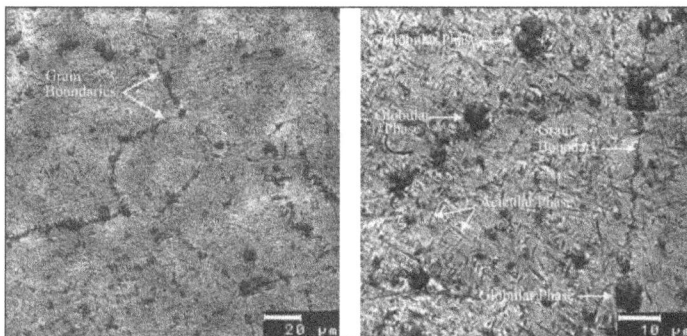

Figure 3.

SEM micrograph of the microstructure observed in an as-cast Pu-60Zr sample.

phase grows primarily at the grain boundaries, although some random precipitates also form within the individual grains. The individual grains display a Widmanstätten structure where plate-shaped particles (the acicular phase) align along specific crystallographic planes of the matrix crystallites. EDS and WDS analyses revealed the globular phase to be depleted in plutonium, while the two-phase (acicular and matrix) mixture is enriched in plutonium. An even distribution of zirconium occurs over both the globular phase and two-phase mixture. The globular phase and the acicular phase seem to contain some amount of oxygen, with the globular phase being more enriched than the acicular phase. The Widmanstätten structure observed in the two-phase region of the 40Pu-60Zr micrographs may form from the alignment of α-Zr plate-shaped particles (the acicular phase) along crystallographic planes of ill-formed δ-Pu, θ-Pu, or κ-Pu matrix crystallites.

A JEOL 7000F field-emission gun SEM instrument, operational since 2008, can achieve magnifications up to 500,000x with a resolution of 3 nm at 30kV. In addition to enhancing the SE, BSE, EDS, and WDS capabilities, the JEOL 7000F also allows electron backscatter diffraction (EBSD) for orientation imaging microscopy (OIM) analysis. The EBSD detector permits acquisition of crystallographic information by scanning the electron beam over a sample and generating an electron diffraction pattern at each point scanned. The pattern is recorded and indexed to identify the specific crystal orientation and phase at that point. This capability is extremely useful, for example, in evaluating the effect of thermomechanical processing on grain texture. In addition, EBSD combined with EDS enables phase identification and more precise determination of phase distributions within complex microstructures. Figure 4 shows an example of a crystal orientation map of an oxide formation on an Inconel 617 sample following long-term aging at elevated temperatures in air. The small, micron-size crystallites in the

oxide, which are difficult to see under standard SEM imaging conditions, are easily discerned from their differences in crystallite orientation.

Fuel cladding chemical interaction is an important aspect of the fuel development programs. The image in Figure 5 was obtained from the interface of a 60U-20Pu-3Am-2Np-15Zr fuel – HT-9 cladding alloy diffusion couple, heated to 650°C for 140 hours. The electron micrograph shows the interdiffusion zone to be a complex mixture of phases developed from the interdiffusion of the cladding constituents into the fuel, and fuel constituents into the cladding. In the base fuel alloy, there is a significant density of zirconium-rich, second-phase precipitates having both globular and stringer-like morphology. X-ray maps illustrated in Figure 6 reveal the distribution of elements within the interdiffusion zone. The maps suggest that the frontal interdiffusion of the fuel into the cladding is actinide dominated, wherein uranium plays a particularly important role and plutonium and neptunium (not shown) track with the uranium. Several distinct phases within the fuel are enriched in iron, chromium, and zirconium; americium appears to concentrate in second-phase particles separate from the other actinide constituents.

The final microscope in the current suite of instruments is a JEOL 2010 transmission electron microscope (TEM) that can operate up to 200kV (LaB$_6$ filament). It is equipped with a Gatan UltraScan 1000 digital

12

NUCLEAR FUELS & MATERIALS SPOTLIGHT

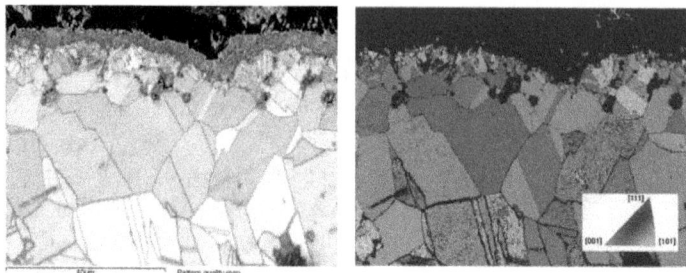

Figure 4.

Image quality map (left) and grain orientation map (right) from an aged Inconel 617 alloy. The image quality map is a representation of the quality of the diffraction pattern formed from each point scanned. Since grain boundaries do not generate indexable patterns, they show up as dark. The grain orientation map is colored to indicate relative grain orientation, as indicated by the legend.

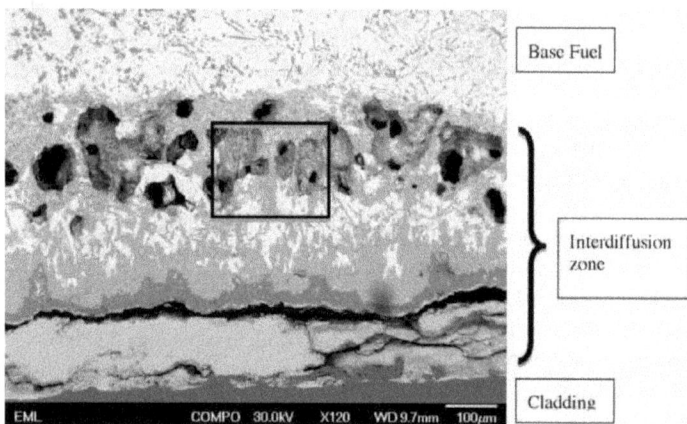

Figure 5.

SEM micrograph of the interaction zone formed from a diffusion couple between 60U-20Pu-3Am-2Np-15Zr fuel and HT-9 stainless-steel cladding heated to 650°C for 140 hours. The rectangular box delineates the region in which a higher resolution elemental X-ray map was collected.

13

ELECTRON MICROSCOPY ON FUELS & MATERIALS

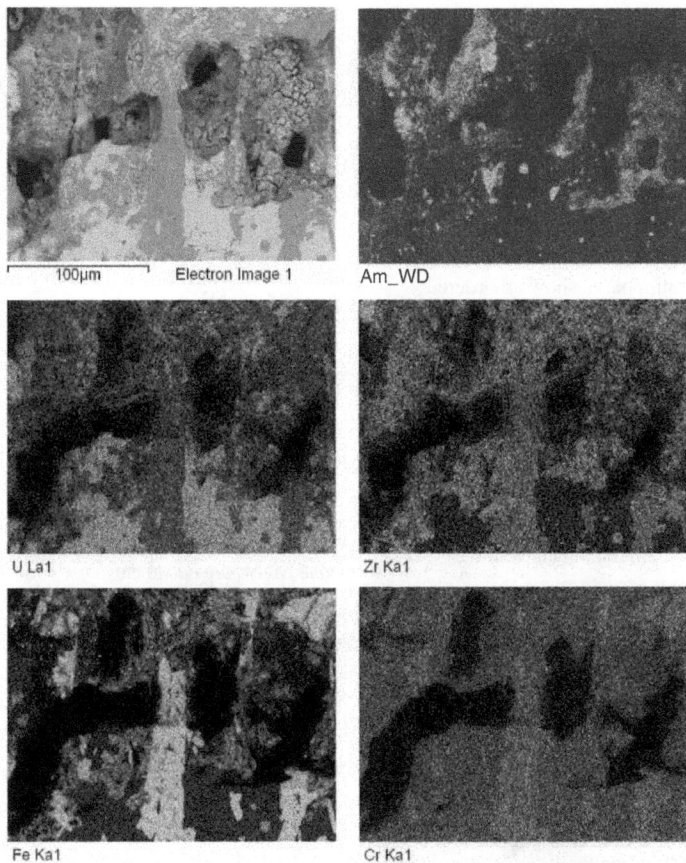

100µm   Electron Image 1   Am_WD

U La1   Zr Ka1

Fe Ka1   Cr Ka1

Figure 6.

Elemental EDS and WDS X-ray maps of the rectangular area in Figure 5 showing the distribution of selected elements within the interdiffusion zone. Bright contrast indicates enrichment in the designated element.

Figure 7.

The [001] zone patterns of a) U(Al,Si)$_3$ and b) (U,Mo)(Al,Si)$_3$ showing ordered Cu$_3$Au type structure (L1$_2$); see indexed major spots in a) and b). The extra spots in the a) U(Al,Si)$_3$ pattern suggest a super lattice structure since no precipitates were identified.

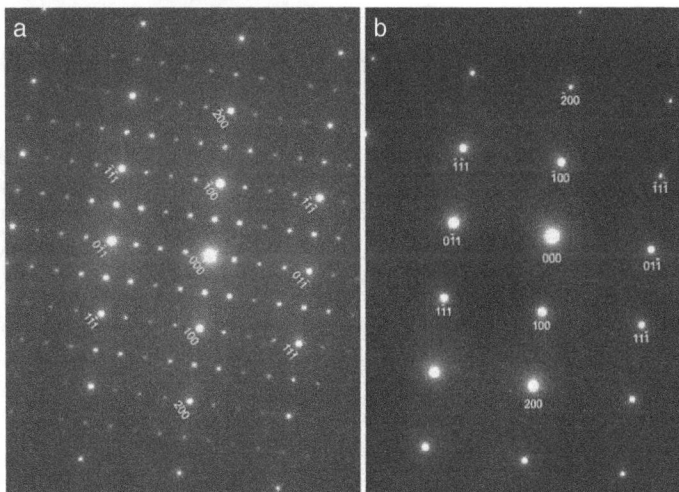

14

N U C L E A R   F U E L S   &   M A T E R I A L S   S P O T L I G H T

camera that produces 2048 x 2048 pixels over a 6.45cm^2 imaging area on the TEM screen and an Oxford EDS detector for chemical analysis. There is a scanning TEM unit attached to the microscope.

TEM characterization is an important part of the RERTR program at the INL. Nanometer scale investigations toward understanding the microstructural features of RERTR fuel are presented in Figures 7 and 8.

A variety of phases can potentially form during irradiation of RERTR fuels as a result of fuel/matrix or fuel/cladding interactions. To study the radiation stability of these phases, three depleted uranium alloys were fabricated with compositions of 67u-5Si-28Al, 48U-5Mo-47Al, and 69U-4Mo-20Al-7Si (wt%). In addition to the excess aluminum phase, microstructural analysis revealed the formation of a single U(Al,Si)$_3$ phase in the first alloy, three UMo$_2$Al$_{20}$, UAl$_4$, and U$_6$Mo$_4$Al$_{43}$ phases in the second alloy, and two (U,Mo)(Al,Si)$_3$ and

UMo$_2$Al$_{20}$ in the third alloy. The microstructural stability of these phases under proton irradiation was investigated. A follow-on study using Kr ion irradiation is in progress. The electron diffraction zone axis patterns of U(Al,Si)$_3$ and (U,Mo)(Al,Si)$_3$ are shown in Figures 7a and 7b, respectively. Both phases show an ordered Cu$_3$Au-type structure. The fine spots in the (Al,Si)$_3$ phase suggest a super lattice structure with a lattice spacing eight times that of U(Al,Si)$_3$. No precipitates were identified in either of the two phases. Proton irradiation at 200°C (up to 3.0 dpa) did not show any discernable changes in the microstructure.

Some interesting features of the UMo$_2$Al$_{20}$ phase are shown in Figure 8. This phase was identified in both the 48U-5Mo-47Al and 69U-4Mo-20Al-7Si alloys. Although both alloys were prepared by arc casting followed by homogenization treatment at 500°C

Figure 8.

Bright-field image near zone [011] showing high-density stacking faults in the UMo$_2$Al$_{20}$ phase in alloy U-5Mo-47Al (a); the streaks in the inset of the diffraction zone pattern are due to the high density of stacking faults on {111} planes. The high-resolution lattice fringe image of the same phase at zone [123] in alloy U-4Mo-20Al-7Si (b) shows the {111} plane projection.

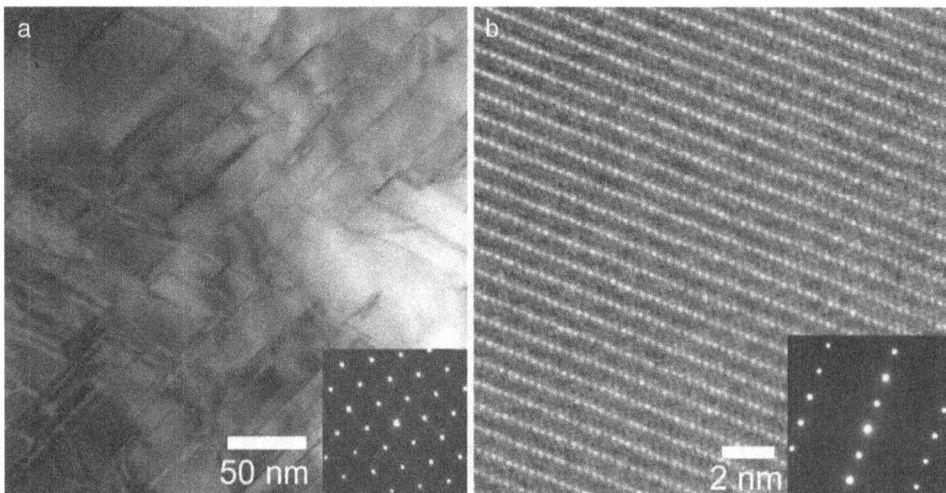

15

for 200 hours, a high-density of stacking faults was observed in the UMo_2Al_{20} phase in the 48U-5Mo-47Al alloy (Figure 8). The weak streaks in the diffraction pattern in the inset are the result of a high concentration of stacking faults along the {111} planes. In contrast, there are only scattered stacking faults at much lower density in the same phase of the 69U-4Mo-20Al-7Si alloy. Clearly, different microstructures can develop for the same phase depending on the details of the alloy fabrication. These differences may subsequently influence the microstructural development under irradiation.

A high-resolution lattice fringe image of the UMo_2Al_{20} phase near the [123] zone axis in the 69U-4Mo-20Al-7Si alloy is shown in Figure 8b. The line features in the image are {111} plane projections and the spacing between the lines corresponds to the spacing between the {111} planes.

The interesting features of the UMo_2Al_{20} phase are shown in Figure 8. This phase is identified in two different depleted-uranium (DU) alloys: U-5Mo-47Al and U-4Mo-20Al-7Si. Although both alloys were prepared by arc casting followed by a homogenization treatment at 500°C for 200 hours,

high-density stacking faults were observed in the UMo_2Al_{20} phase in alloy U-5Mo-47Al (Figure 8[a]). The weak streaks in the diffraction pattern in the inset are the result of high-concentration stacking faults on the {111} planes. There are only scattered stacking faults at a much lower density found in the same phase in alloy U-4Mo-20Al-7Si. This indicates that a different microstructure for the same phase can develop depending on the details of the alloy fabrication, which may influence the microstructural development under irradiation. A high-resolution lattice fringe image of the UMo_2Al_{20}

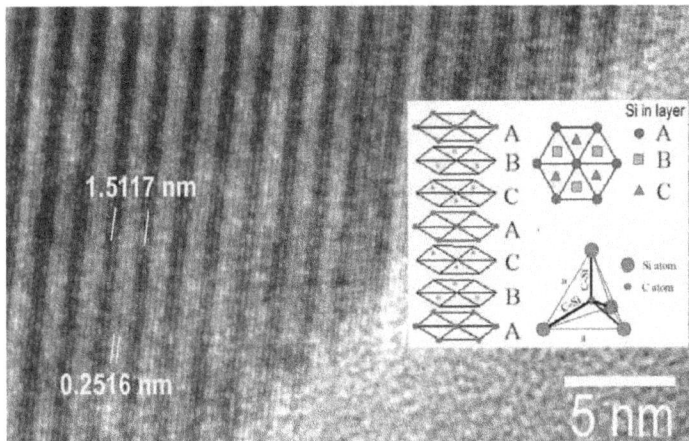

Figure 9.

High-resolution lattice fringe image showing the projection of 6H-SiC basal planes. The six fine fringes between major fringes represent each of the six layers. The inset provides the details on the atomic configuration for the 6H-SiC hexagonal crystal.

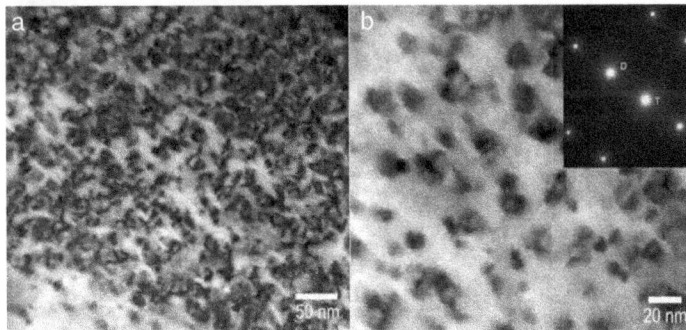

Figure 10.

Low magnification (a) and high magnification (b) bright-field images showing loops in 6H-SiC as a result of Kr ion irradiation at 800°C to a dose of 70 dpa. Dislocation loops were not found in the 10 dpa condition. These loops can affect the mechanical and physical properties of the ceramics.

16

phase at the zone [123] in alloy U-4Mo-20Al-7Si is shown in Figure 8(b). The line features in the image are the {111} plane projection, and the spacing between the lines corresponds to the spacing between {111} planes.

Development of fuel matrices and structural materials for application in the Gas-Cooled Fast Reactor requires detailed understanding of the candidate materials and their response to irradiation. The properties inherent to SiC suggest further studies in its application. The high resolution lattice fringe image in Figure 9 shows the basal plane projection of a 6H-SiC hexagonal crystal with a lattice constant of a = b = 0.3080 nm and c = 1.5117 nm. There are six visible lines between the major fringes representing each of the six layers in 6H-SiC. The details of the atomic configuration in 6H-SiC are shown in the inset. The crystal stacking sequence repeats after every six layers. Irradiation of 6H-SiC produces significant microstructural damage. The images in Figure 10 reveal the dislocation loops produced in the microstructure as a result of selective Kr ion irradiation (1.0 MeV Kr ions, 800°C, 70 dpa). Note that no loops were found in the 10 dpa irradiation condition. This result suggests that high-dose irradiation is required to investigate the development of the radiation-induced defects in SiC at high temperature. The formation of a high-density dislocation loop may have significant impact on both the mechanical and physical properties of the SiC-based ceramic materials.

CONCLUSION

Significant advances have been made in the general materials science community to characterize new and complex materials systems at ever-finer length scales. These characterization techniques are being applied to unique and challenging radioactive fuels and materials at the INL EML. Information gained can be used to support design and modeling efforts that may enhance performance and lead to improved overall life-time prediction capabilities.

17

ELECTRON MICROSCOPY ON FUELS & MATERIALS

JAMES I. COLE (BS, 1989, MS, 1992, PhD 1996, Materials Science, Washington State University) is a Materials Scientist in the Fuel Performance and Design Department at the INL. His research efforts over the last 10 years have focused on radiation effects in reactor structural materials, cladding and structural materials development for advanced reactor systems, and the study of chemical interactions between fast reactor fuels and cladding materials.

J. RORY KENNEDY (BS, 1982, Colorado State University; MS, 1983; and PhD, 1987, Inorganic Chemistry, Northwestern University) is Manager of the Basic Fuel Properties and Modeling Department at the INL. He has been investigating the synthesis and structure/property relations of materials for over 20 years. Over the last 12 years, his interests have been directed to the fabrication and characterization of materials associated with the nuclear fuel cycle, with emphasis on advanced nuclear fuels.

JIAN GAN (BS, 1982, Physics, Fudan University in China; MS, 1992, Physics, Central Michigan University; PhD, 1999, Nuclear Engineering, University of Michigan) is a Research Scientist in the Basic Fuel Properties and Modeling Department at the INL. He has been actively involved in the research on irradiation effects in materials for more than 12 years, with a focus on investigating microstructure using transmission electron microscopy.

DENNIS D. KEISER, JR. (BS, 1987, MS, 1989, Metallurgical Engineering, University of Idaho; PhD, 1992, Materials Engineering, Purdue University) is a member of the Fuel Performance and Design Department at the INL. He has over 20 years experience researching different issues related to the performance of nuclear fuels and materials. His current research relates to developing low-enriched fuels for application in research and test reactors, developing a better understanding of the diffusion behavior of fuel and cladding components in irradiated fuels, and determining the microstructural development of fuels and materials during irradiation.

18

etallic fuels have a history of use that spans the entire nuclear age, from the early plutonium production reactors (1940s) and the first liquid metal fast breeder reactors (1950s) to the most advanced reactor designs of today proposed by industry in support of the Global Nuclear Energy Partnership. Initially, metallic fuels were employed for their ease of fabrication and high heavy-metal density (which led to the most favorable breeding efficiencies in fast reactors). Over subsequent decades of use, metallic fuel technology matured significantly, leading to the realization of major benefits in the areas of fuel reliability and burnup, proliferation-resistant recycling, remote fabrication, and passive reactor safety (Crawford et al., 2007). Such characteristics make metallic fuels a highly competitive technology, especially for use in sodium fast reactors. Research and development activities related to metallic fuel technology have been a major focus of the DOE's research laboratories in Idaho for more than 50 years, and today the INL continues to lead the development of metallic fuels in the United States. This section provides a brief synopsis of the historical progress made on metallic fuel technology, and highlights current activities and recent results in the development of metallic fuels at the INL.

NUCLEAR FUELS & MATERIALS SPOTLIGHT

METALLIC FUELS

Steven L. Hayes, J.Rory Kennedy, Bruce A. Hilton, Randall S. Fielding, Timothy A. Hyde,

Dennis D. Keiser, Jr., and Douglas L. Porter

EARLY EXPERIENCE WITH METALLIC FUELS

TWO YEARS BEFORE President Eisenhower's famous "Atoms for Peace" speech in 1953, the Experimental Breeder Reactor (EBR-I), located in the desert west of Idaho Falls, Idaho, had become the world's first nuclear reactor to produce electricity. EBR-I used metallic fuel and was cooled with a liquid sodium-potassium alloy. The design of a typical metallic fuel pin is illustrated in Figure 1. The fuel slug produces heat by nuclear fission, and the heat is conducted through a liquid metal bond to the cladding and then to the reactor coolant. The reactor coolant flows from the bottom to the top of the fuel pin.

During its 12 years of operation, EBR-I employed several different metallic fuel designs, beginning with an initial loading of unalloyed, 93% enriched uranium. The majority of these fuel pins employed solid cylindrical fuel slugs (9.25 mm in diameter) clad in Type 347 stainless steel. This early metallic fuel pin design used liquid metal NaK in the gap between the fuel and cladding to promote efficient heat transfer and keep fuel temperatures low; it was irradiated to a burnup of only 0.35 at.-% (heavy metal).

The next evolution in metallic fuel design in EBR-I made use of a uranium alloy (U-2Zr), still clad in Type 347 stainless steel and still making use of a liquid metal bond. However, in the Mark-III design, the U-2Zr metallic fuel alloy was coextruded with a Zircaloy-2 cladding tube, producing a fuel pin in which the fuel slug was metallurgically bonded to the cladding. This eliminated the need to put a liquid metal bond in the

fuel-cladding gap, but in time would prove to be incapable of reaching high burnup due to stresses induced in the cladding by fuel swelling. Eventually, EBR-I demonstrated the

Figure 1.

Modern metallic fuel pin design.

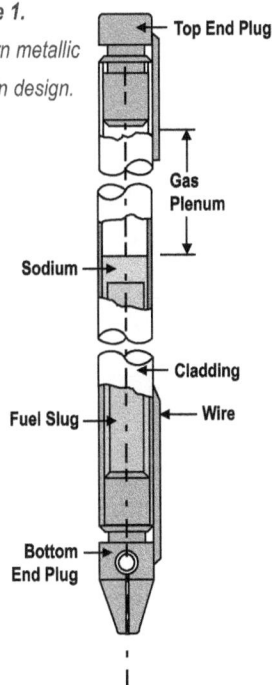

Top End Plug

Gas Plenum

Sodium

Cladding

Fuel Slug

Wire

Bottom End Plug

use of plutonium-based metallic fuel, and the final fuel loading consisted of a Pu-1.25Al metallic alloy clad in zircaloy.

The Dounreay Fast Reactor (DFR) in the United Kingdom, which began full power operation in 1963, also employed a number of metallic fuel designs. These included a U-0.1Cr metallic alloy clad in niobium that employed a sodium bond. Later, U-7Mo and U-9Mo fuel alloys were used. Relatively high burnups of 4 to 9 at.-% were achieved in DFR metallic fuels, considered to be an exceptionally high burnup in early years.

The Enrico Fermi Fast Breeder Reactor, located in Monroe, Michigan, was the first commercial fast reactor. The 200-MWt three-loop design was approved by the Atomic Energy Commission in 1955. Site preparation started in 1956, criticality was achieved in 1963, and operation at 100 MWt began in July 1966. The fuel for the reactor was a U-10Mo metallic alloy, sodium-bonded to zircaloy cladding.

21

METALLIC FUELS

These early metallic fuels were developed using experiments in thermal test reactors (i.e., the Materials Test Reactor in Idaho or CP-5 in Chicago), and the chief fuel performance issue was nonuniform growth of the fuel caused by fuel phase transformations and/or texture created in the fabrication process. These effects sometimes resulted in fuel element bowing or fuel axial growth (causing reactivity anomalies). It was not until these performance issues were understood and corrected/accommodated by design in a subsequent generation of metallic fuel concepts that high burnup became achievable.

Fuel designers at EBR-II engineered fuel designs that could reliably reach 3 at.-% burnup using a U-5Fs (where Fs is a mixture of noble metal elements simulating the fission products carried over in the early melt-refining reprocessing scheme) metallic alloy clad in Type 304 stainless steel and sodium-bonded. By this burnup, however, the fuel began to swell significantly due to gaseous fission products within the fuel. Fuel swelling stressed the cladding, causing breach of the cladding wall.

It was theorized at this time that the swelling problem could be resolved if the fuel had enough space inside the cladding to swell freely to the point where the fission gas could be released through a network of interconnected porosity. The fuel diameter was reduced to 75% smeared density, and the new fuel design achieved 8 at.-% burnup virtually overnight. Figure 2 shows the high percentage of fission gas released as the fuel burnup reaches ~3%, resulting in a dramatically reduced driving force for continued swelling.

Subsequent metallic fuels that used a low smeared density were driven to higher burnups but would eventually fail due to creep rupture of the cladding at or just above the fuel column (where the peak cladding temperature occurred). The much larger volume of released fission gas was overpressurizing the cladding, leading to breach. The gas plenum of these early metallic fuel designs was small. As a final design change, the volume of the gas-collecting plenum above the fuel column was increased, and subsequent metallic fuel pins reliably achieved 20 at.-% burnup without cladding breach.

With the design innovations of a low smeared density fuel pin with a large gas plenum, the final limitation to increased metallic fuel burnup was irradiation-induced swelling of the austenitic stainless steel cladding. Irradiation testing with HT9, a ferritic/martensitic stainless steel, showed it had exceptional resistance to swelling. HT9 was in the process of being qualified as part of a new driver fuel design when EBR-II shut down in 1994. The Fast Flux Test Facility (FFTF) was similarly in the process of qualifying a U-10Zr/HT9 metallic fuel design as the future driver fuel for that reactor at the time of its shutdown.

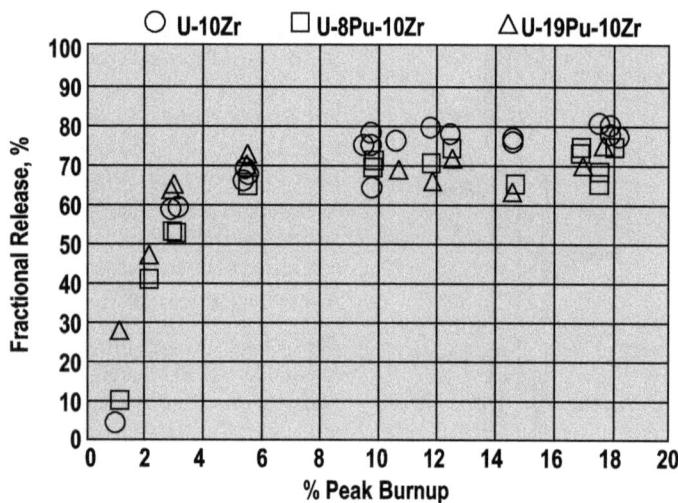

Figure 2.

Fission gas release behavior for a variety of metallic fuels.

22

ADVANCES IN METALLIC FUELS FOR THE INTEGRAL FAST REACTOR

In the 1980s and 1990s, the composition of metallic fuels under serious development evolved from binary to ternary alloys (i.e., U-xPu-10Zr, with x = 20-30 wt%) in support of the Integral Fast Reactor (IFR) program. The IFR program sought to develop a fast reactor that was passively safe and that made use of a fuel that was easily reprocessed and remotely refabricated, but in a way that was inherently proliferation-resistant (i.e., one in which plutonium was never isolated from other recovered actinides). Metallic fuels fit well with such a concept owing to their high thermal conductivity and thermal expansion (which contributed to reactor safety), compatibility with a pyro-metallurgical reprocessing scheme, and their demonstrated fabrication at engineering scale in a remote hot cell environment.

A substantial plutonium content was incorporated into the metallic fuel alloy to simulate an equilibrium, closed fuel cycle in which plutonium was bred in the blanket region of a fast reactor and added back into the driver fuel at the reprocessing stage. It was also important to demonstrate the concept of transmutation (i.e., burning) of plutonium and minor actinide isotopes (americium, neptunium, and curium) that could be recovered from spent light water reactor fuel in order to reduce its radiotoxicity for a geologic repository. Zirconium was selected as the alloying element, as it appeared to inhibit fuel-cladding interdiffusion as well as increase fuel solidus temperatures.

Using the design innovations developed for the metallic fuel pin used in EBR-II for over 20 years, U-Pu-Zr metallic fuel pins were reliably irradiated to 20 at.-% burnup without cladding breach. When the IFR program ended and EBR-II operations were terminated in 1994, EBR-II was on the verge of a total core conversion to U-20Pu-10Zr metallic fuel. Furthermore, experiments were underway to study the effects of adding americium, neptunium, and curium to the fuel. A low-burnup metallic fuels experiment was completed in which U-20Pu-2.1Am-1.3Np-10Zr was irradiated to 7.6 at.-% burnup. A cross-section of fuel from that experiment, where an electron microprobe was used to determine the redistribution of the major fuel constituents during irradiation (Meyer et al., 2008), is shown in Figure 3.

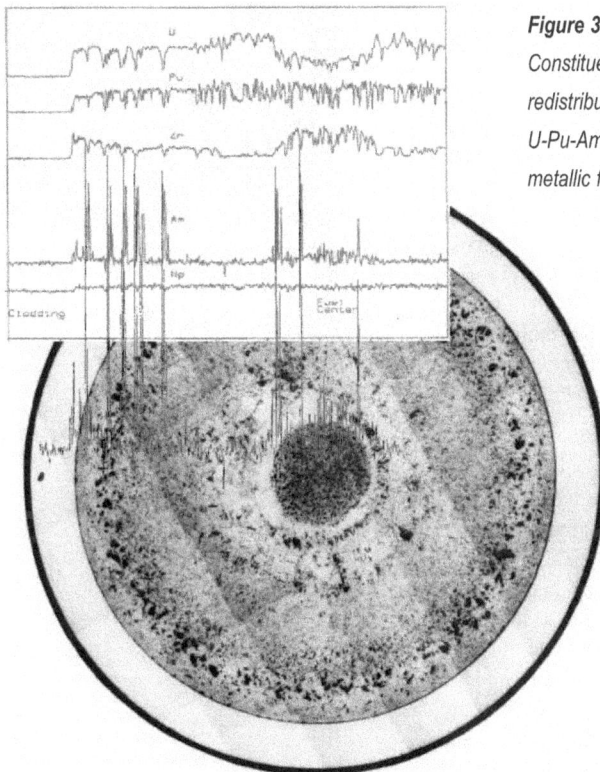

Figure 3. Constituent redistribution in U-Pu-Am-Np-Zr metallic fuel.

23

METALLIC FUELS

CURRENT METALLIC FUELS RESEARCH

Metallic Fuels for Actinide Transmutation

Current research and development activities on metallic fuels focus on their potential use for actinide transmutation in future sodium fast reactors as part of the DOE's Advanced Fuel Cycle research and development (R&D) program. Early factors that led to metallic fuels being identified as a strong candidate for such a mission include the following:

1. Traditional ease of fabrication, especially considering their prior fabrication at an engineering scale in a remote, hot cell environment (Stevenson, 1987)

2. Compatibility with the proliferation-resistant electrochemical recycle scheme

3. Very high burnup potential

4. Passive reactor safety features that derive from attributes inherent to metallic fuels (Chang, 1989).

While the ability to incorporate americium and neptunium into traditional U-Pu-Zr metallic fuel alloys was demonstrated during the IFR program of the 1980s and early 1990s (Trybus et al., 1993), the conventional injection casting fabrication method was not optimized for mitigating loss of the volatile americium constituent in the casting charge. For this reason, significant effort is being applied to the development of an advanced casting system to address this and related issues. The new metallic alloys fabricated as part of the current program are extensively characterized to obtain both fundamental and engineering properties. Finally, the new metallic fuel alloys are incorporated into irradiation experiments to obtain the data needed to assess their in-reactor performance. These activities are summarized in the following subsections.

Fabrication of Metallic Fuels

Arc-casting Experience

A previous attempt at casting metallic fuels with americium using the injection casting furnace that had fabricated hundreds of U-Pu-Zr fuels for EBR-II resulted in significant volatile loss of elemental americium during the process (Trybus et al., 1993). That process held the americium-containing fuel alloy under molten and superheated conditions, with a large melt surface exposed to a vacuum for a substantial period of time. It quickly became apparent that a new casting furnace design would ultimately be needed for fabricating metallic fuels with americium.

Before a new casting furnace became available, however, the immediate need was for a method of fabricating small fuel segments that could be used in characterization studies as well as irradiation experiments using miniature fuel rodlets. An arc-casting technique was developed for this purpose. While this technique was never envisioned as a fabrication method suitable for scale-up, it has provided the small fuel specimens required for current research. In arc-casting, metallic-fuel alloys are synthesized and homogenized using a small, commercial, hand-operated arc-melter inside an inert glovebox, as shown in Figure 4.

Figure 4.
Arc-melter in use fabricating metallic fuel samples.

24

The alloy charge to the arc-melter is considerably less than 50 g, and the total time the fuel alloy is molten is less than a minute. Once homogenization of a fuel alloy is complete, a quartz mold is introduced into the arc-melter, into which the molten metallic fuel alloy is either poured by gravity or drawn by vacuum. Fuel segments up to 0.250 in. in diameter and 1.5 to 2.0 in. in length (see Figure 5) have routinely been cast using this process, with no measurable loss of americium during fabrication.

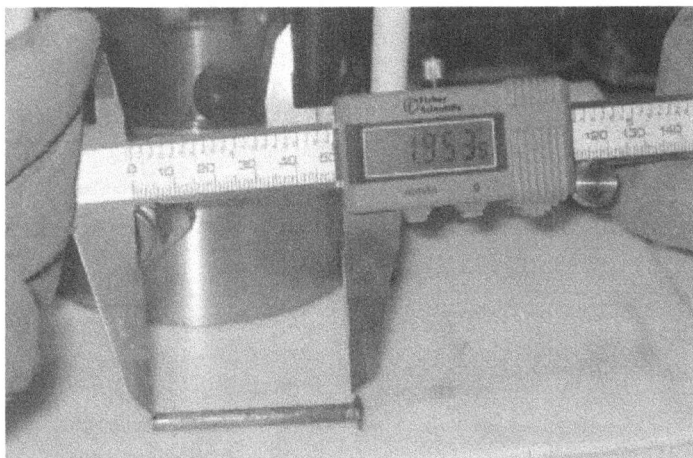

Figure 5.

Typical metallic fuel specimen fabricated by arc-casting.

Advanced Casting Research

Since the late 1950s, metallic fuels have been fabricated at large scale using counter gravity injection casting (CGIC). In CGIC, the fuel material is melted in an induction furnace with glass molds suspended above the crucible. After melting, the furnace and molds are evacuated and the molds are lowered into the molten fuel material. At this time, the furnace is rapidly pressurized, resulting in the molten fuel alloy being forced up into the glass molds where it quickly solidifies. The furnace is then cooled, the molds are removed, and the glass molds shattered to retrieve the metallic fuel slugs.

Well over 100,000 pins have been successfully cast using this method. However, because of the heating cycle duration and the application of a vacuum to a relatively large melt surface, the retention of volatile species (i.e., americium) during casting is a concern for metallic fuel compositions needed for an actinide transmutation mission, as is the amount of waste produced by the one-time use glass molds. For these reasons, an advanced casting proto-typing system (designated as the bench-scale casting system [BCS]) has been fabricated to develop bottom pour casting techniques for minor actinide-bearing metallic fuels, which will ensure volatile species retention. It is also being used as a test bed for the development of reusable molds, which will dramatically reduce the quantity of waste generated in the fabrication process.

The BCS is a bottom-pour, pressure-differential-assisted casting system that has an approximate capacity of 300 g of uranium alloy and can simultaneously cast three fuel slugs approximately 250 mm (10 in.) in length and 4.4 mm (.173 in.) in diameter. The furnace is powered by two 10-KW induction power supplies, one used for crucible heating and the other for mold heating. The power supplies are sized such that rapid processing times are achievable. The furnace shell and coils are passively cooled to avoid complications of cooling systems and possible leaks in radioactive areas, as well as coolant leaks into the furnace. A pressure differential can be applied during the casting process by maintaining a positive pressure on the mold while a vacuum or reduced pressure is applied to the mold. Although a hard vacuum is not applied to the mold, a pressure differential of several hundred torr is achieved. Retention of volatile species is achieved by

METALLIC FUELS

maintaining the melt continuously under pressure and by rapid casting-cycle times. Initial calculations have shown that very significant volatility reductions can be achieved with the application of only modest overpressure with large gas molecules (i.e., argon).

The BCS is currently operated in a radiological hood and tested with uranium alloys. Future development will focus on casting technique optimization, volatility studies, and reuseable crucible/mold material studies. A schematic of the BCS is shown in Figure 6.

Bottom-pour casting has been used successfully in the casting industry generally, although not for fuel manufacturing. While the BCS is still under operational development, an engineering-scale casting system (ECS) is also under design. This system will be very similar to the BCS, although it will have an approximate casting capacity of 60 pins (~5 kg) and be designed for operation in a totally remote environment. A schematic of a preliminary design of the ECS is shown in Figure 7.

Characterization of Metallic Fuels

Thermo-physical Properties

INL research activities aimed at understanding the properties and behavior of metallic fuels focus primarily on characterizing unirradiated, as-fabricated fuels in support of irradiation testing. General characterization includes room temperature density by the Arrhenius method, thermal expansion by the dilatometric method, heat capacity by the differential scanning calorimetry (DSC), thermal diffusivity by the laser flash method, room temperature phase identification by X-ray diffraction coupled with Rietveld analysis, thermally induced phase behavior (transition temperatures and enthalpies of transition) by differential thermal analysis (DTA) techniques, and room temperature microstructure by SEM with EDS and WDS for element identification.

Characterization of the metallic alloys from the most recent irradiation experiment

Figure 6.

Schematic of the assembled BCS (showing internal components).

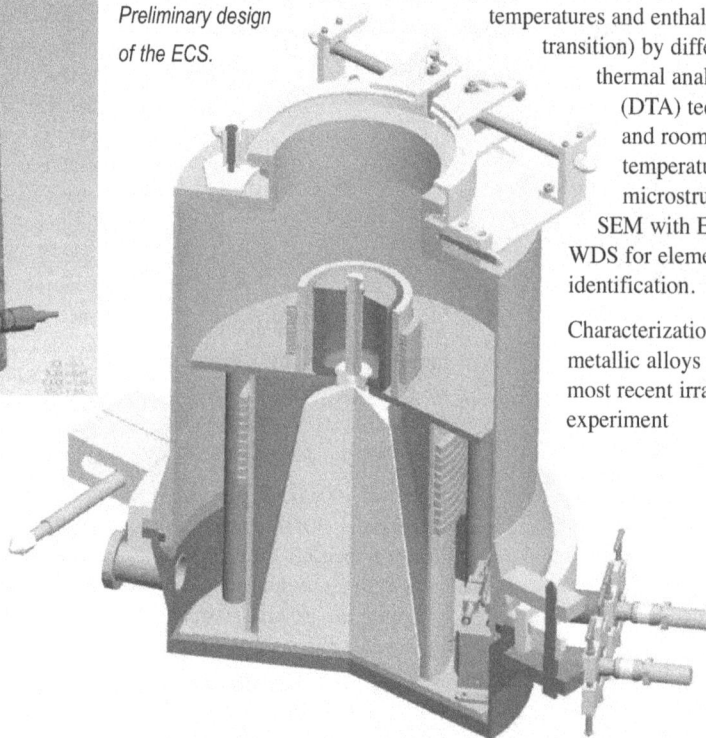

Figure 7.

Preliminary design of the ECS.

26

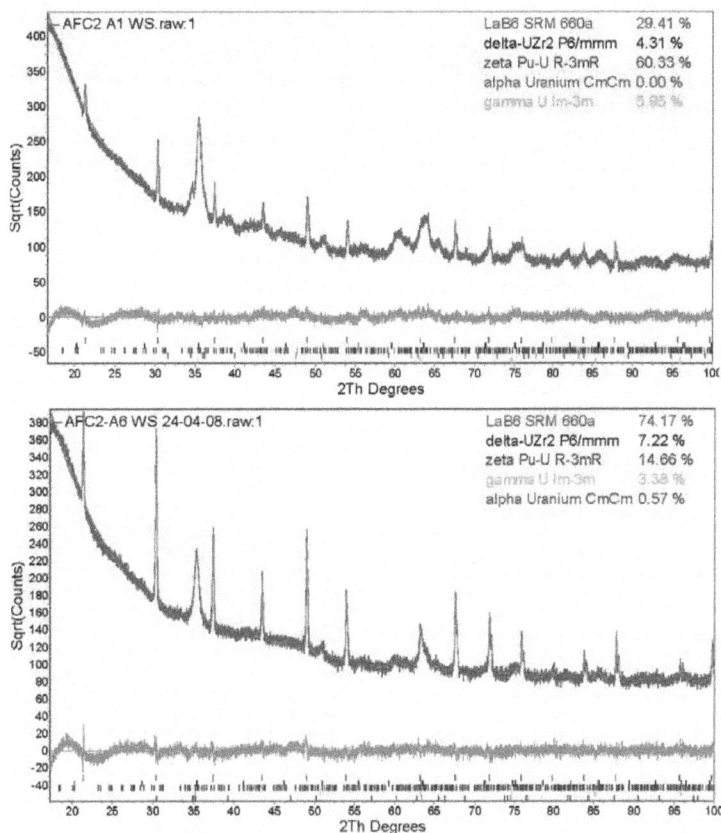

(AFC-2A,B) highlights these capabilities. Two base compositions, 60U-20Pu-3Am-2Np-15Zr and 42U-30Pu-5Am-3Np-20Zr, were fabricated along with companion fuels having additions of lanthanide elements to simulate fission product carryover from electrochemical reprocessing. The X-ray diffraction patterns of the room temperature as-cast, naturally cooled (presumptive non-equilibrium) alloys are shown in Figures 8 and 9, together with the Rietveld refinement analysis results. The samples were prepared as cross-sectional disks from the cast fuel slug and measured only from the fresh cut surfaces. Because the measurements were not performed on powders, the quality of the resulting diffractograms is typically less than optimal. Subtracting out the LaB_6 standard, the percent composition of each of the probable phases within the alloys appears in Table 1 along with the refinement residuals.

Both alloys are composed of primarily two phases: the ζ–phase arising out of the U-Pu binary and the δ–MZr_2 phase arising out of the U-Zr binary (and perhaps the Pu-Zr binary). A small amount of α-U phase may be present, and some residual nonequilibrium γ-U (high temperature sub-solidus phase) remains. The consistency of composition between each of these base alloys and the companion lanthanide addition alloys is very good. Although an experimentally determined room temperature phase diagram is not available for the U-Pu-Zr system, an approximated diagram is provided in Figure 10 for

Figure 8. (top)

Rietveld refinement analysis results for the 60U-20Pu-3Am-2Np-15Zr alloy.

Figure 9. (bottom)

Rietveld refinement analysis results for the 42U-30Pu-5Am-3Np-20Zr alloy.

Alloy	ζ-(U,Pu) (%)	δ-(U,Pu)Zr$_2$ (%)	α-U (%)	γ-U (%)	Refinement Residual R$_{wp}$ (%)
60U-20Pu-3Am-2Np-15Zr	86 ± 0.2	6 ± 0.1	0 ± 0.2	8 ± 1.3	6.47
42U-30Pu-5Am-3Np-20Zr	57 ± 2.3	28 ± 2.4	2.2 ± 0.7	13 ± 1.2	6.29

Table 1.

Phase contents for the 60U-20Pu-3Am-2Np-15Zr and 42U-30Pu-5Am-3Np-20Zr base composition metal alloys.

27

METALLIC   FUELS

visual illustration of the X-ray diffraction results. Because the minor actinide content of the alloys is small, the five component alloys can be reduced to pseudoternary systems for visualization in two dimensions. This can be accomplished by allowing all the americium to be included with the plutonium (uranium and americium anticipated to show limited mutual solubility) and splitting the neptunium content evenly between uranium and plutonium (both uranium and plutonium show mutual solubility with neptunium). The two base alloys reduce as follows:

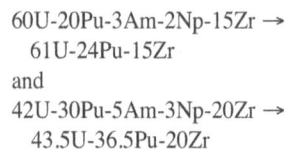

60U-20Pu-3Am-2Np-15Zr →
   61U-24Pu-15Zr
and
42U-30Pu-5Am-3Np-20Zr →
   43.5U-36.5Pu-20Zr

The approximate thermodynamic equilibrium positions for these alloys are indicated in Figure 10. Generally, there is consistency between what is observed and what might be expected, as both alloys are composed of primarily ζ–phase and δ–MZr$_2$ phase with the higher zirconium content alloy having the greater δ–MZr$_2$ content.

The thermally induced phase behavior of the two alloys represented in the traces of Figures 11 and 12 are similar, yet each is distinct. Both show two primary temperature regions through which phase transitions occur: the temperature regions are at about the same temperatures (Table 2), and the transition regions appear to be composed of multiple signals that reveal themselves better in the cooling curves. Included in Table 2 are the determined enthalpies of transition (ΔH) for the general transition regions for the two alloys. The two alloys show that although the higher temperature heats of transition are similar, the lower temperature enthalpies are significantly different.

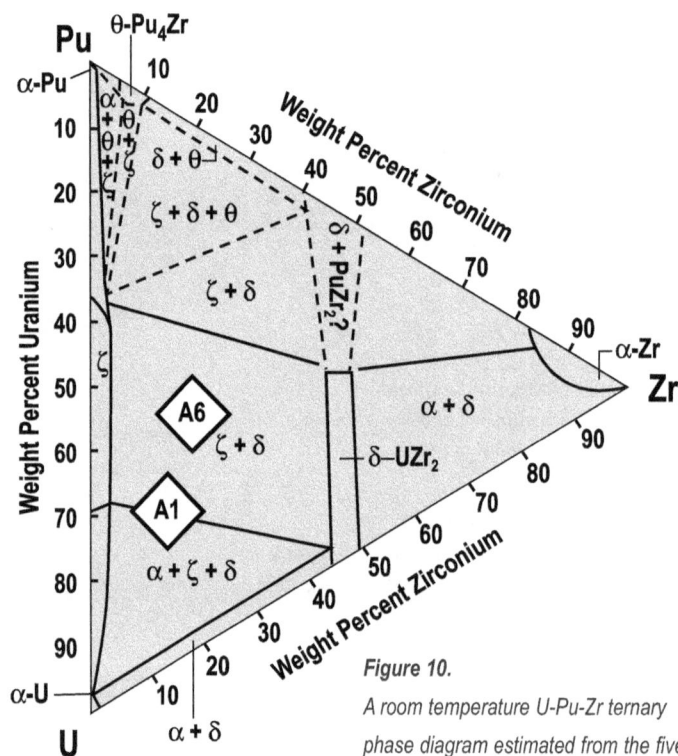

Figure 10.

A room temperature U-Pu-Zr ternary phase diagram estimated from the five isothermal sections from O'Boyle and Dwight (1970) and the three binary phase diagrams of the elements involved. The reduced compositions of alloys 60U-20Pu-3Am-2Np-15Zr (61U-24Pu-15Zr) and 42U-30Pu-5Am-3Np-20Zr (43.5U-36.5Pu-20Zr) are marked as A1 and A6, respectively.

28

NUCLEAR   FUELS   &   MATERIALS   SPOTLIGHT

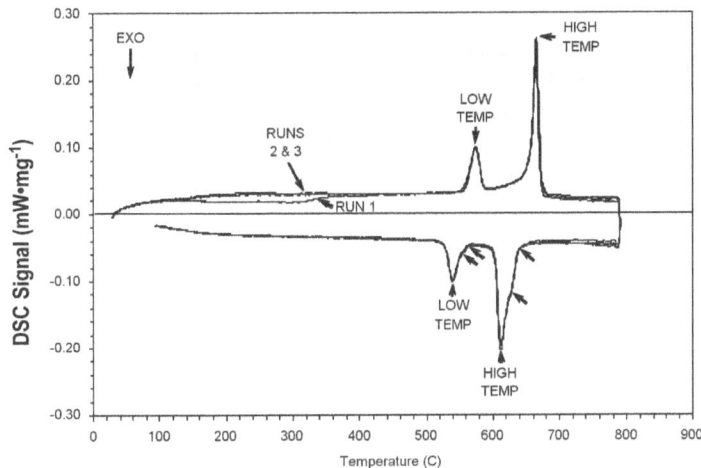

Figure 11.

Heating and cooling traces obtained from the DSC/DTA measurements on the as-cast 60U-20Pu-3Am-2Np-15Zr alloy.

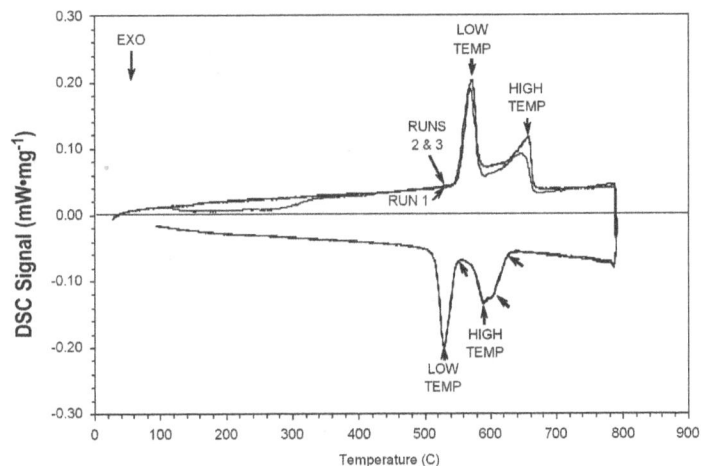

Figure 12.

Heating and cooling traces obtained from the DSC/DTA measurements on the as-cast 42U-30Pu-5Am-3Np-20Zr alloy.

60U-20Pu-3Am-2Np-15Zr	Heating		Cooling	
Transition	Low temp	High temp	Low temp	High temp
T_{tr} (K)	832 ± 1.6	929 ± 0.5	906 ± 5.8	825 ± 0.5
ΔH_{tr} (J•g^{-1})	6.9 ± 0.5	21.8 ± 0.7	-20.2 ± 0.4	-7.1 ± 0.3
42U-30Pu-5Am-3Np-20Zr	Heating		Cooling	
Transition	Low temp	High temp	Low temp	High temp
T_{tr} (K)	824 ± 1.9	900 ± 0.6	898 ± 0.4	820 ± 0.3
ΔH_{tr} (J•g^{-1})	19.6 ± 0.2	19.1 ± 1.5	-18.0 ± 0.2	-19.6 ± 0.4

Table 2.

Transition temperatures (T_{tr}) and enthalpies of transition (ΔH_{tr}) determined from differential scanning calorimetry upon heating and cooling of the AFC2 alloys.

29

M E T A L L I C   F U E L S

Observing the experimental U-Pu-Zr phase diagram data (O'Boyle and Dwight, 1970), the first transition of the 60U-20Pu-3Am-2Np-15Zr alloy can be interpreted to reflect the transition of the ($\delta+\zeta$) phase field directly into the ($\delta+\zeta+\gamma$) field, wherein most of the transition is conversion of δ, which showed to be a minor species from the room temperature X-ray diffraction results and is consistent with the low enthalpy of transition value. This transition is then followed by transition into the ($\gamma+\zeta$) phase field before 600°C and final transition into the sub-solidus γ-phase field at approximately 640°C.

The thermal phase equilibria transitioning for the 42U-30Pu-5Am-3Np-20Zr may be interpreted with a somewhat more complicated path. The endotherm in the heating curve at about 550°C may represent first the $\delta+\zeta \leftrightarrow \delta+\zeta+\eta$ transition with a narrow $\delta+\zeta+\eta$ phase field, followed by the Class II ternary four-phase equilibria reaction $\delta+\zeta+\eta \leftrightarrow \delta+\zeta+\gamma$, wherein approximately 40% of the material transforms ($60\delta+20\zeta+20\eta \leftrightarrow 45\delta+15\zeta+40\gamma$) into γ-phase. The subsequent endotherm signal (signals comprising the saddle portion of total DSC signal) represents transversing the $\delta+\zeta+\gamma$ field (expected 550°C - 565°C) and transitioning into the $\gamma+\zeta$ phase field (expected 565°C), followed by the transition at approximately 630°C into the single phase γ field. No other transitions should occur up to the ~1,100°C solidus.

The importance of the experimental nuclear fuel's thermal conductivity in determining the safety margins and projected fuel performance cannot be overstated. The thermal conductivity of a material at temperature can be computed as the product of its density, heat capacity, and thermal diffusivity. Typically, best results are obtained from a dense material with a homogeneous microstructure. The transmutation metallic fuels produced to date generally show room temperature densities, as determined by the Arrhenius immersion method, that are better than 95% of the theoretical densities (as shown in Table 3). The room temperature density can be combined with thermal expansion data obtained by dilatometric methods to give the material density at temperature.

The temperature dependence of the heat capacity and thermal diffusivity for the fuel alloys is determined and combined with the density to give the fuel thermal conductivity as a function of temperature. Heat capacity is measured by differential scanning calorimetry and thermal diffusivity is obtained with the transient laser flash method. Typically, each of the data sets is fitted to a correlated expression (linear for density and polynomial for heat capacity and thermal diffusivity) and these are carried over to allow the determination of thermal conductivity. The plots in Figure 13 show the averaged measured values for heat capacity (at half-centennial increments) and thermal diffusivity (at centennial increments) for the 60U-20Pu-3Am-2Np-15Zr alloy. The lines representing the polynomial fits are well within the 1σ standard deviations (represented as error bars in plots) determined at each temperature. The plot in Figure 14 illustrates the temperature dependent thermal conductivity result for the 60U-20Pu-3Am-2Np-15Zr test fuel alloy, which itself can be fitted to a polynomial relation (in this case virtually linear). The error bars represented in the plot are shown at the 5% (relative) level, which are taken as conservative. Compounded error estimates obtained from the individual measurements at temperatures generally fall within the 1-6% range at lower temperatures (up to about 600°C) and can increase to 10% or more at higher temperatures (up to 1,200°C). Uncertainty analyses of this type are of increasing importance to the fuel properties and characterization effort at INL as well as the national programs.

Alloy	Average Density (g/cm³) @ 25°C	Theoretical Density (g/cm³)	Std. Dev.
60U-20Pu-3Am-2Np-15Zr	14.05	14.59	0.33
42U-30Pu-5Am-3Np-20Zr	12.88	13.51	0.04

Table 3.

Average room temperature and theoretical ideal solution densities of the 60U-20Pu-3Am-2Np-15Zr and 42U-30Pu-5Am-3Np-20Zr alloys.

30

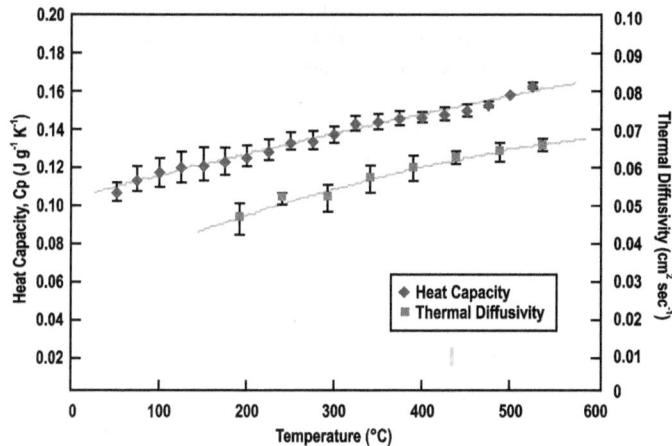

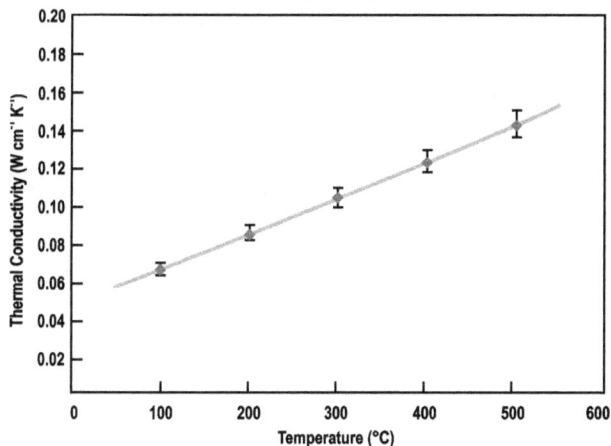

Figure 13. (Top)

Plots showing the temperature dependence of the heat capacity and thermal diffusivity measured for the 60U-20Pu-3Am-2Np-15Zr fuel alloy up to the first phase transition.

Figure 14. (Bottom)

Plot showing the temperature dependence of the thermal conductivity determined from the fitted expressions for density, heat capacity, and thermal diffusivity from the 60U-20Pu-3Am-2Np-15Zr fuel alloy up to the first phase transition. Error bars are at the 5% level.

Fuel-cladding Compatibility

When a metallic nuclear fuel is irradiated, the fuel swells and eventually contacts the cladding. This typically occurs at around 2 at-% burnup. At this point, interdiffusion can occur between the fuel and cladding, resulting in the formation of interdiffusion zones on the inner surface of the cladding. These zones are comprised of phases that are enriched in fuel and cladding constituents as well as fission products. Having interdiffusion zones present on the inner surface of the cladding can affect the performance of a fuel pin, since these zones can potentially be low melting or can be brittle and exhibit poor mechanical properties. Therefore, it is important to develop understanding of the types of phases that will develop during irradiation of metallic fuels and to determine whether the developed phases are low-melting or prone to cracking.

Much experience was gained during the 30 years of operating the EBR-II reactor in developing an understanding of FCCI in irradiated metallic fuels. Many fuel pins that contained U-Pu-Zr or U-Zr metallic fuel alloys were destructively examined to characterize the fuel-cladding interface (Keiser, 2005). Figure 15 shows the fuel-cladding interface in an irradiated fuel element that had U-Pu-Zr fuel and the austenitic steel D9 as the cladding. Composition analyses that have been performed at the fuel-cladding interface in irradiated fuel elements have shown that lanthanide fission products can be especially

31

M E T A L L I C   F U E L S

enriched in these locations. Figure 16 shows a location where hardness indents were generated in a FCCI zone. The interaction zone is observed to be decidedly harder than the unreacted cladding.

Diffusion couple experiments are an effective way to improve understanding of how fuel and cladding components will interdiffuse to form phases. Systematic diffusion couple experiments have been performed using fuel and cladding alloys to determine the interdiffusion behavior of specific fuel and cladding constituents. Some examples of such studies can be found in Keiser and Dayananda (1994) and Keiser and Petri (1996). To study the interdiffusion behavior of fuel components, initial tests used simple binary U-Zr alloys followed by more complex U-Pu-Zr alloys, and then fuel alloys that contained fission products (e.g., molybdenum,

neodymium, ruthenium, etc.) and minor actinides (e.g., americium, neptunium). To look at the behavior of cladding constituents, simple binary iron-containing alloys were first used (e.g., Fe-Cr and Fe-Ni), followed by more complex Fe-Ni-Cr alloys, and then actual cladding steels (e.g., HT-9, 316SS, D9, HCM12A, ODS, or Alloy 800H). An example of a diffusion structure that formed in an annealed diffusion couple is shown in Figure 17. This diffusion structure formed when a diffusion couple between a 35U-29Pu-4Am-2Np-30Zr alloy and an austenitic Type 422 cladding steel was annealed at 550°C for 15 hours. Many complex phases can be found in the interdiffusion zone. The current emphasis with diffusion couple testing is to investigate the compatibility of metallic transmutation fuels, which include minor actinides with HT-9 cladding (Cole et al., 2008).

Also of interest, with regard to FCCI, is the development of lined claddings that can be employed to eliminate the FCCI reaction altogether. If lined claddings can be successfully developed, then the potential operating temperature of the fuel element can potentially be increased. The application of liners has also been investigated for improving fuel performance in light water reactors (Frost, 1982). Diffusion couple tests have demonstrated that zirconium and vanadium have the potential to be effective liner materials for eliminating FCCI (Keiser and Cole, 2007). Ultimately, irradiation tests will be required to demonstrate the true effectiveness for these materials as liners. During the IFR program, V-lined cladding was fabricated for irradiation testing, but due to the shutdown of EBR-II in 1994, reactor testing never transpired. During the same program, casting of a U-Pu-Zr

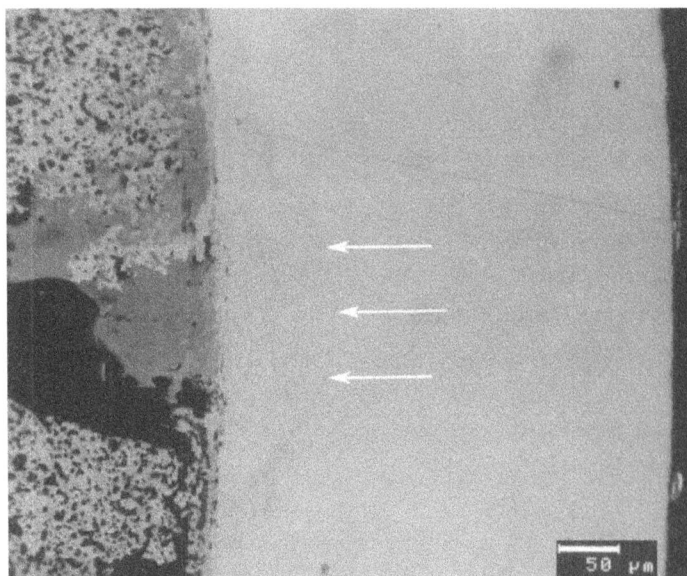

Figure 15.

Optical micrograph showing fuel-cladding interaction (white arrows) adjacent to a large deposit of lanthanide fission products.

32

fuel into a zirconium-sheath was performed to see if zirconium on the fuel surface could act as a barrier to FCCI. It was found that the zirconium-sheath cracked during irradiation and proved not to be an effective barrier (Crawford et al., 1993). Having the zirconium as a liner in intimate contact with the cladding should be a more effective way to employ zirconium as an FCCI barrier.

At this point in time, much progress has been made in improving the understanding of FCCI in irradiated metallic fuels.

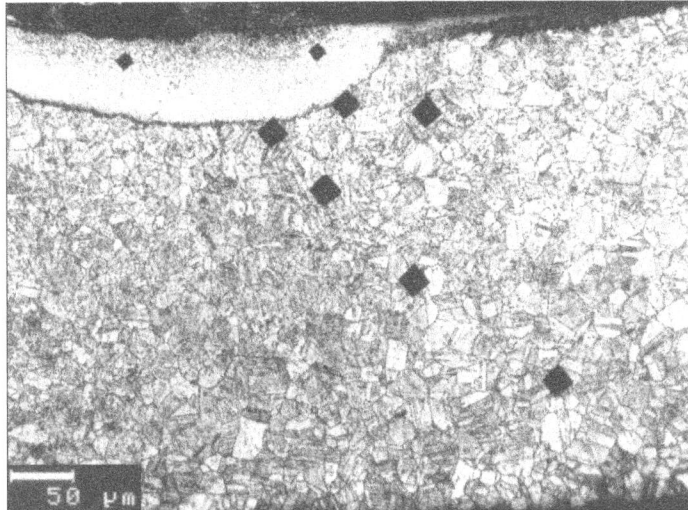

Figure 16. (Top)

Optical micrograph showing the presence of an interaction zone at the fuel-cladding interface. Hardness indents reveal the increased hardness of the interaction zone vis-à-vis the unreacted cladding.

Figure 17. (Bottom)

Backscattered electron micrograph of the interdiffusion structure that formed in a diffusion couple between a 35U-29Pu-4Am-2Np-30Zr fuel alloy and an austenitic Type 422 cladding steel when annealed at 550°C for 15 hours.

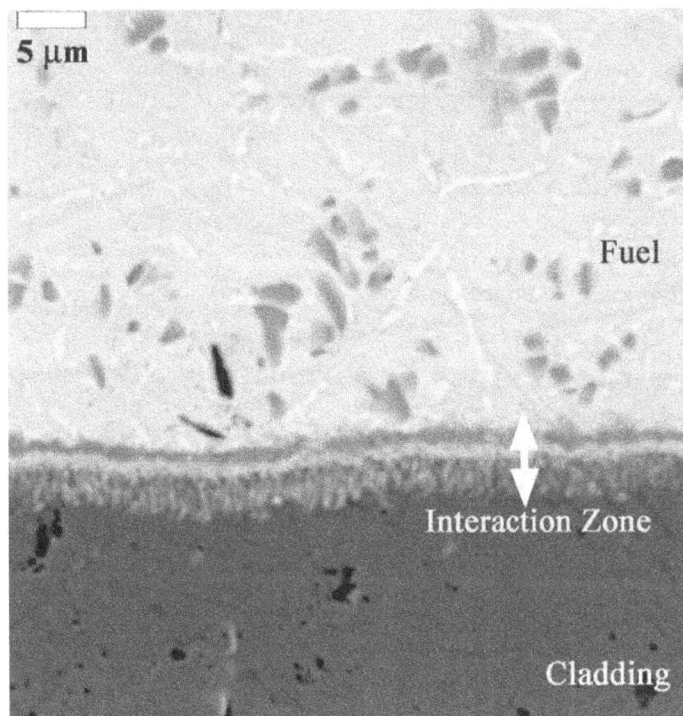

33

METALLIC FUELS

It has been determined that the phases that form uranium, plutonium, zirconium, fission products, and cladding constituents; uranium and plutonium are the original fuel alloy constituents that penetrate furthest into the cladding. Of the fission products, the lanthanide fission products penetrate deepest into the cladding, and palladium is the noble metal fission product to penetrate deepest into the cladding. Iron is the cladding component that penetrates deepest into the fuel. In terms of identifying where the largest interdiffusion zones will form in an irradiated fuel element, the combined high temperature and high-power region of a fuel element is consistently found to contain the widest zones. In this region of a fuel element, the interdiffusion kinetics are increased, and the supply of fission products at the fuel/cladding interface is relatively large.

Future work will focus on using diffusion couple experiments to investigate the kinetics of interaction between metallic fuels appropriate for actinide transmutation and the current cladding to be used, HT9. Furthermore, efforts will continue to identify the best barrier materials and to fabricate lined claddings with these materials so that actual irradiation testing can be performed, which is necessary to find the best material for eliminating FCCI altogether.

Irradiation Testing of Metallic Fuels

Irradiation Testing Strategy

An important aspect of the development of metallic fuels has always been, and continues to be today, the irradiation testing of metallic fuel pins. Historically, such testing was conducted in the experimental fast reactors, EBR-II and the FFTF. Both of these fast reactors, however, were shut down in the early 1990s.

The present lack of a domestic fast flux irradiation capability has of necessity led to a current two-pronged approach to obtaining the necessary irradiation performance data on the new metallic fuel alloys of interest. First, the majority of the fuel performance data are obtained from irradiation experiments conducted in domestically available thermal test reactors, notably the ATR at the INL. Second, a much smaller number of irradiation experiments are conducted in fast reactors available internationally to validate the performance results observed in the thermal-spectrum irradiations and identify any performance differences that may result. In this regard, two metallic fuel pins are currently under irradiation in the Phénix fast reactor in France. The following subsections describe the ongoing irradiation test programs in the ATR and the Phénix and the performance results obtained to date on metallic fuels for actinide transmutation.

ATR Test Program

The ATR has been used to conduct an extensive test program investigating the irradiation performance of metallic fuels with major additions of the minor actinide elements americium and neptunium. When testing began in 2003, accelerator-driven systems were considered to be a primary application for actinide transmutation fuels, so the early metallic fuels irradiated in the ATR were non-fertile (i.e., uranium-free) fuel alloys. As the national emphasis transitioned to sodium fast reactors as the fast-spectrum device to be used for transmutation, uranium-based metallic fuels were included in the test program. Table 4 summarizes both the recently completed metallic fuel tests and those that are still in progress.

Irradiation testing in the ATR is performed on miniature fast reactor fuel pins, or rodlets. The rodlets are fabricated using archive HT9 cladding tubing 0.230 in. in diameter (0.018 in. wall thickness). Finished rodlets are 6.0 in. in length and have a fuel column height of 1.5 in., giving a plenum-to-fuel volume ratio of >2.0. Due to uncertainties regarding fuel swelling associated with the substantial helium production from americium transmutation in a thermal spectrum, the early experiments (AFC-1 test series) made use of a relatively low fuel smear density of ~0.7; subsequent tests (AFC-2 test series) have returned to the traditional value of 0.75 for metallic fuel. Figure 18 gives a schematic of the metallic fuel rodlet design used in the ATR.

34

N U C L E A R   F U E L S   &   M A T E R I A L S   S P O T L I G H T

Up to six of these rodlets are included in a secondary stainless steel capsule for irradiation in the ATR. The secondary capsule provides a safety barrier between the metallic fuel (and bond sodium) and the primary coolant should cladding breach occur, and the gap size between the secondary capsule and rodlet cladding is sized to produce peak cladding temperatures in the 500-550°C range. Additionally, the test capsules are irradiated inside a cadmium shroud to reduce neutron flux levels and produce target linear heat generation rates (LHGR) in the test fuels. A photograph of metallic fuel rodlets (just prior to loading into the secondary capsules) appears in Figure 19.

The important variables under study in these ATR tests include the effects of variable plutonium and minor actinide contents, zirconium content, and lanthanide fission product carry-over (from electrochemical recycle) on the performance of these metallic fuel alloys, particularly as they relate to the large historic database generated on U-20Pu-10Zr metallic fuel from EBR-II and the FFTF.

Figure 18.

Metallic fuel rodlet design used in ATR irradiation tests (dimensions in inches).

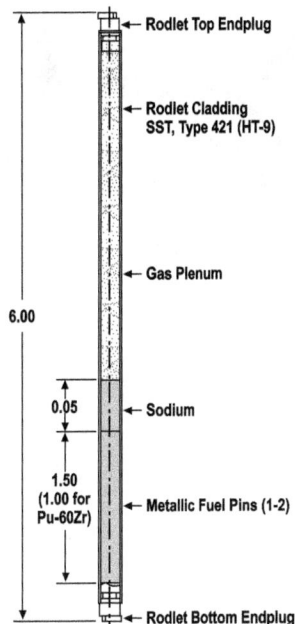

Labels: Rodlet Top Endplug; Rodlet Cladding SST, Type 421 (HT-9); Gas Plenum; 6.00; 0.05 — Sodium; 1.50 (1.00 for Pu-60Zr) — Metallic Fuel Pins (1-2); Rodlet Bottom Endplug

Table 4.

Summary of metallic transmutation fuel irradiation tests in the ATR.

Metallic Fuel Alloy	Experiment	Irradiation Time (EFPDs)	Peak LHGR (W/cm)	Peak Fission Density (fiss/cm³)	Peak Burnup (% fissile)	Status
Pu-40Zr	AFC-1B	93	300	5.26E+20	5.7	In PIE
	AFC-1D	593	300	1.95E+21	33.3	Discharged from ATR
Pu-60Zr	AFC-1B	93	300	3.51E+20	7.0	In PIE
	AFC-1D	593	300	1.33E+21	39.6	Discharged from ATR
Pu-12Am-40Zr	AFC-1B	93	300	4.27E+20	5.9	In PIE
	AFC-1D	593	300	1.71E+21	34.1	Discharged from ATR
Pu-10Np-40Zr	AFC-1G	644	300	1.47E+21	17.6	Discharged from ATR
Pu-10Am-10Np-40Zr	AFC-1B	93	300	3.43E+20	5.5	In PIE
	AFC-1D	593	300	1.35E+21	30.8	Discharged from ATR
U-25Pu-3Am-2Np-40Zr	AFC-1F	94	330	5.89E+20	6.7	In PIE
	AFC-1H	706	330	3.56E+21	38.0	Discharged from ATR
U-28Pu-7Am-30Zr	AFC-1F	94	330	6.38E+20	5.7	In PIE
	AFC-1H	706	330	3.97E+21	33.4	Discharged from ATR
U-29Pu-4Am-2Np-30Zr	AFC-1F	94	330	6.38E+20	5.9	In PIE
	AFC-1H	706	330	3.93E+21	36.2	Discharged from ATR
U-34Pu-4Am-2Np-20Zr	AFC-1F	94	330	5.35E+20	4.5	In PIE
	AFC-1H	706	330	3.48E+21	28.3	Discharged from ATR
U-20Pu-3Am-2Np-15Zr	AFC-2A	219	350	1.33E+21	6.7	In ATR†
	AFC-2B	219	350	1.35E+21	7.0	In ATR†
U-20Pu-3Am-2Np-1.0Ln*-15Zr	AFC-2A	219	350	1.42E+21	9.4	In ATR†
	AFC-2B	219	350	1.44E+21	9.5	In ATR†
U-20Pu-3Am-2Np-1.5Ln*-15Zr	AFC-2A	219	350	1.30E+21	10.7	In ATR†
	AFC-2B	219	350	1.31E+21	10.8	In ATR†
U-30Pu-5Am-3Np-20Zr	AFC-2A	219	350	1.19E+21	8.0	In ATR†
	AFC-2B	219	350	1.23E+21	8.2	In ATR†
U-30Pu-5Am-3Np-1.0Ln*-20Zr	AFC-2A	219	350	1.30E+21	9.9	In ATR†
	AFC-2B	219	350	1.32E+21	10.2	In ATR†
U-30Pu-5Am-3Np-1.5Ln*-20Zr	AFC-2A	219	350	1.40E+21	11.0	In ATR†
	AFC-2B	219	350	1.48E+21	11.3	In ATR†

* Ln = 6% La, 16% Pr, 25% Ce, 53% Nd
† Reported results through end of Cycle 142B

35

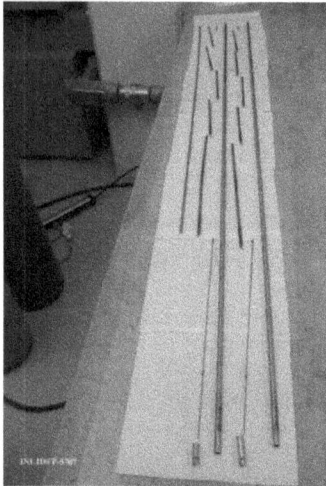

Figure 19.

*Two ATR test capsules
with 12 metallic fuel
rodlets (just before
capsule loading).*

FUTURIX-FTA
Experiment in Phénix

Early in the planning for the ATR test program, two metallic fuel compositions were identified for irradiation in the Phénix fast reactor in France. The compositions selected were Pu-12Am-40Zr and U-29Pu-4Am-2Np-30Zr, and the experiment designed to irradiate them in the Phénix is designated FUTURIX-FTA. Low burnup data on these alloys are already available from ATR irradiations, and experiments to take the same alloys to high burnup are underway. The results from the FUTURIX-FTA metallic fuel pins will provide analogous performance data obtained from a fast-spectrum test. Irradiation of these two metallic fuel pins began in May 2007 and will proceed for 240 effective full-power days, reaching peak burnups in the 10-12 at.-% range; irradiation is anticipated to end in March 2009.

Irradiation Performance Results

A number of ATR experiments (Table 4) have been completed that include a variety of non-fertile and low-fertile metallic fuel alloys irradiated to the intermediate burnup level of 7% (6.4×10^{20} fiss/cm^3). Preliminary postirradiation examination (PIE) of these fuels has been completed (Hilton et al., 2006), although specimens are being retained for more specialized examinations pending the availability of some new instruments (e.g., installation is underway of a shielded Electron Probe Micro-Analyzer to measure constituent redistribution within the fuels).

Figure 20 shows a composite of PIE results from a representative non-fertile (Pu-12Am-40Zr) and low-fertile composition (U-29Pu-4Am-2Np-30Zr), including an image of the irradiated fuel rodlet obtained by neutron radiography, rodlet axial fission product distributions obtained by gamma-ray spectroscopy, and optical metallography performed on a transverse section of the fuel/cladding. Significant fuel swelling has not yet begun for the Pu-12Am-40Zr alloy (a), as is evident from the large fuel-cladding gap that still remains. The U-29Pu-4Am-2Np-30Zr alloy (b), seen at a higher fission density, has swollen considerably, with porosity evident within the fuel and the absence of a discernable fuel-cladding gap. These results are consistent with the historic data for U-Zr and U-Pu-Zr metallic fuels, where the onset of swelling has been observed to begin in the range 3 to 6×10^{20} fiss/cm^3 (Hofman et al., 1990). Axial fission product distributions are relatively uniform. This is to be expected for these short fuel columns, which experienced essentially a uniform axial flux profile during irradiation. Furthermore, the cesium-137 profile is contained within the fuel column region, which indicates that fuel swelling has only just begun. In time, a significant fraction of the fission product cesium will be released from the fuel and alloy with the bond sodium, with a large peak observable above the fuel column.

36

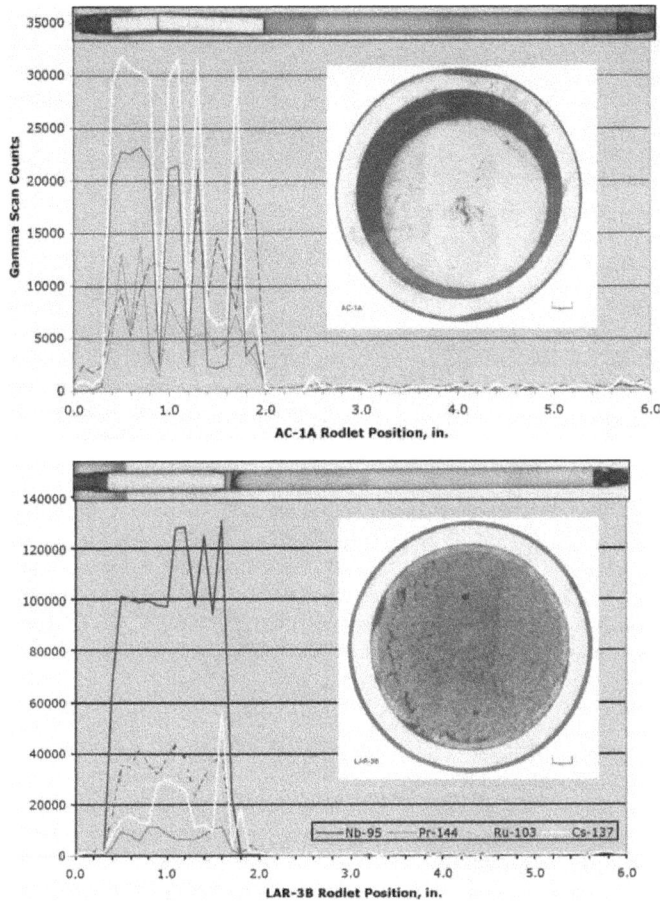

Figure 20.

Fuel swelling and fission product retention in: (a) non-fertile Pu-12Am-40Zr at 4.3 x 10^{20} fiss/cm^3, and (b) low-fertile U-29Pu-4Am-2Np-30Zr at 6.4 x 10^{20} fiss/cm^3.

Figure 21 shows the fission-gas release data from all the metallic fuels alloys. Both non-fertile and low-fertile compositions included in the AFC-1B and AFC-1F irradiation tests are plotted against typical data for U-Pu-Zr irradiations from EBR-II (Hofman and Walters, 1994). The striking feature is that when these data are correlated versus alloy fission density, all the data appear to fall along the same trendline. There may be a slightly longer incubation period to the onset of swelling, and thus, fission gas release for the current metallic fuels with much higher zirconium contents. However, once swelling begins, the fission-gas release behavior appears to converge quickly with the data from previous metallic fuels experiments.

Figure 21.

Fission gas release behavior for metallic fuels.

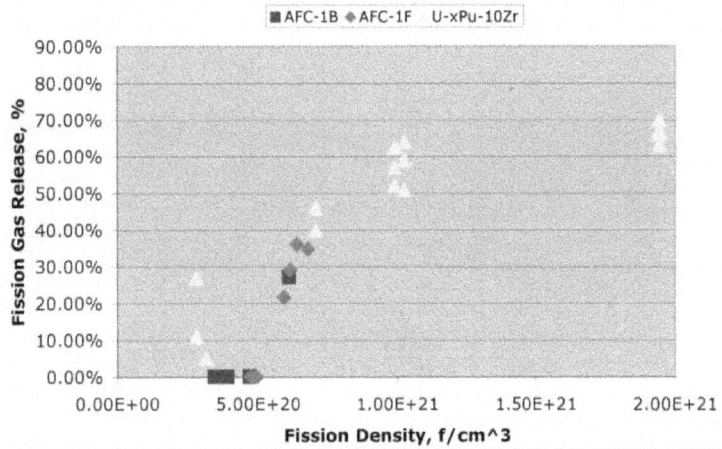

37

703-739-3790 TCNNuclear.com **211**

M E T A L L I C   F U E L S

These results, admittedly at an intermediate burnup level, nevertheless seem to indicate that the metallic fuel alloys currently under development, although having significantly higher levels of plutonium, minor actinides, and zirconium than previous metallic fuels, exhibit behavior under irradiation that is remarkably consistent with the large, historic database on U-Pu-Zr metallic fuels.

CONCLUSIONS AND FUTURE WORK

Metallic fuels have a long history of use in sodium-cooled fast reactors. Continued research, development, and design innovation over a period of decades have demonstrated this fuel technology to be safe and reliable to extremely high burnups over a wide spectrum of compositions. Their simple fabrication has historically given them a significant economic advantage over other fuel forms. Research and development on metallic fuels continues today at the INL, with particular emphasis on showing that this fuel technology can incorporate significant quantities of minor actinides for the purpose of efficiently and economically transmuting these elements in a fast-spectrum reactor. New fabrication methods to improve retention of the volatile fuel constituents during the casting process (and reduce waste generation) are being developed, and characterization of the fundamental and engineering properties of advanced metallic fuel alloys is underway. Irradiation

testing of the new metallic fuels has so far indicated that the new compositions, although they differ significantly (both in the quantities of plutonium and minor actinides included in the zirconium alloying addition), perform very analogously to the historic U-Zr and U-Pu-Zr fuels when behavior is correlated as a function of fission density.

In the next few years, it is anticipated that a new casting furnace prototype will demonstrate that these new metallic fuel alloys can be fabricated at engineering scale, in a remote environment, with no significant fuel loss or waste generation. Irradiation tests currently underway will provide the performance data needed to assess fuel reliability/burnup capability and identify any unique behaviors that are a result of the new compositions. If these future results are consistent with the intermediate data available at present, metallic fuels are expected to continue to be a strong contender for the fuel technology choice of future fast reactors.

38

N U C L E A R   F U E L S   &   M A T E R I A L S   S P O T L I G H T

REFERENCES

Chang, Y. I., 1989, "The Integral Fast Reactor," Nuclear Technology, Vol. 88, pp. 129-138.

Cole, J. I., J. R. Kennedy, and D. D. Keiser, Jr., 2008, "An Investigation of Fuel-Cladding Chemical Interaction in Metallic Transmutation Fuels," Transactions of the ANS Embedded Topical Meeting on Nuclear Fuels and Structural Materials for the Next Generation Nuclear Reactors, Vol. 98, Anaheim, CA, June 8-12, 2008.

Crawford, D. C., C. E. Lahm, H. Tsai, and R. G. Pahl, 1993, "Performance of U-Pu-Zr Fuel Cast into Zirconium Molds," Journal of Nuclear Materials, Vol. 204, pp. 148-156.

Crawford, D. C., D. L. Porter, and S. L. Hayes, 2007, "Fuels for Sodium-cooled Fast Reactors: U.S. Perspective," Journal of Nuclear Materials, Vol. 371, pp. 202-231.

Frost, B. R. T., 1982, Nuclear Fuel Elements: Design, Fabrication, and Performance, Pergamon Press, New York, p. 187.

Hilton, B. A., D. L. Porter, and S. L. Hayes, 2006, "Postirradiation Examination of AFCI Metallic Transmutation Fuels at 8 at.%," Embedded Topical Meeting on Nuclear Fuels and Structural Materials for the Next Generation Reactors," American Nuclear Society, Reno, NV, June 4-8, 2006.

Hofman, G. L., R. G. Pahl, C. E. Lahm, and D. L. Porter, 1990, "Swelling Behavior of U-Pu-Zr Fuel," Metallurgical Transactions A, Vol. 21A, pp. 517-528.

Hofman, G. L., and L. C. Walters, 1994, "Metallic Fast Reactor Fuels", in Materials Science and Technology, Vol. 10A, VCH Publishers, New York, pp. 1-43.

Keiser, D. D., Jr., 2005, "Fuel-Cladding Interaction Layers in Irradiated U-Zr and U-Pu-Zr Fuel Elements," Argonne National Laboratory Report, ANL-NT-240.

Keiser, D. D., Jr. and J. I. Cole, 2007, "An Evaluation of Potential Liner Materials for Eliminating FCCI in Irradiated Metallic Nuclear Fuel Elements," Proceedings of the GLOBAL '07 International Conference on Advanced Nuclear Energy and Fuel Cycle Systems, Boise, ID, September 9-13, 2007.

Keiser, D. D., Jr. and M. A. Dayananda, 1994, "Interdiffusion in U-Zr Fuel and Cladding Steels Couples," Metallurgical Transactions, Vol. 25A, p. 1649.

Keiser, D. D., Jr. and M. C. Petri, 1996, "Interdiffusion Behavior in U-Pu-Zr Fuel Versus Stainless Steel Couples," Journal of Nuclear Materials, Vol. 240, pp. 51-61.

Meyer, M. K., S. L. Hayes, W. J. Carmack, and H. C. Tsai, 2008, "The EBR-II X501 Minor Actinide Burning Experiment," ANS Topical Meeting on Nuclear Fuels and Structural Materials for the Next Generation Nuclear Reactors, Anaheim, CA, June 8-12, 2008.

O'Boyle, D. R., and A. E. Dwight, 1970, Proceedings of the 4th International Conference on Plutonium and Other Actinides, Santa Fe, NM, Session 2.

Stevenson, C. E., 1987, The EBR-II Fuel Cycle Story, American Nuclear Society, La Grange Park, IL, chs. 7-9.

Trybus, C. L., J. E. Sanecki, and S. P. Henslee, 1993, "Casing to Metallic Fuel Containing Minor Actinide Additions," Journal of Nuclear Materials, Vol. 204, pp. 50-55.

39

213

M E T A L L I C   F U E L S

STEVEN L. HAYES (BS, 1988, MS, 1989, PhD, 1992, Nuclear Engineering, Texas A&M University) is Manager of the Fuel Performance and Design Department at the INL. He has been actively involved in the irradiation testing and modeling of fast breeder reactor fuels, fuels for actinide transmutation, and research reactor fuels for more than 15 years.

J. RORY KENNEDY (BS, 1982, Inorganic Chemistry, Colorado State University; MS, 1983, PhD, 1987, Inorganic Chemistry, Northwestern University) is Manager of the Basic Fuel Properties and Modeling Department at the INL. He has been investigating the synthesis and structure/property relations of materials for over 20 years. Over the last 12 years, his interests have been directed to the fabrication and characterization of materials associated with the nuclear fuel cycle, with emphasis on advanced nuclear fuels.

BRUCE A. HILTON (BS, 1990, Physics, Brigham Young University; MS, 1993, PhD, 1995, Nuclear Materials Engineering, Massachusetts Institute of Technology) is a member of the Fuel Performance and Design Department at the INL. He has 13 years experience in nuclear fuels performance and materials, irradiation testing, and postirradiation examination. Current responsibilities at the INL include technical management of Global Nuclear Energy Partnership/Advanced Fuel Cycle Initiative Postirradiation Examinations (GNEP/AFCI PIE) of advanced nuclear fuels, and advanced LWR nuclear fuels development in the LWR Sustainability Program.

RANDALL S. FIELDING (BS, 2000, MS, 2008, Materials Science and Engineering, University of Idaho) is an engineer in the Fuel Fabrication Department at the INL. He has been active in irradiation experiment assembly, metallic fast reactor fuel fabrication, and dispersion fuel-fabrication research since joining the INL in 2001.

TIMOTHY A. HYDE (BS, 1989, Mechanical Engineering, University of Idaho) is an Engineer/Scientist at the INL with 19 years experience in research and development activities. He has worked in a wide range of technical disciplines, including ceramic and metal alloy fuel development, membrane permeability testing, plasma disassociation of hydrocarbons to grow carbon nano-tubes, aerosol particle measurement, magnetic properties measurement, magnetic alloy development, and numerous experimental and mechanical design projects. These activities have resulted in over 20 publications and numerous awards, including a Research and Development 100 award. Mr. Hyde is a coinventor on numerous Invention Disclosure Reports (IDRs) and is coinventor on two issued patents.

40

N U C L E A R   F U E L S   &   M A T E R I A L S   S P O T L I G H T

DENNIS D. KEISER, JR. (BS, 1987, MS, 1989, Metallurgical Engineering, University of Idaho; PhD, 1992, Materials Engineering, Purdue University) is a member of the Fuel Performance and Design Department at the INL. He has over 20 years experience researching different issues related to the performance of nuclear fuels and materials. His current research relates to developing low-enriched fuels for application in research and test reactors, developing a better understanding of the diffusion behavior of fuel and cladding components in irradiated fuels, and determining the microstructural development of fuels and materials during irradiation.

DOUGLAS L. PORTER (BS, 1971, MS, 1973, PhD, 1977, Metallurgy and Materials Science, Case Western Reserve University) has conducted research for more than 30 years at the INL, with major interests in irradiation effects on materials and metallic fuel development for fast reactors.

41

Cross section of an

irradiated U-Mo based

dispersion fuel meat

EML SEI 30.0kV X1,000 WD 9.7mm

Higher magnification

of the fracture surface

shows fission gas

bubble formation along

grain boundaries

within the particle

EML SEI 30.0kV X150 WD 9.7mm

A particle fractured

during sample

preparation showing

cleavage within

the particle

The RERTR program seeks to convert research and test reactor cores throughout the world that currently operate using highly enriched uranium (HEU, ≥20% uranium-235) fuel to low-enriched uranium (LEU, <20% uranium-235) fuel. As the enrichment of the fuel alloy decreases, the volume of fuel material must correspondingly increase, requiring advanced and novel fuel concepts, such as the monolithic fuel form. Monolithic fuel alloys are created by casting a uranium-molybdenum coupon comprised of the desired fuel stoichiometry, hot and/or cold rolling the coupon to a desired thickness (typically between 0.010 and 0.020-in.), and sandwiching the monolith between two aluminum alloy cladding plates. An appropriate joining process is applied to encapsulate the monolithic alloy in the cladding. Development of a reliable, cost effective joining technique is a key challenge in the realization of this fuel system.

216

NUCLEAR   FUELS   &   MATERIALS   SPOTLIGHT

REDUCED ENRICHMENT
FOR RESEARCH AND TEST REACTORS

Daniel M. Wachs

THE RERTR PROGRAM marked its 30th anniversary in 2008. From its inception, the program has been tasked with providing the means necessary to enable research and test reactors worldwide to convert from highly enriched nuclear fuel to low enriched nuclear fuel (which is defined by the International Atomic Energy Agency [IAEA] as <20% uranium-235).

High uranium density fuel development has always been a central component of the RERTR program. In its earliest years, the program was focused on expanding the uranium loading limits for the common fuels of the time to enable enrichment reductions (most notably UZrH but also aluminide and oxide-based dispersion fuels). However, these improvements were not sufficient to provide the basis for widespread LEU conversions, and the program initiated the development of new fuels. The U_3Si_2 based dispersion fuel was subsequently developed and qualified in the 1980s by the program and provided a uranium density of 4.8 g U/cc.

As a result of this accomplishment, only a handful (less than 30) of civilian research and test reactors worldwide are unable to convert from high- to low-enriched nuclear fuel due to the availability of a suitable LEU fuel. However, these reactors are by the far the greatest civilian consumers of HEU, leading to substantial worldwide commerce in this high-risk material. In the early 1990s, the RERTR program embarked on the development of a replacement fuel for these most challenging reactors. In 2004, the program was refocused and accelerated when the terrorist attacks of September 11 added even more urgency, and the program was absorbed into the National Nuclear Security Administration's (NNSA's) Global Threat Reduction Initiative (GTRI). Aggressive timelines for completion of the fuel development program were put in place, and the INL was tasked with leading a multilaboratory team in the execution of a multi-national mission.

Following an investigation of the irradiation performance of several candidate fuel materials, uranium-molybdenum alloys with between 6 and 12 wt% molybdenum were selected to be the base of fuel development. This fuel offered very high uranium densities and demonstrated excellent irradiation behavior under a wide range of relevant conditions. It was originally envisioned as a dispersion-type fuel capable of almost twice the uranium density (8-9 g U/cc) of any other research reactor fuel available; an alternate monolithic version was later proposed to enable uranium loading up to four times the current state of the art (15-16 g U/cc). While the dispersion version of the uranium-molybdenum-based fuel does not come without challenges, it is a relatively modest change compared to the monolithic version, which represents a dramatic shift from conventional fuel designs and challenges the status quo at many levels.

The development of a new nuclear fuel requires that the developer address multiple issues, including fuel fabrication, irradiation testing, characterization of irradiation performance, fuel performance modeling, and final disposal of the fuel. The INL Nuclear Fuels and Materials Division is uniquely equipped to conduct research in all these areas, leading to a holistic development approach that integrates all the relevant areas of fuel development. Since most R&D programs require several iterations on a fuel concept before development is complete, the natural integration of all these functions in a common location often leads to a meaningful acceleration of R&D

43

REDUCED ENRICHMENT FOR RESEARCH & TEST REACTORS

objectives and significant economical advantages. This has clearly been the case for the RERTR uranium-molybdenum fuel development program.

The RERTR fuel development program has faced many challenges along the way that have mobilized the exceptional creativity and technical capability of INL researchers. A sampling of these challenges is described in the following sections. In Subsection I, Dr. Douglas Burkes describes one of the techniques (Friction Bonding) he and his colleagues have developed to fabricate uranium-molybdenum monolithic fuel plates that consist of thin, metallic layers with very different mechanical properties. In Subsection II, Dr. Dennis Keiser describes techniques he and his colleagues developed to investigate the micro-chemical condition of irradiated fuel in order to better understand the interaction of uranium-molybdenum fuels with aluminum and aluminum-silicon alloys during the irradiation of

dispersion fuel designs. In Subsection III, Dr. Daniel Wachs and Mr. Jared Wight outline the development of a device designed to allow irradiation testing of prototypic plate-type fuels under conditions well in excess of their prototypic environment, as well as techniques developed for the high fidelity investigation of fuel performance between irradiation cycles. Subsection IV includes a review of fuel performance modeling led by Dr. Pavel Medvedev to support understanding of integrated fuel performance, including the coupling of thermal, hydraulic, mechanical, and chemical behaviors in the monolithic fuel type.

DANIEL M. WACHS (BS, 1995, Mechanical Engineering, MS, 1997, Nuclear and Mechanical Engineering, Oregon State University; PhD 2000, Mechanical Engineering, University of Idaho) is a member of the Fuel Performance and Design Department at the INL. He is currently the National Technical Director for the RERTR Fuel Development Program. His current research interests and technical background include irradiation testing of nuclear fuel and materials, thermal-hydraulic testing and analysis, and reactor system component design for research reactors and liquid metal cooled fast reactors.

44

N U C L E A R   F U E L S   &   M A T E R I A L S   S P O T L I G H T

FRICTION BONDING PROCESS USED TO FABRICATE FUEL PLATES

Douglas E. Burkes

THE RERTR PROGRAM has identified several techniques for evaluation and development. The first successful approach was based on extension of the friction stir welding (FSW) technique beyond its original applications. FSW has been in existence since the early 1990s and has routinely been used for joining applications in the aerospace and automotive industries. In general, FSW involves rotating a welding tool (comprised of a shank, shoulder, and pin) at a prescribed speed and tilt, and plunging the weld tool into the workpiece material until the tool intimately contacts the workpiece surface and generates friction.

The friction produces intense heat that bonds the workpiece materials as the weld tool traverses the joint line. The FSW process can be employed to join both similar and dissimilar cast and wrought aluminum, titanium, copper, and magnesium alloys, as well as steels, in multiple configurations, including butt, corner, lap, T, spot, fillet, and hem joints. The process typically produces joints with higher strength, increased fatigue life, lower distortion, less residual stress, and less sensitivity to corrosion compared to conventional fusion welding processes.

A modified FSW process, referred to as friction bonding (FB), is being developed at the INL that allows interfacial bonding between multiple layers of thin metallic materials with dissimilar mechanical properties. This process permits the fabrication of nuclear fuel plates for research and test reactors containing the thin, monolithic fuel alloys. Unlike the traditional FSW process, the interface to be bonded is located perpendicular to the plane of the working tool (Figure 1). The rotating tool face is pressed against the surface layer of material and rotates; the combination of force and rotation creates intimate contact between the fuel and cladding layer while the rotation generates the heat required to initiate metallurgical bonding. The tool is rastered across the entire surface of the fuel plate to establish full bonding of all interracial surfaces.

Recent research on this novel FB process has yielded many significant technical and programmatic contributions over the last several years. The process has been utilized to fabricate a number of fuel plates for various irradiation tests, including 35 mini-plates (25 mm x 100 mm x 1.27 mm) and two full-size fuel plates (57 mm x 600 mm x

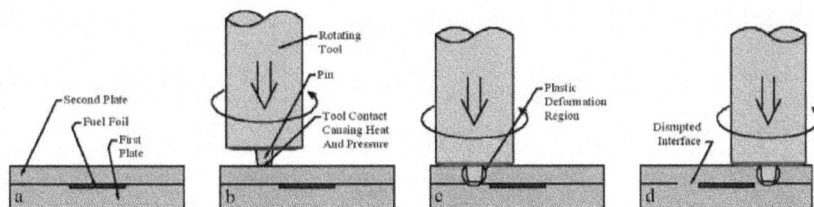

Figure 1.

Schematic of friction-bonding technique developed to enable bonding of aluminum cladding to thin uranium-molybdenum foils.

45

REDUCED ENRICHMENT FOR RESEARCH & TEST REACTORS

1.27 mm). While a few fuel plates in the early experiments showed indications of small delaminations during destructive postirradiation examination (due to fuel chemistry changes during irradiation), most have behaved, and continue to behave, in an expected and normal manner. Techniques to incorporate materials into the fuel/clad interface have successfully demonstrated that they further strengthen the interfacial bond strength. Based on these successes, a prototype FB fabrication unit has been brought online and operation verified for commercial-scale fabrication demonstration.

INTERFACIAL BOND STRENGTH ASSESSMENT

The bond strength between the fuel and cladding after fabrication is a critical performance criterion for the monolithic fuel design. In the absence of a recognized standard for this property, the program is developing examination techniques that will provide a better under-standing of the impact of various fabrication process parameters and, eventually, operational thresholds for interfacial loads.

As an example of this process, several mini-plates were fabricated employing FB for irradiation in the RERTR-9A experiment (Burkes et al., 2007). Four of the plates were fabricated with a standard steel-faced tool, while the remaining three plates were fabricated with a newer, more robust refractory-faced tool – Anviloy. Beginning-of-life (BOL) bonding was suspected to be better for plates fabricated with the

Anviloy-faced tool compared to the steel-faced tool. Pull-tests performed on small samples extracted from the FB fabricated mini-plates confirmed these suspicions, as shown in Figure 2, even though the non-destructive ultrasonic (UT) scans indicated that the interfacial properties were similar (i.e., the sound transmission properties were similar), as shown in Figure 3. However, the stresses occurring during irradiation did not appear to exceed the bond strength for either tool face material or BOL bonding. All seven plates reached the target burn-up without delamination, but the information still helps establish the margins for which the fuel plates can be safely operated.

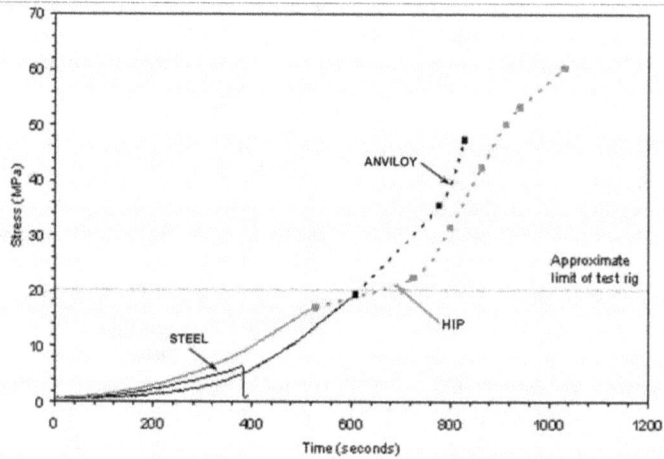

Figure 2.

Pull test results (normal stress) of mini-plates fabricated using HIP, an Anviloy-faced friction-bonding tool, and a steel-faced friction-bonding tool.

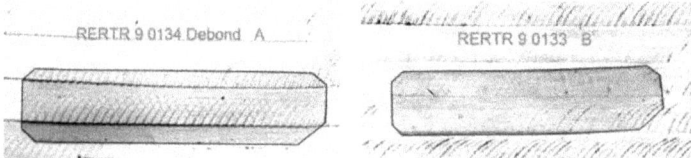

Figure 3.

Ultrasonic testing scans of mini-plates fabricated for the RERTR-9A experiment using a steel-faced tool (left) and an Anviloy-faced tool (right).

46

N U C L E A R   F U E L S   &   M A T E R I A L S   S P O T L I G H T

Additional examination was required to assess the impact of chemical alterations to the fuel/clad interface on the fabrication process and BOL bond conditions. Miniplates were fabricated for the RERTR-9B experiment using FB that included a modified Al-2Si layer to minimize formation of the intermetallic interaction layer that forms between the fuel and the cladding at high temperature during irradiation. The modified Al-2Si layer was applied using thermal-sprayed powders of Al-12Si and aluminum to create the inhomogeneous Al-2Si layer, with two different thicknesses: 13 μm and 25 μm. A cross section of the 13-μm-thick thermal-sprayed layer is shown in Figure 4. These two thicknesses were selected based on the recoil zone of fission products in aluminum: 13 μm. A UT scan of a FB plate with the modified silicon interface is provided in Figure 5. The UT scan reveals that the fabrication process is unaffected by the modification to the interface. A second modification to the fuel system was also proposed and evaluated as part of this testing. Fuel plates employing a thin layer (25 to 50 μm) of zirconium at the fuel-clad interface were also fabricated with FB. A metallographic cross section of an example mini-plate is provided in Figure 6 along with an EDX scan of the interface that revealed a molybdenum enrichment zone at the Zr/U-Mo foil interface. All indications were that the FB process was effective in bonding the interfaces of this system.

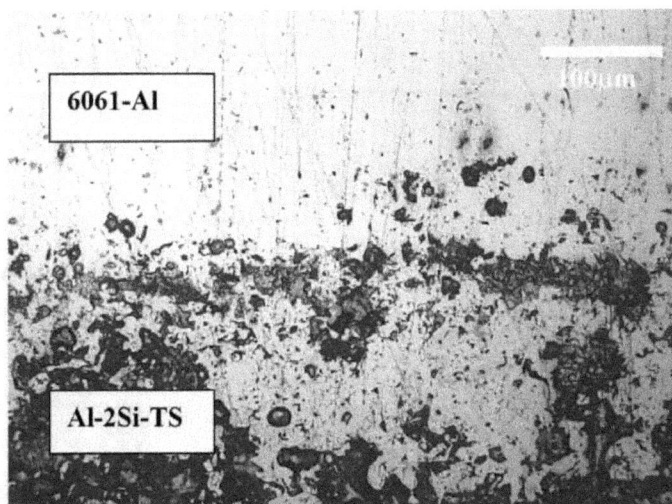

Figure 4.

Cross section of a 13-μm–thick, thermal-sprayed Al-2Si-modified interface.

Figure 5.

Ultrasonic scan of a mini-plate fabricated employing friction bonding with a 25-μm–thick, thermal-sprayed Al-2Si-modified interface.

47

REDUCED ENRICHMENT FOR RESEARCH & TEST REACTORS

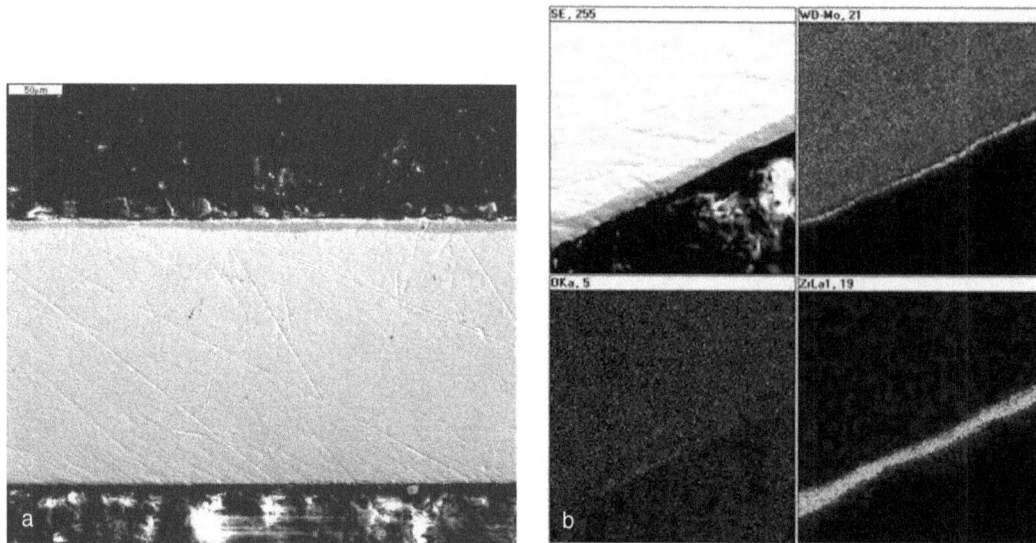

Figure 6.

Metallographic cross section (a) of a mini-plate fabricated using friction bonding with a zirconium interface between the uranium-molybdenum foil and the AA6061 cladding, represented by the dark gray at the top. EDX scans (b) of the interface modification revealed a molybdenum-rich zone between the zirconium diffusion barrier and uranium-molybdenum foil, shown in the top right scan; bottom right is zirconium.

Finally, two full-size fuel plates were fabricated using FB for the AFIP-2 design: one plate containing the modified zirconium interface and the other containing a silicon interface applied using thermal spray. Scale-up of the fabrication process from mini-plates was achieved generally as anticipated; the fuel plates achieved 80% average burnup during irradiation without incident. Not only did this experiment confirm that the selected modifications would not inhibit the FB process, but it also demonstrated, for the first time, scale-up of fuel fabrication from the mini-plate design to the full-size plate design.

Improvements in the overall processing parameters employed for FB and the onset in development of an operational envelope for the process are ongoing. In particular, research into the relationships between process load, temperature, and concomitant microstructure were investigated. The relationship between process load and the temperature measured at the joint interface is illustrated in Figure 7. Results of this study, specifically the large temperature gradient between the advancing edge of the tool and the retreating edge of the tool, initiated the development of a thermodynamic model (Dixon et al., 2007). A snapshot of an early model

run is shown in Figure 8 that basically confirms the temperature gradient across the diameter of the tool face.

Finally, the INL's prototype, full-scale fabrication demonstration unit is now completely online and functional. This piece of equipment, a photograph of which is shown in Figure 9, will allow more fabrication of fuel plates with more consistency and repeatability. Overall, the equipment allows computer controlled feedback, rather than manual operator controlled feedback. The new equipment will also enable the fabrication of long monolithic fuel plates for eventual demonstration of full-scale uranium-molybdenum based monolithic fuel elements in the ATR. The operational envelope established through modeling and experimentation, discussed previously, is currently being validated on the TTI machine.

48

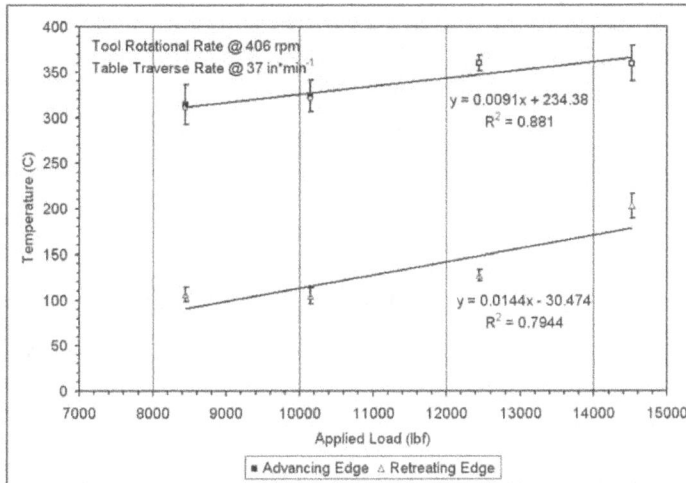

Figure 7.

Temperature measured at the joint interface as a function of applied process load for the advancing edge and retreating edge of the tool. Joint temperature increased with higher process loads, and there is a significant temperature gradient across the diameter of the tool.

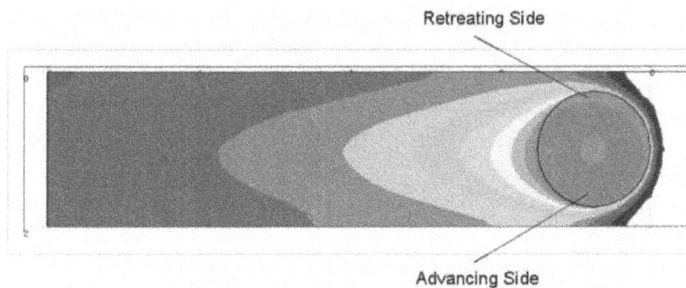

Figure 8.

Thermodynamic model snapshot confirming the temperature gradient across the tool diameter, with the advancing edge having a higher temperature than the retreating edge.

Figure 9.

Photograph of the TTI machine that will be used for full-scale fabrication demonstration of monolithic fuel plates.

49

REDUCED ENRICHMENT FOR RESEARCH & TEST REACTORS

REFERENCES

Burkes, D. E., N. P. Hallinan, C. R. Clark, "Nuclear Fuel Plate Fabrication Employing Friction Bonding," AWS Welding J., accepted for publication (2008) pp. 3-10.

Burkes, D. E., N. P. Hallinan, K. L. Shropshire, P. B. Wells, "Effects of Applied Load on 6061-T6 Aluminum Joined Employing a Novel Friction Bonding Process," Met. Mat. Trans. A, accepted for publication.

Hallinan, N. P., D. E. Burkes, "Friction Stir Weld Tools, Methods of Manufacturing Such Tools, and Methods of Thin Sheet Bonding Using Such Tools," U.S. Patent Application (2007).

Burkes, D. E., N. P. Hallinan, J. M. Wight, M. D. Chapple, "Update on Friction Bonding of Monolithic U-Mo Fuel Plates," Proc. of 2007 RERTR International Meeting, Czech Republic (2007).

Dixon, J., D. Burkes and P. Medvedev, "Thermal Modeling of a Friction Bonding Process," Proceedings of the COMSOL Conference 2007, Boston, MA (2007) pp. 349-354.

DOUGLAS E. BURKES (BS, 2001, Mechanical Engineering, Texas Tech University; PhD, 2005, Materials Science, Colorado School of Mines) is a member of the Fuel Performance and Design Department at the INL. His current research focuses on development of novel nuclear fuel fabrication processes for application in research and fast reactors, characterization of advanced nuclear fuels, and development of advanced fuel characterization methods and techniques. He has been a member of the Fuel Performance and Design Department since 2006.

50

NUCLEAR FUELS & MATERIALS SPOTLIGHT

SILICON AS AN ALLOYING ADDITION TO ALUMINUM FOR IMPROVING THE IRRADIATION PERFORMANCE OF FUEL PLATES

Dennis D. Keiser, Jr.

EARLY TESTING OF RERTR uranium-molybdenum dispersion fuel plates employed plates with pure aluminum as the matrix in the fuel meat. During irradiation of these fuel plates, large pores developed at the interface between a fuel/matrix interaction layer, which developed due to radiation enhanced diffusion (RED), and the unreacted aluminum matrix. The presence of these pores contributed to the eventual failures of some of the fuel plates.

In order to eliminate the development of this type of failure mechanism, possible alloying additions to the fuel and/or matrix of the dispersion fuel meat were evaluated (Kim et al., 2005). It was determined that silicon had the potential to be an effective addition to the aluminum matrix. The idea was that by adding silicon to the aluminum matrix, different, more radiation-stable phases would form during RED that would perform better during irradiation, and in addition, the overall size of the interaction zones would be reduced. Characterization of irradiated uranium-silicide fuels, which were developed early in the RERTR fuel program for converting some of the research reactors, showed that stable interaction layers had formed during RED of the fuel and matrix (Leenaers et al., 2004). If layers similar to these could be generated in the current RERTR fuels, stable irradiation behavior should result.

To develop an understanding of the silicon diffusion behavior in RERTR fuels, out-of-reactor diffusion couple studies are being performed to see

aluminum alloys affects the development of interdiffusion zones in terms of zone width and the types of phases that will form (Keiser, 2007; Perez, 2007). Because these experiments have to be performed at relatively high temperatures (500°C to 600°C) to get measurable diffusion zones in reasonable time periods, these tests are not necessarily representative of the in-reactor phenomena that occur at low temperatures (e.g., 200°C) and under the influence of irradiation (i.e., fission fragments, localized heating, etc). However, these diffusion tests do give an indication of how changing the silicon content in aluminum alloys will affect diffusion behavior. Furthermore, because fuel plates can be exposed to higher temperatures during fabrication, which results in the presence of interaction zones in as fabricated plates, it is important to understand the development of these zones since they will be present in fuel plates going into the reactor.

Diffusion couple experiments that were performed using Al-6061 (0.81 wt% silicon), Al-2Si, Al-5Si, and Al-4043 (4.81 wt% silicon) alloys

indicated that the width of the interdiffusion zones that form decreases as the amount of silicon is increased in the aluminum alloy, up to approximately 2.0 wt% silicon. Concentration additions above this value do not seem to reduce the diffusion zone width over what was the case for the alloy with 2.0 wt% silicon. In terms of the phases that develop in the interaction zones, there can be great differences depending on the silicon-containing aluminum-alloy that is employed for the diffusion couple. The presence of minor alloy constituents in the Al-6061 and Al-4043 alloys can affect the nature of the interaction zones.

Diffusion couple tests have also shown that if not enough silicon is available in an aluminum alloy, the diffusion behaviors that are observed may at some point approximate what occurs in the U-Mo-Al system without silicon (Keiser, 2007). Enough silicon has to interdiffuse to the interaction zone to keep silicon-containing phases stable, or they will break down. Therefore, when adding silicon, a minimum amount should be be added to get good behavior.

51

REDUCED ENRICHMENT FOR RESEARCH & TEST REACTORS

Another research effort associated with diffusion couple experiments has been the characterization of fuel plates after fabrication. Knowing the times that fuel plates were exposed to higher temperatures provides some information about the kinetics of the interaction between fuel alloys and the matrix. Also, the types of phases that are present in interdiffusion zones can be determined. Characterization of as-fabricated fuel plates is crucial for developing a better understanding of in-reactor behavior since comparison of microstructures before irradiation can be made with those after irradiation. Figure 1 shows the microstructure of an Al-4043 matrix-dispersion fuel plate after fabrication. It shows that silicon-rich layers are already present around the fuel particles before the fuel plate is put into the reactor.

Most postirradiation examination of fuel plates for characterizing the fuel microstructures is done using optical metallography. This technique is effective for identifying features such as swelling behavior because the porosity that develops in the fuel can be characterized. However, as noted, the fuel/matrix interaction zones can have large impacts on fuel behavior, and it is necessary to measure the compositions in these zones to determine the phases that are present in irradiated plates. Since silicon is added to change the phases that form in the interaction zones, it is important to see if different, more stable phases are actually present in irradiated fuel plates because of the presence of silicon. Consequently, samples have been taken from actual irradiated RERTR fuel plates in the form of punchings, mounted longitudinally, and then polished before being inserted into a SEM equipped with

an EDS and WDS to conduct microstructural characterization (Janney et al., 2007; Keiser et al., 2008). The samples characterized to date had either low (0.2 wt%) or high (4.81 wt%) silicon. Figure 2 shows the microstructure observed for the low and high silicon samples. The low-silicon plate contained interaction zones that were similar to those that develop in dispersion fuels with pure aluminum as the matrix; the zones in a sample taken from the fuel plate with Al-4043 alloy matrix were relatively narrow and enriched in silicon. The results of these characterizations confirmed that enough silicon has to be present in the matrix to keep silicon-rich layers stable. Therefore, if enough silicon is added to the fuel plate matrix, the interaction zones that are present in a fuel plate irradiated to high burnup will be relatively narrow, and the swelling behavior of these interaction zones will be appreciably improved compared to the zones present in fuel plates without silicon.

Characterization of the surfaces of a fractured sample taken from an irradiated RERTR fuel plate can also be of great value. Such fractured samples are particularly useful because they enable the fine gas bubbles that are present in the fuel to be characterized effectively. These bubbles are typically smeared when polished cross-sections are characterized. Figure 3 shows a few fractured particles that were observed in the fuel meat of an irradiated RERTR fuel plate.

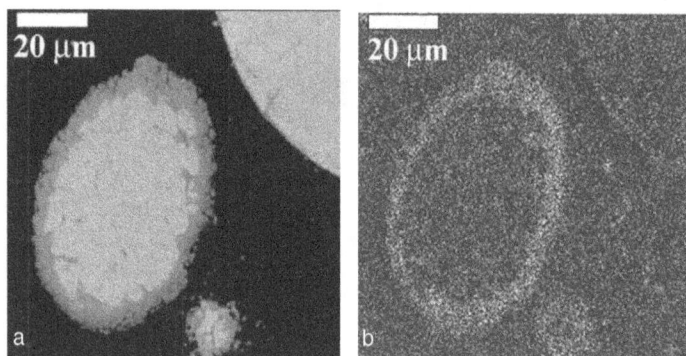

Figure 1.

Backscattered electron micrograph (a) showing white uranium-molybdenum alloy particles in a black Al-4043 matrix and the gray interaction zone around the fuel particles. The silicon X-ray map (b) shows the silicon enrichment in the interaction zone.

52

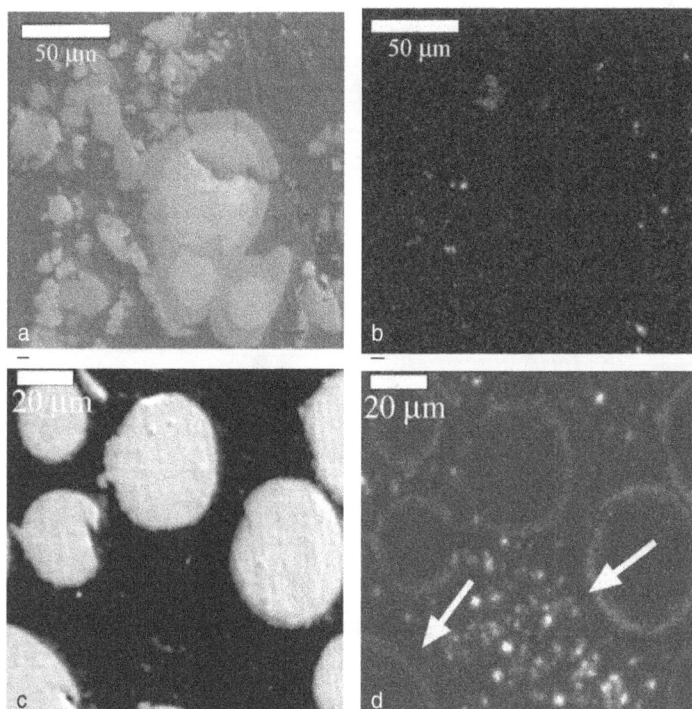

Figure 2.

Backscattered electron images (a) and (c) for an irradiated fuel plate with 0.2 wt% silicon in the matrix and 4.81 wt% silicon in the matrix, respectively, where the bright particles are uranium-molybdenum fuel, the black areas are the aluminum alloy matrix, and the gray regions are the interaction layers. The silicon X-ray map (b) for the low-silicon plate shows negligible silicon in the relatively wide interaction layers, and the silicon X-ray map (d) for the high-silicon fuel plate, where the interaction layers are relatively narrow, contains appreciable silicon The arrows in (d) indicate the areas around fuel particles where recoil zones, which are present around the fuel particles during irradiation, have dissolved the silicon-rich precipitates that are present in the original cladding, creating precipitate-free zones.

Overall, the addition of silicon to the aluminum matrix in RERTR dispersion fuels is an effective technique for improving the overall irradiation performance of the fuel. Based on this optimal behavior, silicon additions are also being employed to improve another type of RERTR fuel, namely monolithic fuel plates. For an early version of the monolithic fuels, where the uranium-molybdenum foil was in contact with plain Al-6061 cladding, interaction zones developed during irradiation that displayed adverse swelling behavior, which was similar to what was observed in dispersion fuels. Based on the positive effect of adding silicon to the aluminum in dispersion fuels, it is anticipated that the addition of silicon can have a similar positive effect for monolithic fuels.

Different approaches have been tried for adding silicon to the uranium-molybdenum foil/cladding interface (e.g., thermal spraying of the silicon onto the cladding), and irradiated samples are in the process of being characterized. Another approach for dealing with fuel/cladding interaction in monolithic fuel plates is to use zirconium (or possibly niobium or molybdenum) as a diffusion barrier for totally eliminating fuel/cladding interactions. Fuel plates with an applied diffusion barrier have also been irradiated, and early results are promising.

Based on the results of recent reactor tests using dispersion and monolithic fuels with added silicon, the goal of qualifying fuel plates in the next few years for use in research and test reactors seems achievable.

53

REDUCED ENRICHMENT FOR RESEARCH & TEST REACTORS

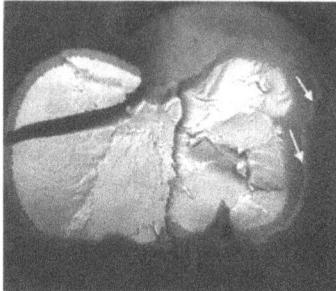

Figure 3.

Backscattered electron micrograph of two uranium-molybdenum particles (white) present in the fuel meat of an irradiated RERTR fuel plate after a sample was fractured for SEM characterization. The white arrows indicate where, based on the fracturing behavior of the interaction zone (gray), an interface may be present between two different phases.

REFERENCES

Janney, Dawn E., et al., 2007, Hot Laboratories and Remote Handling Plenary Meeting, Bucharest, Romania, September 20-21, 2007.

Keiser, D. D., Jr., 2007, Defect and Diffusion Forum, 266 (2007) 131.

Keiser, D. D., Jr. et al., 2008, Proc. of the RRFM 2008 International Meeting on Research Reactor Fuel Management, Hamburg, Germany, March 2-5, 2008.

Kim, Y. S., et al., 2005, in: Proc. of the 25th International Meeting on Reduced Enrichment for Research Reactors, Boston, MA, 2005.

Leenaers, et al., 2004, *J. Nucl. Matter.* 327 (2004) 121.

Perez, E., et al., 2007, Defect and Diffusion Forum, 266 (2007) 149.

DENNIS D. KEISER, JR. (BS, 1987, MS, 1989, Metallurgical Engineering, University of Idaho; PhD, 1992, Materials Engineering, Purdue University) is a member of the Fuel Performance and Design Department at the INL. He has over 20 years experience researching different issues related to the performance of nuclear fuels and materials. His current research relates to developing low-enriched fuels for application in research and test reactors, developing a better understanding of the diffusion behavior of fuel and cladding components in irradiated fuels, and determining the microstructural development of fuels and materials during irradiation.

54

NUCLEAR FUELS & MATERIALS SPOTLIGHT

ADVANCED TEST REACTOR IN-CANAL ULTRASONIC SCANNER: EXPERIMENT DESIGN AND INITIAL RESULTS ON IRRADIATED PLATES

Daniel M. Wachs and Jared.M. Wight

AN IRRADIATION TEST device has been developed to support testing of prototypic scale plate-type fuels in the ATR at the INL. Optimized experiment hardware and operating conditions provide the irradiation conditions necessary to conduct performance and qualification tests on research reactor-type fuels for the RERTR program.

The irradiation device design allows disassembly and reassembly in the ATR spent fuel canal to enable interim inspections on the fuel plates. An ultrasonic scanner was developed to perform dimensional and transmission inspections during these interim investigations. Example results from the AFIP-2 experiment are presented in this section.

Before implementation of any new nuclear fuel design, the fuel must be tested under a wide range of irradiation conditions. These tests provide the basis for evaluating fuel performance and ultimately the data necessary for a regulator to license a reactor to operate with the fuel. In most cases, this testing envelope must meaningfully exceed the anticipated operating conditions to establish appropriate margins for safe operation. This envelope encompasses a number of key environmental variables that include the thermal, hydraulic, nuclear, and geometric conditions of the experiment. The ATR was specifically designed for this purpose, and it provides the experimenter with the ability to control all these variables over a

wide range. The INL Nuclear Fuels and Materials Division (through leadership of the RERTR Fuel Development Program) is utilizing this key asset to develop the nuclear fuels necessary to support the NNSA's GTRI HEU minimization objectives.

The Nuclear Fuels and Materials Division has developed several experimental test trains to enable irradiation testing of nuclear fuel designs in the ATR. The RERTR program focuses on the development of research reactor fuels that operate under conditions that are very similar to the standard ATR coolant environment. Consequently, all the RERTR irradiation test devices are immersed directly in the ATR coolant stream. The first experiments (RERTR-1, -2, and -3) were designed to test the behavior of very small fuel samples (nano- and micro-plates). The size of these samples allowed for testing of a large number of candidate materials and greatly simplified test sample fabrication. The PIE results collected from these screening tests were used to select aluminum clad, uranium-molybdenum alloys for further development. To evaluate performance further, an irradiation

test device was developed to test slightly larger scale plates (mini-plates). This scale allowed the experimenter to account for fabrication variables that might appear at a more representative scale while still accommodating a fairly large number of test specimens at a low cost. The sensitivity of fuel performance to various variables was evaluated by testing multiple samples of each type under different irradiation conditions. After extensive testing at the mini-plate scale, a few fuel designs appeared viable enough to justify prototypic testing. Testing at this scale substantially reduced the number of samples that could be tested while simultaneously increasing the cost of testing for each sample. It was, therefore, desirable to develop an irradiation test device that maximized the amount of data that could be collected from each experiment.

The AFIP (ATR full-size-plates in center flux trap position) test assembly, designed specifically for this purpose, provided a versatile testbed for plate-type fuel experiments. The hardware allows the experimenter to subject experiments of various scales to a broad range of

55

REDUCED ENRICHMENT FOR RESEARCH & TEST REACTORS

nuclear and thermal-hydraulic conditions. The experimental design allowed disassembly in the ATR canal between irradiation cycles. This made it possible to interrogate the experiment behavior at various stages of irradiation. A complementary set of examination tools has also been developed to collect this data.

AFIP Test Hardware

A basic function of the AFIP test assembly is to enable testing of research-reactor fuel designs at a prototypic scale and under prototypic irradiation conditions. The ATR operates a wide range of irradiation test positions capable of supporting all types of nuclear fuels and materials testing; however, only a few positions are appropriate for the type of testing required for this program (large openings with high neutron flux). The ATR center flux trap (CFT) offers a 3-in.-diameter and 48-in.-long cylindrical cavity with a peak (unperturbed) thermal neutron flux of approximately 4.4 x 10^{14} n/cm^2/sec and was selected to house the AFIP experiment. Figure 1 shows a cross-section of the ATR.

An insert was necessary to convert the circular opening into a square channel large enough to accommodate a fuel plate at least 2 in. wide. The hardware also provides the structure necessary to control the experiment configuration during testing (i.e., prevent movement or vibration) and handling (before and after irradiation). The hardware must completely constrain the fuel plates during irradiation (without removable screws, bolts, or pins)

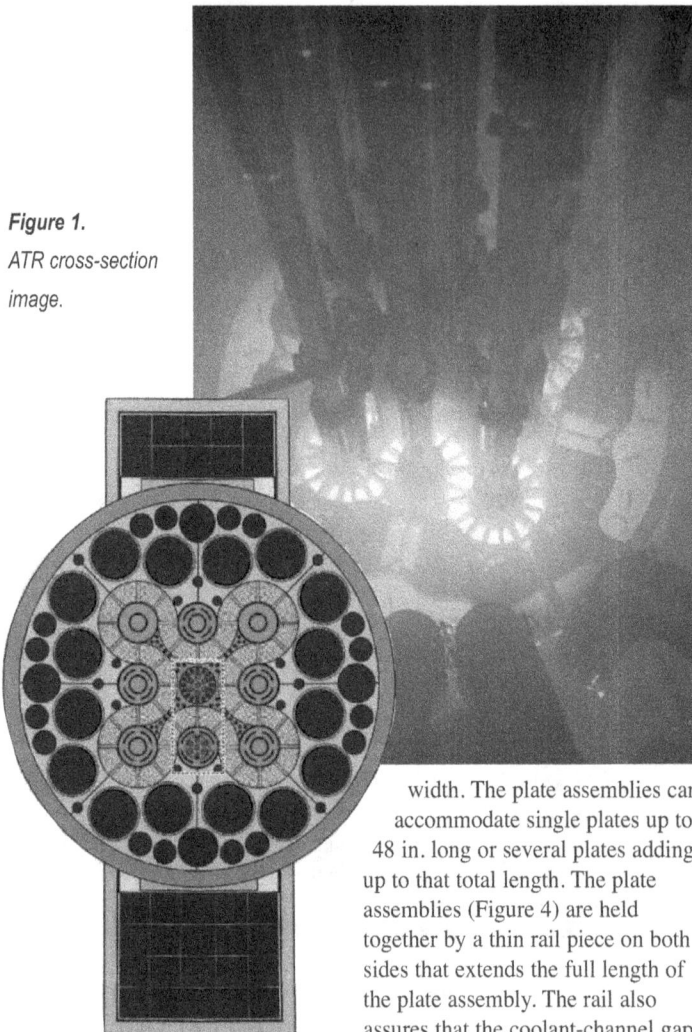

Figure 1.
ATR cross-section image.

while still allowing operators to disassemble the experiment (which is submerged under 20 ft of water). A schematic of the experiment cross-section showing the key components of the hardware is provided in Figure 2. Photographs of the hardware components are shown in Figure 3. The experiment can accommodate two fuel plate assemblies with a 2.2-in. plate width. The plate assemblies can accommodate single plates up to 48 in. long or several plates adding up to that total length. The plate assemblies (Figure 4) are held together by a thin rail piece on both sides that extends the full length of the plate assembly. The rail also assures that the coolant-channel gap width remains fixed during irradiation. The fuel plate assemblies are held in position by a ram that is pinned in place by a tapered ram-rod. Four flux monitor wires are also included in the experiment to baseline the nuclear analysis. The experiment is inserted into the ATR CFT through a long in-pile tube that extends above the reactor core. An extension piece

56

N U C L E A R   F U E L S   &   M A T E R I A L S   S P O T L I G H T

Figure 2.

AFIP test assembly.

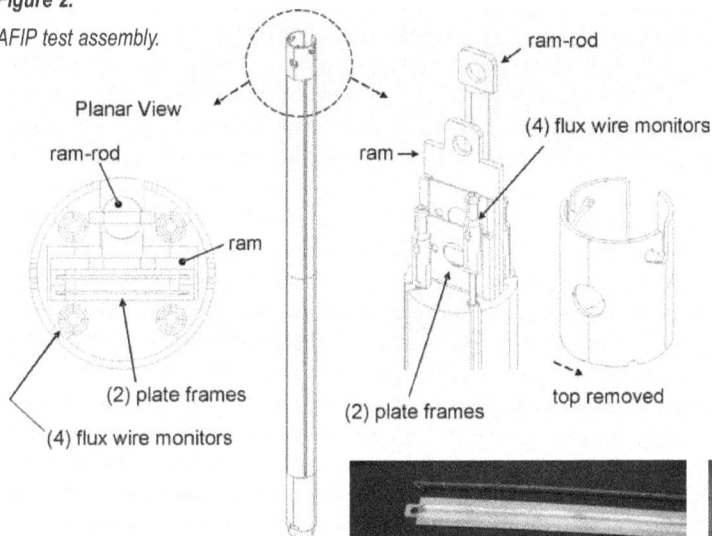

Planar View

ram-rod

ram

(2) plate frames

(4) flux wire monitors

ram-rod

ram

(4) flux wire monitors

(2) plate frames

top removed

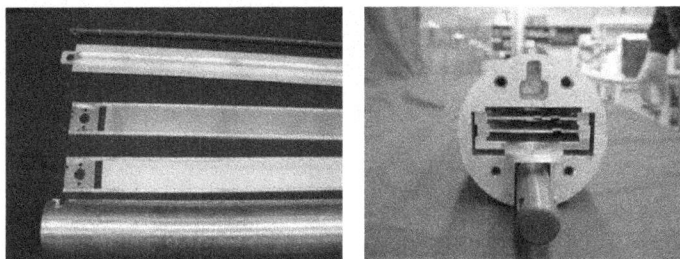

attaches to the top of the experiment hardware to allow loading and unloading of the experiment. The extension also pins the experiment components in place during irradiation.

Figure 3.

Test hardware photographs.

IRRADIATION CONDITIONS

The experiment operating conditions are constrained by the allowable ATR coolant conditions. For ATR experiments, the flow instability ratio (FIR) and departure from nucleate boiling ratios (DNBR) must be larger than 2.0. Hydraulic tests and high fidelity thermal models were developed to establish clearly the experiment-operating envelope. A mock-up of the hardware (including nonfueled dummy plates) was fabricated for the hydraulic tests. The ATR typically operates with a core

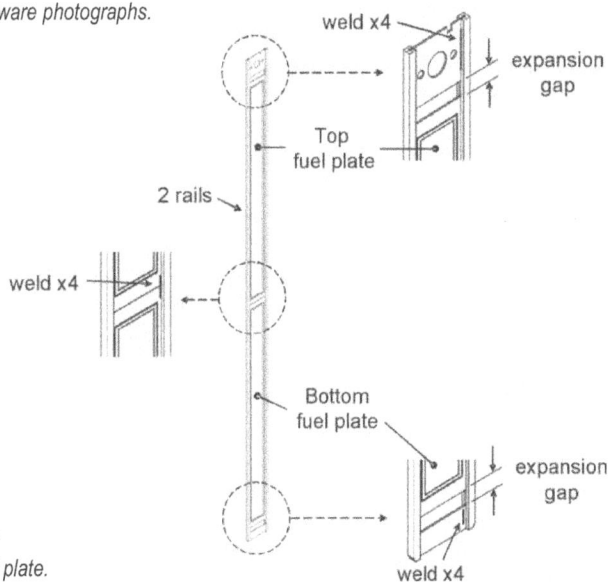

weld x4

expansion gap

Top fuel plate

2 rails

weld x4

Bottom fuel plate

expansion gap

weld x4

Figure 4.

AFIP fuel plate.

57

pressure drop of 77 psi (two-pump mode), where the coolant velocity in the fuel plate channels is roughly 17.4 m/s. The coolant velocity can be increased or decreased by operating in the three-pump mode (100-psi pressure drop) or by inserting an orifice plate at the exit. Thermal analysis confirmed that the experiment could operate at a peak surface-heat flux of at least 500 W/cm^2 without exceeding the FIR or DNBR limits. The fuel plate portion of the thermal model is shown in Figure 5.

IN-CANAL EXAMINATION

The most critical performance measure for nuclear fuels is the ability to maintain a stable geometry during irradiation. The conversion of fissile atoms (typically uranium-235) into two fission fragments inevitably leads to growth of the fuel material during irradiation. All nuclear fuels experience this type of behavior on a microscopic scale. The primary challenge of the fuel developer is to identify materials that accommodate fission fragments in a manner that minimizes swelling, and that are geometrically stable at the macroscopic scale during irradiation. This swelling is typically manifest by growth in the plate thickness as a function of fission density. It is also possible that local differences in fission density (and thus, local swelling) could lead to the formation of significant nonuniformities in stress throughout the plate. These stresses could result in overall plate distortion or even delamination of the interface between the fuel and cladding.

Macroscopic shifts in fuel geometry are routinely evaluated after irradiation. However, periodic nondestructive examination that allows this data to be collected several times during the experiment can enhance the value of an individual experiment substantially. This effectively allows the experimenter to transform a single experiment into the equivalent of several experiments, which translates into substantial time and money savings for large experiments.

UT scanning has long been used to inspect fresh fuel plates before irradiation; thus, techniques to determine both the cladding thickness and level of contact (bonding) between fuel and clad layers have been well established. It is also possible to use UT technology to map the surface profile of the plates. Since the UT scanning requires that the fuel plates be submersed in water initially, implementation of the technique for use on irradiated fuel plates in the ATR spent fuel canal is ideal for interim examination. A UT test stand that can be operated remotely with radiation-hardened detectors was developed at the INL for this purpose (Figure 6).

The in-canal UT scanner was first demonstrated as an integral part of the AFIP-2 experiment on full-size monolithic fuel plates. Although significant work had been successfully performed at the miniplates scale, the experiment provided the first opportunity to evaluate the performance of prototypic scale monolithic uranium-molybdenum fuel plates. The examinations yielded unprecedented dimensional resolution for irradiated fuel. Examinations were conducted at

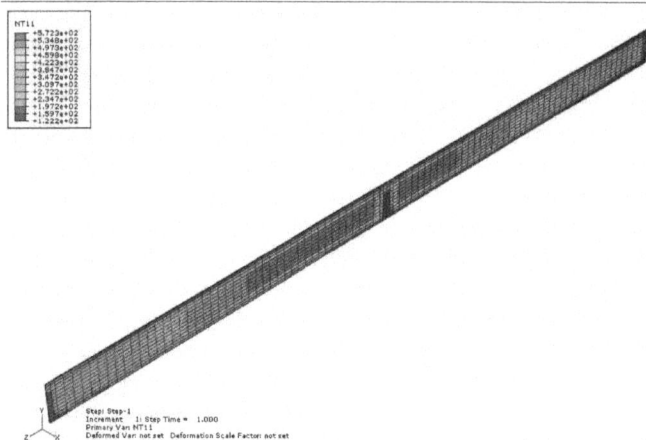

Figure 5.

Fuel surface temperature estimates from thermal analysis.

58

N U C L E A R   F U E L S   &   M A T E R I A L S   S P O T L I G H T

Figure 6.

ATR in-canal UT scanner (prior to installation).

Figure 7.

Thickness change of one AFIP-2 plate after each irradiation cycle. The right edge of the plate was located at the reactor core centerline and was, therefore, at the highest power/fission density. This location is where the maximum swelling condition occurs. (The highest peak on the far right is from the upper plate.)

four points: 1) prior to irradiation, 2) after 35 days of irradiation, 3) after 77 days of irradiation, and 4) after 133 days of irradiation. The results clearly show the relationship between swelling and fission density, as the plate thickness mirrors the plate power profile (Figure 7). The measurements (Figure 8) showed that the plate contours remain relatively unchanged during the irradiation (i.e., plate buckling or warping does not occur). Transmission scans were also completed to reveal the formation of any delaminations (Figure 9). However, the scans identified small regions where void densities were large enough to diminish sound transmission in one of the fuel designs. These indications will be an area of focus during destructive examination in the hot cell.

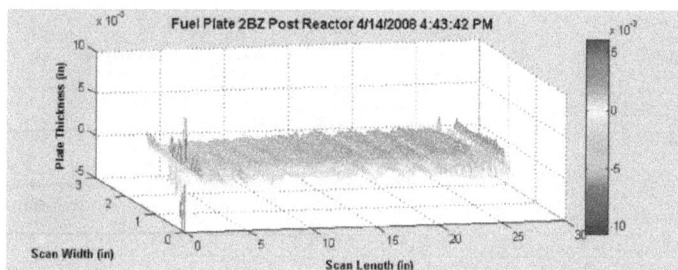

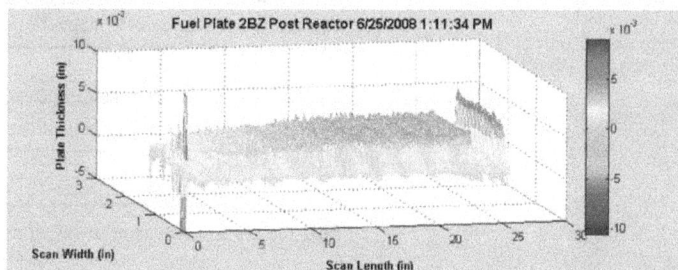

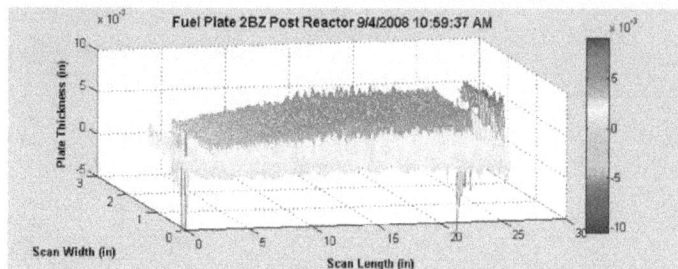

59

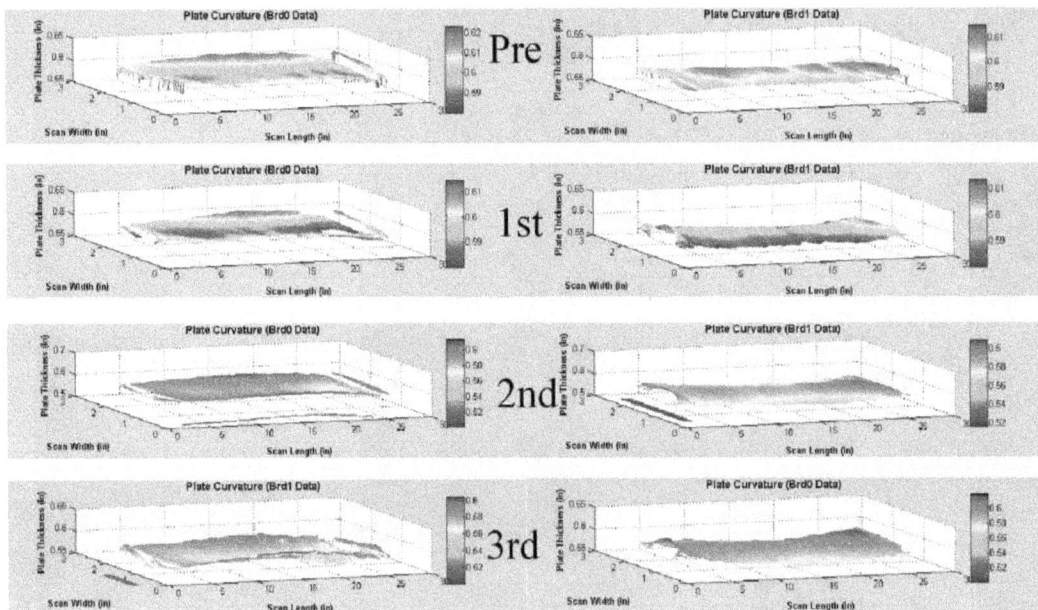

Figure 8.

Surface profiles for one AFIP-2 plate after each irradiation cycle. The images show that the general shape of the plates remains relatively unchanged during irradiation.

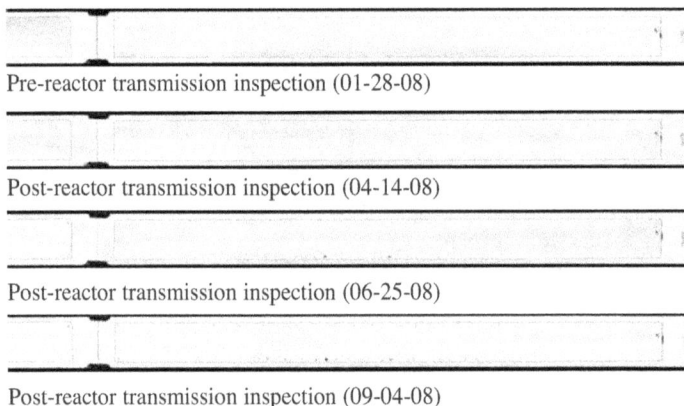

Pre-reactor transmission inspection (01-28-08)

Post-reactor transmission inspection (04-14-08)

Post-reactor transmission inspection (06-25-08)

Post-reactor transmission inspection (09-04-08)

Figure 9.

UT transmission images of one AFIP-2 plate taken from each irradiation cycle. The images show the formation of several small indications in the final cycle.

60

NUCLEAR FUELS & MATERIALS SPOTLIGHT

Conclusions

The AFIP experiment hardware design enables detailed testing of prototypic scale research reactor fuels designs. The flexible design allows the experimenter to configure the test to achieve a wide range of desired irradiation conditions (i.e., nuclear, thermal-hydraulic, and geometric). The test apparatus also enables collection of a substantial amount of performance data previously unavailable to the experimenter, substantially reducing the cost and length of time required to conduct a fuels development program. Ultimately, this experiment design will be the workhorse capability that enables qualification of new research-reactor fuel designs.

DANIEL M. WACHS (BS, 1995, Mechanical Engineering, MS, 1997, Nuclear and Mechanical Engineering, Oregon State University; PhD 2000, Mechanical Engineering, University of Idaho) is a member of the Fuel Performance and Design Department at the INL. He is currently the National Technical Director for the RERTR Fuel Development Program. His current research interests and technical background include irradiation testing of nuclear fuel and materials, thermal-hydraulic testing and analysis, and reactor system component design for research reactors and liquid metal cooled fast reactors.

JARED M. WIGHT (BS, 2005, Mechanical Engineering, Brigham Young University - Idaho) is an engineer in the Fuel Fabrication Department at the INL. He has been actively involved in the project coordination for irradiation testing of research reactor fuels, hardware design and fabrication of irradiation devices, and engineering of devices needed for fabrication of irradiation samples for 4 years.

61

N U C L E A R   F U E L S   &   M A T E R I A L S   S P O T L I G H T

FUEL PERFORMANCE MODELING

Steven L. Hayes and Pavel G. Medvedev

THE INL'S FUEL performance code Plate Lifetime Accurate Thermal Evaluation (PLATE) simulates multiple simultaneous coupled physical phenomena to predict evolution of the fuel temperature and dimensions with burnup. PLATE is used by INL's engineers to analyze fuel behavior during irradiation, and as a testbed for the new fuel performance models, leading to a better understanding of fuel behavior. While the ultimate objective of the RERTR fuel performance modeling effort is to predict the service lifetime of the fuel plates, PLATE represents a critical step towards this goal.

Plate-type fuels are widely used in research and test reactors worldwide. The development of U_3Si_2 as the dispersed fuel phase in an aluminum matrix during the 1980s and the current interest and work on uranium alloys as the fuel phase has resulted in fuel systems that undergo substantial chemical interaction between the fuel and matrix phases during irradiation. Results from the PIE of RERTR fuels indicate that the interaction between the fuel and matrix phases occurs readily during irradiation and is a sensitive function of temperature, as illustrated in Figure 1. As the interaction proceeds, a low conductivity reaction-product phase accumulates, with a corresponding depletion of the high-conductivity aluminum matrix phase. This leads to a substantial degradation of fuel meat thermal conductivity with time, and fuel centerline temperatures can increase substantially with burnup even if plate powers decrease. This strong interrelationship between fuel temperature and fuel-matrix interaction makes the development of a simple empirical correlation between the two difficult, since it is unclear what temperature to employ; without a correlation for interaction thickness, it is impossible to calculate fuel temperatures during irradiation. For this reason, a sophisticated thermal model has been developed to calculate fuel temperatures, taking into account the changing volume fractions of fuel meat constituents, including fuel, matrix, and reaction-product phases within the fuel meat, gas generation/swelling in the fuel and

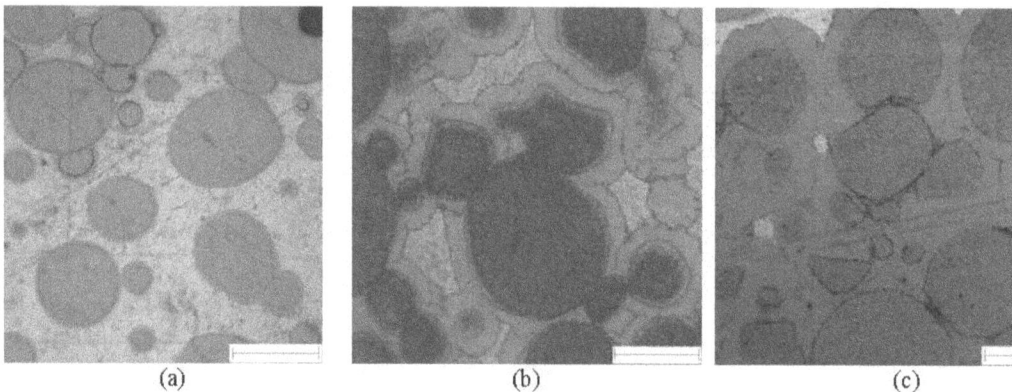

Figure 1.

Increasing U-Mo/Al fuel-matrix interaction with temperature; (a) U-10Mo at 139°C and 30% burnup, (b) U-10Mo at 203°C and 38% burnup, and (c) U-6Mo at 224°C and 40% burnup (scale=100μm).

63

reaction-product phases, and cladding corrosion. Within the framework of this best-estimate temperature calculation, an empirical fuel-matrix interaction-rate correlation has been developed in an integral way. The resulting interaction rate correlation and other associated behavior correlations and models have been implemented within a computer code designated as PLATE, which is now in use to evaluate RERTR fuel plate irradiation performance.

During irradiation, the fuel phase reacts readily with the aluminum matrix. U-Mo/Al fuels loaded at 8 g-U/cm^3 and irradiated at high temperature (>200°C) resulted in some cases of the aluminum matrix phase near the fuel centerline being totally depleted due to this reaction by approximately 40% burnup. The stoichiometry of the reaction product has not yet been determined definitively; however, it now appears to be much more aluminum-rich than previously thought. Though it has not been experimentally determined, it is assumed that this aluminide reaction product has a low thermal conductivity (e.g., UAl_4 has a very low thermal conductivity of ~0.06 W/cm-°C). Thus, the depletion of the high conductivity aluminum matrix phase and the buildup of a significant quantity of low conductivity reaction-product phase leads to a dramatic decrease in fuel meat thermal conductivity (i.e., reduction by a factor of 2 to 3). Since the reaction rate is an Arrhenius function of temperature, this fuel-matrix interaction can drive

an escalating positive feedback thermal response at high temperatures.

Fissions occur within both the fuel and reaction product phases and produce fission product atoms having a larger volume than the fissioned uranium (or plutonium) atoms that produced them. Furthermore, approximately 25% of the fission products are either xenon or krypton gas atoms, some of which coalesce into bubbles over time. These phenomena result in swelling in both the fuel and reaction product phases that increase with burnup. Swelling of the uranium-containing aluminide reaction product in U-Mo/Al fuels is relatively low and does not appear to be very different from the swelling of the reaction product phase in U_3Si_2/Al fuels; the swelling of the uranium-molybdenum alloys as a function of burnup, however, is larger than that observed in U_3Si_2/Al fuels.

Mechanical constraint minimizes dimensional changes in the width and length dimensions of a fuel plate; as a result, fuel plate swelling is essentially manifest as a thickness increase. Because the coolant channels between fuel plates in most research reactors are normally very small, any fuel plate thickness increases that could substantially decrease the coolant channel size are important to predict for safety evaluations. Fuel plate thickness increases are a composite of three phenomena: 1) cladding corrosion and 2) fuel-matrix interaction, which both produce a reaction product of

lower density than the reactants; and 3) the fission product swelling of the uranium-containing fuel and reaction product phases. Fission product swelling is the largest of these components for U-Mo/Al fuels.

The corrosion of the cladding outer surface by the reactor coolant, when extensive, is an important phenomenon to include in fuel modeling. Aluminum cladding corrosion by water results in a reaction product that is a mixture of alumina hydrates, but is often assumed to be either totally or predominantly boehmite ($Al_2O_3 \cdot H_2O$). Boehmite has a low thermal conductivity (~0.002 W/cm-°C), so even relatively thin layers on the cladding surface can lead to significant increases in fuel temperatures; thick boehmite layers (>50μm) can result in fuel temperature increases of 50°C or more.

The fuel performance code PLATE has been developed to model the irradiation performance of high density RERTR fuel plates. Many of its performance predictions are based on empirical correlations developed using data obtained from the extensive fuel testing campaign conducted by the RERTR program during the past decade. However, the heart of the PLATE code is an analytical model for estimating the fuel meat thermal conductivity, as influenced by the changing volume fractions of fuel meat constituents during irradiation. Comparison of PLATE calculations for experimental fuel plates from

64

NUCLEAR FUELS & MATERIALS SPOTLIGHT

various RERTR tests (with measured data obtained during PIE) shows positive overall agreement. Development work on PLATE continues with goals to extend performance correlations outside these parameter ranges, develop more mechanistic irradiation performance models (where possible), extend PLATE to the monolithic RERTR fuels, and couple PLATE with ABAQUS finite element software to analyze the structural integrity of the RERTR fuels.

Analyses made using PLATE have led to a much better understanding of the fabrication and irradiation variables important for acceptable RERTR fuel performance. First, the total fuel surface area should be minimized in order to minimize fuel-matrix interaction. In this regard, spherically shaped fuel particles have an advantage over non-spherical shapes; for any shape, however, the use of small fuel particles (fines) should be avoided. Calculations have also shown that knowledge of the fuel particle size and shape distribution will be necessary to make accurate predictions of fuel plate performance. Second, fuel-matrix interaction is a sensitive function of temperature; thus, efforts should be taken to keep temperatures as low as possible for RERTR fuels. For this reason, cladding surface corrosion must be kept low for this fuel form.

STEVEN L. HAYES (BS, 1988, MS, 1989, PhD, 1992, Nuclear Engineering, Texas A&M University) is Manager of the Fuel Performance and Design Department at the INL. He has been actively involved in the irradiation testing and modeling of fast breeder reactor fuels, fuels for actinide transmutation, and research reactor fuels for more than 15 years.

PAVEL G. MEDVEDEV (PhD, 2004, Nuclear Engineering, Texas A&M University) is a staff member of the Fuel Performance and Design Department at the INL. His interests include nuclear fuel fabrication, testing, and performance.

65

U.S. DEPARTMENT OF ENERGY | Nuclear Energy

The Advanced Fuel Cycle Initiative

Science Based Fuel Cycle Research and Development

Phillip Finck
Idaho National Laboratory

June 9, 2009

U.S. DEPARTMENT OF ENERGY
Nuclear Energy

Former Programmatic Approach

- **Incremental improvement of existing technologies to allow for short-term (~20 years) deployment, driven by better utilization of Yucca Mountain**
  - Specific choice of technologies and integrated system (dictated by time frame and Yucca Mountain characteristics)
  - Challenges were well identified
  - Engineering approaches were chosen to address these challenges
  - Fundamental challenges had also been identified (2006 workshops), but were marginally acted upon (e.g., modeling and simulation)
- **The industrial approach resulted in very limited investment in the tools needed to develop a better understanding of the fundamentals**

2

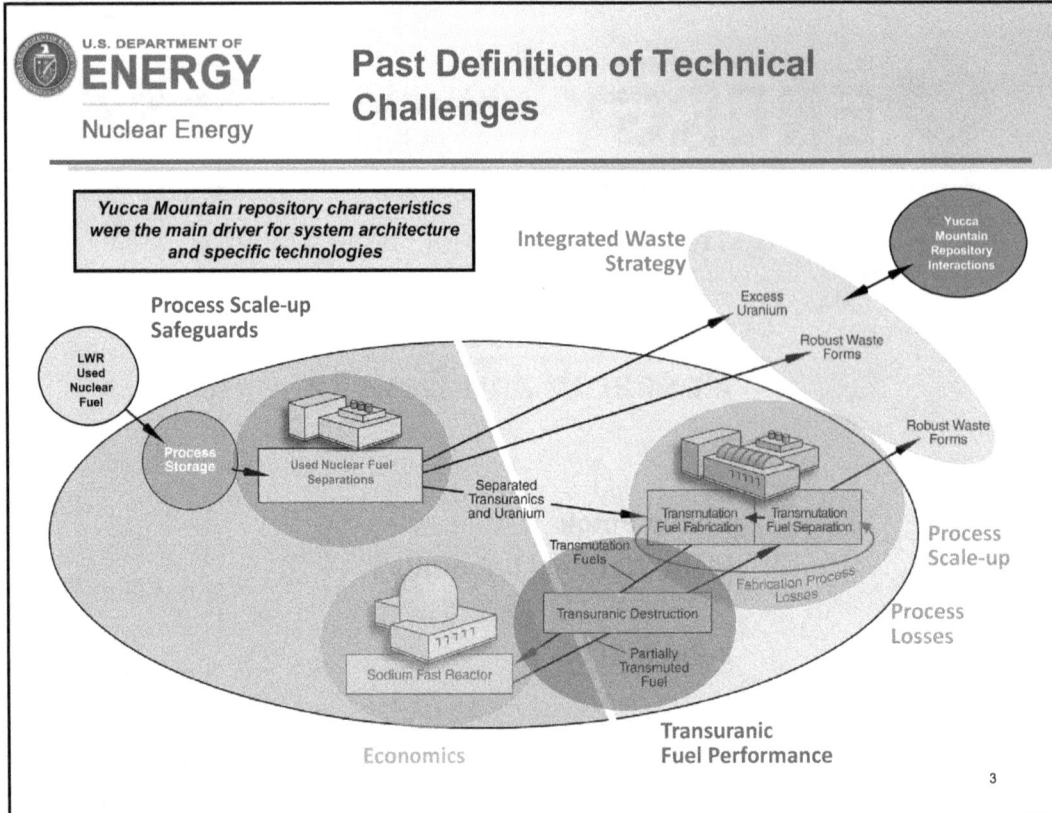

U.S. DEPARTMENT OF
ENERGY
Nuclear Energy

Past Definition of Technical Challenges

Yucca Mountain repository characteristics were the main driver for system architecture and specific technologies

U.S. DEPARTMENT OF
ENERGY
Nuclear Energy

Proposed New Approach

1. **Long term deployment of fuel cycle technologies**

2. **Based on an initial analysis of a broad set of options**

3. **Based on the use of modern science tools and approaches designed to solve challenges and develop better performing technologies**

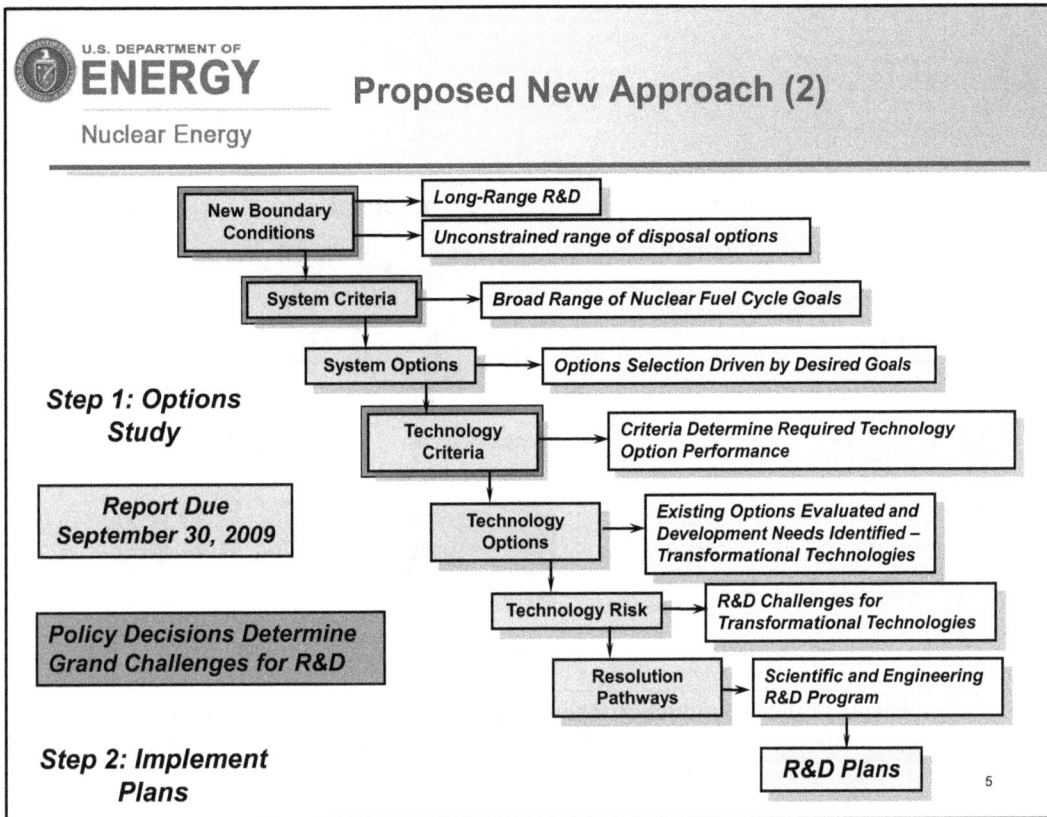

U.S. DEPARTMENT OF
ENERGY
Nuclear Energy

Proposed New Approach (2)

New Boundary Conditions → Long-Range R&D

New Boundary Conditions → Unconstrained range of disposal options

System Criteria → Broad Range of Nuclear Fuel Cycle Goals

System Options → Options Selection Driven by Desired Goals

Technology Criteria → Criteria Determine Required Technology Option Performance

Technology Options → Existing Options Evaluated and Development Needs Identified – Transformational Technologies

Technology Risk → R&D Challenges for Transformational Technologies

Resolution Pathways → Scientific and Engineering R&D Program

R&D Plans

Step 1: Options Study

Report Due September 30, 2009

Policy Decisions Determine Grand Challenges for R&D

Step 2: Implement Plans

5

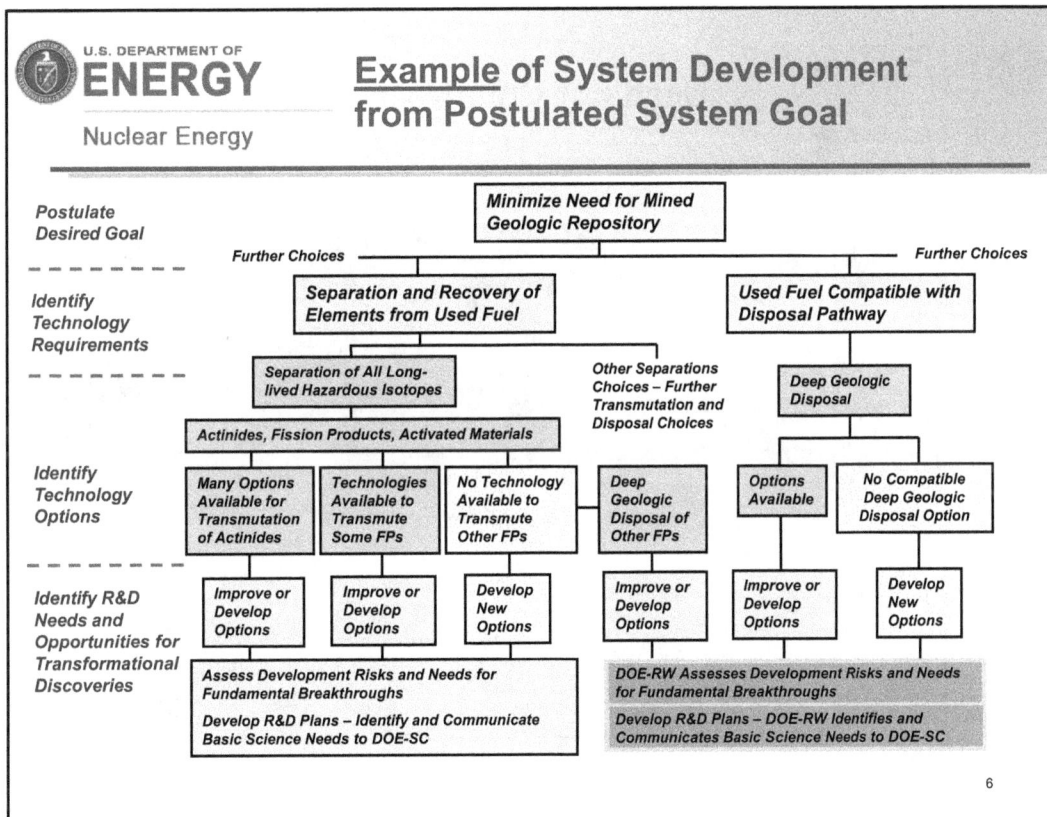

U.S. DEPARTMENT OF
ENERGY
Nuclear Energy

Example of System Development from Postulated System Goal

Postulate Desired Goal

Minimize Need for Mined Geologic Repository

Further Choices — Further Choices

Identify Technology Requirements

Separation and Recovery of Elements from Used Fuel | Used Fuel Compatible with Disposal Pathway

Separation of All Long-lived Hazardous Isotopes | Other Separations Choices – Further Transmutation and Disposal Choices | Deep Geologic Disposal

Actinides, Fission Products, Activated Materials

Identify Technology Options

Many Options Available for Transmutation of Actinides | Technologies Available to Transmute Some FPs | No Technology Available to Transmute Other FPs | Deep Geologic Disposal of Other FPs | Options Available | No Compatible Deep Geologic Disposal Option

Identify R&D Needs and Opportunities for Transformational Discoveries

Improve or Develop Options | Improve or Develop Options | Develop New Options | Improve or Develop Options | Improve or Develop Options | Develop New Options

Assess Development Risks and Needs for Fundamental Breakthroughs | DOE-RW Assesses Development Risks and Needs for Fundamental Breakthroughs

Develop R&D Plans – Identify and Communicate Basic Science Needs to DOE-SC | Develop R&D Plans – DOE-RW Identifies and Communicates Basic Science Needs to DOE-SC

6

Deep Geologic Disposal

U.S. DEPARTMENT OF **ENERGY**
Nuclear Energy

■ **There are many options for deep geologic disposal**

Mined geologic repositories –
Saturated rock
Hard rock – granite, basalt, …
Soft rock – shale, sedimentary
rocks, …
Unsaturated rock
Volcanic tuff
Clay (saturated)
Salt (dry)

Isolation is provided by engineered systems, chemical environment, geologic stability, …

Deep boreholes
Isolation is provided by depth

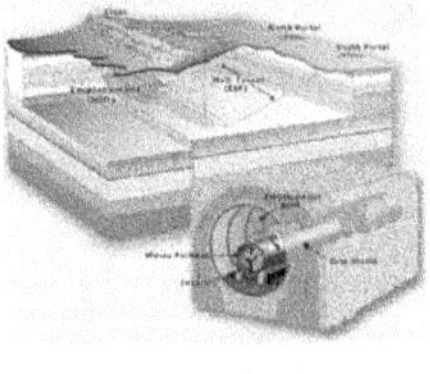

Yucca Mountain 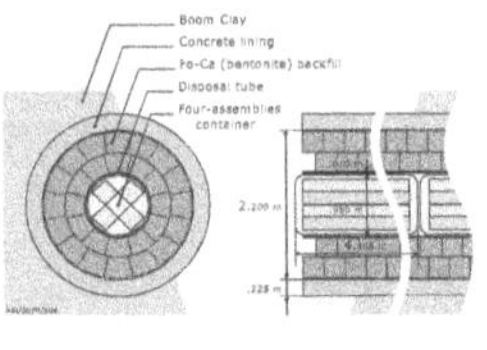   Saturated Clay Layer

Granite 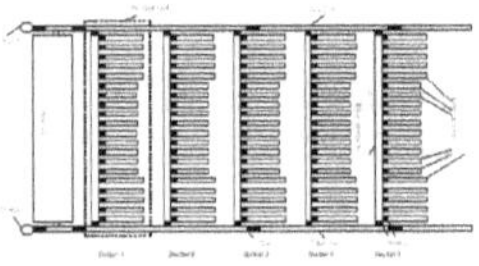   Salt

Seabed/sub-seabed, Subduction Zone, Rock melt, Island (intentional dilution in ocean), Ice sheet, Space, … Many issues - isolation potential, international law, geology, …

7

Need For Creativity

U.S. DEPARTMENT OF **ENERGY**
Nuclear Energy

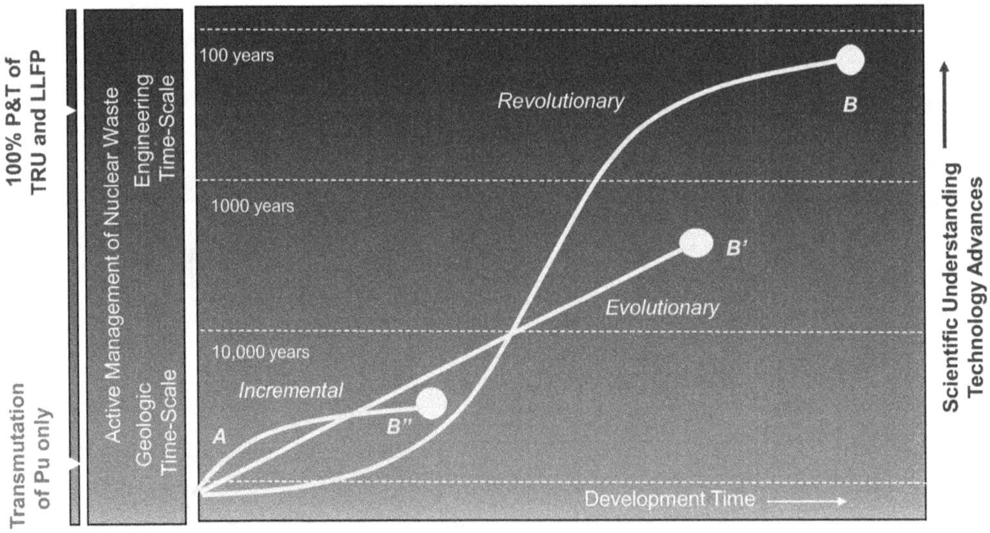

8

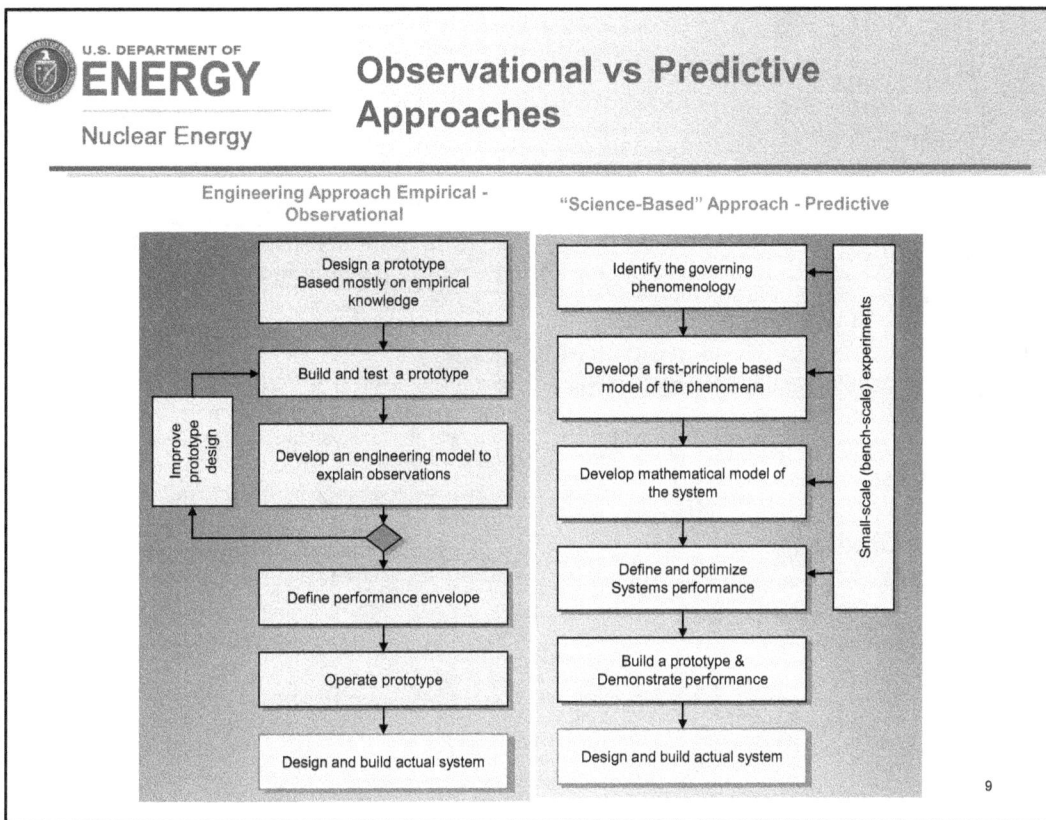

U.S. DEPARTMENT OF **ENERGY**
Nuclear Energy

Observational vs Predictive Approaches

Engineering Approach Empirical - Observational

- Design a prototype Based mostly on empirical knowledge
- Build and test a prototype
- Develop an engineering model to explain observations
- Improve prototype design
- Define performance envelope
- Operate prototype
- Design and build actual system

"Science-Based" Approach - Predictive

- Identify the governing phenomenology
- Develop a first-principle based model of the phenomena
- Develop mathematical model of the system
- Define and optimize Systems performance
- Build a prototype & Demonstrate performance
- Design and build actual system

Small-scale (bench-scale) experiments

9

U.S. DEPARTMENT OF **ENERGY**
Nuclear Energy

Transformational Nuclear Fuels Scientific Research and Development

Today's Technology Challenges

For fuels with variable compositions

- Understanding and predicting fuel behavior and performance
- Reliably fabricating fuel with zero defects and with zero losses

Grand Challenge

- Zero loss and zero defect fuel fabrication
- Ultra-high burnup fuel operation with zero clad-breach

Development Path

Develop a μ-structural understanding of fuels and materials

- Closure of combined transport and phase-field equations
- Separate effect testing and properties measurement at sub-grain scale
- Effect of nano-scale implantations
- Innovative clean and reliable fabrication techniques with tightly controlled microstructures tailored to desired performance

Transformational Result

- Predictive capability for fuel process and in-pile behavior for a variety of initial and boundary conditions
- Novel fuel forms

10

U.S. DEPARTMENT OF ENERGY
Nuclear Energy

Waste Storage and Disposal Scientific Research and Development

Today's Technology Challenges

Storing and disposing spent fuel, HLW, GTCC, and LLW from a range of fuel cycles

- Understanding and predicting geologic repository performance
- Safe, secure, and cost effective storage and disposal

Grand Challenge

Integrated waste management with near zero radionuclide release from storage and disposal system

Development Path

Develop an understanding of geologic repository performance

- Review extensive technical basis developed in the U.S. and internationally over the past several decades
- Explore range of geologic settings, including granite, salt, clay, and tuff, and range of disposal concepts, including shaft-room, ramp-drift, borehole, and shallow land burial
- Investigate storage concepts for a range of waste streams
- Develop an integrated waste management strategy applicable to a range of fuel cycle options

Transformational Result

Predictive capability for performance of storage and disposal options for a range of fuel cycles

11

U.S. DEPARTMENT OF ENERGY
Nuclear Energy

Transmutation Systems Scientific Research and Development

Today's Technology Challenges

- Fast reactors have not been commercially deployed – perception of higher system cost of electricity
- Licensing regime is based on light water reactor technology
- Ability to design and assess other systems

Grand Challenge

- Risk to public health and safety prohibited by inherent safety
- Cost of fast spectrum systems less than current ALWR
- Ability to model new systems

Development Path

Develop key cost reduction features

- Modeling and simulation for optimized design and performance, and safety assurance
- Advanced materials for performance, reliability, longevity, and safety
- Energy conversion innovations for improved efficiency and component cost
- R&D facilities for validation of innovative features and exploration of options

Transformational Result

- Revolutionary improvements in fast spectrum system performance (and cost) to enable transmutation and economic fuel cycle closure
- Novel transmutation systems

12

244 703-739-3790 TCNNaturalGas.com

U.S. DEPARTMENT OF ENERGY
Nuclear Energy

Separations and Waste Form Scientific Research and Development

Today's Technology Challenges

Recycling used nuclear fuel:

- Meeting current air emission requirements
- Economical recovery of transuranic elements for recycle/transmutation
- Minimal waste generation

Grand Challenge

- Near-zero radioactive off-gas emissions
- Simplified, single-step recovery of transuranic elements
- Significantly less waste produced

Development Path

Develop fundamental understanding of separation process and waste form thermodynamics

- Understand underlying separation driving forces
- Exploit thermodynamic properties to effect separations
- Elucidate microstructural waste form corrosion mechanisms

Transformational Result

- Predictive capability for separation and waste form performance over a broad range of operational conditions
- Novel separations technologies

13

U.S. DEPARTMENT OF ENERGY
Nuclear Energy

Materials Protection, Accounting, and Control for Transmutation Scientific Research and Development

Today's Technology Challenges

- Large throughput facilities require shutdown for periodic inventory
- New reactor designs require new nuclear material management approach
- Move from reactive to preventive systems approach

Grand Challenge

Develop online, real-time, continuous, accountability instruments and techniques that permit an order of magnitude improvement in the ability to inventory fissile materials in domestic fuel cycle systems, in order to detect diversion and prevent misuse

Development Path

- **Next generation instrumentation**
  - High sensitivity and specificity
  - Enabled by new physics data
  - New sensor materials
- **Integration of disparate data in quantitative manner**
  - Real time assessments
  - Probability basis with uncertainties
- **Predictive modeling and simulation at atomistic and plant level**

Transformational Result

Real time nuclear materials management with continuous inventory

14

U.S. DEPARTMENT OF ENERGY
Nuclear Energy

Modeling and Simulation
Scientific Research and Development

Today's Technology Challenges

- Theory drives experiment design
- Experiments provide discoveries to drive theory
- Empirically based modeling and simulation heavily dependent on staying close to experimental basis

Grand Challenge

Develop process/methodologies to enable the use of computer simulation in a fundamentally new way for operation, design, and licensing of nuclear systems

Development Path

- Treat simulations as numeric experiments
- Focus on simulating physics vs characterizing specific devices
- Numerically solve governing equations of motion in detailed 3-D grids
- Carry out simulations prior to experiments
- Leverage massive computing power (petascale) + HPC expertise
- Combine single-effects validation to infer behavior of integrated systems

Transformational Result

Modeling and simulation tools that are based on fundamental understanding of physical processes and capable of predicting performance of fuel cycle technologies

15

U.S. DEPARTMENT OF ENERGY
Nuclear Energy

Example: Mixing in upper plenum

- **New CFD TH-validation experiment**
- **BG/P simulations supported through separate INCITE award**
- **Comparing LES & RANS results**
  - *First experimental data this summer*

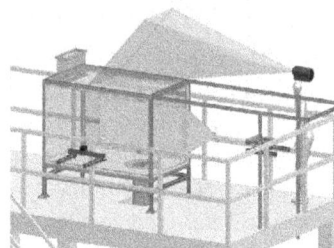

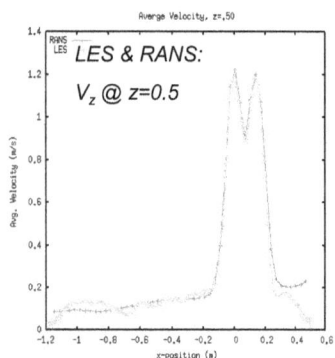

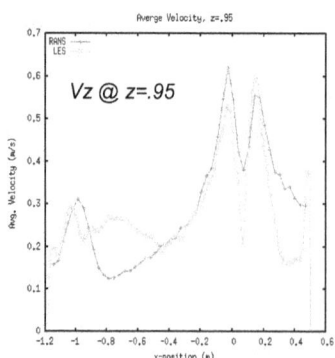

Average Velocity, z=.50

LES & RANS:

V_z @ z=0.5

Average Velocity, z=.95

Vz @ z=.95

16

247

Nuclear Fuel Reprocessing: U.S. Policy Development

Anthony Andrews
Specialist in Energy and Energy Infrastructure Policy

March 27, 2008

Congressional Research Service

7-5700

www.crs.gov

RS22542

CRS Report for Congress
Prepared for Members and Committees of Congress

Summary

As part of the World War II effort to develop the atomic bomb, reprocessing technology was developed to chemically separate and recover fissionable plutonium from irradiated nuclear fuel. In the early stage of commercial nuclear power, reprocessing was thought essential to supplying nuclear fuel. Federally sponsored breeder reactor development included research into advanced reprocessing technology. Several commercial interests in reprocessing foundered due to economic, technical, and regulatory issues. President Carter terminated federal support for reprocessing in an attempt to limit the proliferation of nuclear weapons material. Reprocessing for nuclear weapons production ceased shortly after the Cold War ended. The Department of Energy now proposes a new generation of "proliferation-resistant" reactor and reprocessing technology.

*R*eprocessing refers to the chemical separation of fissionable uranium and plutonium from irradiated nuclear fuel. The World War II-era Manhattan Project developed reprocessing technology in the effort to build the first atomic bomb. With the development of commercial nuclear power after the war, reprocessing was considered necessary because of a perceived scarcity of uranium. Breeder reactor technology, which transmutes non-fissionable uranium into fissionable plutonium and thus produces more fuel than consumed, was envisioned as a promising solution to extending the nuclear fuel supply. Commercial reprocessing attempts, however, encountered technical, economic, and regulatory problems. In response to concern that reprocessing contributed to the proliferation of nuclear weapons, President Carter terminated federal support for commercial reprocessing. Reprocessing for defense purposes continued, however, until the Soviet Union's collapse brought an end to the Cold War and the production of nuclear weapons. The Department of Energy's latest initiative to promote new reactor technology using "proliferation-resistant" reprocessed fuel raises significant funding and policy issues for Congress. U.S. policies that have authorized and discouraged nuclear reprocessing are summarized below.

1946

The Atomic Energy Act of 1946 (P.L. 79-585) defined *fissionable materials* to include plutonium, uranium-235, and other materials determined to be capable of releasing substantial quantities of energy through nuclear fission.[1] The act also created the Atomic Energy Commission (AEC) and transferred production and control of fissionable materials from the Manhattan Project. As the exclusive producer of fissionable material, the AEC originally retained title to all such material for national security reasons.

1954

Congress amended the Atomic Energy Act, authorized the AEC to license commercial reactors, and eased restrictions on private companies using special nuclear material (fissionable material). Section 183 (Terms of Licenses) of the act, however, kept government title to all special nuclear material utilized or produced by the licensed facilities in the United States.

1956

Lewis Strauss, then chairman of the AEC, announced a program to encourage private industry's entry into reprocessing spent nuclear fuel.[2]

1957

The AEC expressed its intent to withdraw from providing nuclear reprocessing services for spent nuclear fuel in a *Federal Register* notice of March 22, 1957.

---

[1] In the amended Atomic Energy Act of 1954 (P.L. 83-703), the term *special nuclear material* superseded the term *fissionable material* and included uranium enriched in isotope 233, material the AEC determined to be special nuclear material, or any artificially enriched material. Laws of 83rd Congress, 2nd Session, 1118-21.

[2] U.S. House of Representatives, Committee on Science and Technology, Subcommittee on Investigations and Oversight, West Valley Cooperative Agreement Hearing, p. 233, July 9, 1981.

1959

The Davison Chemical Company, later called Nuclear Fuel Services, began extensive discussions with the AEC on commercial reprocessing.

1963

The AEC-sponsored Experimental Breeder Reactor (EBR II), constructed at the Argonne National Laboratory West near Idaho Falls, began operating. Irradiated fuel was reprocessed by "melt-refining."

1964

The AEC was authorized to issue commercial licenses to possess special nuclear material subject to specific licensing conditions (P.L. 88-489).

1966

The AEC granted an operating permit for commercial reprocessing to Nuclear Fuel Services for the West Valley plant, near Buffalo, NY. The plant operated from 1966 until 1972, reprocessing spent fuel from the defense weapons program.[3] Commercial spent fuel was never reprocessed. Stricter regulatory requirements forced the plant's shutdown for upgrades. The plant was permanently shut down in 1976 after it was determined that the stricter regulatory requirements could not be met.[4]

1967

The AEC authorized General Electric Company (GE) to construct a spent fuel reprocessing facility in Morris, IL.[5]

1969

The AEC invited public comment on a proposed policy in the form of Appendix F to 10 C.F.R. Part 50 on siting a fuel reprocessing plant.[6]

1969

EBRII fuel reprocessing and refabrication operations were suspended.

---

[3] U.S. Department of Energy, Plutonium Recovery from Spent Fuel Reprocessing by Nuclear Fuel Services at West Valley, New York from 1966 to 1972, February 1996, http://www.osti.gov/opennet/document/purecov/nfsrepo.html.

[4] Congressional Budget Office, *Nuclear Reprocessing and Proliferation: Alternative Approaches and their Implications for the Federal Budget*, May 1977.

[5] 65 *Federal Register* 62766-62767, October 19, 2000: *General Electric Company, Morris Operation; Notice of Docketing, Notice of Consideration of Issuance, and Notice of Opportunity for a Hearing for the Renewal of Materials License SNM-2500 for the Morris Operation Independent Spent Fuel Storage Installation.*

[6] *Federal Register*, June 3, 1969.

1970

Allied-General Nuclear Services began constructing a large commercial reprocessing plant at Barnwell, SC.

1972

GE halted construction and decided not to pursue an operating license for its Morris reprocessing facility. Instead, GE applied for and received a license to store spent fuel.[7]

1974

The AEC determined that any decision to permit nuclear fuel reprocessing on a large scale would require an environmental impact statement under Section 101(2)(c) of the National Environmental Policy Act (U.S.C. 4332(2)(c)).

1974

The Energy Reorganization Act (P.L. 93-438), October 11, 1974, split the AEC into the Nuclear Regulatory Commission (NRC) and the Energy Research and Development Administration (ERDA). The responsibility for licensing nuclear facilities was transferred to the NRC.

1976

Exxon applied for a license to construct a large reprocessing plant but received no final action on its license application.

1976

In an October 28 nuclear policy statement, President Ford announced his decision that

> the reprocessing and recycling of plutonium should not proceed unless there is sound reason to conclude that the world community can effectively overcome the associated risks of proliferation ... that the United States should no longer regard reprocessing of used nuclear fuel to produce plutonium as a necessary and inevitable step in the nuclear fuel cycle, and that we should pursue reprocessing and recycling in the future only if they are found to be consistent with our international objectives.[8]

With that announcement, agencies of the executive branch were directed to delay commercialization of reprocessing activities in the United States until uncertainties were resolved.

---

[7] Comptroller General, *Federal Facilities for Storing Spent Nuclear Fuel—Are they Needed?* June 27, 1979.

[8] Gerald R. Ford Presidential Documents, vol. 12, no. 44, pp. 1626-1627, 1976.

1977

In an April 7 press statement, President Carter announced, "We will defer indefinitely the commercial reprocessing and recycling of plutonium produced in the U.S. nuclear power programs."[9] He went on to say, "The plant at Barnwell, South Carolina, will receive neither federal encouragement nor funding for its completion as a reprocessing facility." (It was actually Carter's veto of S. 1811, the ERDA Authorization Act of 1978, that prevented the legislative authorization necessary for constructing a breeder reactor and a reprocessing facility.)[10]

1977

The NRC issued an order terminating the proceedings on the Generic Environmental Statement on Mixed Oxide Fuel and most license proceedings relating to plutonium recycling.[11] It stated, however, that it would reexamine this decision after two studies of alternative fuel cycles were completed.

1978

The Nuclear Nonproliferation Act (P.L. 95-242), March 10, 1978, amended the Atomic Energy Act of 1954 to establish export licensing criteria that govern peaceful nuclear exports by the United States, including a requirement of prior U.S. approval for re-transfers and reprocessing; and a guaranty that no material re-transferred will be reprocessed without prior U.S. consent.

1980

President Carter signed Executive Order 12193, Nuclear Cooperation With EURATOM (45 *Federal Register* 9885, February 14, 1980), which permitted nuclear cooperation with the European Atomic Energy Community (EURATOM) to continue to March 10, 1981, despite the agreement's lack of a provision consistent with the intent of the Nuclear Nonproliferation Act requiring prior U.S. approval for reprocessing. This cooperation was extended through December 31, 1995, by a series of executive orders.[12] It has since expired and been replaced by a new agreement.

1981

President Reagan announced he was "lifting the indefinite ban which previous administrations placed on commercial reprocessing activities in the United States."[13]

---

[9] Jimmy Carter Library, Records of the Speech Writer's Office, *Statement on Nuclear Power Policy*, April 7, 1977.

[10] J. Michael Martinez, *The Journal of Policy History*, "The Carter Administration and the Evolution of American Nuclear Nonproliferation Policy, 1977—1981," vol. 14, no. 3, 2002.

[11] Allied-General Nuclear Services v. United States, no. 87-1902 in the Supreme Court of the United States, Petition for Writ of Certiorari, October term, 1988.

[12] EO 12295, February 24, 1981; EO 12351, March 9, 1982; EO 12409, March 7, 1983; EO 12463, February 23, 1984; EO 12506, March 4, 1985; EO 12554, February 28, 1986; EO 12587, March 9, 1987; EO 12629, March 9, 1988; EO 12670, March 9, 1989; EO 12706, March 9, 1990; EO 12753, March 8, 1991; EO 12791, March 9, 1992; EO 12840, March 9, 1993; EO 12902, March 8, 1994; EO 12955, March 9, 1995.

[13] "Announcing a Series of Policy Initiatives on Nuclear Energy," Pub. Papers 903 (October 8, 1981) in *Allied-General* (continued...)

1981

Convinced that the project could not proceed on a private basis and that reprocessing was commercially impracticable, Allied halted the Barnwell project.[14]

1982

President Reagan approved the *United States Policy on Foreign Reprocessing of Plutonium Subject to U.S. Control* as National Security Decision Directive 39 (June 4, 1982). Although specific details of the directive have not been declassified, the policies approved pertain to the nonproliferation and statutory conditions for safeguards and physical security for a continued commitment by Japan to nonproliferation efforts.

1990

In the National Defense Authorization Act for Fiscal Year 1991 (P.L. 101-510, Sec. 3142), Congress declared under Findings and Declaration of Policy that

> [a]t the present time, the United States is observing a de facto moratorium on the production of fissile materials, with no production of highly enriched uranium for nuclear weapons since 1964. While the United States has ceased operation of all of its reactors used for the production of plutonium for nuclear weapons, the Soviet Union currently operates as many as nine reactors for the production of plutonium for nuclear weapons." Also, under Sec. 3143—Bilateral Moratorium on Production of Plutonium and Highly Enriched Uranium for Nuclear Weapons and Disposal of Nuclear Stockpiles, the law urged "an end by both the United States and the Soviet Union to the production of plutonium and highly enriched uranium for nuclear weapons.

(In its fullest sense, plutonium production implies reprocessing.)

1992

President G. H. W. Bush disapproved Long Island Power Authority's attempt to enter into a contract with the French firm Cogema to reprocess the slightly irradiated initial core from the decommissioned Shoreham reactor.

1992

President G. H. W. Bush halted weapons reprocessing in a policy statement on nuclear nonproliferation declaring: "I have set forth today a set of principles to guide our nonproliferation efforts in the years ahead and directed a number of steps to supplement our existing efforts. These steps include a decision not to produce plutonium and highly enriched uranium for nuclear explosive purposes...."[15]

(...continued)

Nuclear Services v. United States.

[14] Letter to James B. Edwards, Secretary of Energy, from Brian D. Force, Allied Corporation, October 15, 1981.

[15] President's statement, The White House, Office of the Press Secretary, Washington, DC, July 13, 1992.

1992

Energy Secretary Watkins announced the permanent closure of the Hanford, WA, PUREX reprocessing plant in December.

1993

President Clinton issued a policy statement on reprocessing stating that "[t]he United States does not encourage the civil use of plutonium and, accordingly, does not itself engage in plutonium reprocessing for either nuclear power or nuclear explosive purposes. The United States, however, will maintain its existing commitments regarding the use of plutonium in civil nuclear programs in Western Europe and Japan."[16]

1995

On November 29, 1995, a new nuclear cooperation agreement with EURATOM was submitted to Congress. Although the Clinton Administration determined it met all the requirements of Section 123 a. of the Atomic Energy Act, some Members believed it did not meet the requirement of prior consent for reprocessing. The agreement entered into effect in 1996 without a vote.

2001

President Bush's National Energy Policy included the recommendation that "[t]he United States should also consider technologies (in collaboration with international partners with highly developed fuel cycles and a record of close cooperation) to develop reprocessing and fuel treatment technologies that are cleaner, more efficient, less waste intensive, and more proliferation-resistant."[17]

2006

As part of the ongoing Advanced Fuel Cycle Initiative (AFCI), the Department of Energy announced that it will initiate work toward conducting an engineering scale demonstration of the UREX+ separation process (operation planned for 2011) and developing an advanced fuel cycle facility capable of laboratory development of advanced separation and fuel manufacturing technologies. UREX refers to the process of chemically separating uranium from spent nuclear fuel. The AFCI is intended to develop proliferation resistant nuclear technologies in association with the Global Nuclear Energy Partnership (GNEP) for expanding nuclear power in the United States and around the world. The Department of Energy later requested an expression of interest from domestic and international industry in building a spent nuclear fuel recycling and transmutation facility that would meet GNEP goals.[18]

---

[16] Fact Sheet—Nonproliferation And Export Control Policy, The White House, Office of the Press Secretary, September 27, 1993.

[17] Report of the National Energy Policy Development Group, May 2001.

[18] 71 *Federal Register* 44673-44676, August 7, 2006, *Notice of Request for Expression of Interest in a Consolidated Fuel Treatment Center to Support the Global Nuclear Energy Partnership.*

2007

In July 2007, DOE announced that four consortia had been selected to receive up to $16 million for technical and supporting studies to support GNEP (AREVA Federal Services, LLC; EnergySolutions, LLC; GE-Hitachi Nuclear Americas, LLC; and General Atomics). DOE followed with an August announcement that it was making $20 million available to conduct detailed siting studies for public or commercial entities interested in hosting GNEP facilities. The original GNEP partnership—China, France, Japan, Russia and the United States—expanded to include Australia, Bulgaria, Ghana, Hungary, Jordan, Kazakhstan, Lithuania, Poland, Romania, Slovenia, Ukraine, South Korea, Italy, Canada, and Senegal by year end.

Author Contact Information

Anthony Andrews
Specialist in Energy and Energy Infrastructure
Policy
aandrews@crs.loc.gov, 7-6843

INL/MIS-08-14918
Revision 2

Light Water Reactor Sustainability Research and Development Program Plan

Fiscal Year 2009–2013

December 2009

DOE Office of Nuclear Energy

DISCLAIMER

This information was prepared as an account of work sponsored by an agency of the U.S. Government. Neither the U.S. Government nor any agency thereof, nor any of their employees, makes any warranty, expressed or implied, or assumes any legal liability or responsibility for the accuracy, completeness, or usefulness, of any information, apparatus, product, or process disclosed, or represents that its use would not infringe privately owned rights. References herein to any specific commercial product, process, or service by trade name, trade mark, manufacturer, or otherwise, do not necessarily constitute or imply its endorsement, recommendation, or favoring by the U.S. Government or any agency thereof. The views and opinions of authors expressed herein do not necessarily state or reflect those of the U.S. Government or any agency thereof.

INL/MIS-08-14918
Revision 2

**Light Water Reactor Sustainability Research and Development
Program Plan**

Fiscal Year 2009–2013

2009

**Idaho National Laboratory
Idaho Falls, Idaho 83415**

http://www.inl.gov

**Prepared for the
U.S. Department of Energy
Office of Nuclear Energy
Under DOE Idaho Operations Office
Contract DE-AC07-05ID14517**

EXECUTIVE SUMMARY

Nuclear power has reliably and economically contributed almost 20% of electrical generation in the United States over the past two decades. It remains the single largest contributor (more than 70%) of non-greenhouse-gas-emitting electric power generation in the United States.

By the year 2030, domestic demand for electrical energy is expected to grow to levels of 16 to 36% higher than 2007 levels. At the same time, most currently operating nuclear power plants will begin reaching the end of their 60-year operating licenses. Figure E-1 shows projected nuclear energy contribution to the domestic generating capacity. If current operating nuclear power plants do not operate beyond 60 years, the total fraction of generated electrical energy from nuclear power will begin to decline—even with the expected addition of new nuclear generating capacity.

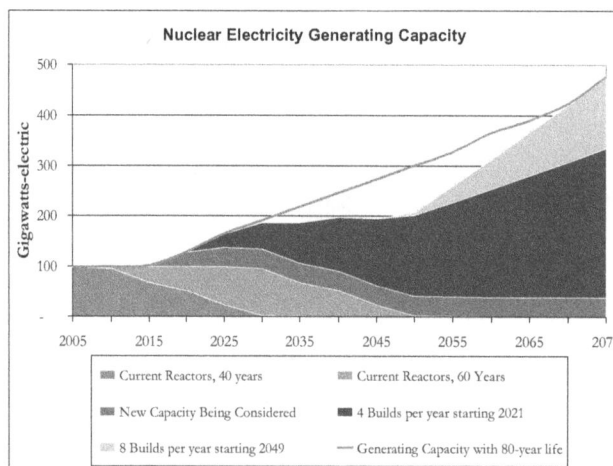

The red line represents the total generating capacity of current and planned nuclear power plants, assuming extended operation to 80 years. The unshaded area below the line represents lost capacity if the current nuclear power plant fleet is decommissioned after 60 years.

Figure E-1. Projected nuclear power generation.

The oldest commercial plants in the United States reached their 40th anniversary this year. U.S. regulators have begun considering extended operations of nuclear power plants and the research needed to support long-term operations. The Light Water Reactor Sustainability (LWRS) Research and Development (R&D) Program, developed and sponsored by the Department of Energy, is performed in close collaboration with industry R&D programs. The purpose of the LWRS R&D Program is to provide technical foundations for licensing and managing long-term, safe and economical operation of the current operating nuclear power plants.

The LWRS R&D Program vision is captured in the following statements:

iii

Existing operating nuclear power plants will continue to safely provide clean and economic electricity well beyond their first license-extension period, significantly contributing to reduction of United States and global carbon emissions, enhancement of national energy security, and protection of the environment.

There is a comprehensive technical basis for licensing and managing the long-term, safe, economical operation of nuclear power plants. Sustaining the existing operating U.S. fleet also will improve its international engagement and leadership on nuclear safety and security issues.

The following five R&D pathways have been identified to achieve the program vision:

Nuclear Materials Aging and Degradation. Research to develop the scientific basis for understanding and predicting long-term environmental degradation behavior of materials in nuclear power plants. Provide data and methods to assess performance of systems, structures, and components essential to safe and sustained nuclear power plant operation.

Advanced LWR Nuclear Fuel Development. Improve scientific knowledge basis for understanding and predicting fundamental nuclear fuel and cladding performance in nuclear power plants. Apply this information to development of high-performance, high burn-up fuels with improved safety, cladding integrity, and improved nuclear fuel cycle economics.

Advanced Instrumentation, Information, and Control Systems Technologies. Address long-term aging and obsolescence of instrumentation and control technologies and develop and test new information and control technologies. Develop advanced condition monitoring technologies for more automated and reliable plant operation

Risk-Informed Safety Margin Characterization. Bring together risk-informed, performance-based methodologies with scientific understanding of critical phenomenological conditions and deterministic predictions of nuclear power plant performance, leading to an integrated characterization of public safety margins in an optimization of nuclear safety, plant performance, and long-term asset management.

Economics and Efficiency Improvement. Improve economics and efficiency of the current fleet of reactors while maintaining excellent safety performance. Develop methodologies and scientific basis to enable more extended power uprates or ultra high power uprates. Improve thermal efficiency by developing advanced cooling technologies to minimize water usage. Study the feasibility of expanding the current fleet into nonelectric applications.

With the 60-year licenses beginning to expire between the years 2029 and 2039, utilities are likely to initiate planning baseload replacement power by 2014 or earlier. Research for addressing nuclear power plant aging questions must start now and is likely to extend through 2029. The LWRS R&D Program represents the timely collaborative research needed to retain the existing nuclear power plant infrastructure in the United States.

iv

CONTENTS

EXECUTIVE SUMMARY iii

ACRONYMS ix

1. PURPOSE OF THE PROGRAM 1

    1.1 Introduction 1

    1.2 Sustainability 5

    1.3 Critical Path for Nuclear Power Plants..6

2. DESCRIPTION 7

    2.1 Vision 7

    2.2 Program Goals 8

         2.2.1 Scientific Basis..8
         2.2.2 Economic Viability 8

    2.3 Implementing Strategy 9

3. RESEARCH AND DEVELOPMENT PATHWAYS 11

    3.1 Nuclear Materials Aging and Degradation...11

         3.1.1 Background and Introduction...11
         3.1.2 Vision and Goals 12
         3.1.3 Highlights of Research and Development..13
         3.1.4 Products and Implementation Plan..14

    3.2 Advanced Light Water Reactor Nuclear Fuel Development 16

         3.2.1 Background and Introduction...16
         3.2.2 Vision and Goals 16
         3.2.3 Highlights of Research and Development..16
         3.2.4 Products and Implementation Plan..18

    3.3 Advanced Instrumentation, Information, and Control Systems Technologies...................20

         3.3.1 Background and Introduction...20
         3.3.2 Vision and Goals 21
         3.3.3 Highlights of Research and Development..21
         3.3.4 Products and Implementation Plan..23

    3.4 Risk-Informed Safety Margin Characterization 24

         3.4.1 Background and Introduction...24

v

FIGURES

vii

TABLES

viii

ACRONYMS

CO_2	carbon dioxide
DOE	U.S. Department of Energy
DOE-ID	U.S. Department of Energy-Idaho Office
DSA	deterministic safety analysis
EPRI	Electric Power Research Institute
FY	fiscal year
II&C	instrumentation, information, and control
INL	Idaho National Laboratory
LWR	light water reactor
LWRS	light water reactor sustainability
NRC	U.S. Nuclear Regulatory Commission
PRA	probabilistic risk analysis
R&D	research and development
RISMC	risk-informed safety margin characterization
SiC	silicon carbide
SiC/SiC_f	silicon carbide/silicon carbide fiber (reinforced)
SSC	systems, structures, and components
TIO	Technical Integration Office

ix

Light Water Reactor Sustainability Research and Development Program Plan

1. PURPOSE OF THE PROGRAM

1.1 Introduction

The electric energy sector is entering a time of serious challenge and tremendous opportunity. Expanding energy demand and a growing awareness of the environmental impact caused by various forms of electricity generation have prompted debate on how best to achieve a sustainable, affordable, and environmentally sensitive energy solution. Nuclear power is integral to meeting that objective.

The Light Water Reactor Sustainability (LWRS) Program is a research and development (R&D) program sponsored by the U. S. Department of Energy (DOE), performed in close collaboration with industry R&D programs, with the aim to provide technical foundations for licensing and managing the long-term safe and economical operation of current nuclear power plants. DOE's program focus is on the longer term and higher risk/reward research that contributes to the national policy objectives of energy security and reduction of carbon dioxide (CO_2) emissions.

Electric power is a vital component of the nation's economy and way of life. As the energy needs of the United States grow over the coming decades, the national energy supply faces growing pressures on global and domestic scales. In 2006, 70% of domestic electricity generation relied on fossil fuels. Greenhouse gas emissions from these fossil fuels are a mounting problem that threatens the future production of electricity from coal and natural gas. President Obama has called for a reduction of CO_2 emissions to the 1990 levels by the year 2020, with a further 80% reduction by the year 2050. Meeting these aggressive goals while gradually increasing the overall energy supply requires that all nonemitting technologies must be advanced.

Nuclear power is the largest contributor of non-greenhouse-gas-emitting electric power generation, comprising nearly three-quarters of the nonemitting sources as shown in Figure 1-1. Energy efficiency and carbon storage are expected to play increasing roles in providing clean, reliable energy; however, nuclear power will be depended on well into the 21st century for a large-scale supply of dependable clean electricity.

Other forms of low CO_2-emitting and renewable energy production also have the potential to produce environmentally friendly energy. Among the most promising forms of energy production are hydroelectric, wind, geothermal, and solar power. Hydroelectric power is the most widely used renewable energy in the United States; however, there is limited opportunity for expansion. Wind, geothermal, and solar power have demonstrated promise in producing

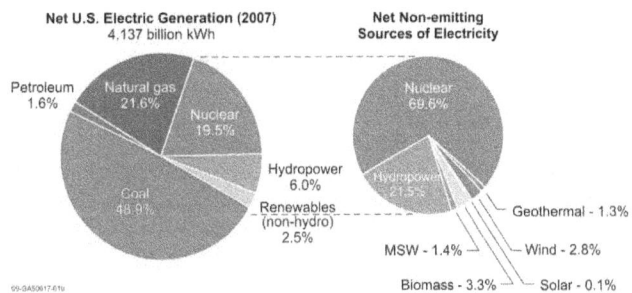

Figure 1-1. Current electric generating portfolio showing dominance of nuclear as low carbon emission power source.

environmentally friendly energy to meet the nation's growing demand. These sources of power have been

1

deployed only recently and they currently contribute only a small fraction of the nation's rapidly growing energy demands. In addition, wind and solar power, by nature, are dilute with low power density and a low capacity factor. Figure 1-2 provides a graph of current capacity factors by energy source. The very high capacity factor for nuclear power makes it the only reliable and nearly non-CO_2-emitting source of baseload power available.

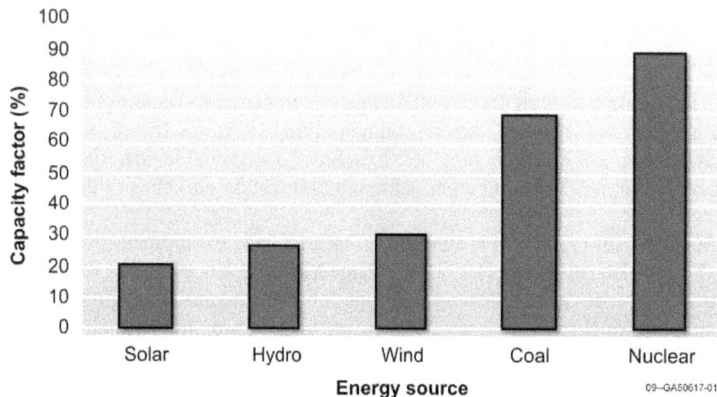

Figure 1-2. United States electrical generation capacity factors by energy source showing high operating performance.

The National Energy Policy Act of 2005 established and authorized the DOE's Nuclear Power 2010 Program to stimulate construction of new nuclear power plants with demonstration of streamlined but unproven licensing processes and facilitating "first mover" new nuclear power plants. Construction of new nuclear power plants is a clear option for new, emission-free, electrical generating capacity.

As of July 2009, 18 combined operating license applications have been submitted to construct 28 new nuclear power plants.[a] Over 30 proposed nuclear power plants currently are in the planning or licensing stage, making it clear that new plant construction is an option that is being pursued seriously. However, bringing new nuclear power plants online is facing substantial challenges and uncertainties, including formidable high capital cost, high financing cost, long construction time, limitations in domestic fabrication capacity, and small market values. It is anticipated that there will be a modest pace of construction of new nuclear power plants. Only a fraction of the planned new nuclear power plants might be built, as evidenced by the fact that, as of today, no utility has committed to constructing a new advanced reactor. Each nuclear power plant currently under consideration is expected to be capable of producing between 1.1 and 1.7 GWe, depending on design.

On the other hand, 104 nuclear power plants currently operate in 31 states (Figure 1-3). The existing, operating fleet of U.S. nuclear power plants has consistently maintained outstanding levels of nuclear safety, reliability, and operational performance over the last two decades and operates with an average capacity factor above 90%, far superior to the 70% capacity factor a decade ago.[b] This significant improvement in performance has made nuclear power plants considerably more economical to operate. Major improvements were made in all areas of plant operation, including operations, training, equipment maintenance and reliability, technological improvements, and improved understanding of component degradation. More broadly, these improvements reflect effective management practices, advances in technology, and the sharing of safety and operational experience. Today, nuclear production costs are the lowest among major U.S. power-generating options.

---

[a] U.S. Department of Energy Nuclear Power 2010, *Nuclear Power Deployment Scorecard*, http://nuclear.energy.gov/np2010/neScorecard/neScorecard.html, web page updated July 14, 2009, web page visited July 28, 2009.

[b] Blake, Michael E., "U.S. capacity factors: Another small gain, another new peak," *Nuclear News*, May 2008, pp. 28-34.

2

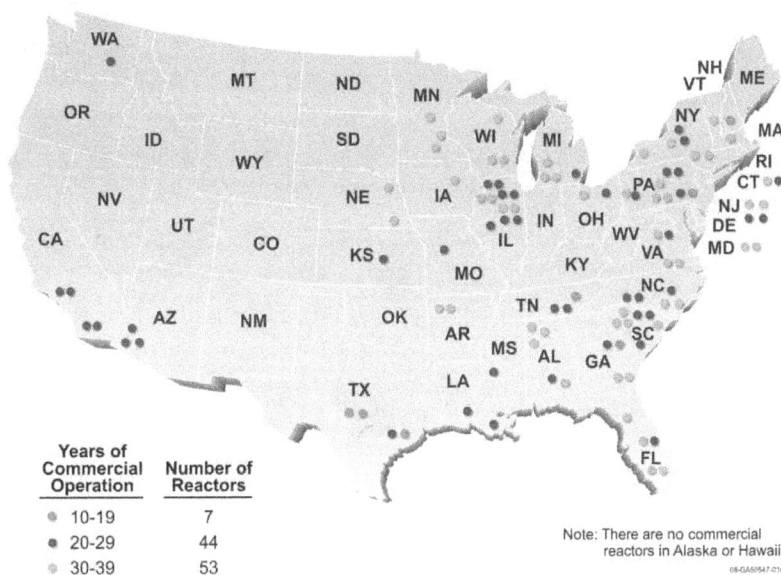

Figure 1-3. National distribution of operating nuclear power plants.

Years of Commercial Operation	Number of Reactors
10-19	7
20-29	44
30-39	53

Note: There are no commercial reactors in Alaska or Hawaii.

Most operating nuclear power plants have obtained, are applying, or intend to apply for license extension. Figure 1-4 shows the following: (1) the oldest nuclear power plant started operation in 1969 and the newest plant started operation in 1996, (2) the first group of nuclear power plants were brought online between 1969 and 1979 and the second group between 1980 and 1996, and (3) all most all operating nuclear power plants have been issued, are applying for, or plan to apply for a 20-year license extension. This license extension will result in a licensed operating life of 60 years.

In about the year 2030, unless further licensing renewal occurs the current fleet of nuclear power plants will start decommissioning. Absent additional research to address critical plant-aging issues, these valuable generating stations may be retired after reaching 60 years of operation. Furthermore, with the state of present research, degradation and obsolescence threaten to decrease power production from these nuclear power plants even before their scheduled end of licensed lifetimes. Over the next three decades, this would result in a loss of 100-GWe, emission-free generating capacity and is comparable to electrical generation of new nuclear power plants built over the same time period, leaving a gap in projections of required emission-free generating capacity. This gap might be filled with higher construction rates of new nuclear power plants or with other technologies. However, continued safe and economical operation of current reactors for an even longer period of commercial operation, beyond the current license renewal lifetime of 60 years, is a low-risk option to fill the gap and to add new power generation at a fraction of the cost of building new plants.

In order to receive a 20-year license extension, a nuclear power plant operator must ensure that the plant will operate safely for the duration of the license extension. The 40-year operating license period established in the Atomic Energy Act was based on antitrust considerations, not technical limitations. The 20-year license extension periods are presently authorized under the governing regulation of 10 CFR Part 54, "Requirements for Renewal of Operating Licenses for Nuclear Power Plants." This rule places no limit on the number of times a plant can be granted a 20-year license renewal as long as the licensing basis is maintained during the renewal term in the same manner and to the same extent as during the original licensing term.

3

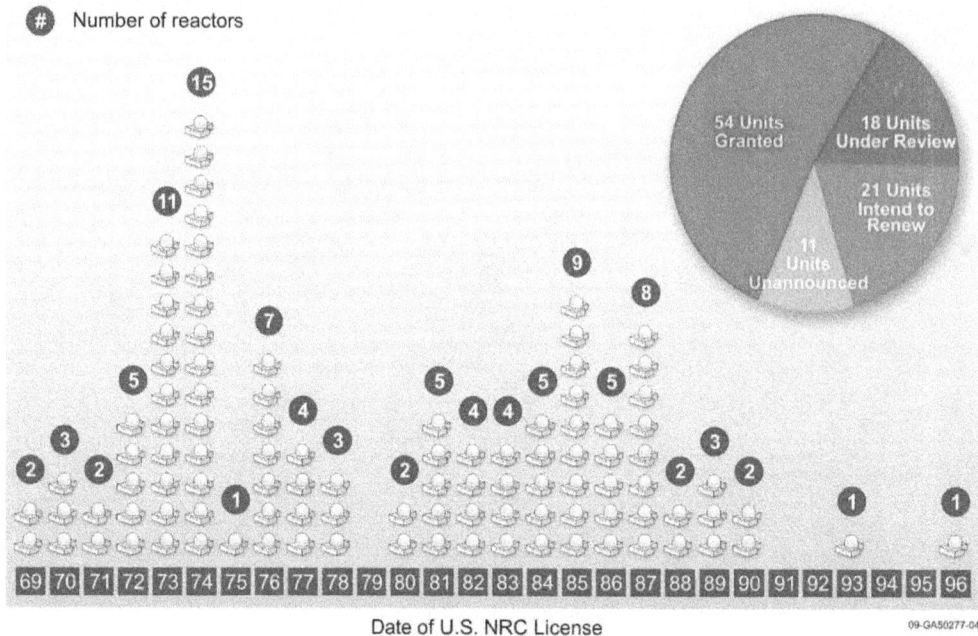

Figure 1-4. Nuclear power plant initial license date and license extension plans.

This regulatory process ensures continued safety of all currently operating nuclear power plants during future renewal periods. The license extension process requires a safety review and an environmental review, with multiple opportunities for public involvement. The applicant must demonstrate safety issues through technical documentation and analysis, which the U.S. Nuclear Regulatory Commission (NRC) confirms before granting a license extension. A solid technical understanding of how systems, structures, and components (SSCs) age is necessary for nuclear power plants to demonstrate continued safety. A well-established knowledge base for the current period of licensed operation exists; however, additional research will be needed to obtain the same robust technical basis required for continued operational evaluations beyond 60 years.

In early 2007, DOE, with the Idaho National Laboratory (INL) engaging the Electric Power Research Institute (EPRI) and other industry stakeholders, initiated planning that lead to the LWRS R&D Program. The aim was to develop an R&D strategy that addresses nuclear energy issues within the framework of the National Energy Policy and the Policy Act of 2005. Based on considerable analysis and information gathering, the "*Strategic Plan for Light Water Reactor Research and Development*," was developed and reviewed by an independent committee of experts. The plan, which recommended ten top priority areas for a government-industry, cost-shared R&D program, was issued in November 2007.[c]

Building on the strategic plan and collaborative relationships that were developed while preparing it, DOE and INL immediately started developing the LWRS R&D Program. In February 2008, DOE and NRC sponsored a workshop, which identified necessary R&D for long-term operation and licensing of

---

[c] NL/EXT-07-13543, *Strategic Plan for Light Water Reactor Research and Development*, Idaho National Laboratory, November 2007.

4

nuclear power plants.[d] Input from a large set of stakeholders provided important definition of needs and focused program objectives on long-term operation of existing nuclear power plants.

In developing the strategic plan and more specific program plans, it has become apparent that a government/industry cost sharing arrangement for R&D is desirable for addressing the long-range, policy-driven goals of government and the acceptability and usefulness of derived solutions to industry. The LWRS R&D Program requires the long-term vision and support of national laboratories to address strategic reliability and safety requirements of existing nuclear power plants that could not be addressed by more inherently tactical organizations. The long-term, higher-risk research required to construct a scientific basis to understand the complex effects of plant aging is not likely to be carried out by industry alone.

While industry is likely to invest in applied research programs that are directed toward enhancing operations or in developing incremental improvements, industry is unlikely to invest significantly in research programs that focus on longer-term or higher-risk gains. Additionally, because research necessary for nuclear power plant life extension is of a broad nature that provides benefits to the entire industry, it is unlikely that a single company will make the necessary investment on its own. Government cost sharing and involvement will be required to promote the necessary programs that are of crucial long-term importance. The LWRS R&D Program, by incorporating long-term collaborative industry stakeholder inputs and shared costs, will support the strategic national interest of maintaining nuclear power as an available resource.

Over the past several decades, academia and national laboratories have made enormous advances in the area of general materials science and modeling of fundamental structures. Applications of these sciences, although not specifically nuclear in nature, have the potential to bring tremendous advances over the narrowly focused, step-wise improvements the nuclear industry has realized thus far. Additionally, because of their unique resources (such as experimental irradiation and post-irradiation examination facilities), the national laboratory infrastructure is positioned to bridge the nuclear industry, R&D, and demonstration infrastructures. The LWRS R&D Program serves to facilitate use of this knowledge with further R&D that is specific to the current fleet of nuclear power plants in understanding ongoing and complex challenges to long-term operations.

In summary, the electrical energy sector is challenged to supply increasing amounts of electricity in a dependable and economical manner and with reduced CO_2 emissions. Consistent with the National Energy Policy, nuclear power is an important part of answering the challenge through long-term safe and economical operation of current nuclear power plants and with building new nuclear power plants. The LWRS R&D Program is designed to provide, in collaboration with industry programs, the sound technical basis for licensing and managing the long-term safe operation of existing operating nuclear power plants.

1.2   Sustainability

Sustainability in the context of LWRs is defined as the ability to maintain safe and economic operation of the existing operating fleet of nuclear power plants for a longer than licensed lifetime. It has two facets with respect to long-term operations: (1) manage the aging of hardware so the nuclear power plant lifetime can be extended and the plant can continue to operate safely, efficiently and economically;

---

[d] "Life Beyond 60 Workshop Summary Report, NRC/DOE Workshop U.S. Nuclear Power Plant Life Extension Research and Development," U.S. Nuclear Regulatory Commission and U.S. Department of Energy, Prepared by Energetics Inc., Feb. 19-21, 2008.

5

and (2) provide science-based solutions to the industry to implement technology to exceed the performance of the current labor-intensive business model.

Programmatically, LWRS is dependent on a sequence of four successful phases: (1) utility's decision to invest in extending the nuclear power plant life beyond 60 years; (2) licensing and public confidence in the nuclear power plant life extension; (3) implementation of nuclear power plant refurbishment and upgrade to meet the licensing and enhanced performance requirements; and (4) safe and economic nuclear power plant operation for the intended period of the nuclear power plant life extension. While tightly coupled, each of the four sequential phases is critical to nuclear power generation on its own with a specific set of challenges. The four phases span over several decades, a feature important for planning and implementation of the supporting R&D program.

The industry must also have the confidence that these sustainability critical technologies and processes will be acceptable with the regulators. On the technical side, the key is to establish the availability of an adequate body of knowledge (e.g., data and methods) to assess characterizing nuclear power plant SSC, their aging behavior, and plant safety margins. On the regulatory side, it is important to account for evolution of the regulatory paradigm. Because of the long-term character of investing in LWRS, it is in the best interest of the industry that a predictable, science-based regulatory framework be established. Application of the new predictable science-based licensing concepts will provide additional confidence for the public and industry.

Through technological innovation, existing operating nuclear power plants have established a remarkable performance track record. However, as a plant ages, performance normally drops. The May 2009 issue of *Nuclear News* provides evidence that reactors in their fourth decade have not achieved results quite as impressive as those of newer reactors.[c] Without innovation, performance most likely will deteriorate even more when the older nuclear power plants enter their fifth and sixth decades. Therefore, new innovations and business models are needed in order to significantly enhance performance from today's high standard.

The new business model can be achieved through enabling transforming technology advancements and by leveraging the resources of the entire industry through seamless integration of plant owner/operator, suppliers, service providers and regulators. Nuclear power plant data and analysis tool interfaces would be standardized across the industry.

1.3   Critical Path for Nuclear Power Plants

Ultimately, extending the life of an existing asset is an individual utility business and risk decision. A utility anticipates that, in most situations, extending the life of an existing nuclear power plant is likely to cost less than building a new plant; however, operating costs must remain competitive. Individual owner-operators are likely to seriously consider extending the life of their existing nuclear power plants well in advance of committing to new construction, assuming existing assets can economically meet anticipated demand growth and assuming that the option to do so is still available. It is also likely that decisions of extending nuclear power plant lifetimes will be accompanied by facility upgrading and uprate assessments, thus helping to manage the operational risks of aging and taking advantage of technical enhancement opportunities. These capital-spending decisions will require a thorough business case and a technical understanding and predictability of aging and degradation risks.

Decisions to develop, construct, and license baseload generation must be made far in advance of power demands outgrowing supply capacities. Actions for retaining existing nuclear power infrastructure

---

[c] "U.S. Capacity Factors: Can Older Reactors Keep Up the Pace?," *Nuclear News*, May 2009.

6

in the United States must begin in a timely manner. Given the risk-adverse influences of financial markets, state public utility/service commissions, and NRC, power-generating utilities must use all available information in carrying out these decisions. With extended operational lifetimes, aging-related technical or operational questions that did not exist previously have now become important decision factors.

Extending nuclear power plant life beyond 60 years is expected to remain a technically viable option for filling the power-generation gap between license expiration of older nuclear power plants and having newer nuclear power plants come online. In addition to the environmental benefits, extending the life of highly efficient existing nuclear power plants defers the up-front costs of building new nuclear power plants.

With the present 60-year licenses beginning to expire between the years of 2029 and 2039 for the first group of nuclear power plants that came online between 1969 and 1979 (as shown in Figure 1-4) utilities are likely to initiate planning of baseload replacement power by 2014 or earlier. If the option to extend current plant lifetimes is not available, strategic planning and investment required to maintain the current LWR fleet may not happen in a sustainable manner. The research window for supporting the utility's decisions to invest in lifetime extension and to support NRC decisions to extend the license must start now and is likely to extend through the following 20-year period (i.e., 2010 to 2029), with higher intensity for the first 10 years. The LWRS R&D Program represents the beginning of timely collaborative research needed to retain the existing nuclear power infrastructure of the United States.

2.   DESCRIPTION

2.1   Vision

Today's commercial nuclear power plant fleet has reliably produced environmentally friendly power in the United States for decades. As these nuclear power plants reach the end of their original 40-year operating license and enter their first 20-year extended license, sound engineering principles used in designing and building them are being applied to demonstrate their continued safety for a possible second license extension. In order to preserve the option of continued safe and economical operation of these nuclear power plants, a technical basis is required for the utility to evaluate investments in life-extending improvements and for the regulator to accept license extension applications. This program plan identifies R&D activities for enhancing scientific understanding of aging mechanisms important to the SSCs in nuclear power plants and to develop methods and technologies for managing plant aging and evaluating safety of nuclear power plants for long-term operation.

The LWRS R&D Program vision is captured in the following statements:

Existing operating nuclear power plants will continue to safely provide clean and economic electricity well beyond their first license-extension period, significantly contributing to reduction of United States and global carbon emissions, enhancement of national energy security, and protection of the environment.

There is a comprehensive technical basis for licensing and managing the long-term, safe, economical operation of nuclear power plants. Sustaining the existing operating U.S. fleet also will improve its international engagement and leadership on nuclear safety and security issues.

7

Extending the life of nuclear power plants is a vital step in meeting the electrical needs of the United States today and in decades to come. By keeping these plants safely in service, the nation will retain valuable infrastructure and allow additional time to construct new sources of clean, reliable, and secure energy. Until other reliable sources of power are built and placed on the electrical grid, the existing fleet of nuclear power plants is a vital component of the economy.

2.2   Program Goals

The LWRS R&D Program is designed to help achieve its vision by addressing long-term operational challenges that face nuclear utilities in the United States. Program goals are to develop scientific understanding, tools, processes, and technical and operational improvements to do the following:

1.   Support long-term licensing and operation of the existing operating nuclear power plants to successfully achieve planned lifetime extension up to 60 years and lifetime extension beyond 60 years

2.   Support maintenance and enhancement of performance of the existing operating fleet of LWRs to ensure superior safety, high reliability, and economic performance throughout their full lifetime.

2.2.1   Scientific Basis

Ensuring public safety and environmental protection is a prerequisite to all nuclear power plant operating decisions. For extended operating periods, it must be shown that adequate aging management programs are present or planned and that appropriate safety margins exist throughout the subsequent license renewal periods. Through research, this program will seek to contribute to the technical foundation on which licensees can base their analyses to determine if these adequate safety margins and superior economic performance can be maintained or even enhanced. In order to make the technically justified case when deciding to apply for a subsequent license extension, the nuclear industry will require definitive knowledge into the effects of aging. The scientific means (such as sound fundamental understanding) and transformative technologies (such as advanced analytical and computational tools and state-of-the-art diagnostic tools and leading expertise) will be employed to address practical challenges facing the nuclear industry.

2.2.2   Economic Viability

Once scientific research establishes how nuclear power plants will age and aging management programs are identified, operators must demonstrate that the costs associated with continuing to maintain and operate their nuclear power plants are justified and remain in the best interest of their owners. It is likely that as nuclear power plants operate beyond their original license periods, significant component replacements will become necessary, thereby increasing costs. Each utility will need to be able to accurately predict such costs in order to make sound business decisions regarding continued long-term plant operation.

Technology, in combination with effective plant management programs, is expected to support new opportunities for further cost savings in areas such as aging management, information technologies, operations and maintenance, training, fuel design, and management. Some of these cost improvements will be within the scope of a regulatory license renewal process (e.g., reactor pressure boundary materials issues), while others may be important to continued economic viability but not have regulatory significance. Safety and economic viability are considered complementary goals. Developed properly, programs that enhance economics also are likely to benefit plant safety.

8

2.3 Implementing Strategy

Three diverse, yet interrelated sequential strategies will be implemented in the program:

1. Develop the scientific basis to understand, predict, and measure changes in materials and SSCs as they age in environments associated with continued long-term operation of existing LWRs

2. Apply this fundamental knowledge in collaborative public-private and international partnerships, developing and demonstrating methods and technologies that support safe and economical long-term operation of existing LWRs

3. Identify and verify the efficacy of new technology to address obsolescence while enhancing plant performance and safety.

Because of the scale, cost, and time horizons involved in sustaining the current operating fleet of LWRs, achieving the strategic goals of the LWRS R&D Program will require extensive collaboration with the industry, NRC, and international R&D institutions of extensive technical expertise. In addition, recognizing the need to support education and training of the next generation of scientists and engineers, the following strategic guidelines were established to guide organization and implementation of the program:

- Leverage institutional knowledge and collaborative opportunities between the nuclear industry, national laboratories, universities, and the federal government in developing the basic scientific understanding in predicting key materials and safety margin characterizations

- Using the LWRS R&D Program vision and goals, build relationships across established relevant research interests, both at international and domestic levels

- Integrate Nuclear Energy University Program projects with selected R&D pathways

- Ensure the LWRS R&D Program is accountable to sponsors, partners, and other stakeholders.

The 60-year lifetime license for the first group of nuclear power plants will expire between the years 2029 and 2039. The LWRS R&D Program can be divided into four sequential, yet interconnected, phases that correspond to the four phases of sustainability (Section 1.2). The following describes the main objectives of each phase and the timeframe applicable to those nuclear power plants with the 60-year license expiring in 2029 and beyond:

- Phase I: Build confidence for the industry to proceed with 80-year license renewal, using data and tools (the timeframe for this phase is 2010 to 2015)

- Phase II: Enable the industry to make the decision to invest in plant refurbishments, modernizations, and licenses for 80-year operations (the timeframe for this period is 2015 to 2020)

- Phase III: Apply scientific solutions and continuing technology development to support NRC review and plant capital investment (the timeframe for this period is 2020 to 2030)

- Phase IV: Enable safe and economic operations with the 80-year license (the timeframe for this phase is 2030 and beyond).

9

275

On a more abstract level, this program can be broken into the following two periods:

1. Period of license application and review for 80-year operation (Phases I, II, and III fall into this period)

2. Period during which the nuclear power plant fleet operates beyond 60 years of life (Phase IV falls into this period).

The implementation schedule (Figure 2-1) is structured to support the following high-level milestones:

*   2010: Ensure that long-term operation is an accepted high priority option for power generation by industry, DOE, and NRC

*   2015: Build confidence in long-term operation with data and tools

*   2020: Enable industry decision to invest and license for long-term operation

*   2025: Acceptance of advanced tools, methods, and technologies

*   2030: Commence licensed long-term operations.

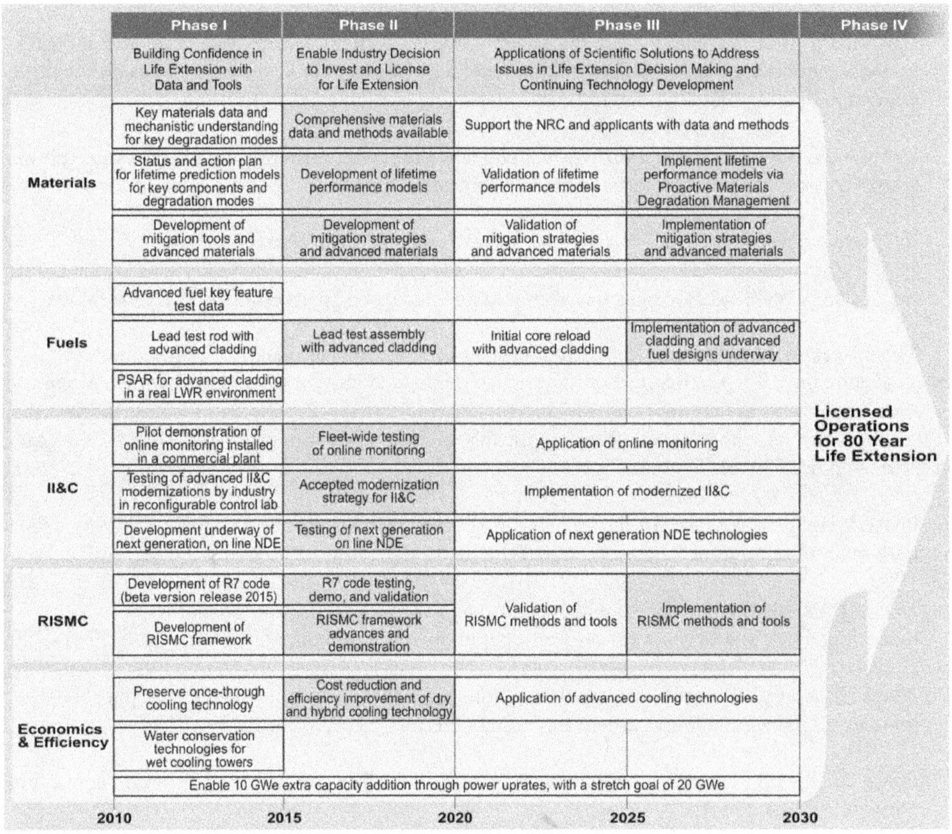

	Phase I	Phase II	Phase III		Phase IV
	Building Confidence in Life Extension with Data and Tools	Enable Industry Decision to Invest and License for Life Extension	Applications of Scientific Solutions to Address Issues in Life Extension Decision Making and Continuing Technology Development		
Materials	Key materials data and mechanistic understanding for key degradation modes	Comprehensive materials data and methods available	Support the NRC and applicants with data and methods		
	Status and action plan for lifetime prediction models for key components and degradation modes	Development of lifetime performance models	Validation of lifetime performance models	Implement lifetime performance models via Proactive Materials Degradation Management	
	Development of mitigation tools and advanced materials	Development of mitigation strategies and advanced materials	Validation of mitigation strategies and advanced materials	Implementation of mitigation strategies and advanced materials	
Fuels	Advanced fuel key feature test data				
	Lead test rod with advanced cladding	Lead test assembly with advanced cladding	Initial core reload with advanced cladding	Implementation of advanced cladding and advanced fuel designs underway	Licensed Operations for 80 Year Life Extension
	PSAR for advanced cladding in a real LWR environment				
II&C	Pilot demonstration of online monitoring installed in a commercial plant	Fleet-wide testing of online monitoring	Application of online monitoring		
	Testing of advanced II&C modernizations by industry in reconfigurable control lab	Accepted modernization strategy for II&C	Implementation of modernized II&C		
	Development underway of next generation, on line NDE	Testing of next generation on line NDE	Application of next generation NDE technologies		
RISMC	Development of R7 code (beta version release 2015)	R7 code testing, demo, and validation	Validation of RISMC methods and tools	Implementation of RISMC methods and tools	
	Development of RISMC framework	RISMC framework advances and demonstration			
Economics & Efficiency	Preserve once-through cooling technology	Cost reduction and efficiency improvement of dry and hybrid cooling technology	Application of advanced cooling technologies		
	Water conservation technologies for wet cooling towers				
	Enable 10 GWe extra capacity addition through power uprates, with a stretch goal of 20 GWe				
	2010	2015	2020	2025	2030

09-GA50277-05d

Figure 2-1. Program implementation schedule.

10

3. RESEARCH AND DEVELOPMENT PATHWAYS

Safety is a fundamental requirement for reliable economic operation; therefore, most of the knowledge and methodologies developed in this program are expected to serve the regulator and the utility. This commonality is a key consideration in defining the R&D pathways and individual R&D projects. The LWRS R&D Program currently is comprised of the following five principal R&D pathways, each of which focuses on a key technical element that ensures the safe, economic, and reliable operation of the existing nuclear power plant fleet:

1. Nuclear Materials Aging and Degradation

2. Advanced LWR Nuclear Fuel Development

3. Advanced Instrumentation, Information, and Control Systems Technologies

4. Risk-Informed Safety Margin Characterization

5. Economics and Efficiency Improvement.

The objective of these R&D pathways is to create a greater level of safety through application of increased knowledge and an enhanced economic understanding of nuclear power plant operational risk beyond the first license extension period. These R&D pathways also provide possible solutions to future challenges and will ensure safe and economic extended nuclear power plant operation.

3.1 Nuclear Materials Aging and Degradation

3.1.1 Background and Introduction

Nuclear reactors present a very harsh environment for components service. Components within a reactor core must tolerate high temperature water, stress, vibration, and an intense neutron field. Degradation of materials in this environment can lead to reduced performance, and in some cases, sudden failure.

Materials degradation in a nuclear power plant is extremely complex due to the various materials, environmental conditions, and stress states. Over 25 different metal alloys can be found within the primary and secondary systems; additional materials exist in concrete, the containment vessel, instrumentation and control equipment, cabling, buried piping, and other support facilities. Dominant forms of degradation may vary greatly between different SSCs in the reactor and can have an important role in the safe and efficient operation of a nuclear power plant. When this diverse set of materials is placed in a complex and harsh environment, coupled with load and degradation over an extended life, an accurate estimate of the changing material behaviors and lifetime is complicated. A small sampling of these metals for a pressurized water reactor is shown in Figure 3-1.

Clearly, materials degradation will impact reactor reliability, availability, and, potentially, safe operation. Routine surveillance and component replacement can mitigate these factors; however, failures still occur. With reactor life extensions up to 60 years or beyond and power uprates, many components must tolerate more demanding reactor environments for even longer times. This may increase susceptibility to degradation for different components and may introduce new degradation modes. While all components (except perhaps the reactor pressure vessel) can be replaced, it may not be economically favorable. Therefore, understanding, controlling, and mitigating materials degradation processes and a technical basis for long-range planning for necessary replacements are key priorities for reactor operation,

11

power uprate considerations, and life extensions. Appendix A contains detailed information on research tasks for nuclear materials aging and degradation.

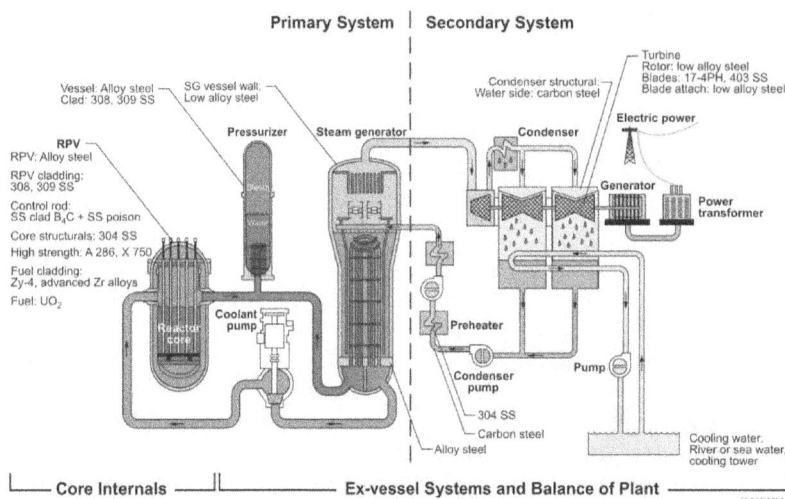

Figure 3-1. Light water reactor metals.

3.1.2   Vision and Goals

Materials research provides an important foundation for licensing and managing the long-term, safe, and economical operation of nuclear power plants. Aging mechanisms and their influence on nuclear power plant SSCs are predictable with sufficient confidence to support planning, investment, and licensing for necessary component repair, replacement, and relicensing. Understanding, controlling, and mitigating materials degradation processes are key priorities. While our knowledge of degradation and surveillance techniques are vastly improved, unexpected degradation can still occur. Proactive management is essential to help ensure that any degradation from long-term operation of nuclear power plants does not affect the public's confidence in the safety and reliability of those nuclear power plants.

The strategic goals of the Nuclear Materials Aging and Degradation R&D pathway are to develop the scientific basis for understanding and predicting long-term environmental degradation behavior of materials in nuclear power plants and to provide data and methods to assess performance of SSCs essential to safe and sustained nuclear power plant operations.

Specific outputs from this pathway will include improved mechanistic understanding of key degradation modes and sufficient experimental data to provide and validate operational limits and development of advanced mitigation techniques to provide improved performance, reliability, and economics. Mechanistic and operational data also will be used to develop performance models for key material systems and components in later years.

12

3.1.3 Highlights of Research and Development

The Nuclear Materials Aging and Degradation R&D pathway activities have been organized into five areas: (1) reactor metals, (2) concrete, (3) cable aging, (4) buried piping, and (5) mitigation strategies. These research areas cover material degradation in SSCs that were designed for service without replacement throughout the life of the plant. Management of long-term operation of these components can be difficult and expensive. As nuclear power plant licensees seek approval for extended operation, the way in which these materials age beyond 60 years will need to be evaluated and their capabilities reassessed in order to ensure that they maintain the required design functions safely and economically. In addition to the five research areas, a Materials Aging and Degradation Assessment also will be conducted to provide a comprehensive assessment of materials degradation.

3.1.3.1 Reactor Metals. Numerous types of metal alloys can be found throughout the primary and secondary systems. Some of these materials, particularly the reactor internals, are exposed to high temperatures, water, and neutron flux. This creates degradation mechanisms that may be unique or environmentally exacerbated. Research programs in this area will provide a foundation upon which a safe regulatory environment can be established for life beyond 60 years. The following eight activities will encompass the reactor metals area (see Appendix A for detailed information about the activities):

- Mechanisms of irradiation-assisted stress corrosion cracking in stainless steels

- High-fluence effects on reactor pressure vessel steels

- Crack initiation in Ni-alloys

- High-fluence effects on irradiation-assisted stress corrosion cracking of stainless steels

- Irradiation-assisted stress corrosion cracking of alloy X-750

- Evaluation of swelling effects in high-fluence core internals

- Irradiation-induced phase transformations in high-fluence core internals

- Surrogate and attenuation effects on reactor pressure vessel steels.

3.1.3.2 Concrete. Currently, there is little or no data on long-term concrete performance in nuclear power plants. Long-term stability and performance of concrete structures within a nuclear power plant is a concern. The objective of this task is to assess the long-term performance of concrete. Research task evaluation and prioritization will be performed on an ongoing basis. Plans for research will continue to be evaluated by collaborators at EPRI and NRC to ensure complementary and cooperative research. In addition, formation of an Extended Service Materials Working Group will provide a valuable resource for additional and diverse input.

3.1.3.3 Cabling. Cable aging is a concern that currently faces the operators of existing nuclear power plants. Utility companies carry out periodic cable inspections using nondestructive examination techniques to measure degradation and determine when replacement is needed. Degradation of these cables is primarily caused by long-term exposure to high temperatures. Additionally, stretches of cables that have been buried underground are frequently exposed to groundwater.

13

3.1.3.4 ***Buried Piping.*** Maintaining the many miles of buried piping is an area of concern when evaluating the feasibility of continued plant life. While much of the buried pipes comprise either secondary plant or other non-safety-related cooling systems, some buried piping serves a direct safety function. Maintaining the integrity and reliability of all of these systems is necessary for continued plant operation. These systems must be maintained to ensure predictable plant operation and to maintain plant efficiency.

3.1.3.5 ***Mitigation Technologies.*** Welding is widely used for component repair. Weld-repair techniques must be resistant to long-term degradation mechanisms. Extended lifetimes and increased repair frequency welds must be resistant to corrosion, irradiation, and other forms of degradation. The purpose of this research area is to develop new techniques for weldments, weld analysis, and weld repair. A critical assessment of the most advanced methods and their viability for LWR repair weld applications is needed.

3.1.3.6 ***Integrated Research Activities.*** This research element includes (1) international collaboration to conduct coordinated research with international institutions such as the Materials Aging Institute in order to provide more collaboration and cost sharing, (2) coordinated irradiation experiments to provide a single integrated effort for irradiation experiments, (3) advanced characterization tools to increase materials testing capability, improve quality, and develop new methods for materials testing, and (4) additional research tasks based on results and assessments of current research activities (see Appendix A for more details on these research activities).

3.1.4   Products and Implementation Plan

The main products from the Nuclear Materials Aging and Degradation R&D pathway are (1) mechanistic understanding of key degradation modes, (2) lifetime performance models, (3) advanced mitigation strategies, and (4) advanced replacement materials. The implementation schedule shown in Figure 3-2 is structured to support the following high-level milestones:

- 2010:

    - Complete the first iteration of reactor material degradation matrix

    - Identify the status and potential magnitude of key degradation modes for materials systems and issues.

- 2015:

    - Develop materials data and mechanistic understanding for key degradation modes in hand:

        – Determination of mechanisms of stress corrosion cracking underway

        – Bounding data for reactor pressure vessel embrittlement

        – Concrete degradation

        – Cabling

    - Develop status and action plan for lifetime prediction models for key components and degradation modes

    - Develop mitigation tools and advanced materials options underway:

        – Validation of post-irradiation annealing

14

– Development of advanced replacement materials.

- 2020:

  - Ensure materials data and methods are available to support high confidence of successful operation to 80 years and predictable service times (replacement times) for major components

    – Validation of lifetime performance models

    – Development of mitigation strategies.

- 2025: Support applicants and NRC with data and methods for materials degradation issues and limitations via proactive materials degradation management.

- 2030: Implement lifetime performance models, mitigation strategies, and advanced replacement materials.

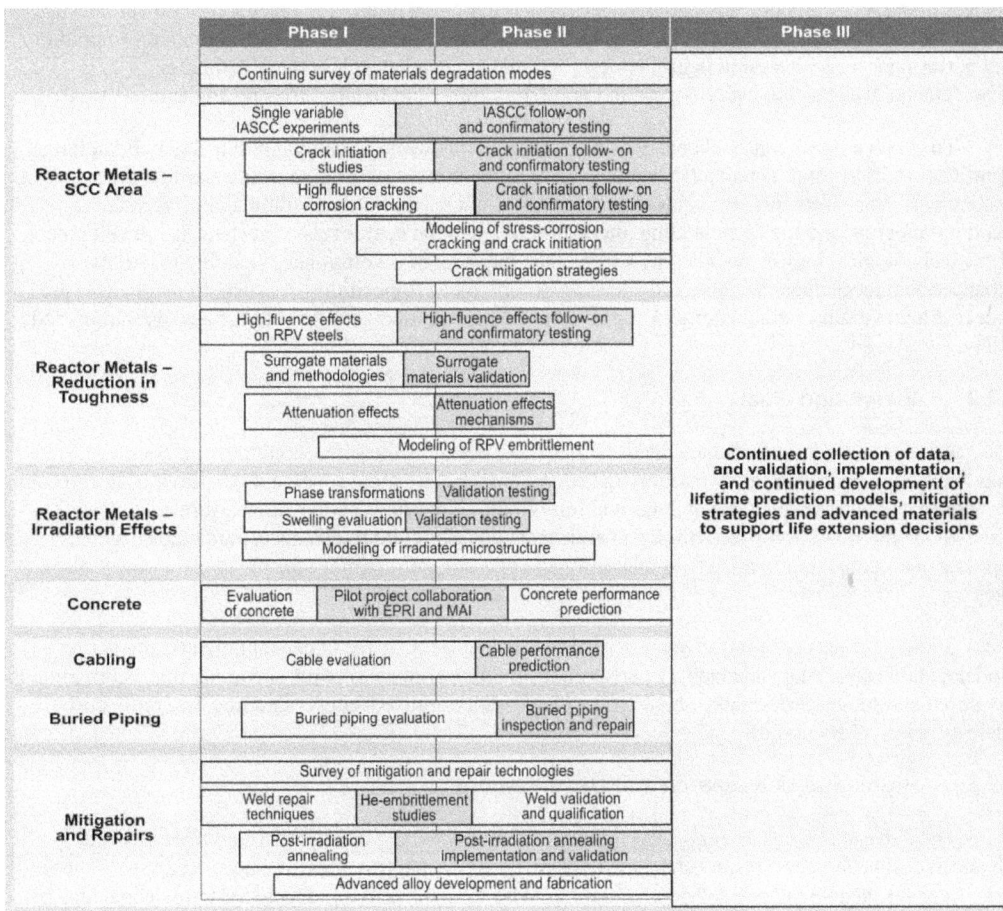

	Phase I	Phase II	Phase III
	Continuing survey of materials degradation modes		Continued collection of data, and validation, implementation, and continued development of lifetime prediction models, mitigation strategies and advanced materials to support life extension decisions
Reactor Metals – SCC Area	Single variable IASCC experiments	IASCC follow-on and confirmatory testing	
	Crack initiation studies	Crack initiation follow- on and confirmatory testing	
	High fluence stress-corrosion cracking	Crack initiation follow- on and confirmatory testing	
	Modeling of stress-corrosion cracking and crack initiation		
	Crack mitigation strategies		
Reactor Metals – Reduction in Toughness	High-fluence effects on RPV steels	High-fluence effects follow-on and confirmatory testing	
	Surrogate materials and methodologies	Surrogate materials validation	
	Attenuation effects	Attenuation effects mechanisms	
	Modeling of RPV embrittlement		
Reactor Metals – Irradiation Effects	Phase transformations	Validation testing	
	Swelling evaluation	Validation testing	
	Modeling of irradiated microstructure		
Concrete	Evaluation of concrete	Pilot project collaboration with EPRI and MAI / Concrete performance prediction	
Cabling	Cable evaluation	Cable performance prediction	
Buried Piping	Buried piping evaluation	Buried piping inspection and repair	
Mitigation and Repairs	Survey of mitigation and repair technologies		
	Weld repair techniques	He-embrittlement studies / Weld validation and qualification	
	Post-irradiation annealing	Post-irradiation annealing implementation and validation	
	Advanced alloy development and fabrication		

09-GA50277-00c

Figure 3-2. Nuclear Materials Aging and Degradation pathway implementation schedule.

15

281

3.2   Advanced Light Water Reactor Nuclear Fuel Development

3.2.1   Background and Introduction

Nuclear fuel performance is a significant driver of nuclear power plant operational performance, safety, operating economics, and waste disposal requirements. Over the past two decades, the nuclear power industry has improved plant capacity factors with incremental improvements in fuel reliability and use or "burnup." However, these upgrades are reaching their maximum achievable impact within the constraints of existing fuel design, materials, licensing, and enrichment limits. Although the development, testing, and licensing cycle for new fuel designs is typically long (about 10 years from conception through utility acceptance), these improvements are often used with only an empirical understanding of the fundamental phenomena limiting their long-term performance.

Continued development of high-performance nuclear fuels through fundamental research focused on common aging issues can enable plant operators to extend plant operating cycles and enhance the safety margins, performance, and productivity of existing nuclear power plants. The Advanced LWR Nuclear Fuel Development R&D pathway performs research on improving reactor core power density, increasing fuel burnups, advanced cladding, and developing enhanced computational models to predict fuel performance. This research is further designed to demonstrate each of these technology advancements while satisfying all safety and regulatory limits through rigorous testing and analysis.

To achieve significant fuel cost and use improvements while remaining within safety boundaries, significant steps beyond incremental improvements in the current generation of nuclear fuel are required. Fundamental improvements are required in the areas of nuclear fuel composition and performance, cladding integrity, and the fuel/cladding interaction to reach the next levels of nuclear fuel development. These technological improvements are likely to take the form of revolutionary cladding materials, enhanced fuel mechanical designs, and alternate isotope fuel compositions. As such, these changes are expected to have substantial beneficial improvements in nuclear power plant economics, operation, and safety.

3.2.2   Vision and Goals

Advanced, high-performance fuels are an essential part of the safe, economic operation of LWRs. New fuels have improved safety margins and economics and are more reliable. Fuel provides head-room for additional power uprates and high burnup limits. The scientific basis for fuel performance is well understood, and its response to changing operational conditions and transients is predictable, which supports continuous improvements to reliability and operational flexibility for the nuclear power plant fleet.

Strategic goals are to improve the scientific knowledge basis for understanding and predicting fundamental nuclear fuel and cladding performance in nuclear power plants, and apply this information to development of high-performance, high burnup fuels with improved safety, cladding, integrity, and nuclear fuel cycle economics.

3.2.3   Highlights of Research and Development

The Advanced Nuclear Fuels Development Program element is separated into three R&D tasks: advanced design and concepts, mechanistic understanding of fuel behavior, and advanced tools. These tasks were selected to balance development of new knowledge, verifying developed knowledge, and creation of new advanced fuel technology. The scope of the pathway includes all aspects important to fuel design and performance, including fuel design, exposure effects, and cladding material performance and

16

development. Figure 3-3 shows a typical pressurized water reactor fuel assembly. A boiling water reactor assembly is of different design; however, the fuel rods are quite similar.

3.2.3.1   *Advanced Designs and Concepts.* The purpose of this task area is to increase the understanding of advanced fuel design concepts, including use of new cladding materials, increases to fuel lifetime, and expansions to the allowable fuel performance envelope. These improvements will allow the fuel performance related plant operating limits to be optimized in areas such as operating temperatures, power densities, power ramp rates, and coolant chemistry. Accomplishing these goals leads to improved operating safety margins and improved economic benefits. Detailed information on the Advanced Designs and Concepts task can be found in Appendix B.

3.2.3.2   *Mechanistic Understanding of Fuel Behavior.* This task area will involve testing and modeling of specific aspects of LWR fuel, cladding, and coolant behavior. Examples include pellet cladding interaction, fission gas release, coolant chemistry effects on corrosion, and crud (oxide) formation. Improved understanding of fuel behavior can be used in fuel design, licensing, and performance prediction.

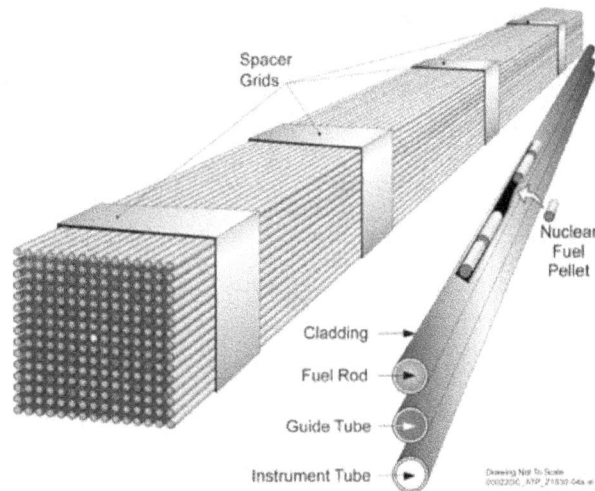

Figure 3-3. Nuclear fuel assembly.

An improved fundamental understanding of phenomena that impose limitations on fuel performance will allow fuel designers, fabricators, plant chemists, and code developers to optimize the performance of current fuels and the designs of advanced fuel concepts. A life-cycle concept will be applied so that optimization will be applied to fabrication, in-reactor use, and performance as spent fuel in storage. Fundamental mechanistic models will provide a foundation for supporting the LWRS R&D Program strategic objectives in developing advanced fuels. The following models will be included in this task (see Appendix B for detailed information about the following models):

1.   Fuel mechanical property change model as a function of exposure

2.   Pellet cladding interaction model development

3.   Chemistry coolant model development

17

4. Mesoscale models of microstructure fuel behavior

5. Hydrogen uptake behavior of Zr cladding.

3.2.3.3 Advanced Tools. This task area will use increased understanding of specific fuel performance phenomena that will be integrated into encompassing fuel performance advanced tools. These advanced tools, including modeling and simulation codes, advanced experimental capabilities, and real-time performance monitoring, will be developed to enhance plant and repository efficiency. In addition, the advanced tools developed will be used to minimize the time required to realize the gains made through this R&D effort by decreasing the amount of time needed for materials development and fuel qualification. The following activities will be included in this task (see Appendix B for detailed information about the following activities):

1. Engineering design and safety analysis tool

2. Mechanical models of composite cladding

3. Irradiation design studies of advanced silicon carbide (SiC) cladding

4. Experimental campaign to verify design and safety margin calculation tool

5. Advanced mathematical tools to support advanced nuclear fuels calculations.

3.2.4 Products and Implementation Plan

The main product produced from this pathway is development of SiC/silicon carbide fiber (reinforced) (SiC/SiC$_f$) fuel cladding. This activity allows direct product development and development of the supporting enabling technology and understanding required to design and license a new generation of fuel. Without the specific SiC/SiC$_f$ cladding development, another high value fuel development activity would be used to focus fuel development activities toward the roll out of a specific product. The implementation schedule shown in Figure 3-4 is structured to support the following high-level milestones:

- 2010:
    - Design and planning of SiC/SiCf rodlet irradiation campaign
    - Rodlet testing planning/design with SiC
    - Rodlet irradiation with SiC
    - Mechanical modeling of SiC/SiCf matrix
    - Licensing case for SiC applications in commercial applications
    - Out-of-core testing, repeated stress, thermal cycles, and failure modes for SiC.
- 2015:
    - Initial lead test rod design with SiC and planning
    - Rod testing planning/design with SiC
    - Rod irradiation with SiC.
- 2020:
    - Initial SiC lead test assembly licensing
    - Reload testing planning/design with SiC

18

- Reload irradiation with SiC.
- 2025:
    - Initial SiC reload design
    - Initial core reload with SiC
    - Irradiation program for increased enrichment bundles
    - Irradiation program for increased exposure bundles.
- 2030:
    - Fleetwide implementation of SiC reload and advanced fuel designs under way
    - Lead test assembly for increased enrichment fuel
    - Lead test assembly for increased exposure fuel.
- 2040:
    - Advanced fuel designs
    - Advanced uprated cores using SiC cores.

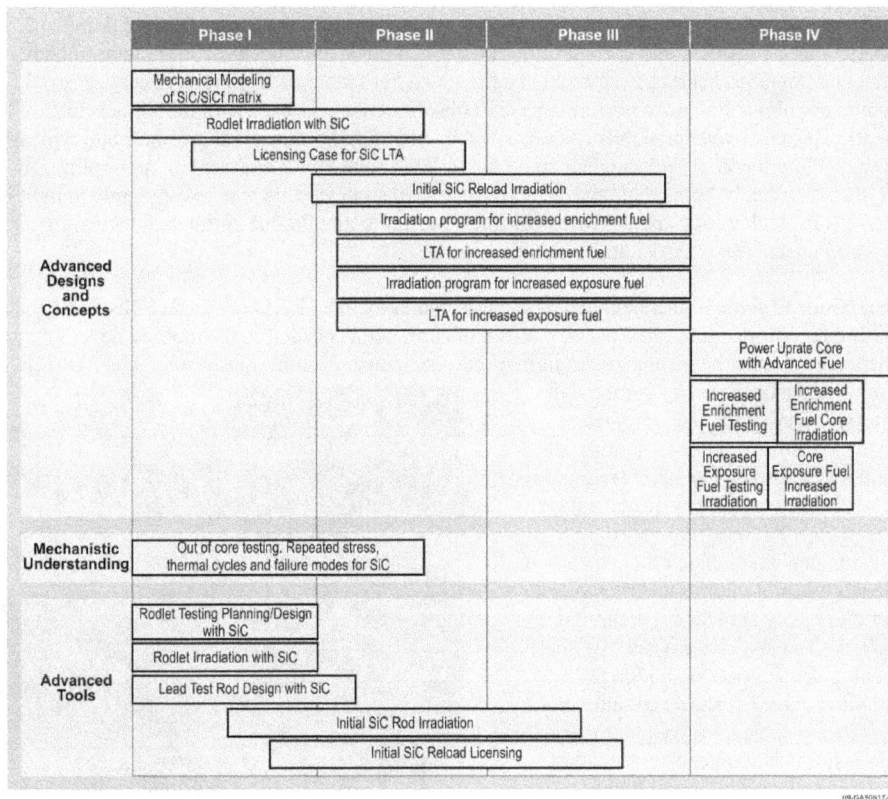

Figure 3-4. Advanced Light Water Reactor Nuclear Fuels Development pathway implementation schedule.

19

3.3 Advanced Instrumentation, Information, and Control Systems Technologies

3.3.1 Background and Introduction

Instrumentation, information, and control (II&C) systems technologies are essential to ensuring delivery and effective operation of nuclear power systems. They are enabling technologies that affect every aspect of nuclear power plant and secondary plant operations – analogous to a central nervous system. In 1997, the National Research Council conducted a study concerning the challenges involved in modernization of digital instrumentation and control systems in nuclear power plants. Their findings identify the need for new II&C technology integration. Unfortunately, this report, issued in 1997, still reflects the current state of affairs at nuclear power plants. Numerous issues that must be addressed in order to implement new types of II&C systems in commercial nuclear power plants have not been satisfactorily demonstrated in the commercial nuclear power industry of the United States. Without new types of II&C systems, today's nuclear power plants II&C systems will become antiquated and unreliable, unfamiliar to a future workforce, and a liability on the corporate balance sheet.

Digital II&C technologies are deployed in a number of power generation settings worldwide. The situation in the United States nuclear power sector differs from these other settings in several key respects: analog systems that have been operated beyond their intended service lifetimes dominate II&C systems in place today; regulatory uncertainty and associated business risk concerns are dominant contributors to the status quo; and current utility business models have not evolved to take full advantage of digital technologies to achieve performance gains. As a consequence, digital technologies are implemented as point solutions to performance and obsolescence concerns with individual II&C components. This reactive approach is characterized by planning horizons that are short and typically only allow for 'like-for-like' replacements to be made. This results in a fragmented, non-optimized approach that is driven by immediate needs. As a long-term strategy, this is not sustainable in light of the evolution of II&C technology, availability of skills needed to maintain this antiquated technology, and high costs and uncertainties associated with doing so.

In addition to some of the technical challenges and associated R&D needs, in order to be successful in supporting long-term operational goals, a different approach is needed to technology development and deployment. These must be recognized in light of current industry trends and factors. The first is that the nuclear power generation sector is rarely an early adopter of new II&C technologies. Consequently, the nuclear power industry does not drive the development of II&C technology needs or availability in the power generation sector as a whole. Rather, it reacts to developments implemented in other sectors of power generation and implements them some time after others. Second, digital technologies are deployed on an as-needed basis to replace failing analog devices that are no longer maintainable. Figure 3-5 shows a contemporary control room at a nuclear power plant that relies on analog instrumentation and controls that require extensive procedures and highly trained operators. Because these technologies replace like-for-like capability – analog with digital –

Figure 3-5. A contemporary control room at a nuclear power plant.

20

the planning horizon for such activities is typically short, which tends to marginalize the potential benefits that can be achieved through digital II&C technology development and deployment (see Appendix C for more detailed information).

Individual force-fitting approaches to digital technology deployment and ever increasing obsolescence, long-term safety, and reliability of analog devices necessitate reconsideration of potential solutions involving digital technologies for nuclear energy systems. This reconsideration must include the long-term issues associated with monitoring and managing aging and degradation of plant systems and initiatives that must be undertaken to ensure long-term sustainability of II&C systems in a way that achieves availability of a cost-competitive, reliable nuclear energy supply.

A technology-driven approach in this R&D area alone will be insufficient to yield the type of transformation that is needed to secure a long-term source of nuclear energy base load; a new approach is needed. An effective R&D initiative must engage the perspectives of stakeholders (i.e., asset owners, regulators, vendors, and R&D organizations) in order to articulate and initiate relevant R&D activities.

3.3.2 Vision and Goals

Maintaining the reliability and safety of II&C systems used for process measurement and control is crucial in meeting the licensing basis of nuclear power generation assets. Aging and obsolescence of the installed technologies is a continuing concern for asset owners. Advances are needed to support crucial characterization and monitoring activities that will become increasingly important as materials age. The aim of collaborations, demonstrations, and approaches envisioned by this pathway are intended to lessen the inertia that sustains the current status quo of today's II&C systems technology and to motivate transformational change and a shift in strategy – informed by business objectives – to a long-term approach to II&C modernization that is more sustainable.

One of the goals of this program is to ensure the issues do not become a limiting factor in the decisions on long-term operation of these assets. Goals for technology introduction are to enhance efficiency, safety, and reliability; improve characterizations of the performance and capabilities of passive and active components during periods of extended operation; and to facilitate introduction of other advanced II&C systems technologies by reducing regulatory uncertainties. The R&D activities of this program are intended to set the agenda for a long-term vision of future operations, including fleetwide integration of new technologies.

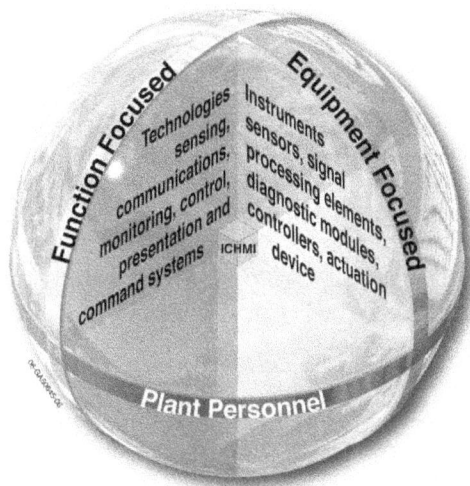

3.3.3 Highlights of Research and Development

A program element of R&D activities is proposed to develop some of the needed critical capabilities of digital technologies to support long-term nuclear asset operations and management. This includes comprehensive programs intended to do the following:

- Develop national capabilities at the university and laboratory level to support R&D

21

- Create or renew infrastructure needed for long-term research, education, and testing

- Support creation of new technologies that can be deployed to address the sustainability of today's II&C systems technologies

- Improve understanding of, confidence in, and facilitate transition to these new technologies

- Support development of the technical basis needed to achieve technology deployments.

3.3.3.1   *Centralized Online Monitoring and Information Integration.* As nuclear power systems begin to be operated during periods longer than originally anticipated, the need arises for more and better types of monitoring of material and system performance. This includes the need to move from periodic, manual assessments and surveillances of physical systems to online condition monitoring. This represents an important transformational step in the management of physical assets. It enables real-time assessment and monitoring of physical systems and better management of active components based on their actual performance. It also provides the ability to gather substantially more data through automated means and to analyze and trend performance using new methods to make more informed decisions about asset management and safety management.

3.3.3.2   *New Instrumentation and Control and Human System Interface Capabilities.* R&D activities are aimed at the eventual modernization of II&C systems technologies used in nuclear energy production. Asset owners and regulators view these as enabling in the dialogue of long-term asset and safety management. The evidence of aged and obsolete technologies is abundant in the control centers of nuclear power plants. The analogy of control rooms as the tip of the iceberg for aging analog technology is particularly apt because it typifies both the problem and a substantial opportunity for R&D to impact systems on a plant scale much larger than what can be readily observed.

Through long-term collaborations with leading international research institutes and capitalizing on new national capabilities for simulation-based technology development and testing, research in visualization, process control, and automation is planned. The long-term objectives of these research activities are to demonstrate new concepts of operations for nuclear power generation assets that address the need for technology modernization, improved state awareness, improved safety, and optimized asset management. These objectives will be achieved by a series of multiyear pilot programs aimed at developing and demonstrating new technologies and concepts for information and control technologies, including the following:

1.   Advanced instrumentation and information pilot projects

2.   Future concept of operations pilot projects

3.   Advanced automation pilot projects.

3.3.3.3   *Nondestructive Examination Technologies.* Activities are proposed to develop and test sensors and characterization methods and technologies for a range of nondestructive examination applications. Working closely with the Nuclear Materials Aging and Degradation R&D pathway, this pathway will develop sensors and accompanying technologies to detect and characterize the condition of material parameters needed to assess the performance of SSC materials during long-term operation, including sensors for measuring material properties to derive parameter estimates of specific aging and performance features and analytic capabilities and methods for characterizing the state and condition of material properties in order to obtain 'diagnostic' accuracy about material aging and degradation. This will provide the ability to move from identification of damage and incipient change to more precise

22

descriptions about the underlying mechanisms of change, their progression in materials, and a description of the specific transformations that affect a material or system's ability to achieve its design function.

Activities also are proposed to build on sensors, characterization, and more refined diagnostics to enable prognostic assessments of materials and performance to be made. These capabilities will aid in answering the 'so what' types of questions that arise in connection with material assessments. This entails extending our knowledge and models of materials and material change processes to include predictions about the eventual consequences of change. This requires the need to incorporate information from material science studies and from other R&D pathways and research programs, including international consortia, to develop interim prognostic models that can be validated and improved through bench scale, engineering scale, and accelerated testing to yield models for predicting the effects of different aging mechanisms and associated phenomena.

3.3.3.4 *Regulatory Engagement.* Ongoing working group activities between the staff of NRC and the nuclear power industry on digital technologies for advanced LWR design submittals underscore the need for a process of engagement within this pathway. Research results and data are needed that can be used for establishment of a regulatory technical basis to support rulemaking and reviews and to provide necessary confidence in the tailored application of these technologies for asset owners. This program includes a specific engagement activity to support interactions with the regulator in order to derive the greatest benefit from these research activities and to achieve goals for eventual deployments.

3.3.3.5 *Industry Working Groups.* Nuclear asset owner engagement is a necessary and enabling activity to obtain data and accurate characterization of long-term operational challenges, assess suitability of proposed research for addressing long-term needs, and gain access to data and representative infrastructure needed to assure success of the proposed R&D activities.

3.3.4 Products and Implementation Plan

The main products of the Advanced II&C Systems Technologies R&D pathway are as follows:

- Technologies for and demonstrations of highly integrated control and display technologies that address long-term objectives of nuclear power plant operation, including the following:

  - Fleetwide management of asset information to support integrated operations

  - Improved visualization and use of information to support decision making and actions

  - Greater automation of functions and availability of operator support systems to improve efficiencies and reduce errors

- Online monitoring of active and passive components to reduce demands for unnecessary surveillance, testing, and inspection; minimize forced outages; and provide monitoring of physical performance of critical SSCs

- Nondestructive examination technologies for characterizing performance of physical systems in order to monitor and manage the effects of aging on SSCs.

The program activities occur in three phases (see Figure 3-6). Phase I, from FY 2010 to FY 2015, R&D activities are intended to create technologies with new functional capabilities. The objectives of this phase are to create and demonstrate new capabilities to achieve the objectives and vision of long-term asset operation. Phase II, from FY 2016 to FY 2020, R&D activities will create more mature technologies that are capable of some field deployments, pilot projects with asset owners, and consortia. Phase III,

23

from FY 2021 to FY 2029, the technology maturity and success with initial deployments will lead to and motivate a shift in the technology base for II&C systems used during long-term operation. Fleet wide deployments and standardization of technology will be ongoing and more R&D activities will lead to greater regulatory engagement and acceptance. Appendix C contains detailed information on the three phases.

Projects	Phase I	Phase II	Phase III
Centralized Online Monitoring	• Algorithm development • Scale studies • Field studies • Industry participation	• Technology maturity • Fleet-wide tests • Industry leadership • Industry standards • License amendments	• Technology standardization • Industry-wide implementation • Regulatory acceptance
New I&C and HSI	• Advanced visualization technology development • System integration concept development • New automation	"First movers" • Individual plant deployment • Industry demonstration	"Modernized Industry" • Fleet-wide deployments • Industry deployment • Standardization
NDE Technologies	• SSC characterization needs defined • Characterization methods and technologies developing	• SSC characterization demonstrated • Characterization methods refinement • License applications using NDE methods	• SSC characterization needs being met • Characterization methods & technologies standard • Industry-wirde and international trending
Advanced I&C Inputs	• Regulatory engagement • Industry participation • International collaboration • University engagement	• Joint regulatory research • Industry working groups and standards bodies • International coordination • University infrastructure	• Integration of R&D with regulatory technical bases • Industry-wide meetings • International standards

09-GA50617-01-15

Figure 3-6. Advanced II&C Systems Technologies pathway implementation schedule.

3.4   Risk-Informed Safety Margin Characterization

3.4.1   Background and Introduction

The Risk-Informed Safety Margin Characterization (RISMC) pathway focuses on advancing the state-of-the-art in safety analysis and risk assessment to support decision making on nuclear power plant life extension beyond 60 years. A comprehensive approach involves four questions that need to be addressed and resolved from the risk and safety perspectives (Figure 3-7). With the plant life extension well beyond the originally designed lifetime, the safety questions take on additional significance due to plant aging (namely how plant aging affects our answer to the four questions), and how confident we are about the answers. In

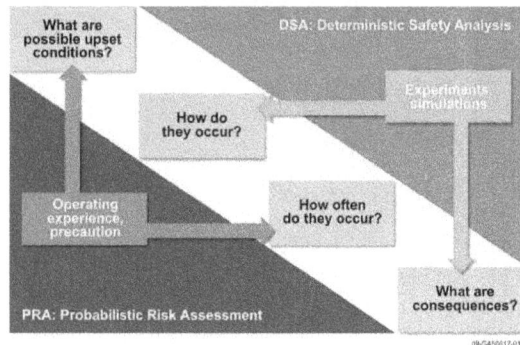

Figure 3-7. Nuclear plant safety analysis.

24

particular, aging of SSCs has potential to increase frequency of initiating events of certain safety transients; create new sequences associated with previously-not-considered SSC failures; and increase severity of safety transients due to cascading failures of SSCs.

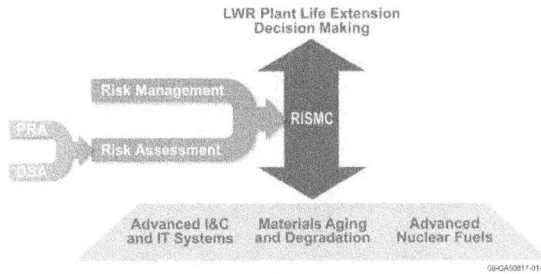

Figure 3-8. RISMC in context of the LWRS R&D Program (PRA = probabilistic risk assessment; DSA = deterministic safety analysis).

The decision on life extension requires us to scrutinize and quantify the uncertainty by which we predict the safety envelope of the aging plant and the efficacy of measures undertaken to manage the aging effect. In this context, the main objective of RISMC R&D is to establish science-based, risk-informed methodology and tools to determine the safety margin envelope with high confidence. Within the LWRS R&D Program, the RISMC pathway provides the bridge from physics and technology-driven pathways to life extension decision-making (Figure 3-8).

The concept of safety margins as a cornerstone in nuclear reactor design emerged during the early days of nuclear power as a part of the defense-in-depth approach to ensuring nuclear safety. Defined as the minimum distance between the system's "loading" and "capacity," safety margin is expressed in terms of safety-significant parameters (e.g., cladding temperature and containment pressure) and determined for a range of anticipated system operating conditions (Figure 3-9). Traditionally, in nuclear power plant design and licensing, availability of safety margins must be demonstrated for a prescribed set of design-basis accidents.

In parallel with a deterministic safety analysis (DSA) approach, probabilistic risk analysis (PRA) methods have been developed and applied to analyze the safety of nuclear power plants. Notably, safety margins calculated by the DSA methods (e.g., accident simulation codes and structural capacity codes) are used to support the specification of "success criteria" in the plant's PRA. Pioneered by the "Reactor

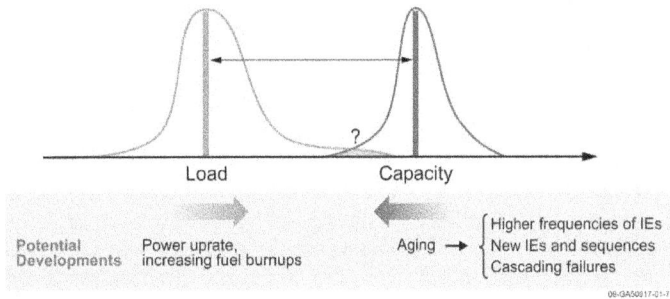

Figure 3-9. Safety margin trend relevant to light water reactor sustainability.

Safety Study" (WASH-1400, 1975), the PRA technology has matured and currently provides the nuclear power industry and the regulator with powerful tools to analyze plant safety, identify system vulnerabilities, provide a framework for effective resource allocation, and focus research and plant operations on risk-significant safety threats. (Appendix D provides more information on PRA methods and the DSA approach.)

The state-of-the-art and trends in R&D related to risk-informed safety analysis topics can be viewed in three interrelated groups: (1) advanced PRA techniques, (2) advanced DSA techniques, and (3) methods and tools for analysis integration and visualization of results that support effective decision making. Overarching themes in all three groups are analysis completeness, uncertainty treatment, and computational efficiency.

25

With respect to methodology for integrated safety assessments, quantification and utilization of plant safety margins and their regulatory implication have received increased attention during recent years, paving way to formulation of RISMC as an R&D area. A comprehensive review of the state-of-the-art and discussion of open issues related to RISMC can be found in the CSNI Safety Margin Action Plan group report, NEA/CSNI/2007(9). Beyond the still-open formidable questions on RISMC framework, it is widely recognized that success of the risk-informed approach requires enhanced simulation tools (computer codes) to enable system analysis with high fidelity and treatment of uncertainties, which can be significant (e.g., in non-design-basis accidents and beyond- design-basis accident situations). These challenges will increase as plant operational life is extended further.

Figure 3-10 depicts technical elements of RISMC in the context of the LWRS R&D Program. In the spirit of defense-in-depth, margin is considered to be significant to the degree that it exceeds uncertainties and variability associated with a given comparison between "load" and "capacity." This idea applies to the success of active functions and passive SSC integrity, which is instrumental to characterization, mechanistic understanding, prediction, and monitoring of the plant aging and degradation behaviors and their impact on plant life extension decision making.

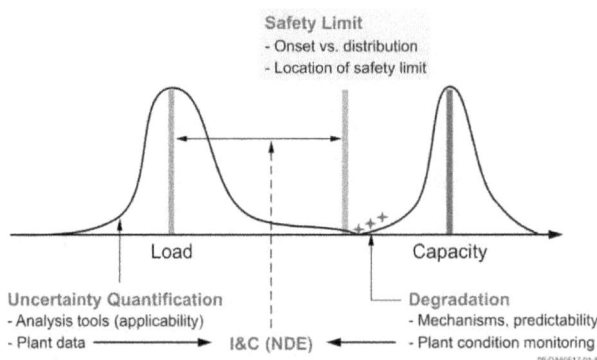

Figure 3-10. Elements of the RISMC model for light water reactor sustainability.

3.4.2   Vision and Goals

Safety is central to design, licensing, operation, and economy of nuclear power plants. As the current LWR nuclear power plants age beyond 60 years, there are possibilities for increasing the frequency of equipment failures that initiate safety-significant events and for creating new failure modes. Accurate characterization of plant safety margins can play an important role in facilitating decision making related to LWRS. In addition, as R&D in the LWRS R&D Program and other collaborative efforts obtain new data and improve scientific understanding of physical processes that govern materials aging and degradation and develop technological advances in nuclear reactor fuels and plant II&C, there are needs and opportunities to manage plant safety, performance, and assets in an optimal way.

Advanced analysis methods and simulation tools for predicting and managing plant response and safety margins are an accepted and essential part of operating and licensing nuclear power plants. Using the science-based models and databases, RISMC provides effective support and guidance to plant operations, maintenance, major components replacement, and plant licensing decisions.

The strategic objectives of the RISMC R&D pathway are to bring together risk-informed, performance-based methodologies with scientific understanding of critical phenomenological conditions and deterministic predictions of nuclear power plant performance, leading to an integrated characterization of public safety margins in an optimization of nuclear safety, plant performance, and long-term asset management. The RISMC research pathway aims to develop an integrated framework and advanced tools for safety assessment that enables more accurate characterization and visualization of the plant's safety margins.

26

3.4.3 Highlights of Research and Development

The RISMC R&D pathway is driven by recognition that risk-informing plant safety margins present an avenue for enhancing operational flexibility and safety benefits obtained from the transition toward risk-informed and performance-based regulation. Existing methods and tools used today in deterministic and probabilistic safety analysis, by themselves and within the current assessment framework, are not adequate to cost-effectively manage the risk and operability significance of aging of SSC. Therefore, there are conceptual and technical "capability gaps" (in frameworks, tools, and data that need to be filled to enable integrated and defensible decision-making regarding the continued operation of nuclear power plants after their current license terms.

Once matured and established, RISMC developments will benefit LWRS R&D Program objectives by (1) creating a strong technical basis for an enhanced risk-informed regulatory structure that enables optimization of plant operation, inspection, maintenance, and replacement of plant SSCs; (2) enabling effective long-term management of plant resources (for which accurate characterization and prediction of safety margins are a prerequisite); and (3) helping guide R&D planning toward maximum payoff from both resource utilization and risk perspectives.

The RISMC R&D pathway is built on the vision that long-term operation of the existing fleet of nuclear power plants requires continued demonstration of their high-level of performance in plant reliability, safety, and economy, and that such objectives require advanced methods and tools to support analysis of plant safety margins and input into operational decision-making. While RISMC pathway planning does not exclude theoretical considerations and generic developments in a broad context, the programmatic approach is driven by the need to ensure effective use of limited resources to meet the anticipated time window (i.e., 2014 through 2019) for investment decision-making of nuclear power plant operators to support plant life extension beyond 60 years. This narrowing down of focus is necessary to develop necessary methods and tools to address the highest priority issues in a topic as broad as RISMC, which involves the whole domain of PRA, DSA, and their short and long-term developmental needs.

Given the LWRS R&D Program focus, the RISMC R&D pathway devised strategy is shown in Figure 3-11 (the RISMC pathway facilitates integration and visualization of R&D contributions in other pathways on sustainability critical information and sustainability critical analytical tools). Areas marked

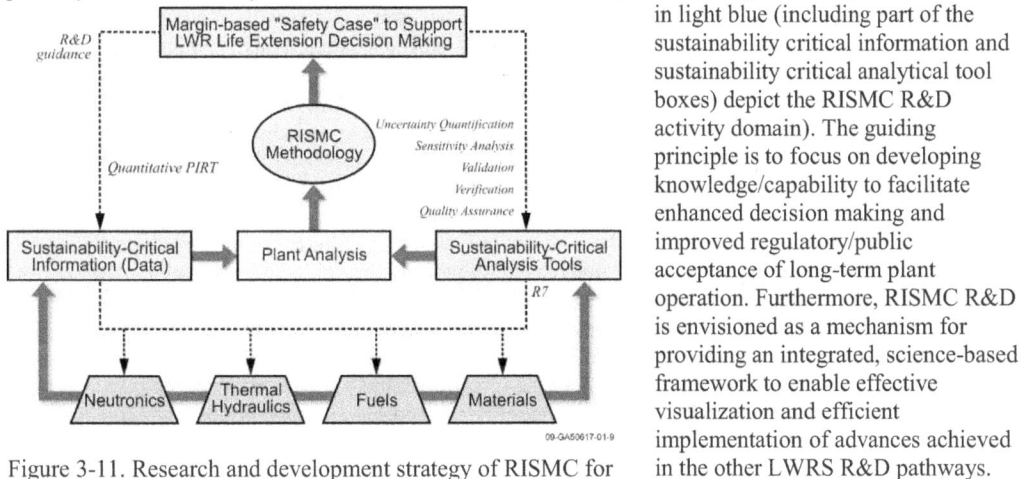

Figure 3-11. Research and development strategy of RISMC for LWR sustainability.

in light blue (including part of the sustainability critical information and sustainability critical analytical tool boxes) depict the RISMC R&D activity domain). The guiding principle is to focus on developing knowledge/capability to facilitate enhanced decision making and improved regulatory/public acceptance of long-term plant operation. Furthermore, RISMC R&D is envisioned as a mechanism for providing an integrated, science-based framework to enable effective visualization and efficient implementation of advances achieved in the other LWRS R&D pathways.

27

The RISMC study effort in FY 2009 resulted in further clarification of the RISMC concept and formed the basis for the project planning as outlined in Figure 3-12 (see Appendix D for more detailed information on the RISMC R&D pathway).

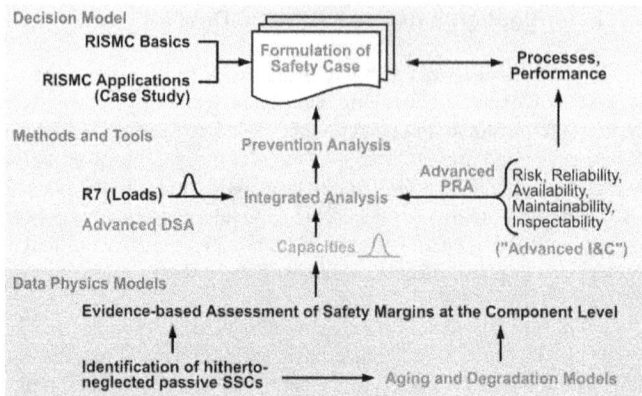

Figure 3-12. RISMC project hierarchy and information flow.

3.4.3.1   RISMC Framework Development. Although LWRs are a mature and successful technology in the United States with an impressive track record in nuclear power plant safety and performance over the past two decades, the next wave of new plant deployment and life extension of the existing operating LWR fleet beyond 60 years is anything but certain. There is broad consensus that technical, cost, and schedule uncertainties in certification and licensing are a significant hindrance to prospective applicants for new licenses, especially for technologies other than LWR. Many discussions tacitly assign a great deal of blame for this to NRC processes.

Part of the traditional approach to licensing is to invest very substantially in margin. The concept of margin has enormous benefits in decision-making, but traditional implementation of the concept has proven to be enormously expensive. A comprehensive set of plausible safety margins will make the sustainability decision easier for both licensees and NRC. Thus, it is important to formulate a margin-based safety case framework aimed at streamlining NRC review and subsequent licensee implementation. The actual technical content of a safety case is necessarily plant-specific; the framework will establish a set of plant-specific technical demonstrations whose integrated presentation to NRC should help to reduce regulatory uncertainty.

Development and demonstration of a new technology-neutral paradigm of science-based safety case development, evaluation, and acceptance will ensure predictable, streamlined, and cost-effective licensing of nuclear installations. It will be achieved through (1) a set of advanced simulation and analysis tools to enable accurate quantification of the system's margins to safety, (2) a formalized (computerized) technology-neutral framework for safety case development, and (3) a knowledge center of previous license applications. A comprehensive, high-quality, and defensible safety analysis submitted by the license applicant is paramount to ensuring the effectiveness of the application's regulatory review.

The proposed research is driven by the idea that reducing uncertainties facing applicants can be achieved not only by working on improved understanding of the technical factors governing particular margins, but by proactively establishing the character and rigor of the portfolio of tests, demonstrations, and commitments to be comprised in the licensing safety case. In the abstract, this idea is not new, but in the United States, previous implementations of it have defaulted to licensing tradition, rather than proactively trying for an improved formulation of the safety case. In short, the proposed task will take up, from a DOE perspective, where NRC left off and identify and address technical issues within the RISMC framework.

28

3.4.3.2 *Technology Integration.* This task was formulated with the objective of identifying crosscut case studies that support formulation and demonstration of the RISMC framework for LWRS. The work scope is accomplished largely within the RISMC working group activity.

3.4.3.3 *Enabling Methods and Tools.* With focus on the effect of plant aging on life extension decision making, characterization of the nuclear power plant safety margin is hindered by large uncertainties that exist in modeling and predicting behaviors of aging SSCs in a broad range of nuclear power plant operating and abnormal conditions and nuclear power plant system dynamics in accident scenarios involving SSC failure modes not studied before. Existing PRA and DSA methods ignore reliability of the plant's passive SSCs and their failure physics, making them unsuited for capturing the essence of aging impact. Of particular interest is identification of catastrophic system degradation scenarios (e.g., cascading failures that cannot be ruled out as "physically unreasonable"). These scenarios require measures (in nuclear power plant inspection, maintenance, and modification) to eliminate system vulnerability. This thrust focuses on advancing the PRA and DSA methods to enable their use in assessing the aging effects on nuclear power plant safety margin.

3.4.3.3.1 *Deterministic Safety Analysis*—Although incremental advances were made continuously over the past two decades to improve modeling of plant components and transient/accident phenomena, the system (plant) analysis tools used in industry's engineering applications remain based on the decades-old modeling framework and computational methodology that have not taken advantage of modern developments in computer/computational science and engineering. Fundamental limitations in the current generation of system analysis codes are well known to the community. Although the codes have served as an adequate basis to address traditional safety margin analysis, significant enhancements will be necessary to support the challenges of extended and enhanced plant operations.

The new generation of system analysis codes (i.e., R7) provides critical capabilities not available in the legacy codes (e.g., RELAP5), which were developed in the 1970s and were used to analyze design-basis accidents. Notably, enhanced capability for simulation of plant dynamics is central to quantification of safety margins in postulated sequences with aging-induced (new and cascading) failure modes. More broadly, the new DSA capability would help address, in a risk-informed manner, a number of safety and licensing issues facing the nuclear power industry. Together, advanced deterministic and probabilistic modeling capabilities would greatly enable RISMC to the benefit of both the regulator and the nuclear power plant operator.

We envision that the new generation production code will build on the decade-old and tested legacy codes (like RELAP5) while capitalizing on extraordinary advances in computing power and computational science (including computational fluid dynamics, neutron diffusion/transport, and fluid-structure interactions) of the past decades. The high-order accurate schemes, modern software architecture, and rigorous procedures for verification and validation are critical in implementing algorithms for sensitivity analysis and performing uncertainty

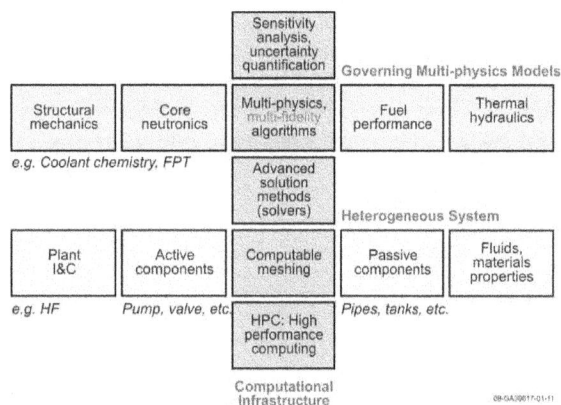

Figure 3-13. Composition of a next-generation production code for nuclear system analysis and safety margin quantification.

29

quantification, which are essential components to improve understanding and utilization of safety margins (Figure 3-13).

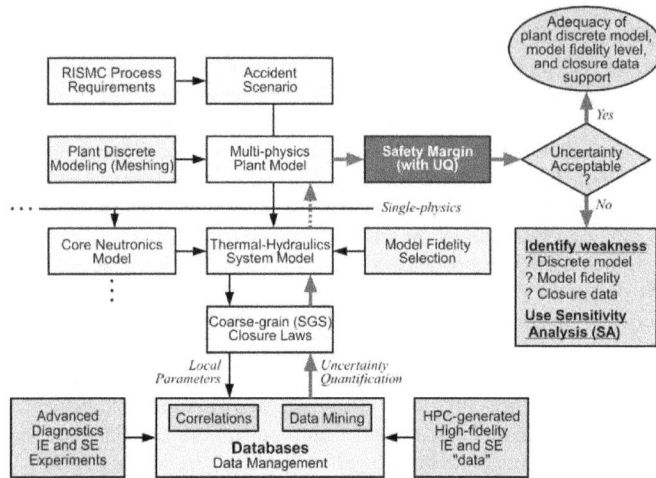

Figure 3-14. Elements of a next generation system code to support the RISMC process.

The new DSA capability would help address, in a risk-informed manner, a number of safety and licensing issues facing the nuclear power industry. Multiphysics coupled treatment offers potential to qualify and improve fidelity of the prediction of safety margins in design-limiting scenarios (Figure 3-14). These same advances and their promulgation in engineering applications would allow identification and quantification of risk-significant transients (both operating and accident sequences), on a scale never before achieved in probabilistic risk analyses. Together, advanced deterministic and probabilistic modeling capabilities would greatly enable RISMC to the benefit of the regulator and the regulated.

3.4.3.3.2 *Probabilistic Risk Analysis*—Like the state-of-practice thermal hydraulic analysis tools that are still used for licensing, the traditional PRA paradigm was formulated in the mid-1970s to resolve certain issues of that time. It, too, is based on simplifications and approximations that are not adequate to support certain decisions today. Although state-of-practice PRA makes some high-level use of certain thermal hydraulic analyses, the usual coupling between thermal hydraulic and scenario-based risk modeling is nowhere near to being close enough to support evaluation of RIMSC. Efforts to transcend the 1970s PRA paradigm are underway; these efforts incorporate dynamical considerations that are all but suppressed in existing PRAs and try to couple directly to mechanistic codes like RELAP. The RISMC R&D pathway needs this development and needs for it to be formulated in a particular way to support RISMC objectives. Participants in this activity will be chosen in a way that leverages the capabilities other institutions in this area.

The computational situation continues to improve at a rapid rate, and the field of dynamic PRA continues to develop. The present subtask is aimed at joining this development and bringing it to bear within the RISMC framework. The capability will require integration of three major components: (1) the simulation engine itself, (2) an internal facility for decision-making that reflects operational procedures based on the current plant state, and (3) coupling to the mechanistic code(s). In addition to all of the inputs to today's PRA, the RISMC application will require consideration of many passive components that are neglected today.

3.4.3.3.3 *Prevention Analysis*—Optimal development of a safety case calls for optimal selection of a set of SSCs and associated levels of performance as the backbone of that safety case. Prevention analysis is the name that has been given to one specific way of doing this. Prevention analysis works by driving a risk model backward. Most applications of risk models proceed by estimating SSC performance a priori and using that information to synthesize plant risk for comparison with objectives.

30

This supports a trial-and-error approach to optimization. In contrast to that approach, prevention analysis starts with a desired top-level safety objective and determines what level of SSC performance would need to be credited in the risk model in order to optimally satisfy that safety objective (in this case, optimality means crediting a complement of equipment that is necessary and sufficient to do the job). The solution to this is not unique; correspondingly, prevention analysis presents the decision-maker with alternative strategies for satisfying top-level objectives. These strategies can be ranked with respect to difficulty and expense of implementation. In short, prevention analysis identifies a complement of nuclear power plant capabilities that, taken together, serve to prevent accidents to the degree specified by the top-level safety objective.

3.4.3.4   *Technology Inputs.* Apart from specialized application areas (such as seismic PRA), most current PRA methodology takes most passive SSCs for granted because it is believed that these components do not contribute significantly to offsite risk. Within the LWRS R&D Program, it is important to challenge that presumption and to examine whether margin issues could emerge for SSCs whose performance is presently taken for granted.

The point of this task is to incorporate, into risk models, passive SSCs whose performance has previously been taken for granted in PRA, but whose loss of physical margin may need to be analyzed. Ultimately, the risk model that these SSCs are added to is the same risk model to be quantified in the enhanced PRA paradigm described above.

3.4.4   Products and Implementation Plan

The main products of the RISMC R&D pathway are as follows:

- R7 code – A system code for mechanistic description and effective simulation of plant transient behavior under a broad range of upset conditions and sequences of risk importance under life extension operation

- RISMC framework – A comprehensive methodology that brings together advanced modeling, simulation and analysis tools, and relevant data to characterize nuclear power plant safety margins, including the effect of plant aging to support plant life extension decision making

- Enabling methods and tools for advanced PRA and advanced prevention analysis to support life extension decision making.

The implementation schedule (Figure 3-15) is structured to support the following high-level milestones:

- 2010

  - Initiate R&D on technology that potentially transforms safety and economics of operating LWRS

  - Formulation of RIMSC methodology

  - Next generation safety analysis tools and R7 code development.

- 2015

  - Release R7 beta version for testing and validation

31

- - Initiate demo of R7-enabled safety analysis that supports life extension decision.

- 2020

  - - High confidence and acceptance by industry and NRC for RISMC process and tools to support power uprate and long-term operations evaluations.

- 2025

  - - Validation of RISMC methods and tools for life extension applications.

- 2030

  - - Industry's broad implementation of RISMC to support plant life extension licensing and enhanced performance.

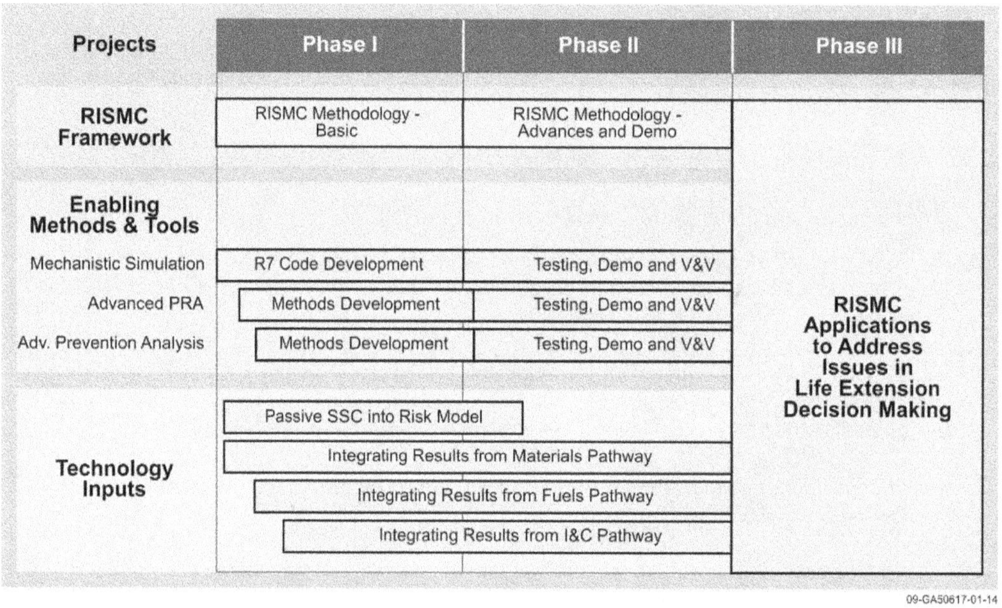

Projects	Phase I	Phase II	Phase III
RISMC Framework	RISMC Methodology - Basic	RISMC Methodology - Advances and Demo	**RISMC Applications to Address Issues in Life Extension Decision Making**
Enabling Methods & Tools			
Mechanistic Simulation	R7 Code Development	Testing, Demo and V&V	
Advanced PRA	Methods Development	Testing, Demo and V&V	
Adv. Prevention Analysis	Methods Development	Testing, Demo and V&V	
Technology Inputs	Passive SSC into Risk Model		
	Integrating Results from Materials Pathway		
	Integrating Results from Fuels Pathway		
	Integrating Results from I&C Pathway		

09-GA50617-01-14

Figure 3-15. RISMC pathway implementation schedule.

3.5   Economics and Efficiency Improvement

3.5.1   Background and Introduction

Improving the economics and efficiency of current LWRs while maintaining excellent safety performance is one of the primary objectives of the LWRS R&D Program. Power uprates have been the most important methods that enable enhancement of the economic performance of the current operating fleet of LWRs. Cooling capability influences thermal efficiency and reliable operation. Increased reactor power and climate change concerns place more burdens on cooling requirements. Expanding the current fleet into nonelectric applications would further increase the value of LWR asset owners. This R&D pathway will focus on three activities: (1) alternative cooling technologies, (2) nonelectric applications

32

(process heat), and (3) power uprate (more detailed information on each activity can be found in Appendix E).

3.5.1.1 *Alternative Cooling.* Water consumed by thermoelectric power plants (such as those fueled by coal, natural gas, and nuclear) continues to receive increasing scrutiny as new power plants are proposed and existing power plants encounter water shortages. Climate change may exacerbate the situation through hotter weather and disrupted precipitation patterns that promote regional droughts. Before 1970, thermoelectric power plants addressed their need for cooling with either fresh or saline water withdrawals for once-through cooling. Since that time, closed-cycle systems (evaporative cooling towers or ponds) have become the dominant choice, with certain impacts on water usage. Figure 3-16 shows the Limerick nuclear power plant in Pennsylvania, which uses mine pool water for a substantial fraction of its cooling.

Figure 3-16. Limerick nuclear power plant.

3.5.1.2 *Nonelectric Application (Process Heat).* Nuclear power plants have very high capital investment and low operating costs. Therefore, to minimize the cost of electricity, these nuclear power plants are typically operated at full power to provide base load needs. With the potential extended power uprates for these nuclear power plants in the future and the eventual construction of new nuclear power plants in the United States, some of the nuclear power plants may need to be operated at reduced power levels when electricity demand is low at off-peak times, such as during the night. Operating nuclear power plants at a reduced power level is certainly not desirable. On the other hand, only about one-fifth of the world's energy consumption is used for electricity generation. Most of the world's energy consumption is for heat and transportation. The existing LWR fleet in the United States has no experience in nonelectric applications. However, the existing LWR fleet might have considerable potential to penetrate into the heat and transportation sectors, which are currently served by fossil fuels that are characterized by price volatility, finite supply, and, more importantly, environmental concerns. There are a wide variety of purely thermal applications of a reactor's output, which may be integrated with an electrical generating plant. These nonelectric applications of nuclear energy include nuclear hydrogen production, providing heat and steam to industrial processes, seawater desalination, and district heating (see Appendix E for more detailed information about these applications). The desalination of seawater using nuclear energy has been demonstrated and nearly 200 reactor-years of operating experience have been accumulated worldwide. District heat involves the supply of heating and hot water through a distribution system, which is usually provided in a cogeneration mode in which waste heat from power production is used as the source of district heat. Several countries have district heating using heat from nuclear power plants.

3.5.1.3 *Power Uprates.* The nuclear industry has been using power uprate since the 1970s as a way of increasing the power output of its nuclear power plants. The primary methods of producing more power are changes in the fuel design, operational changes in reactor thermal-hydraulic parameters, and upgrade of the balance of plant capacity by component replacement or modification (such as replacing a high-pressure turbine). Other changes may include replacing selected feedwater and condensate motors that are already operating at capacity, providing additional cooling for some plant systems, various electrical upgrades to accommodate the higher currents and to improve electrical stability, modifications to accommodate greater steam and condensate flow rates, and instrumentation upgrades that include replacing parts, changing set points, and modifying software. As of today, NRC has approved 127 power

33

uprate submittals. The total extra power generated from power uprate is equivalent to building five 1,000-MWe new nuclear power plants, which significantly enhanced the asset value of the plant owners. There are three types of power uprates[f] (descriptions of the power uprates are provided in Appendix E): (1) measurement uncertainty recapture power uprates are less than 2% and are achieved by implementing enhanced techniques for calculating reactor power; (2) stretch power uprates are typically up to 7% and are within the design capacity of the plant; and (3) extended power uprates are greater than stretch power uprates and have been approved for increases as high as 20%.

3.5.2   Vision and Goals

The commercial nuclear power industry will undertake more extended power uprate and ultra power uprate activities, have alternative cooling technology options ready to maximize water usage and accommodate uprated power output, and expand to nonelectric applications within the framework of plant life extension to minimize the cost of production and maximize return on investment.

The programmatic goals for this R&D pathway are captured in the following statements:

1.   Power Uprates: Provide scientific and engineering solutions to facilitate extended power uprates and ultra high power uprates for all operating LWRs in a cost-effective manner. Specific goals are to enable boiling water reactors to achieve above 20% extended power uprate and pressurized water reactors to achieve up to 20% power uprate by the year 2030.

2.   Alternative Cooling Technology: Conceive, develop, and establish deployable technologies for optimizing use in the nuclear energy thermocycle while minimizing reliance on water resources at the same time.

3.   Nonelectric Application (Process Heat): Penetrate the applications of existing LWRs to low temperature process heat and hydrogen production market.

3.5.3   Highlights of Research and Development

3.5.3.1   Alternative Cooling. Alternatives to closed-cycle cooling (wet cooling tower) are generally dry cooling (waste heat rejected to the atmosphere) or hybrid cooling (using aspects of both wet and dry cooling), as well as replacing freshwater supplies with degraded water sources. Degraded water is polluted water that does not meet water-quality standards for various uses such as drinking, fishing, or recreation. Existing operating LWRs in the United States use either once-through cooling or wet cooling tower.

It is essential to provide adequate and timely cooling for safe and economic operation of nuclear power plants. With more stringent regulation on the temperature of the discharged cooling water from a nuclear power plant, the availability of clean cooling water, increased cooling load with the power uprates, and potential warmer weather in the summer season due to global climate change, alternative and potentially advanced cooling technology has to be developed in order to ensure the reactors can be safely and economically operated without being forced to shut down or reduce the power output due to cooling water issues. R&D activities will focus on: (1) technology development such as advanced condenser design, reducing water losses in the wet cooling tower system, improving dry cooling and hybrid cooling technology, and ice thermal storage system; (2) evaluating applicability of alternative water-conserving cooling technologies (such as dry cooling, hybrid cooling, and ice thermal storage system) to improve LWR plant efficiency and relieve the cooling water requirement, as well as expand use of alternative

---

f.   http://www.nrc.gov/reactors/operating/licensing/power-uprates/type-power.html.

sources of water; and (3) improving analysis methodology and performing actual analysis to identify optimal designs and developing water resource assessment and management decision support tools (more detailed information on these technologies is found in Appendix E).

3.5.3.2   *Nonelectric Application (Process Heat).* Nuclear power plants produce 1,500 to 4,500 MW of steam. Very few markets exist for such large quantities of steam. Usually, it is not economical to modify a nuclear power plant to produce a few megawatts of heat to meet a local industry or district-heating need; therefore, district heating will not be considered. Under current circumstances, seawater desalination using existing LWRs also is a very remote possibility. However, biomass-to-fuel-ethanol plants require very large quantities of low-temperature steam. Using nuclear energy for transportation indirectly through transportation fuel ethanol production has the potential to open new markets for existing LWRs. For example, low-temperature steam from nuclear power plants can be extracted to help produce ethanol from starch.

Steam from nuclear power plants also can be used to provide process heat to a Fischer-Tropsch chemical process (or similar processes) to produce synthetic fuel. Coal gasification has the advantage of the reduction of air emissions from coal combustion, an increased thermal efficiency of combustion, and use of a large resource base. If coal gasification becomes widespread, economic and environmentally benign technologies for the supply of gasification energy will be required. Nuclear energy, being an industrially proven and nonpolluting technology, is a valid candidate for this purpose.

Using nuclear energy to produce hydrogen is likely to facilitate another application of nuclear energy. The share of nuclear energy in a hydrogen-based system will depend on its competitiveness with other energy options such as natural gas. Successful demonstration projects (such as use of surplus nuclear capacity for hydrogen production using cheap off-peak electricity) would help promote the nuclear-hydrogen link.

The technical and economic viability of different applications will be studied. One key issue to be addressed is interface design and plant modifications. Appendix E provides further details on low-temperature distillation, nuclear hydrogen production, and heat source for synthetic fuel production.

3.5.3.3   *Power Uprates.* R&D activities will be focused on enabling safe and cost-effective plant modifications and modernizations required to gain margins by enhancing the plant power limiting equipment capability. Consistent with the main themes currently identified in this R&D pathway, activities are planned in the following main areas to significantly uprate the current LWR power levels (details on each of these activities can be found in Appendix E):

1.   Collaboration with Nuclear Materials Aging and Degradation R&D Pathway

2.   Fuel performance and loading management

3.   Reactor thermal hydraulics

4.   Safety assessment under high power

5.   Balance of plant, including steam generators for pressurized water reactors

6.   Operation with higher core outlet temperature

7.   Instrumentation and control systems and software reliability

35

3.5.4   Products and Implementation Plan

The main products of this pathway are as follows:

- Advanced cooling technologies that would reduce cooling water requirements and improve the plant's thermal efficiency

- Tools, methods, and technologies (collaborating with the other four pathways) to enable more extended power uprates or even ultra high power uprates

- Feasibility studies of the technical and economic viability of expanding the existing fleet into nonelectric applications.

The implementation schedule (Figure 3-17) is structured to support the following high-level milestones:

- 2015: Preserve the once-through cooling technologies (advanced water conservation technologies for wet cooling tower).

- 2015: Complete feasibility studies for nuclear hydrogen production and low temperature distillation applications.

- 2020: Ensure significant cost reduction of dry cooling technology and thermal efficiency improvement in the hot summer timeframe.

- 2020: Ensure next generation safety analysis tools available to support extended power uprates and ultra power uprates.

- 2025: Apply advanced cooling technologies.

- 2030: Enable 10-GWe extra capacity additions through more extended power uprates or even ultra high power uprates.

	Phase I	Phase II	Phase III
Alternative Cooling Technology	Preserve once-through technology	Cost reduction and efficiency improvement for dry and hybrid cooling technology	Application of advanced cooling technologies
	Development of water conservation technology for wet cooling tower		
Non-electirc Applications	Technology and economics viability	Interface design	Applications
Power Uprate	Collaborate with other pathways to enable 10 GWe extra capacity addition through power uprates, with a stretch goal of 20 GWe		

09-GA50277-05f

Figure 3-17. Economics and Efficiency Improvement pathway implementation schedule.

36

3.6   Pathway Crosscutting and Integration

The overall focus of the R&D activities will be on practically advancing the ability of the owner of nuclear assets to manage the effects of the aging of passive components and increase the efficiency and economics of operations. Transformational activities initially should be developed as limited-scope pilots that provide obvious, value-driven return for the asset owner. In selecting projects, it is vital that all consideration be given to how each of the pathways can support achievement of safety and economic sustainability for existing LWRs by ensuring that each pathway is appropriately coordinated with the desired outcomes of the other pathways. Technical integration is an important and significant part of the LWRS R&D Program. R&D within the program is integrated across scientific and technical disciplines in the five R&D pathways. The LWRS R&D Program is integrated with outside sources of information and parallel R&D programs in industry, universities, and other laboratories, both domestic and international. Different methods of integration are used depending on the situation and goals.

3.6.1   Technical Integration

Interfaces between R&D pathways and the required integration across them are naturally defined by common objectives for materials and fuel performance and the system monitoring of their performance. Similarly, interface and integration of the pathways with the RISMC R&D pathway is defined by data and models, which affect performance, monitoring, and control (Figure 3-18).

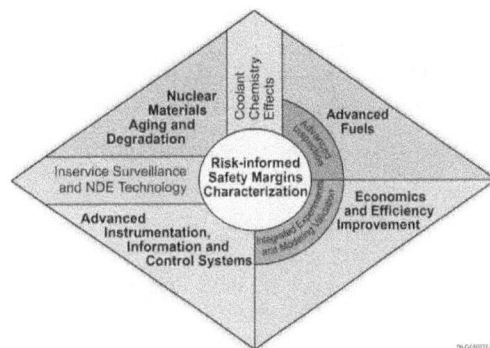

Figure 3-18. Integration of five research and development pathways.

Data and information from the Nuclear Materials Aging and Degradation, Advanced LWR Nuclear Fuel, and Economics and Efficiency Improvement R&D pathways will be fed into the RISMC models. Results of the RISMC analysis will guide development of advanced fuels; materials aging and degradation mitigation; advanced II&C systems; and economics and efficiency improvement. Table 3-1 includes examples of some crosscutting areas in the LWRS R&D Program.

3.6.2   Enhanced Modeling as a Crosscutting Activity

The most common theme from all five R&D pathways is use of computer modeling of physical processes or development of a larger system computer model. Extensive use of computer modeling by all five R&D pathways is intended to distill the derived information so that it can be used for further research in other pathways and as the basis for decision making.

Computer modeling occurs in three forms with many overlapping aspects within the LWRS R&D Program. Modeling a physical behavior (such as crack initiation in steel) is an example of direct computer modeling. The resulting model is used to store information for use in other pathways and to use in its own right for further research.

A second computer modeling activity is development of more detailed computer modeling tools capable of encoding more complex behaviors. One of the intended outcomes from Advanced LWR Nuclear Fuels Development research are new modeling tools that can describe behavior of such complexity that current computer models are incapable of producing adequate results for the LWRS R&D Program. The increased accuracy will allow improved results to be incorporated into other pathways.

Table 3-1. Program crosscutting areas.

Crosscutting Area	Materials Aging and Degradation	Advanced Fuels	Advanced II&C	RISMC	Economics and Efficiency Improvement
Coolant chemistry effects	X	X	X	X	X
Crack growth mitigation effects	X	X		X	X
Irradiation testing	X	X			
Irradiation source term changes		X		X	
Improved online monitoring of reactor chemistry	X	X	X	X	X
Advanced instrumentation for the study of system degradation	X		X	X	X
Fuel failure mechanisms		X		X	X
Creation of SSC aging database	X		X	X	X
Advanced measurement techniques	X	X	X	X	X
Field testing and data collection/capture	X	X	X	X	X
Nondestructive evaluation/assay tools	X	X	X	X	X
Advanced inspection techniques	X	X	X	X	X

The final computer modeling improvement is creation of larger integrated databases that roll up results and allow decision-making. The large system-wide, integrated models allow complex behavior to be understood in new ways and new conclusions to be drawn. These integrated databases can be used to further guide physical and modeling research, improving the entire program.

Because of their overlapping nature and personal interfaces, these modeling activities tend to be natural crosscutting activities between R&D pathways. Computer modeling will remain an integrating and crosscutting element of the LWRS R&D Program.

3.6.3   Coordination with other Research Efforts

In order to encourage communication and coordination with outside experts and parallel programs, the LWRS R&D Program will be aware of issues and changes of technical needs that affect long-term, safe, and economical operation of existing operating LWRs, and share information and resources with other professionals and programs that can assist the LWRS R&D Program to provide timelier, less expensive, and better solutions to the needs and issues.

Primarily, coordination will be with the EPRI Long-Term Operation Program. At the program level, formal interface documents will be used to coordinate planning and management of the work. This will provide a ready source of information from EPRI's Nuclear Power Council and through their contact with utilities. At the R&D project level, both programs encourage frequent communication and collaboration.

Consistent with the vision of the LWRS R&D Program, working relationships have been established with international organizations in FY 2009 and will continue in FY 2010 and beyond. The goal is to facilitate communication and cooperative R&D with international R&D organizations.

38

3.6.4   Performance of Technical Integration and Coordination

The LWRS R&D Program will lead and encourage technical integration and coordination of issues affecting the LWR long-term operation program using methods that best match the issue. For known gaps in data, understanding, or technology, the LWRS R&D Program will plan and manage integrated R&D projects through the LWRS R&D Program Technical Integration Office (TIO) and its multiple interfaces.

To accommodate currently unknown issues or gaps in technology that may arise as result of ongoing R&D or nuclear power plant operations, a broader approach is necessary. This approach should include active internal and external communication with professional organizations, industry groups, and interdisciplinary teams for project and program reviews. The Steering Committee is an essential part of this process. The LWRS R&D Program encourages participation in professional technical societies and national standards committees.

4.   PROGRAM MANAGEMENT

4.1   Organization Structure

The entire LWRS R&D Program falls within the DOE Office of Nuclear Energy. Program management and oversight, including programmatic direction, project execution controls, budgetary controls, and TIO performance oversight is provided by the DOE Office of LWR Technologies in conjunction with the DOE Idaho Operations Office (DOE-ID) (Figure 4-1).

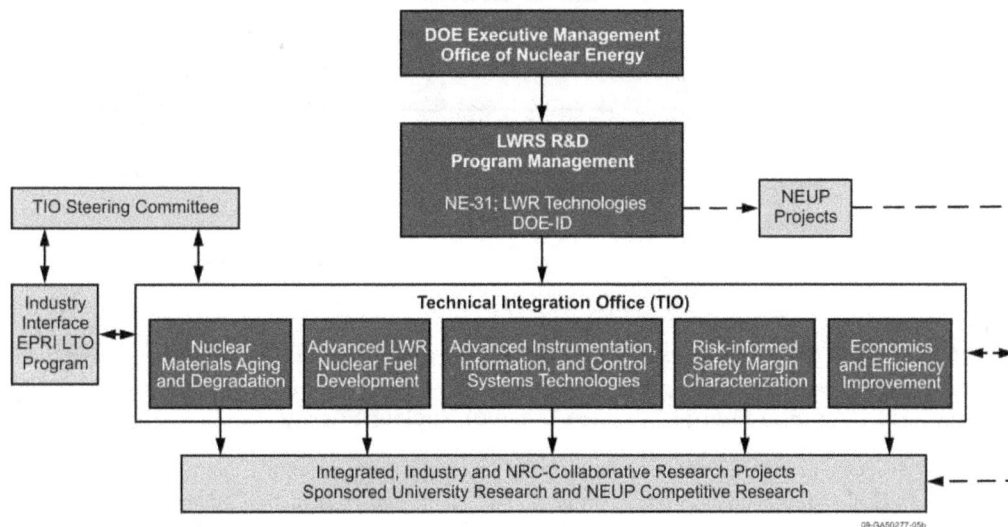

Figure 4-1. Program organization.

DOE-ID will provide technical and administrative support to the LWRS R&D Program. This support includes activities such as assisting in development of administrative requirements in support of contracting actions, conducting merit reviews and evaluations of applications received in response to program solicitations, performing all contracting administration functions, and providing technical project management and monitoring of assigned projects.

39

The TIO basic organizational structure is used to accommodate the crosscutting nature of the proposed research pathways. This organization is responsible for developing and implementing integrated research projects consistent within the LWRS R&D Program vision and objectives. Additionally, the TIO is responsible for developing suitable industry and international collaborations appropriate to individual research projects and acknowledging industry stakeholder inputs to the program.

Within the TIO structure is the TIO Director, each of the five R&D pathway leads, and an external Steering Committee. Nuclear industry interfaces and stakeholders' contributions are accommodated in program development and project implementation actions through the TIO management structure. Recognition of continuing industry collaborations reflecting issues and concerns necessary to extend plant licenses are incorporated through the same program development and implementation actions.

The functional organization, reporting relationships, and roles and responsibilities for the TIO are explained in the following sections and are shown in Figure 4-2.

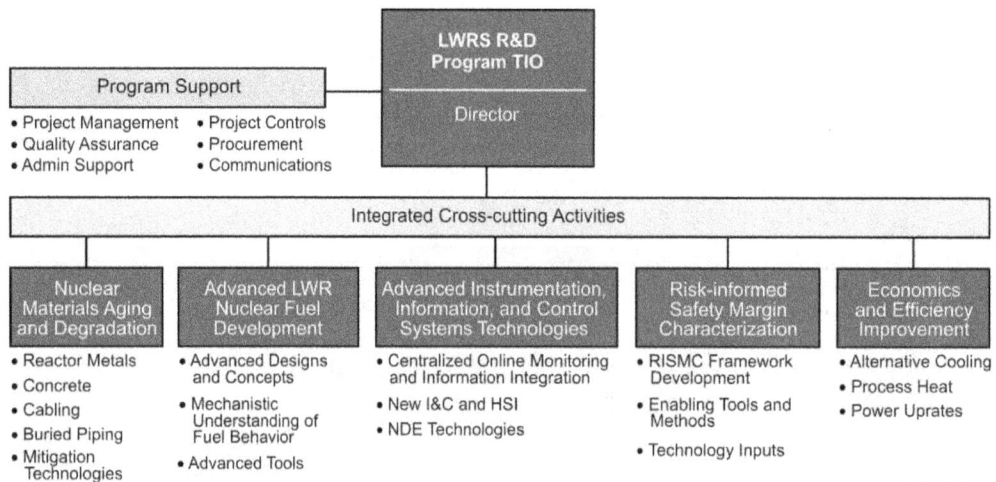

Figure 4-2. Technical Integration Office organization.

4.2 Roles, Responsibilities, Accountabilities, and Authorities

4.2.1 Department of Energy Program Office of Nuclear Energy

DOE is responsible for the Federal government's investments in nuclear power research, development, demonstration, and incentive programs, which all further the nation's supply of clean, dependable nuclear-generated electricity. The LWRS R&D Program conducts research that enables licensing and continued reliable, safe, long-term operation of current nuclear power plants beyond their initial license renewal period. The DOE Office of LWR Deployment directs the program, establishes policy, and approves scope, budget, and schedule for the program through the LWRS R&D Program Manager. The DOE LWRS R&D Program Manager is assisted with program management and oversight by DOE-ID.

The essential programmatic DOE functions include, but are not limited to, the following:

•   Establish program policy and issue program guidance

40

- Establish requirements, standards, and procedures

- In cooperation with the TIO, establish requirements and develop strategic and project plans

- Establish performance measures and evaluate progress

- Represent the DOE program to other government agencies.

4.2.2   Technical Integration Office

TIO supports the DOE Program Manager. The program is a cost-shared, collaborative program aimed to meet the needs of a diverse set of stakeholders. In addition to supporting national policy (energy and environmental security needs), the program supports agreed upon technical needs of NRC in assessing safety and relicensing requests for nuclear power plant extended life operation. It also supports industry needs for data and planning tools for long-term safe economical operation of their nuclear power plants. TIO is staffed with a director, R&D pathway leads, and program management staff. The director and leads are all well-known technical and management experts from DOE laboratories. The TIO is structured and staffed to provide the program director with strong interfaces and communications with stakeholders, R&D plans based on stakeholder needs, proposals for R&D-specific projects and budgets, management of the projects, including funding, and communication of the results.

The LWRS R&D Program TIO is a national organization and is expected to have international participants as the LWRS R&D Program evolves. The intent of the organization is to staff the program with the right people to accomplish the work, regardless of location or affiliation. As appropriate, the technology integration and execution activities will use facilities and staff from multiple national laboratories, universities, industrial alliance partners, consulting organizations, and research groups from cooperating foreign countries.

TIO functions include the following:

- Maintaining the long-range technical strategy plan for the LWRS R&D Program

- Maintaining the LWRS R&D Program Plan

- Developing annual project scope statements

- Developing and implementing the project execution plan

- Monitoring authorized project work

- Coordinating weekly/monthly status meetings

- Coordinating periodic technical review meetings

- Providing formal status reporting

- Maintaining baseline change control

- Performing project closeout planning and completion.

41

4.2.2.1   Technical Integration Office Director. The TIO director provides general program management for the LWRS R&D Program. This position leads the planning, performance, and communication of results from the R&D pathways. The TIO director works with the program support team and R&D pathway leads to integrate and ensure all requirements are well defined, understood, and documented through long-range planning. The TIO director works with the project support staff to ensure proper annual financial planning, scoping, oversight, and scheduling of the project work. The TIO director and the Steering Committee oversee assignment of appropriate resources and evaluate and resolve R&D needs of the LWRS R&D Program. The TIO director reports to DOE Program Manager.

4.2.2.2   R&D Pathway Leads. The TIO currently includes five R&D pathway leads for the major R&D areas currently developed. The leads are the technical managers for their pathways and are responsible for ensuring that technical planning, project management, and leadership is provided for each pathway. R&D pathway leads are the primary interface between technically diverse organizations that form the structure of the LWRS R&D Program. They are responsible for integration and translation of project requirements into an overall program plan tailored to accomplish their assigned R&D mission. They are responsible for establishing scope, cost, and schedule of the R&D activities. They interface with other R&D pathway leads to ensure effectiveness of crosscutting activities.

4.2.2.3   Program Support Team. The program support staff is responsible for contractual operations of TIO and assists other parts of TIO to execute work. The team provides personnel with expertise in project management, quality assurance, procurement, project controls, and communications. They provide tools, structure, oversight, and rigor to maintain R&D schedules and interfaces to the LWRS R&D Program; provide financial information to management (through the TIO director's office); monitor technical progress and earned value; and track milestones.

4.2.3   Project Monitoring and Evaluation

DOE and TIO use a variety of methods to provide oversight of their projects, including semiannual project reviews, periodic progress reports, and scheduled evaluations, invoice reviews, and participation in periodic project meetings and conference calls.

4.2.3.1   Project Reviews. DOE and TIO conduct semiannual and annual project progress review meetings with the project participants, including all research pathway leaders. During these project review meetings, project activities, schedule progress, and cost are discussed in detail. Status of deliverables, funding, or schedule concerns and potential changes in scope are discussed. Performance expectations for the remainder of the budget period and project are reviewed. On an annual basis, DOE staff reviews the work scope, budget requirements, schedule, deliverables, and milestones for the subsequent budget periods as part of the approval of project continuation requests. Review of these continuation requests often requires face-to-face meetings with project participants to fully understand the future planned work.

4.2.3.2   Periodic Project Status Meetings and Conference Calls. DOE, TIO, and pathway leaders participate in periodic project status meetings and conference calls. Typically, project conference calls are the method of choice because of the number and location of participants; they are held at least twice a month. In addition, DOE staff participates in TIO conference calls on specific tasks.

4.2.3.3   Monthly Progress and Earned Value Reporting. DOE personnel review and evaluate project monthly progress reports for the project task and activity progress, accomplishment of deliverables, and budget and cost status. Because of the size, cost, and complexity of integrated LWRS projects and collaborative projects, earned value will be reported on a monthly basis. This earned value reporting provides project participants and DOE staff with a monthly snapshot of overall project cost and schedule performance against the project baseline.

42

4.3   INTERFACES

The LWRS R&D Program TIO is intended as a national organization and is expected to have multiple national laboratory, governmental, industrial, international, and university partnerships. As appropriate, the LWRS R&D Program technology development and execution activities will use facilities and staff from national laboratories, universities, industrial alliance partners, consulting organizations, and research groups from cooperating foreign countries.

TIO is responsible for ensuring the necessary memorandum purchase orders, interagency work orders, or contracts are in place to document work requirements, concurrence with work schedules and deliverables, and transfer funds to the performing organizations for R&D activities.

4.3.1   Steering Committee

A standing TIO Steering Committee advises TIO on the content, priorities, and conduct of the program. The committee is comprised of technical experts selected and agreed upon by the TIO director and the DOE Program Manager. The committee, as a group, is knowledgeable of the various R&D needs of DOE, industry, and NRC; ongoing and planned research as related to nuclear power technology; and policies and practices in public and private sectors that are important for the collaborative R&D program. The TIO director, in consultation with the Steering Committee, may form ad hoc subcommittees to review specific technical issues.

4.3.2   Industry Partners

Planning, execution, and implementation of the LWRS R&D Program are done in coordination with U.S. industry and NRC to assure relevance and good management of the work. The LWRS R&D Program addresses some of the most pressing R&D needs identified in the Strategic Plan, including R&D needed by currently operating LWRs to extend their safe economical lifetime to significantly contribute to the long-term energy security and environmental goals of the United States. EPRI has established the Long-Term Operations Program to run in parallel with the DOE LWRS R&D Program. The Long-Term Operations Program is based on the LWR R&D Strategic Plan and focuses on the long-term operations of the current fleet. EPRI and industry's interests are applications of the scientific understanding and the tools to achieve safe, economical, long-term operation. Therefore, the government and private sectors' interests are similar and interdependent, leading to strong mutual support for technical collaboration and cost sharing. Formal interface agreements between EPRI and the TIO will be used to coordinate collaborations. Contracts with EPRI or other businesses may be used as appropriate for some work.

4.3.3   International Partners

TIO has made contact with several international organizations with interests and programs in long-term operation of LWR technology and the R&D to support those interests and programs. We expect to continue to develop these contacts to provide timely awareness of emerging issues and their scientific solutions. A close working relationship with the Organization for Economic Cooperation and Development's Halden Project and with Electricite de France's Materials Aging Institute are particularly important to the LWRS R&D Program. As funding is available, the LWRS R&D Program intends to initiate formal R&D agreements with both institutions.

43

4.3.4   University Partners

Universities will participate in the program in at least two ways: (1) through the Nuclear Energy University Program and (2) with direct contracts. In addition to contributing funds to the Nuclear Energy University Program, the LWRS R&D Program will provide to the Nuclear Energy University Program descriptions of research from universities that would be helpful to LWRS R&D Program. In some cases, R&D contracts will be let to key university researchers.

4.3.5   NRC Partnership

DOE's mission to develop the scientific basis to support both planned lifetime extension up to 60 years and lifetime extension beyond 60 years and to facilitate high-performance economic operations over the extended operating period for the existing LWR operating fleet in the United States is the central focus of the LWRS R&D Program. Therefore, more and better coordination with industry and NRC is needed to ensure that there is a single national strategic plan, shared objectives, and efficient integration of collaborative work for LWRS. This coordination requires that articulated criteria for the work appropriate to each group be defined in memoranda of understanding that is executed among these groups. NRC has a memorandum of understanding[g] in place with DOE, which specifically allows for collaboration on research in these areas. Although the goals of NRC and DOE research programs differ in many aspects, fundamental data and technical information obtained through joint research activities is recognized as potentially of interest and useful to each agency under appropriate circumstances. Accordingly, to conserve resources and to avoid needless duplication of effort, it is in the best interest of both parties to cooperate and share data and technical information and, in some cases, the costs related to such research, whenever such cooperation and cost sharing may be done in a mutually beneficial fashion.

---

[g] "Memorandum of Understanding Between U.S. Nuclear Regulatory Commission and U.S. Department of Energy on Cooperative Nuclear Safety Research," dated April 22, 2009, and signed by Brian W. Sheron, Director, Office of Nuclear Regulatory Research, U.S. Nuclear Regulatory Commission and Rebecca Smith-Kevern, Acting Deputy Assistant Secretary for Nuclear Power Deployment, Office of Nuclear Energy, U.S. Department of Energy.

5. BUDGET SUMMARY

Table 5-1. Five-year program budget profile by work breakdown structure ($K).

		FY-09	FY-10	FY-11	FY-12	FY-13	>FY-13[1]
1.0	Light Water Reactor Sustainability Program						
1.1	Management	481	3,100	8,600	14,500	22,000	28,000
	1.1.1 Technical Integration Office (TIO)						
	1.1.2 DOE Headquarters Program Management[2]						
	1.1.3 Program Controls						
1.2	Materials	602	2,000	6,000	10,000	15,000	20,000
	1.2.1 Project Management at Oak Ridge National Laboraroty						
	1.2.2 Reactor Metals						
	1.2.3 Concrete						
	1.2.4 Cabling						
	1.2.5 Buried Piping						
	1.2.6 Mitigation Technologies						
	1.2.7 Integrated Research Activities						
1.3	Fuels	480	1,900	5,000	9,000	15,000	18,000
	1.3.1 Project Management at INL						
	1.3.2 Advanced Designs and Concepts						
	1.3.3 Mechanistic Understanding of Fuel Behavior						
	1.3.4 Advanced Tools						
1.4	Instrumentation Information Systems Technologies	208	900	4,000	6,000	7,000	9,000
	1.4.1 Project Management at INL						
	1.4.2 Centralized Online Monitoring and Information Integration						
	1.4.3 New Instrumentation and Control and Human System Interfaces and Capabilities						
	1.4.4 Nondestructive Examination Technologies						
1.5	Risk-Informed Safety Margin Characterization	229	2,100	5,400	8,000	11,000	15,000
	1.5.1 Project Management at INL						
	1.5.2 RISMC Framework						
	1.5.3 Technology Integration						
	1.5.4 Enabling Methods and Tools						
	1.5.5 Technology Inputs						
1.6	Economics and Efficiency Improvements		1,000	2,500	5,000	10,000	
	1.6.1 Project Management at INL						
	1.6.2 Alternative Cooling						
	1.6.3 Process Heat						
	1.6.4 Power Uprates						
Grand Totals		2,000	10,000	30,000	50,000	75,000	100,000

1. Steady-state, long-term funding levels.
2. Includes Nuclear Energy University Program (20% of total budget) and SBIR/STTR (2.8% of total budget) after FY 2009. For FY 2009, a Nuclear Energy University Program project was funded under the Gen-IV Program and is not included here.

45

6. APPENDIXES

Appendix A, Nuclear Materials Aging and Degradation

Appendix B, Advanced Light Water Reactor Nuclear Fuel Development

Appendix C, Advanced Instrumentation, Information, and Control Systems Technologies

Appendix D, Risk-Informed Safety Margin Characterization

Appendix E, Economics and Efficiency Improvement

46

Congressional
Research
Service

Renewable Energy R&D Funding History: A Comparison with Funding for Nuclear Energy, Fossil Energy, and Energy Efficiency R&D

Fred Sissine
Specialist in Energy Policy

April 9, 2008

Congressional Research Service

7-5700

www.crs.gov

RS22858

CRS Report for Congress ————————————
Prepared for Members and Committees of Congress

Summary

Energy research and development (R&D) intended to advance technology played an important role in the successful outcome of World War II. In the post-war era, the federal government conducted R&D on fossil fuel and nuclear energy sources to support peacetime economic growth. The energy crises of the 1970s spurred the government to broaden the focus to include renewable energy and energy efficiency. Over the 30-year period from the Department of Energy's inception at the beginning of fiscal Year (FY) 1978 through FY2007, federal spending for renewable energy R&D amounted to about 16% of the energy R&D total, compared with 15% for energy efficiency, 25% for fossil, and 41% for nuclear. For the 60-year period from 1948 through 2007, nearly 11% went to renewables, compared with 9% for efficiency, 25% for fossil, and 54% for nuclear.

Contents

Figures

Tables

Contacts

Introduction

This report provides a cumulative history of Department of Energy (DOE) funding for renewable energy compared with funding for the other energy technologies—nuclear energy, fossil energy, and energy efficiency. Specifically, it provides a comparison that covers cumulative funding over the past 10 years (FY1998-FY2007), a second comparison that covers the 30-year period since DOE was established at the beginning of fiscal year 1978 (FY1978-FY2007), and a third comparison that covers a 60-year funding history (FY1948-FY2007).

Guide to Tables and Charts

Table 1 shows the cumulative funding totals in real terms for the past 10 years (first column), 30 years (second column), and 60 years (third column). **Table 2** converts the data from **Table 1** into relative shares of spending for each technology, expressed as a percentage of total spending for each period.

Figure 1 displays the data from the first column of **Table 2** as a pie chart. That chart shows the relative shares of cumulative DOE spending for each technology over the 10 years from FY1998 through FY2007. **Figure 2** provides a similar chart for the period from FY1978 through FY2007. **Figure 3** shows a chart for FY1948 through FY2007.

Background

The availability of energy—especially gasoline and other liquid fuels—played a critical role in World War II. Another energy-related factor was the application of research and development (R&D) to the atomic bomb and other military technologies. During the post World War II era, the federal government began to apply R&D to the peacetime development of energy sources to support economic growth. At that time, the primary R&D focus was on fossil fuels and new forms of energy derived from nuclear fission and nuclear fusion.

From FY1948 through FY1977 the federal government provided an extensive amount of R&D support for fossil energy and nuclear power technologies.[1] Total spending on fossil energy technologies over that period amounted to about $15.4 billion, in constant FY2008 dollars. The federal government spent about $46.4 billion (in constant FY2008 dollars) during that period for nuclear fission and nuclear fusion energy R&D.[2]

The energy crises of the 1970s spurred the federal government to expand its R&D programs to include renewable (wind, solar, biomass, geothermal, hydro) energy and energy efficiency technologies. Modest efforts to support renewable energy and energy efficiency began during the early 1970s. From FY1973 through FY1977 the federal government spent about $1.5 billion (in

---

[1] DOE. Pacific Northwest Laboratory. *An Analysis of Federal Incentives Used to Stimulate Energy Production.* 1980. The spending for fossil energy included coal, oil, and natural gas technologies.

[2] DOE (Pacific Northwest Laboratory), *An Analysis of Federal Incentives Used to Stimulate Energy Production,* 1980.

constant FY2008 dollars) on renewable energy R&D, $140 million on energy efficiency R&D, and $170 million on electric systems R&D.[3]

The Department of Energy was established by law in 1977. All of the energy R&D programs—fossil, nuclear, renewable, and energy efficiency—were brought under its administration. DOE also undertook a small program in energy storage and electricity system R&D that supports the four main energy technology programs.[4] DOE's funding support for those technologies began in FY1978. Funding for all four of the main technologies skyrocketed initially, and then fell dramatically in the early 1980s.

Table 1. DOE Energy Technology Cumulative Funding Totals
(billions of 2008 dollars)

Technology	Period		
	FY1998-FY2007 (10 years)	FY1978-FY2007 (30 years)	FY1948-FY2007 (60 years)
Renewable Energy	$ 3.94	$ 15.43	$ 16.96
Energy Efficiency	6.02	14.18	14.32
Fossil Energy	5.36	24.22	39.60
Nuclear Energy	6.41	38.62	85.01
Electric Systems	0.93	2.85	3.02
Total	$22.66	$95.30	$158.91

Sources: DOE Budget Authority History Table by Appropriation, May 2007; DOE Congressional Budget Requests (several years); DOE (Pacific Northwest Laboratory), An Analysis of Federal Incentives Used to Stimulate Energy Production, 1980. Deflator Source: The Budget for Fiscal Year 2009. Historical Tables. Table 10.1. Gross Domestic Product and Deflators Used in the Historical Tables, 1940-2013. p. 194-195.

Table 2. DOE Energy Technology Share of Funding
(percent; derived from **Table 1**)

Technology	Period		
	FY1998-FY2007 (10 years)	FY1978-FY2007 (30 years)	FY1948-FY2007 (60 years)
Renewable Energy	17.4%	16.2%	10.7%
Energy Efficiency	26.6%	14.9%	9.0%
Fossil Energy	23.7%	25.4%	24.9%
Nuclear Energy	28.3%	40.5%	53.5%
Electric Systems	4.1%	3.0%	1.9%
Total	100.0%	100.0%	100.0%

---

[3] *DOE Conservation and Renewable Energy Base Table.* February 1990.

[4] This program includes R&D on advanced batteries to store electricity and transmission equipment to transfer electricity with less heat loss (i.e. at higher levels of energy efficiency).

Renewable Energy R&D Funding History: A Comparison

Sources: DOE Budget Authority History Table by Appropriation, May 2007; DOE Congressional Budget Requests (several years); DOE (Pacific Northwest Laboratory), An Analysis of Federal Incentives Used to Stimulate Energy Production, 1980; DOE Conservation and Renewable Energy Base Table. February 1990. Deflator Source: The Budget for Fiscal Year 2009. Historical Tables. Table 10.1. Gross Domestic Product and Deflators Used in the Historical Tables, 1940-2013. p. 194-195.

Figure 1. DOE Energy Technology Share of Funding, FY1998-FY2007

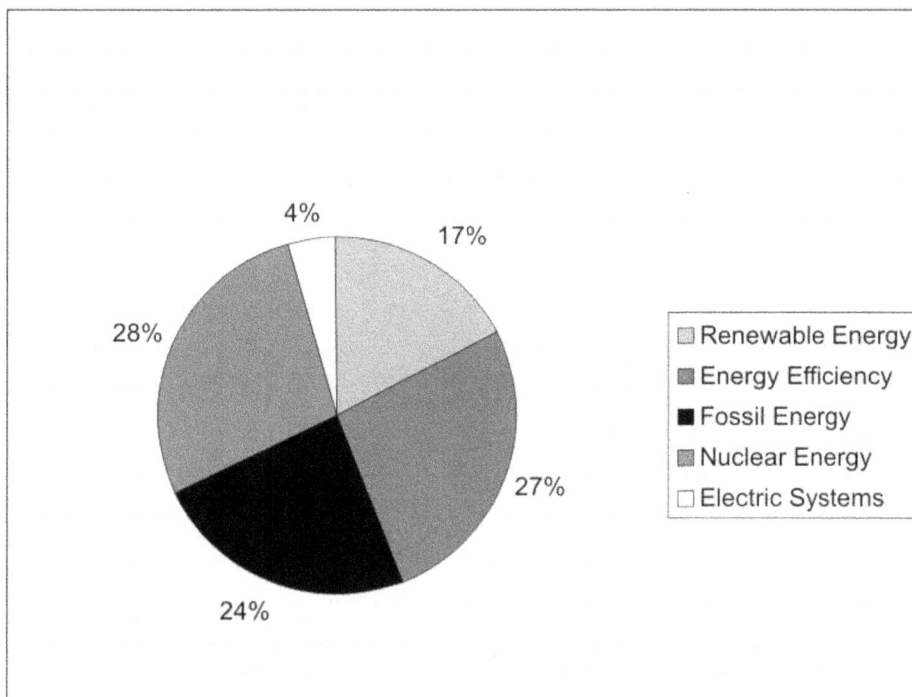

Sources: *DOE Budget Authority History Table by Appropriation*, May 2007; *DOE Congressional Budget Requests* (several years); Deflator Source: *The Budget for Fiscal Year 2009*. Historical Tables. Table 10.1. Gross Domestic Product and Deflators Used in the Historical Tables, 1940-2013. p. 194-195.

Figure 2. DOE Energy Technology Share of Funding, FY1978-FY2007

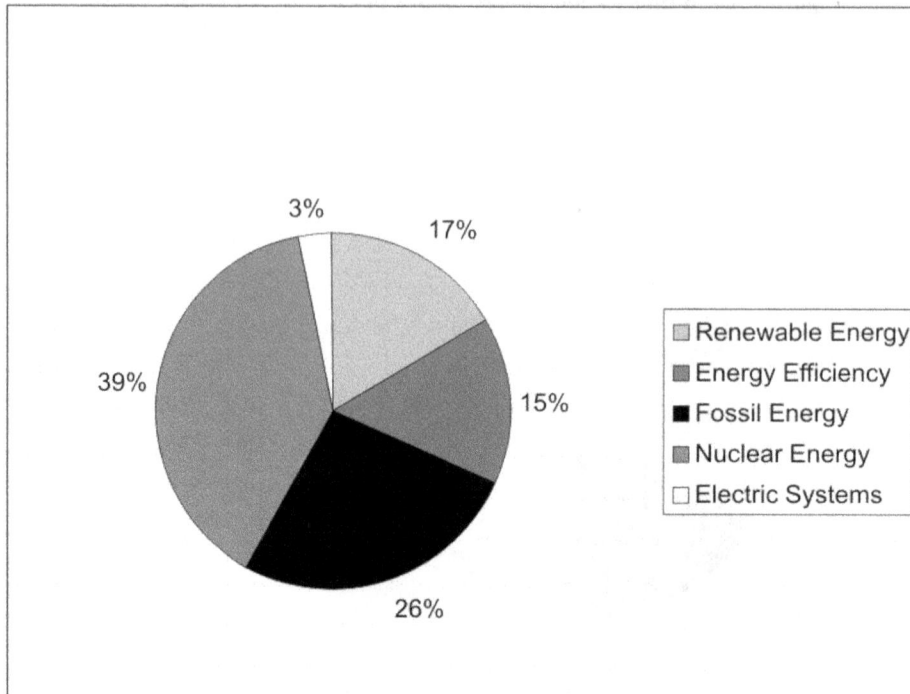

Pie chart values: 3%, 17%, 15%, 26%, 39%

Legend:
- Renewable Energy
- Energy Efficiency
- Fossil Energy
- Nuclear Energy
- Electric Systems

Sources: *DOE Budget Authority History Table by Appropriation*, May 2007; *DOE Congressional Budget Requests* (several years); Deflator Source: *The Budget for Fiscal Year 2009*. Historical Tables. Table 10.1. Gross Domestic Product and Deflators Used in the Historical Tables, 1940-2013. p. 194-195.

Figure 3. DOE Energy Technology Share of Funding, FY1948-FY2007

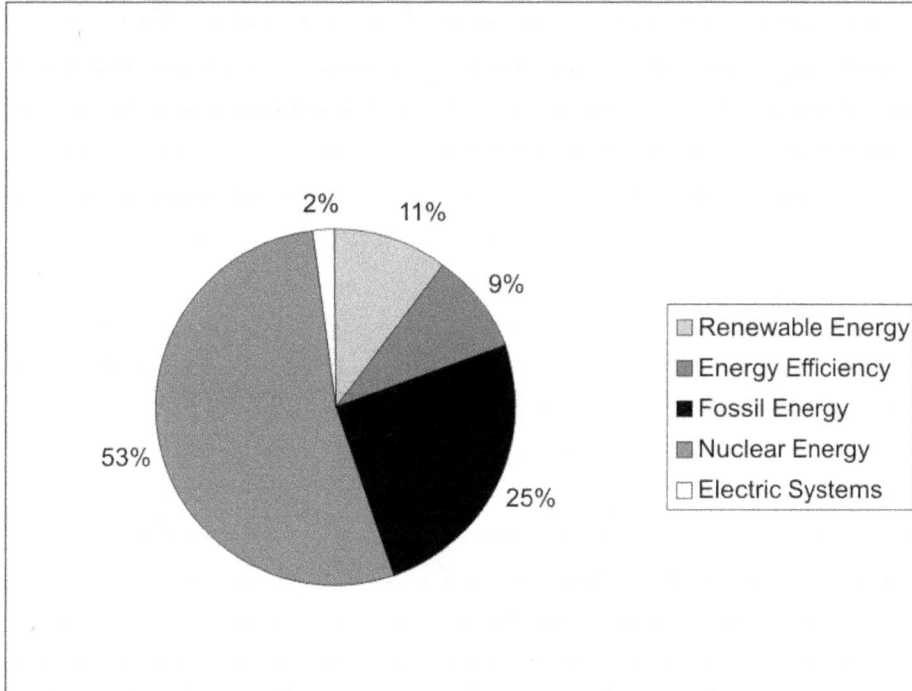

Sources: DOE Budget Authority History Table by Appropriation, May 2007; DOE Congressional Budget Requests (several years); DOE (Pacific Northwest Laboratory), An Analysis of Federal Incentives Used to Stimulate Energy Production, 1980; DOE Conservation and Renewable Energy Base Table. Feb. 1990. Deflator Source: The Budget for Fiscal Year 2009. Historical Tables. Table 10.1. p. 194-195.

Author Contact Information

Fred Sissine
Specialist in Energy Policy
fsissine@crs.loc.gov, 7-7039

LIGHT WATER REACTOR SUSTAINABILITY

The U.S. Department of Energy's Office of Nuclear Energy

The current fleet of nuclear power plants provides almost 20 percent of the total U.S. electricity production and more than 70 percent of the U.S. non-greenhouse-gas-emitting electric power generation.

The Light Water Reactor Sustainability (LWRS) Program is developing the scientific basis to extend existing nuclear power plant operating life beyond the current 60-year licensing period and ensure long-term reliability, productivity, safety, and security. The program is conducted in collaboration with national laboratories, universities, industry, and international partners. Idaho National Laboratory serves as the Technical Integration Office and coordinates the Research and Development (R&D) projects in the following pathways: Nuclear Materials Aging and Degradation; Advanced Light Water Reactor Fuel Development; Advanced Instrumentation, Controls, and Information Systems Technology; Risk-Informed Safety Margin Characterization, and Efficiency Improvements.

BENEFITS OF THE INITIATIVE

Nuclear power has reliably and economically contributed almost 20 percent of electrical generation in the United States over the past 2 decades. It remains the single largest contributor (more than 70 percent) of non-greenhouse-gas-emitting electric power generation in the United States. By the year 2030, domestic demand for electrical energy is expected to grow to levels of 16 to 36 percent higher than 2007 levels. At the same time, most currently operating nuclear power plants will begin reaching the end of their 60-year operating licenses.

If current operating nuclear power plants do not operate beyond 60 years, the total fraction of generated electrical energy from nuclear power will begin to decline — even with the expected addition of new nuclear generating capacity. The oldest commercial plants in the United States reached their 40th anniversary this year.

Continued safe and economical operation of current reactors for an even longer period of commercial operation, beyond the current license renewal lifetime of 60 years, is a low-risk option to fill the gap and to add new power generation at a fraction of the cost of building new plants. The cost to replace the current fleet would require hundreds of billions of dollars. Replacement of this 100 GWe generating capacity with traditional fossil plants would lead to significant increases in greenhouse gas emissions. Extending operating licenses beyond 60 to perhaps 80 years would enable existing plants to continue providing safe, clean, and economic electricity.

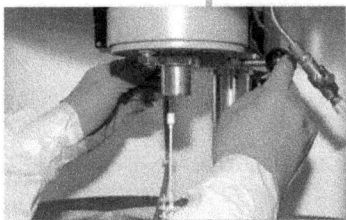

www.nuclear.energy.gov
February 2010

LIGHT WATER REACTOR SUSTAINABILITY

Program Budget

Light Water Reactor Sustainability
($ in Millions)

	FY 2010 Actual	FY 2011 Request
	$9.7*	$25.8

appropriated under GEN IV programs

To provide the technical basis for this life extension, the following five R&D pathways have been identified:

- **The Nuclear Materials Aging and Degradation** pathway will conduct research to develop the scientific basis for understanding and predicting long-term environmental degradation behavior of materials in nuclear power plants. Data and methods to assess performance of systems, structures, and components essential to safe and sustained nuclear power plant operation will be developed.

- **The Advanced LWR Nuclear Fuel Development** pathway will improve the scientific knowledge basis for understanding and predicting fundamental nuclear fuel and cladding performance in nuclear power plants. This information will be applied to the development of high-performance, high burn-up fuels with improved safety, cladding integrity, and improved nuclear fuel cycle economics.

- **The Advanced Instrumentation, Information, and Control Systems Technologies** pathway addresses long-term aging and obsolescence of instrumentation and control technologies and the development and testing of new information and control technologies. Advanced condition monitoring for more automated and reliable plant operation will be developed.

- **The Risk-Informed Safety Margin Characterization** pathway will bring together risk-informed, performance-based methodologies with scientific understanding of critical conditions and predictions of nuclear power plant performance, leading to an integrated characterization of public safety margins in an optimization of nuclear safety, plant performance, and long-term asset management.

- **The Economics and Efficiency Improvement** pathway will improve the economics and efficiency of the current fleet of reactors while maintaining excellent safety performance. Methodologies and the scientific basis to enable more extended power uprates or ultra-high power uprates will be developed. In addition, improving thermal efficiency by developing advanced cooling technologies to minimize water usage will be explored.

With the 60-year licenses beginning to expire between the years 2029 and 2039, utilities are likely to initiate planning baseload replacement power by 2014 or earlier. Research for addressing nuclear power plant aging questions must start now. The LWRS Program represents the timely collaborative research needed to retain the existing nuclear power plant infrastructure in the United States.

www.nuclear.energy.gov
February 2010

The U.S. Department of Energy's Office of Nuclear Energy

PLANNED PROGRAM ACCOMPLISHMENTS

FY 2010

- Complete a review of current information and technology related to concrete durability and aging relevant to nuclear power plant environments.
- Complete a technology development plan for Silicone Carbide clad fuel development.
- Complete a workshop and research and development plan for online monitoring and non-destructive examination technologies.
- Complete a report on the architectural and algorithmic requirements for a next-generation system analysis code.
- Complete the development of the initial test code version of the R7 system analysis computer code.

FY 2011

- Address high-fluence neutron irradiation effects on reactor metals including the reactor pressure vessels and core internals (stainless steels and high strength alloys), radiation-induced swelling effects, and phase transformation of core internals.
- Evaluate long-term aging of concrete structures.
- Investigate crack initiation in nickel-based alloys (steam generator tubing).
- Examine advanced mitigation techniques such as welding and weld repair techniques, post-irradiation annealing and modern replacement alloys.
- Develop a risk-informed simulation-driven methodology to guide safety system analysis and uncertainty quantification.
- Enhance the deterministic safety analysis capability to simulate plant dynamics and compute safety margin.
- Incorporate passive structures, systems, and components into a probabilistic safety analysis at one plant type.
- Develop alternative and new cooling technologies that can be applied in the near term to reactors impacted by insufficient cooling water supplies and innovative technologies that lessen the environmental impacts of removing large volumes of cooling water from naturally occurring sources.
- Develop plant control and monitoring systems to improve plant efficiency, facilitate power up-rates, and enable remote monitoring and support.
- Develop a model for fuel cracking at the mesoscale level with sufficient understanding to develop a predictive model for fission gas release.
- Begin the development of new long-life fuel designs with advanced fuel and cladding materials.

Fuel testing increases our understanding of material performance and is essential to safe and sustained nuclear power plant operation.

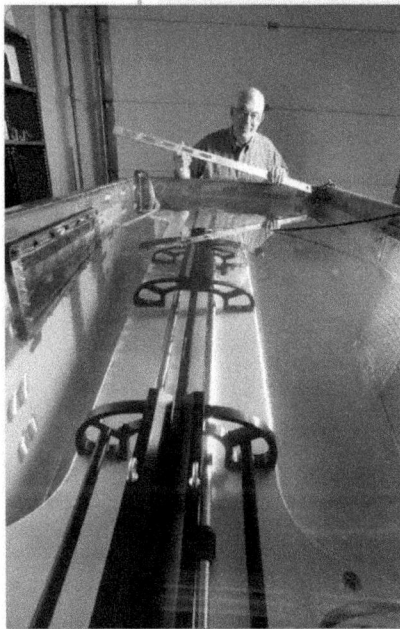

www.nuclear.energy.gov
February 2010

LIGHT WATER REACTOR SUSTAINABILITY

The U.S. Department of Energy's Office of Nuclear Energy

www.nuclear.energy.gov
February 2010

ENERGY INNOVATION HUB FOR MODELING AND SIMULATION

The U.S. Department of Energy's Office of Nuclear Energy

The Energy Innovation Hub for Modeling and Simulation will provide new ways to address safety, waste management, and nonproliferation.

The Energy Innovation Hub for Modeling and Simulation will be modeled after highly successful endeavors such as Bell Labs and the Bioenergy Research Centers. It will utilize existing advanced modeling and simulation capabilities (*e.g.,* computational fluid dynamics) developed by the Department of Energy's (DOE) Office of Science, National Nuclear Security Administration (NNSA), and other DOE research and development programs to fundamentally change how the U.S. designs and manages its nuclear facilities. Over the past decade and a half, a new capability has been added to theory and experimentation to create and demonstrate scientific insight about complex physical systems. With the advent of very high-powered computing, advanced modeling and simulation can provide faster and more detailed insights into the operation of these complex physical systems. The goals are to find ways to improve waste management, reduce proliferation risk, and lower the cost of nuclear facilities.

NEW WAYS OF UNDERSTANDING

Using the world's most powerful computers, the modeling and simulation hub will provide new ways for scientists and engineers to advance nuclear energy technologies. This will provide a faster technology innovation cycle while reducing risk and cost. Today's immersive visualization technology will allow scientists and engineers to stand in the center of an operational "virtual" reactor, observing coolant flow, nuclear fuel performance, and even the reactor's response to changes in operating conditions, accident conditions, or design parameters. This approach will provide the critical research and development insight needed to develop nuclear power systems that are better understood, more readily licensed, and even safer to operate.

The design and licensing of current reactors was based on conventional engineering that relied on a series of incremental steps moving from prototypes to demonstrations to commercial power plants. Without significant experimentation, validation and verification, the engineering development processes had to ensure the designs were sufficiently conservative to cover a lack of precise models to simulate system behaviors under steady state, transient, and potential accident conditions. This process led to a technology development and licensing process that was long and expensive, and resulted in overly conservative designs. Modern, science-based, integrated modeling and simulation such as what the Hub will provide, will allow engineers to understand how systems will perform beyond what historically was directly measured through experiments. Modeling and simulation tools will now enable them to engineer facilities that have even greater margins of safety than we know of today.

www.nuclear.energy.gov
February 2010

ENERGY INNOVATION HUB FOR MODELING AND SIMULATION

Program Budget

EIH – Modeling and Simulation
($ in Millions)

	FY 2010 Actual	FY 2011 Request
	$21.4	$24.3

GRAND CHALLENGES

Over the last two decades, the world has seen an explosive growth in the power of computers. This is the result of faster computer processors and the ability to have up to hundreds of thousands of these processors working on a single problem. Scientists and engineers are now using this unprecedented computing power to put modeling and simulation on par with theory and experiment. Now, scientists and engineers are able to create entirely new levels of understanding about both the results of physical processes and insights into the physical processes themselves.

The challenge of building this new level of science-based modeling and simulation is bringing together the large number of disciplines needed for a successful outcome. These include physics, chemistry, mathematics, computer science, electrical engineering, and software engineering. In addition, these disciplines must address issues of multi-physics across multiple length and time scales on parallel computers running thousands of program threads across up to 100,000 processors.

SEEKING THE BEST TEAM AND THE BEST APPROACH

As envisioned, the Modeling and Simulation Hub will be a competitively awarded partnership with national laboratories, industry, and academia. It is intended to use the best existing, relevant modeling and simulation capabilities, or develop new ones, to deliver new levels of science-based understanding of nuclear energy technologies. The Hub will employ a cross-disciplinary team of nuclear engineers and scientists, computer scientists, mathematicians, verification and validation experts, and sociologists and psychologists to change the existing user environment.

A well-considered competitive process is a critical step to ensuring success and a Funding Opportunity Announcement was released in early 2010 to that end. Once that competition is completed and the Modeling and Simulation Hub is established, the winning team will begin to address the intermediate and long-term goals aimed at removing the barriers to transforming advanced nuclear systems into commercially deployable materials, devices, and systems. Exact deliverables will be determined by the winning proposal.

In addition to the Hub, advanced modeling and simulation efforts for the associated research areas within the Office of Nuclear Energy will continue to develop new computational science capabilities from the micro behavioral level of fuels and materials to the macro behavioral level of entire systems such as reactors, repositories, and even entire fuel cycles. This effort, known as Nuclear Energy Advanced Modeling and Simulation (NE-AMS), along with the Modeling and Simulation Hub will provide complementary and essential capabilities to advance nuclear energy technologies.

PLANNED PROGRAM ACCOMPLISHMENTS

FY 2010

- Develop and issue a Funding Opportunity Announcement.

- Select an applicant and award a Cooperative Agreement contract for five years with the possibility of a five-year extension if a high standard of performance is achieved.

- Stand-up the Energy Innovation Hub for Modeling and Simulation, including:

www.nuclear.energy.gov
February 2010

- Hire subject matter experts for required core capabilities and relocate personnel as needed for optimum Hub operating efficiency.

- Prepare the Hub infrastructure including any required renovation of existing buildings, leasing buildings, purchase of research equipment and instrumentation, and installation of state-of-the-art Hub communications and interface capabilities for long distance collaboration.

- Initiate robust interaction with private industry for the collection of requirements from expected users of the nuclear energy engineering environment.

- Initiate educational/training programs for students, post-doctoral fellows, and scientists.

2011

- Continue to work toward achieving the Hub's goals and objectives presented in the winning proposal, consistent with the Hub's funding plan, including cost sharing if applicable.

- Establish an Energy Innovation Hubs Oversight Board to review the progress of the Hub's scientific program and its management structure, policies, and practices.

- Provide ongoing review of the Hub's deliverables and performance.

The new 1.64-petaflop Cray XT Jaguar features more than 180,000 processing cores, each with 2 gigabytes of local memory. The resources of the ORNL computing complex provide scientists with a total performance of 2.5 petaflops.

www.nuclear.energy.gov
February 2010

ENERGY INNOVATION HUB FOR MODELING AND SIMULATION

The U.S. Department of Energy's Office of Nuclear Energy

www.nuclear.energy.gov
February 2010

NUCLEAR ENERGY UNIVERSITY PROGRAMS

The U.S. Department of Energy's Office of Nuclear Energy

DOE is committed to strengthening the Nation's nuclear education infrastructure.

The Office of Nuclear Energy (NE) supports the Department of Energy's (DOE) development of advanced nuclear science and technology through its Nuclear Energy University Programs (NEUP). NEUP funds nuclear energy research, helps educate and train the next generation of the nuclear-energy workforce through equipment and instrumentation upgrades and curriculum development at U.S. colleges and universities, and provides scholarships and fellowships to nuclear science and engineering students.

NEUP's goal is to support outstanding and innovative nuclear energy research at U.S. universities. The program aims to:

- Fund creative research ideas that can potentially produce breakthroughs in nuclear reactor technology;

- Attract the brightest students to the nuclear professions and support the Nation's intellectual capital in nuclear engineering and relevant nuclear science, such as Health Physics, Nuclear Materials Science, Radiochemistry, and Applied Nuclear Physics;

- Integrate research and development (R&D) at universities, national laboratories, and industry to revitalize nuclear education;

- Improve university and college infrastructures for conducting R&D and educating students; and

- Facilitate the transfer of knowledge from an aging nuclear workforce to the next generation of workers.

To accomplish these goals, NE will support:

- Nuclear energy-related R&D;
- Scholarships and fellowships;
- Curriculum development; and
- Infrastructure and equipment upgrades.

NUCLEAR ENERGY R&D

NEUP seeks to conduct nuclear energy research at U.S. colleges and universities to further DOE's mission and goals. The program supports R&D projects focused on the research needs and priorities of the Reactor Concepts Research, Development and Demonstration (RCRD&D), NE Fuel Cycle R&D (FCR&D), and Nuclear Energy Enabling Technologies (NEET) programs.

www.nuclear.energy.gov
February 2010

NUCLEAR ENERGY UNIVERSITY PROGRAMS

SCHOLARSHIPS AND FELLOWSHIPS

Increasing the number of students entering the nuclear science and engineering fields is a key goal for NE. While the demand for engineers and scientists in these areas is growing, about half of the nuclear industry's workforce will be eligible to retire in the next 10 years. If the United States is to stay competitive, it must keep the pipeline of key personnel filled. The average NEUP undergraduate scholarship is $5,000 for 1 year, and graduate fellowship awards can be as high as $150,000 over 3 years.

INFRASTRUCTURE AND EQUIPMENT

NE recognizes that U.S. universities require proper support to conduct cutting-edge research and educate the next generation of nuclear science and engineering students. Infrastructure grants will be available for equipment and instrumentation for research reactors; other specialized facilities; classrooms and laboratories; non-reactor nuclear science and engineering research; and for developing the curriculum required to advance nuclear energy education and to support the university departmental missions.

Center for Advanced Energy Studies at Idaho National Laboratory — a public/private partnership that integrates resources, capabilities and expertise to create new research capabilities.

AWARD PROCESS

In FY 2011, NE will designate up to 20 percent of funds appropriated to its R&D programs, as well as funds from the RE-ENERGYSE Initiative, for work to be performed at university and research institutions. These funds, competitively awarded, will support:

- Mission-specific applied R&D activities;
- Investigator-initiated basic research;
- Human capital development activities, such as fellowship and scholarships awards; and
- Curriculum development and infrastructure and equipment upgrades for universities, colleges, and other post-secondary institutions.

PLANNED PROGRAM ACCOMPLISHMENTS

FY 2010

- Award NEUP R&D projects in support of NE Fuel Cycle R&D, Next Generation Nuclear Plant (NGNP), and Light Water Reactor Sustainability (LWRS) activities;
- Award grants to U.S. universities, colleges, and other post-secondary institutions for infrastructure and equipment to support nuclear energy-relevant education and R&D;
- Award scholarships and fellowships to students attending U.S. universities and colleges who major in nuclear science and engineering fields of study;
- Conduct workforce and infrastructure assessments to baseline program;

www.nuclear.energy.gov
February 2010

332 703-739-3790 TCNNaturalGas.com

The U.S. Department of Energy's Office of Nuclear Energy

- Complete research on 33 R&D projects initiated in FY 2007 in the NGNP and FCR&D areas; and

- Conduct an NEUP workshop to plan for FY 2011 program solicitations.

FY 2011

- Solicit and competitively award new mission-specific NEUP R&D projects in support of the Reactor Concepts Research, Development and Demonstration (RCRD&D), NE Fuel Cycle R&D (FCR&D), and Nuclear Energy Enabling Technologies (NEET) programs;

- Solicit and competitively award university infrastructure and equipment support relevant to the R&D needs; and

- Solicit and competitively award scholarships and fellowships in nuclear engineering and science areas to students attending U.S. universities and colleges.

Fuel rods for a university reactor.

www.nuclear.energy.gov
February 2010

333

NUCLEAR ENERGY UNIVERSITY PROGRAMS

The U.S. Department of Energy's Office of Nuclear Energy

www.nuclear.energy.gov
February 2010

NEXT GENERATION NUCLEAR PLANT DEMONSTRATION PROJECT

The U.S. Department of Energy's Office of Nuclear Energy

NGNP will extend the benefits of nuclear energy by providing carbon-free, high-temperature process heat to industry.

The Department of Energy (DOE) is laying the groundwork for a lower emissions future, free of reliance on imported energy. The Next Generation Nuclear Plant Demonstration Project (NGNP) is a vital part of this vision.

DOE's Next Generation Nuclear Plant (NGNP) Demonstration Project supports a transformative application of nuclear energy to address the President's goals for reducing greenhouse gas emissions and enhancing energy security.

BENEFITS OF THE NGNP PROGRAM

Through scientific and international collaboration, NGNP supports the development of gas-cooled nuclear reactor technology that promises improved performance in sustainability, economics, and proliferation resistance. As a result of these efforts, nuclear energy will increase its contribution to the reduction of CO_2 emissions when it is used to replace conventional sources of process heat, such as the burning of fossil fuels.

By investing in gas-cooled reactor technology that makes possible more efficient electricity production and the production of nuclear process heat for industry, NGNP is endeavoring to jump start a new application for nuclear energy with potential benefits to the environment that rival the reduction in greenhouse-gas emissions credited to current generation nuclear power plants.

The NGNP Demonstration Project will provide:

High-temperature gas-cooled reactor technology — Gas-cooled reactors are a revolutionary advance in reactor technology. They are inherently safe, efficient, and can use less fuel than the current generation of light-water reactor designs. Gas reactors can be used to extend the benefits of nuclear energy beyond the electrical grid by providing industry with low carbon, high-temperature process heat for a variety of applications, including petroleum refining, biofuels production, and production of chemical feedstocks for use in the fertilizer and chemical industries

Underlying Technologies — Underlying technologies (fuels, materials, neutronic and thermofluid modeling) benefit the majority of reactor concepts and sizes that are being investigated as part of the overall NE R&D portfolio.

www.nuclear.energy.gov
February 2010

NEXT GENERATION NUCLEAR PLANT DEMONSTRATION PROJECT

Program Budget

Next Generation Nuclear Plant
($ in Millions)

	FY 2010 Actual	FY 2011 Request
	$164.3	$103.0

PROGRAM BACKGROUND

The NGNP Demonstration Project was formally established by the Energy Policy Act of 2005 (EPAct 2005) to demonstrate the generation of electricity and/or hydrogen with a high-temperature nuclear energy source. The Project is executed in collaboration with industry, DOE national laboratories, and U.S. universities. The U.S. Nuclear Regulatory Commission (NRC) is responsible for licensing and regulatory oversight of the demonstration nuclear reactor.

The NGNP Demonstration Project includes design, licensing, construction, and research and development conducted in two phases. Phase 1 is pre-conceptual and conceptual design leading to the selection of a single technology for NGNP. Phase 2 is preliminary and final design necessary for licensing and construction of a demonstration plant. R&D as well as licensing activities are included in both Phase 1 and Phase 2. Licensing scope supports the development of a licensing framework for high-temperature gas reactors and includes the preparation and submission of a Combined Operating License Application (COLA) for the NGNP. R&D activities are organized into four major technical areas: (1) Fuel Development and Qualification, (2) Graphite Qualification, (3) High-Temperature Materials Qualification, and (4) Design and Safety Methods Validation.

INTERNATIONAL COOPERATION

The United States collaborates with the international community via the Generation IV International Forum (GIF), the International Atomic Energy Agency (IAEA), and through a number of bilateral agreements pioneered under the International Nuclear Energy Research Initiative.

PLANNED PROGRAM ACCOMPLISHMENTS

In FY 2010, the Department engaged with industry to complete conceptual designs for the NGNP. In the first quarter of FY 2011, the Nuclear Energy Advisory Committee will review the conceptual design reports along with the state of NGNP R&D and licensing activities and make recommendations on whether or not to proceed to Phase 2 of the project.

FY 2010

- Begin cost-sharing with industry to complete the conceptual design of one or two gas-cooled reactor concepts for the NGNP.

- Complete the first Advanced Gas Reactor fuel irradiation experiment, perform post-irradiation examinations of the irradiated fuel specimens, and commence irradiation of the first fuel produced in large-scale production equipment.

- Continue irradiation of the first Advanced Graphite Creep test experiment to provide data for nuclear graphite qualification.

- Continue the development of advanced gas reactor system-simulation software and initiate bilateral cooperation with Japan on the use of its experimental gas reactor as a test facility for code validation, operational experience, and as an instrumentation and controls test bed.

- Continue the study of liquid salt properties and materials issues associated with its use as a high-temperature medium for intermediate energy transport loops.

- Continue collaboration with the NRC on scale reactor tests to benchmark thermal-fluid reactor system modeling tools.

www.nuclear.energy.gov
February 2010

- Submit white papers to NRC on key licensing topics and resolve any comments.
- Continue international R&D collaborations through the Generation IV International Forum (GIF), IAEA, and bi-lateral agreements.

FY 2011

- Sponsor an independent review of NGNP activities by the Nuclear Energy Advisory Committee to inform a decision on readiness to proceed with Phase 2 of the NGNP Project.
- Enter into a cost-sharing public-private partnership to conduct the R&D, design, and licensing activities leading to NRC issuance of a Combined Operating License.
- Continue the irradiation in the Advanced Test Reactor (ATR) of the first NGNP fuel produced in commercial scale production equipment and complete post-irradiation examination of the first NGNP fuel removed from ATR in November 2009.
- Continue selection and characterization of NGNP graphite and composite materials, including the irradiation of the first graphite test experiment in ATR and the assembly of High-Temperature Vessel irradiation experiments planned for Oak Ridge National Laboratory.
- Continue environmental, mechanical property, and joining method (*e.g.* welding) studies for selected heat exchanger and reactor pressure vessel materials for code-case data package development and qualification.
- Continue topical report analysis and respond to Requests for Additional Information from the NRC, pending the availability of design data.
- Complete Regulatory Gap Analysis that will review existing NRC rules and regulations and identify their applicability to gas reactors.
- Continue international R&D collaborations through the GIF on fuels, materials, modeling, and process heat applications.
- Use existing test facilities (High-Temperature Test Reactor in Japan) and construction of separate effects and integral effects test experiments for validating NGNP thermal-fluid behavior and the capability of the passive system to remove decay heat.
- Conduct research on process heat applications, including system interface requirements and materials compatibility issues, for coupling NGNP to various non-electric applications.
- Continue fuel performance modeling, fabrication modeling, and fission product transport.

R&D focuses on enabling technologies such as high-temperature metal alloys, nuclear-grade graphites, and coated-particle fuels.

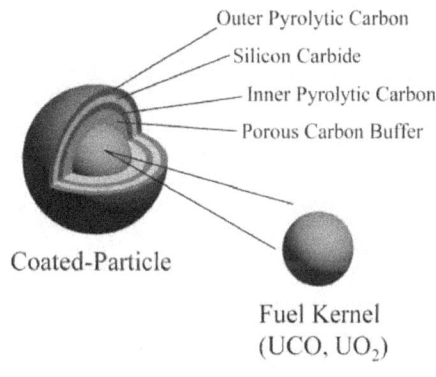

Outer Pyrolytic Carbon
Silicon Carbide
Inner Pyrolytic Carbon
Porous Carbon Buffer

Coated-Particle

Fuel Kernel
(UCO, UO$_2$)

www.nuclear.energy.gov
February 2010

NEXT GENERATION NUCLEAR PLANT DEMONSTRATION PROJECT

The U.S. Department of Energy's Office of Nuclear Energy

www.nuclear.energy.gov
February 2010

PLUTONIUM-238 PRODUCTION PROJECT

The U.S. Department of Energy's Office of Nuclear Energy

The production of Pu-238 is vital to the continued exploration of space and our national security.

The United States is initiating activities to reestablish the domestic production of plutonium-238 (Pu-238), which is required for radioisotope power systems used for National Aeronautics and Space Administration's (NASA) space missions, and for national security needs.

In the past, the Department of Energy obtained Pu-238 from the Savannah River Site K Reactor, which was taken off-line in the late 1980s. More recently, the Department has augmented its available inventory for non-national security missions by purchasing Pu-238 from Russia, but those stocks are limited, and Russia is no longer producing Pu-238. Only a limited amount of Pu-238 remains available for U.S. purchase under the current contract with Russia, which expires in January 2013.

The production of Pu-238 is vital to the continued exploration of space, and our national security. NASA's plans for an outer planets flagship mission to be launched in 2020 will exhast the remaining available supply of Pu-238. Additional Pu-238 is required. NASA has established Pu-238 requirements to meet the power and heating needs of planned missions to explore the outer planets and a range of other solar system destinations for the next two decades. Its requirement for Pu-238 is expected to remain constant in the more distant outyears.

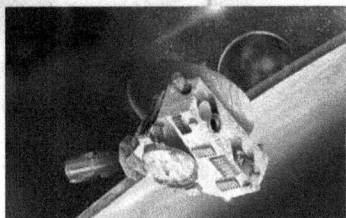

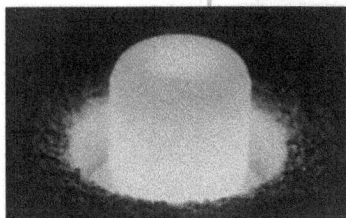

WHY Pu-238?

Pu-238 was chosen as the heat source material because of its inherent characteristics. This isotope combines a high heat output with a long half-life, which allows radioisotope power systems to remain useful over long mission durations.

HOW WILL Pu-238 BE PRODUCED?

The Pu-238 production process consists of the fabrication of targets, irradiation of the targets in a nuclear reactor, and recovery of Pu-238 from the irradiated targets through chemical extraction. Facilities will have to be modified or constructed to reestablish target fabrication and Pu-238 recovery capabilities. Existing reactors, the Advanced Test Reactor (ATR) at Idaho National Laboratory and the High Flux Isotope Reactor (HFIR) at Oak Ridge National Laboratory, are suitable for target irradiation. As the primary user of Pu-238, NASA will share with DOE the cost of reestablishing this production capability.

www.nuclear.energy.gov
February 2010

PLUTONIUM-238 PRODUCTION PROJECT

The U.S. Department of Energy's Office of Nuclear Energy

Program Budget

Pu-238 Production
($ in Millions)

	FY 2010 Actual	FY 2011 Request
	$0.0	$15.0

PLANNED PROGRAM ACCOMPLISHMENTS

FY 2011

- Complete National Environmental Policy Act requirements.
- Prepare a Conceptual Design.
- Develop a Safety Design Strategy.
- Prepare a Preliminary Security Vulnerability Assessment.
- Conduct a Technical Independent Project Review.
- Initiate target fabrication using existing laboratory facilities and equipment.
- Initiate target irradiations in ATR and HFIR.
- Finalize separations flow sheets using existing laboratory facilities and equipment.

The Advanced Test Reactor (ATR) has the capability to support production of Pu-238.

www.nuclear.energy.gov
February 2010

SMALL MODULAR REACTORS

The U.S. Department of Energy's Office of Nuclear Energy

Small Modular Reactors (SMRs) are nuclear power plants that are smaller in size (300 MWe or less) than current generation base load plants (1,000 MWe or higher). These smaller, compact designs are factory-fabricated reactors that can be transported by truck or rail to a nuclear power site.

The Department of Energy (DOE) believes that there is a need and a market in the United States for SMRs. The DOE Office of Nuclear Energy's Small Modular Reactor program will advance the licensing and commercialization of SMR designs.

BENEFITS OF SMRs

The term "modular" in the context of SMRs refers to a single reactor that can be grouped with other modules to form a larger nuclear power plant. Even though current large nuclear power plants incorporate factory-fabricated components (or modules) into their designs, a substantial amount of field work is required to assemble components into an operational power plant. SMRs are envisioned to require limited on-site preparation as they are expected to essentially be ready to "plug and play" when they arrive from the factory. SMRs provide simplicity of design, economies and quality of factory production, and offer more flexibility (financing, siting, sizing, and end-use applications) compared to larger nuclear power plants.

SMRs can reduce a nuclear plant owner's capital outlay or investment due to the lower plant capital cost. Modular components and factory fabrication can reduce construction costs and duration. Additional modules can be added incrementally as demand for energy increases. SMRs can provide power for applications where large plants are not needed or sites lack the infrastructure to support a large unit. This would include smaller electrical markets, isolated areas, smaller grids, limited water and acreage sites, or unique industrial applications. SMRs can replace aging fossil plants or complement existing industrial processes or power plants with an energy source that does not emit greenhouse gases. Some reactor designs will produce a higher temperature process heat for either electricity generation or industrial applications.

SMRs also provide potential nonproliferation benefits to the United States and the wider international community. Some SMRs will be designed to operate for decades without refueling. These SMRs would be fabricated and fueled in a factory, sealed and transported to sites for power generation or process heat, and then returned to the factory for defueling at the end of the life cycle. This approach could help to minimize the transportation and handling of nuclear material. There is both a domestic and international market for SMRs and U.S. industry is well positioned to compete for these markets.

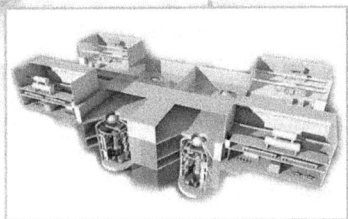

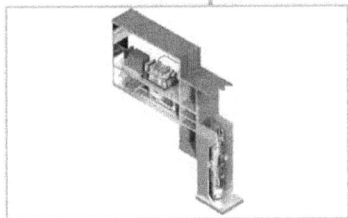

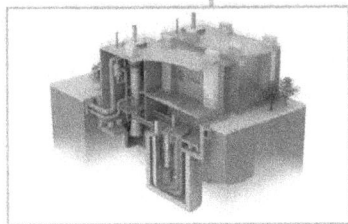

www.nuclear.energy.gov
February 2010

SMALL MODULAR REACTORS

The U.S. Department of Energy's Office of Nuclear Energy

Program Budget

Small Modular Reactors ($ in Millions)		
	FY 2010 Actual	FY 2011 Request
	$0.0	$38.9

DOE SMR PROGRAM

The SMR program supports two activities:

- Public/private partnerships to advance mature SMR designs; and
- Research and development (R&D) activities to advance the understanding and demonstration of innovative reactor technologies and concepts.

Although several light water reactor (LWR) SMR concepts are based on proven reactor technologies, they have not been previously designed, licensed, or built for commercial deployment. DOE believes that these LWR SMRs can be commercially deployed within the next decade. NE's SMR program will establish competitive cost-shared projects to support NRC design certification of new LWR SMR designs.

Other SMRs are based on advanced and innovative concepts — designs based on fast spectrum neutrons, or high-temperature reactors — that offer added functionality and affordability. DOE will support R&D activities at its national laboratories and universities to develop and prove the proposed design concepts. Emphasis will be on advanced reactor technologies that offer simplified operation and maintenance for distributed power and load-following applications and increased proliferation resistance and security.

R&D activities will focus on:

- Basic physics and materials research and testing;
- State-of-the-art computer modeling and simulation of reactor systems and components;
- Probabilistic risk analyses of innovative safety designs and features;
- High-temperature and radiation effects on fuels and materials; and
- High efficiency power conversion systems.

PLANNED PROGRAM ACCOMPLISHMENTS

FY 2011

- Solicit, competitively select, and award project(s) with industry partners for cost-sharing the U.S. Nuclear Regulatory Commission (NRC) review of design certification document for up to two of the most promising LWR SMR concept(s) for near-term licensing and deployment.
- Conduct research, development, and testing of innovative technologies, structures, systems, and components necessary for licensing.
- Establish and support national laboratory and university R&D activities to advance innovative technologies.
- Support the development of new/revised nuclear industry codes and standards necessary to support licensing and commercialization of innovative designs.
- Develop recommendations, in collaboration with NRC and industry, for changes in NRC policy, regulations or guidance to license and enable SMRs for deployment in the United States.
- Collaborate with Department of Defense (DoD) and Idaho National Laboratory (INL) to assess the feasibility of SMR designs for energy resources at DoD domestic installations.

www.nuclear.energy.gov
February 2010

GENERATION IV NUCLEAR ENERGY SYSTEMS

The U.S. Department of Energy's Office of Nuclear Energy

Generation IV systems concepts excel in safety, sustainability, cost effectiveness, and proliferation resistance.

The Department of Energy (DOE) is laying the groundwork for a zero emissions future, free of reliance on imported energy. The Generation IV (Gen IV) program is a vital part of this vision.

The goal of the Gen IV Nuclear Energy Systems Initiative is to address the fundamental research and development (R&D) issues necessary to establish the viability of next-generation nuclear energy system concepts to meet tomorrow's needs for clean and reliable electricity, and non-traditional applications of nuclear energy. Successfully addressing the fundamental R&D issues will allow Gen IV concepts that excel in safety, sustainability, cost-effectiveness, and proliferation risk reduction to be considered for future commercial development and deployment by the private sector.

The program focuses on Gen IV reactor concepts with an emphasis on very high-temperature reactor technologies and on the underlying Gen IV technologies that will improve the economic and safety performance of the previous generations of existing light-water reactors (LWRs).

BENEFITS OF THE INITIATIVE

Through scientific R&D and international collaboration, Gen IV supports the development of next-generation nuclear reactor technologies that promise improved performance in sustainability, economics, and proliferation resistance. As a result of these efforts, nuclear energy will be able to increase its contribution to the reduction of CO_2 emissions when nuclear energy is used to replace conventional sources of process heat, such as burning fossil fuels.

By investing in Gen IV technologies that make possible more efficient electricity production and the production of nuclear process heat for industry, Gen IV R&D holds the potential to match, or even exceed, the reduction in greenhouse-gas emissions credited to current generation nuclear power plants.

Generation IV R&D will provide:

- **High-temperature gas-cooled reactor technology** — Gas-cooled reactors are a revolutionary advance in reactor technology. They are inherently safe, efficient, and can use less fuel than the current generation of light-water reactor designs. Gas reactors can be used to extend the benefits of nuclear energy beyond the electrical grid by providing industry with low carbon, high-temperature process heat for a variety of applications, including petroleum refining, bio-fuels production, and production of feedstock for use in the fertilizer and chemical industries.

www.nuclear.energy.gov
May 2009

GENERATION IV NUCLEAR ENERGY SYSTEMS

Program Budget

Gen IV Nuclear Energy Systems Initiative
($ in Millions)

	FY 2009 Actual	FY 2010 Request
	$180.0	$191.0

- **Underlying Technologies** — Underlying technologies (fuels, materials, neutronic and thermofluid modeling) benefit the majority of reactor concepts and sizes. These technologies will receive limited but sustained Gen IV support in cooperation with international R&D.

- **Hub for Modeling and Simulation** — The Modelling and Simulation Hub will focus on providing validated advanced modeling and simulation tools necessary to enable fundamental change in how the U.S. designs and manages nuclear facilities. The goal is to find ways to improve waste management, reduce proliferation risk, and lower the cost of nuclear facilities.

INTERNATIONAL COOPERATION

Key to all Gen IV research and development is the multiplication effect on investment derived from international collaboration. By coordinating U.S. efforts with partner nations, our funding is leveraged by a factor of two to ten.

The United States collaborates with the international community via the Generation IV International Forum (GIF), the International Atomic Energy Agency (IAEA), and through a number of bilateral agreements pioneered under the International Nuclear Energy Research Initiative.

PLANNED PROGRAM ACCOMPLISHMENTS

FY 2009

- Continue the gas reactor fuel development, manufacturing, and qualification program. The first of eight irradiation experiments (AGR-1) has survived more than two years without any evidence of failure, and some of the samples will reach 18 percent burnup before the experiment is removed early next fiscal year.

- Complete preparations for the next fuel irradiation test, AGR-2, which contains fuel that was fabricated with production scale equipment, and complete the equipment design for the next series of tests, AGR-3 and -4, that will test fuel with hypothetical fuel manufacturing defects.

- Insert the first of six irradiation capsules to determine the properties of candidate nuclear-grade graphites, while under simulated conditions. Graphite is the primary reactor core structural material, and the Advanced Graphite Creep (AGC) experiment series is designed to qualify one or more commercial graphites for this mission.

- Complete environmental and mechanical property tests for potential Intermediate Heat Exchanger (IHX) metal alloys and continue to support industry-code committees in the development of test standards and design rules for the use of these alloys under Gen IV conditions.

- Continue analytical computer modeling methods development in both physics and heat transport, and continue benchmarking of computer codes for gas reactor design concepts.

www.nuclear.energy.gov
May 2009

The U.S. Department of Energy's Office of Nuclear Energy

- Upgrade the laboratory-scale components and setup for demonstrating a promising advanced Brayton cycle energy-conversion system that uses supercritical carbon dioxide as the working fluid, and obtain data on the component behavior under supercritical conditions.

- Continue, in collaboration with international partners, the development of crosscutting benchmarking methodologies (economics, proliferation resistance and physical protection, and reactor safety).

FY 2010

- Perform post-irradiation examinations of unique, highly-irradiated metallic material samples obtained from the Fast Flux Test Facility and the Phoenix Fast Reactor in France.

- Continue development of advanced materials such as ceramics, composite materials, and nano-structured ferritic materials for use in structural systems, fuel claddings, and other high-temperature applications.

- Complete the first Advanced Gas Reactor fuel irradiation experiment (AGR-1), perform post-irradiation examinations of the irradiated fuel specimens, and commence irradiation of the first fuel produced in large-scale production equipment (AGR-2).

- Continue irradiation of the first Advanced Graphite Creep (AGC-1) test experiment to provide data for nuclear graphite qualification.

- Continue the development of advanced gas reactor system-simulation software and initiate bilateral cooperation with Japan on the use of their experimental gas reactor as a test facility for code validation, operational experience, and as an instrumentation and controls test bed.

- Continue study of the use of liquid salt as a circulating fluid in primary and intermediate cooling loops, and demonstrate the technical and economic viability of an advanced Brayton-cycle energy-conversion system using supercritical carbon dioxide as the working fluid.

- Continue international R&D collaborations through the Generation IV International Forum, IAEA, and bi-lateral agreements.

- Continue advanced modeling techniques utilizing the Department's high-speed, parallel computers for the development of close-coupled neutronic and thermofluid codes.

R&D focuses on enabling technologies such as high-temperature metal alloys, nuclear-grade graphites, and coated-particle fuels.

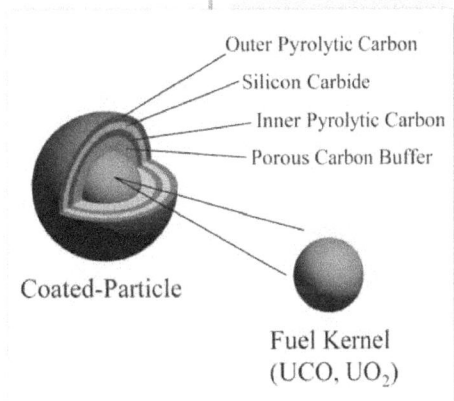

Outer Pyrolytic Carbon
Silicon Carbide
Inner Pyrolytic Carbon
Porous Carbon Buffer

Coated-Particle

Fuel Kernel (UCO, UO_2)

www.nuclear.energy.gov
May 2009

GENERATION IV NUCLEAR ENERGY SYSTEMS

The U.S. Department of Energy's Office of Nuclear Energy

www.nuclear.energy.gov
May 2009

NUCLEAR POWER 2010

The U.S. Department of Energy's Office of Nuclear Energy

New baseload nuclear generating capacity is required to support the U.S. energy objectives of enhancing energy supply, diversity, and security.

Nuclear Power 2010 (NP 2010) is a government-industry, 50-50 cost-shared initiative with two main goals: removing the technical, regulatory, and institutional barriers to building new nuclear power plants in the United States, and securing industry decisions to construct and operate those plants. With most of the objectives and activities planned under NP 2010 accomplished or near completion, the program is winding down, before phasing out entirely in FY 2011.

Through NP2010, the Department of Energy actively engages with industry to address issues affecting future expansion of nuclear generation. The NP 2010 program, initiated in 2002, has worked to:

- Demonstrate untested regulatory processes,
- Identify sites for new nuclear power plants,
- Develop and bring to market advanced, standardized nuclear plant technologies, and
- Evaluate the business case for building new nuclear power plants.

Accomplishing these program objectives paves the way for industry decisions to build advanced, light-water reactor nuclear plants in the United States that would begin operation by the middle of the next decade. The NP 2010 program is based on expert industry recommendations documented in *A Roadmap to Deploy New Nuclear Power Plants in the United States by 2010* and *The Business Case for New Nuclear Power Plants in the United States*.

MEETING ENERGY DEMAND

Electricity demand in the United States is expected to grow sharply in the 21st century: nearly 26 percent by 2030, according to the Energy Information Administration (EIA). Global electricity demand is expected to almost double by 2030. These projections could go even higher if electricity demand continues to grow at the rates experienced in recent years. This growth could require building a significant number of new power plants over the next two decades.

Today, nuclear power plants generate 19.4 percent of the electricity produced in this country. Notwithstanding projections in energy demand, new nuclear power plants must be licensed and built to maintain nuclear power's contribution to the national energy portfolio and to increase the supply of low carbon energy.

www.nuclear.energy.gov
May 2009

NUCLEAR POWER 2010

FUTURE NEED FOR ADDITIONAL GENERATING CAPACITY

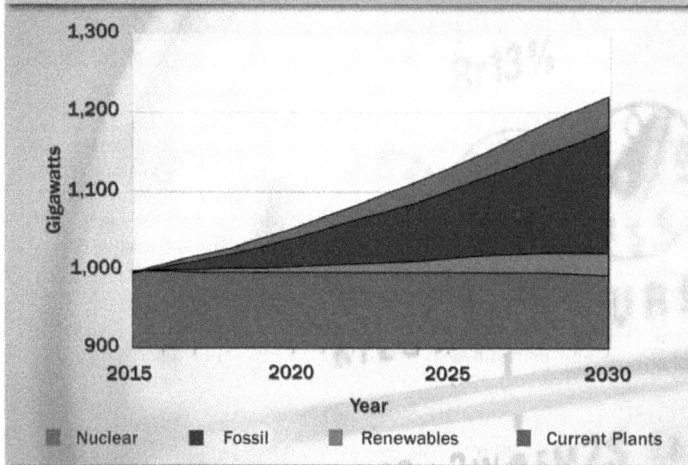

Nuclear ■ Fossil ■ Renewables ■ Current Plants

Capacity projections are based on the premise that nuclear energy will continue to provide 19.4 percent of the Nation's energy mix through 2030.

Despite the excellent and safe performance of current nuclear power plants, no new plant has been ordered in this country for nearly three decades, although nuclear plant owners have obtained license renewals and electrical power generation increases. For the past 30 years and as recently as the early 2000s, electric utilities chose to build new electric-generating plants that used inexpensive natural gas as fuel. However, natural gas prices are experiencing substantial volatility, making planning difficult.

Over-reliance on a single fuel source, such as natural gas, is a vulnerability to the long-term security of our Nation's energy supply. Bringing new nuclear plants into operation will:

- Reduce energy supply vulnerability,
- Address increasing concerns over air quality and climate change,
- Ease the pressures on natural gas supply, and
- Provide a source of abundant, affordable, reliable and low carbon energy to complement the current baseload capacity.

For the reasons identified above, power companies are submitting combined Construction and Operating License (COL) applications to the Nuclear Regulatory Commission (NRC) for new advanced nuclear power plant designs. Despite lingering financial and licensing concerns, reactor vendors and plant owners are proceeding to finalize design and licensing, and it is likely that construction will start on several new plants within the next few years.

DEPLOYING ADVANCED NUCLEAR POWER

The NP 2010 program focuses on reducing the technical, regulatory, and institutional barriers to deployment of new nuclear power plants. The technology focus of the NP 2010 program is on Generation III+ advanced, light-water reactor designs, which offer advancements in safety and economics over current nuclear plant designs and the nuclear plant designs certified by the NRC in the 1990s.

Regulatory Issues and Licensing —

To enable deployment of the new Generation III+ nuclear power plants in the United States in the relatively near term, it is essential to complete first-of-a-kind Generation III+ reactor design activities and to demonstrate the untested NRC regulatory and licensing processes for the siting, construction, and operation of new nuclear plants. The Early Site Permit (ESP) process is a licensing process to approve sites for new nuclear plants prior to a power company's commitment to build. The combined COL, is a 'one-step' licensing process by which the NRC approves and issues a license to build and operate a new nuclear power plant.

www.nuclear.energy.gov
May 2009

The U.S. Department of Energy's Office of Nuclear Energy

Cooperative Projects —

In 2002, the Department initiated cooperative projects with industry to obtain the NRC's approval of three sites for construction of new nuclear power plants under the NRC's ESP process. In 2003, three ESP applications were submitted by power companies to the NRC for review. They were approved in FY 2007 and early FY 2008. One additional ESP application is currently under review with the NRC as of FY 2009; three more are expected in the FY 2010–2012 timeframe.

In 2005, the Department, in cooperation with industry teams, initiated two new Nuclear Plant Licensing Demonstration Projects to demonstrate the COL licensing process and complete the certification of first-of-a-kind designs for Generation III+ reactor technologies. In early FY 2008, these industry consortia developed and submitted COL applications for two commercial nuclear plant sites for the Westinghouse Advanced Passive Pressurized Water Reactor (AP1000) and the General Electric (GE) Economic Simplified Boiling Water Reactor (ESBWR) technologies. The two industry consortia involve power companies currently operating more than two-thirds of the existing U.S. commercial nuclear power plants. As of FY 2009, these demonstration projects are proceeding successfully toward their respective goals. Design Certifications for both reactor technologies are expected in the early to mid-2011 timeframe, with COL approvals in late 2011.

Standby Support —

To mitigate some of the financial risk associated with new nuclear power plants—thus encouraging the construction of new nuclear plants—the Energy Policy Act of 2005 (EPAct 2005) allows the Secretary of Energy to pay certain costs to project sponsors if construction or full-power operation of an advanced nuclear facility is delayed. This standby support provision covers costs attributed to regulatory delays or litigation that delays full-power operations of the new nuclear plants. The Secretary is permitted to pay the delay costs for six reactors, up to a total of $2 billion.

UPCOMING DECISIONS

Notwithstanding the phase-out of NP 2010 by FY 2011, decisions are expected as a result of ongoing activities in the NP 2010 program:

- Issuance of conditional agreements for standby support is expected in FY 2009.
- An industry decision to build a new nuclear power plant is anticipated as early as 2010.
- Execution of final Standby Support contracts is expected in FY 2011.
- Issuance of nuclear loan guarantees is anticipated in FY 2011.
- NRC is expected to issue its rulemaking on the revised AP1000 certification in April 2011.
- NRC is expected to issue its rulemaking on the ESBWR design certification by June 2011.
- NRC decisions on issuance of two COLs are expected in 2011.

EMERGING REACTOR DESIGNS

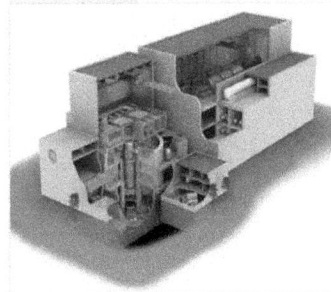

ESBWR

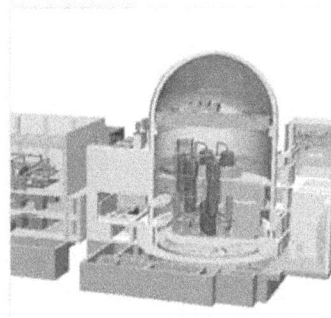

US-APWR

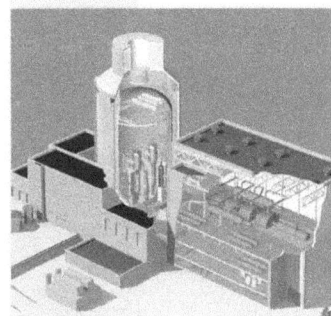

AP1000

www.nuclear.energy.gov
May 2009

NUCLEAR POWER 2010

The U.S. Department of Energy's Office of Nuclear Energy

Program Budget

Nuclear Power 2010
($ in Millions)

	FY 2009 Actual	FY 2010 Request
	$177.5	$20.0

PLANNED PROGRAM ACCOMPLISHMENTS

FY 2009

- Continue industry interactions with NRC on the COL applications, including issuance of Safety Evaluation Reports (SERs) and Final Environmental Impact Statements.

- Continue first-of-a-kind design finalization activities for the standardized AP1000 and ESBWR designs.

- Accelerate design finalization activities necessary to permit vendors and utilities to specify and procure components and equipment.

- Initiate additional First-of-a-Kind Engineering and design details to increase standardization of component design, specification and qualification.

- Resolve open ESBWR certification items to allow the NRC to issue the Final Design Approval and initiate the design certification rulemaking.

- Complete review of application requests and issue conditional agreements for standby support.

- Review documentation for final standby support contracts.

FY 2010

- Continue NP 2010 Licensing Demonstration projects.

- Continue efforts to close NRC open items on selected COL applications so that NRC Safety Evaluation Reports and Final Environmental Impact Statements can be issued on schedule.

- Support industry decisions to start construction of a new nuclear power plant beginning in 2010.

www.nuclear.energy.gov
May 2009

Nuclear Power 2010
Meeting Tomorrow's Energy Needs

The **Nuclear Power 2010** (NP 2010) program is a government-industry, 50-50 cost-shared initiative aimed at reducing the technical, regulatory, and institutional barriers to building new nuclear power plants the United States. These new plants are needed to meet an expected increase in electricity demand and to replace older power plants with innovative, more efficient designs.

Authorized by the **Energy Policy Act of 2005**, the NP 2010 program focuses on deployment of **Generation III+ advanced light-water reactor designs** that offer advancements in safety, efficiency, and economics over existing U.S. nuclear plant designs.

NP 2010 program goals:

- Develop and bring to market advanced, standardized nuclear plant technologies.
- Demonstrate streamlined Federal regulatory and licensing processes for siting, building, and operating new nuclear power plants.

NP 2010 is managed by the U.S. Department of Energy (DOE) Office of Nuclear Energy (NE).

Nuclear Power 2010
Moving toward Deployment
www.nuclearenergy.gov

January 2010

Nuclear Power Deployment Status

Early Site Permit (ESP)
- Four ESPs issued by NRC.
- No ESP applications under NRC review.
- Four ESP applications expected in 2010–2012.

Design Certification (DC)
- Two advanced reactor designs certified by NRC.
- Four reactor designs undergoing NRC review.

Combined Construction & Operating License (COL)
- Eighteen COL applications submitted to NRC.
- Thirteen COL applications under NRC review.
- Five COL applications are suspended, pending technology decision or for financial reasons.

New Nuclear Plant Orders
- Nine utilities have ordered large, long-lead nuclear component forgings from three reactor vendors.
- Four Engineering, Procurement, and Construction Contracts signed (Vogtle, V.C. Summer, STP, and Progress).
- TVA resumed construction of Watts Bar 2; construction permits reinstated for Bellefonte 1 & 2.

Federal Financial Incentives
Nuclear Power Loan Guarantees — DOE authorized to guarantee $18.5 billion in loans for nuclear power projects.
Standby Support (Risk Insurance) — DOE authorized to issue insurance to six reactors to cover delays in operations attributed to NRC licensing reviews or litigation.
Production Tax Credits — 1.8 cents/kw tax credit for the first 6,000 MWe of deployed nuclear power.

DESIGN · SIMULATE · LICENSE · BUILD · OPERATE · SUPPLY

Quarterly NEWS

January 2010

- Areva signed a letter of intent with the Fresno Nuclear Energy Group December 31, 2009, to investigate a possible EPR in California's Central Valley region.
- On December 23, Alternate Energy Holdings Inc. announced a delay in its COL application for a nuclear plant in Elmore County, Idaho to the fourth quarter of 2011. They also announced plans to submit applications for two new plants:
  - Payette County, ID, 2nd quarter 2011
  - Pueblo, CO, 2nd quarter 2012.
- General Electric-Hitachi announced on December 17, that it had signed an agreement with Detroit Edison on site planning for an ESBWR at Detroit Edison's Fermi site.
- NRC held a a meeting in Waynesboro, Georgia on December 16, to discuss the agency's inspection plans under the Limited Work Authorization issued for the Vogtle nuclear plant site.
- Unistar asked the NRC in a December 1 letter to suspend review of their COL application for an EPR at Nine Mile due to uncertainties in loan guarantee funding.
- The Maryland Public Service Commission has granted Electricité de France (EdF) conditional rights to take over 49.99% of Constellation Energy's nuclear generation on November 2, including the proposed new EPR at Calvert Cliffs.

Updates available at http://www.nuclear.gov

Planned Reactors

Number of Planned Reactors—36 Total

14

COLs Submitted   Expected

2   5   4   4   2   3   2

Proposed Sites of New U.S. Commercial Nuclear Power Plants

Payette County US-EPR - 1 Unit
Pueblo Colorado US-EPR - 1 Unit
River Bend Unknown- 1 Unit
Grand Gulf Unknown - 1 Unit
Callaway US-EPR - 1 Unit
Fermi ESBWR - 1 Unit
Nine Mile Point US-EPR - 1 Unit
Bell Bend US-EPR - 1 Unit
Calvert Clifs US-EPR - 1 Unit
North Anna ESBWR - 1 Unit
Harris AP1000 - 2 Units
William Lee AP1000 - 2 Units
Turkey Point AP1000 - 2 Units
Levy County AP1000 - 2 Units
V.C. Summer AP1000 - 2 Units
Vogtle AP1000 - 2 Units
Bellefonte AP1000 - 2 Units
South Texas ABWR - 2 Units
Victoria County ABWR - 2 Units
Comanche Peak US-APWR - 2 Units
Amarillo US-EPR - 2 Units
Blue Castle Unknown - 1 Unit
Hammet US-EPR - 1 Unit

Construction and Operating License (COL) Status
☆ Under Nuclear Regulatory Review (NRC)   ◆ Planned Future Submittal   ● NRC Review Suspended

Emerging Nuclear Reactor Designs

- **Advanced Passive Pressurized Water Reactor (AP1000)** — Twin units, 1,117 MWe each (Westinghouse International)
- **Advanced Boiling Water Reactor (ABWR)** — 1,356 MWe (General Electric)
- **Economic Simplified Boiling Water Reactor (ESBWR)** — 1,560 MWe (General Electric)
- **United States Advanced Power Reactor (US-APWR)** — 1,700 MWe (Mitsubishi Heavy Industry)
- **United States Evolutionary Power Reactor (US-EPR)** — 1,600 MWe (AREVA)

IDAHO NUCLEAR INFRASTRUCTURE

The U.S. Department of Energy's Office of Nuclear Energy

ADVANCED TEST REACTOR

The Department of Energy supports nuclear science and technology through one of the world's most comprehensive research infrastructures.

The Idaho National Laboratory (INL) serves as the center for U.S. nuclear energy research and development (R&D) efforts. INL combines the expertise of government, industry, and academia in a single laboratory dedicated to the development of advanced reactor and fuel-cycle technologies.

A MULTI-PROGRAM NATIONAL LABORATORY

INL employs more than 3,900 personnel located primarily at the Idaho Site and in the city of Idaho Falls. In addition to its broad spectrum of nuclear energy and national security programs, the laboratory provides essential site services to DOE and other governmental agencies and private-sector companies doing business on the Idaho Site. INL conducts science and technology research across a wide range of disciplines. Its core missions include:

- Developing advanced, next-generation reactor and fuel-cycle technologies;

- Promoting nuclear technology education; and

- Applying technical skills and unique features of the laboratory site to enhance the Nation's security.

Under the oversight of the Department's Office of Nuclear Energy (NE), INL provides technical leadership to support long-term nuclear science and engineering R&D activities to address the Nation's energy and nuclear security goals. Key technical areas include nuclear fuel cycle science-based research, the development of alternative radioactive waste management strategies for the United States, and technology programs that support nuclear nonproliferation and other critical infrastructure protection.

INL also supports NE by conducting R&D and technical integration support for the new Reactor Concepts Research, Development and Demonstration program. INL is the lead laboratory for the Next Generation Nuclear Plant (NGNP) program and, together with Oak Ridge National Laboratory (ORNL), is the principal laboratory responsible for the development of advanced gas reactor fuel and materials R&D. INL is also responsible for staffing the Technical Secretariat for the Generation IV International Forum.

INL provides technical support for cross-cutting technologies including advanced fuels, fabrication and construction methods, and proliferation risk assessment within the new Nuclear Energy Enabling Technologies program. INL has the lead on the development of advanced instruments and sensors for the existing light water reactor fleet.

www.nuclear.energy.gov
February 2010

IDAHO NUCLEAR INFRASTRUCTURE

INL also provides the facilities and expertise needed to fuel and test radioisotope power systems for space and defense applications, and to accomplish national and homeland security missions, including critical infrastructure protection and nuclear nonproliferation.

NUCLEAR ENGINEERING AND SCIENCE EDUCATION

The Center for Advanced Energy Studies (CAES) is a public-private partnership including the State of Idaho and its academic research institutions, DOE, and INL. CAES serves to advance the educational opportunities at the Idaho universities in energy-related areas, creating new capabilities within its member institutions and delivering technological innovations leading to technology-based economic development for the intermountain region. CAES also provides students and professors from across the country with access to the Laboratory's unique capabilities.

CAES also administers the Nuclear Energy University Program (NEUP), which includes three components — R&D, scholarships and fellowships, and research infrastructure enhancements. The goal is to improve America's competitiveness and develop more effective collaborations among universities, national laboratories, and industry in direct support of DOE's NE R&D programs.

INL'S NUCLEAR INFRASTRUCTURE

Two programs support the nuclear infrastructure at INL:

- **Idaho Facilities Management (IFM) Program.** Through IFM, NE maintains its research facilities in a safe, reliable, and environmentally compliant condition to support national nuclear programs.

- **Idaho Site-Wide Safeguards and Security Program.** Through this program, NE supports activities that are required to protect the assets of the Idaho complex from theft, diversion, sabotage, espionage, unauthorized access, compromise, and other hostile acts.

The Department manages and operates three main engineering and research complexes at INL:

Advanced Test Reactor (ATR) Complex —

This is the site of the ATR, a 250-megawatt test reactor used to provide irradiation services for a range of users. ATR is the largest and most versatile thermal test reactor in the world. Its current primary mission is to provide irradiation and testing services to the Naval Reactors Program.

ATR supports the NE R&D programs, as well as National Nuclear Security Administration (NNSA) programs. ATR also provides irradiation and testing services on a cost-reimbursable basis to other national and international nuclear energy research groups and medical and industrial isotope producers.

In April 2007, DOE designated ATR as a National Scientific User Facility (NSUF).

IDAHO NATIONAL LABORATORY MAP

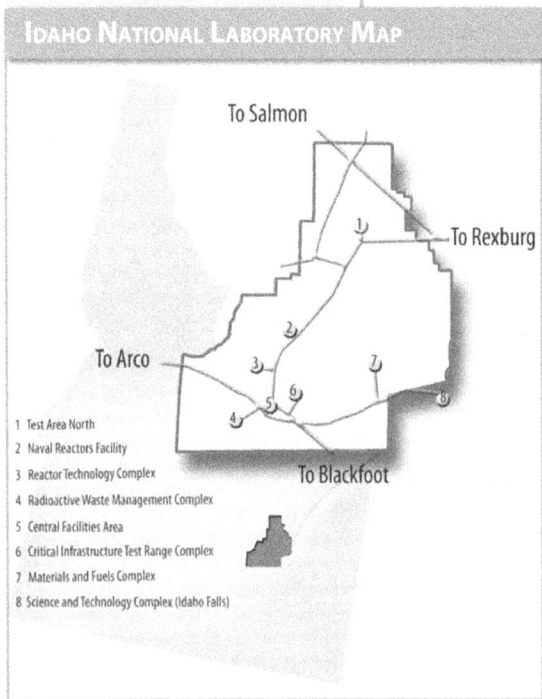

To Salmon

To Rexburg

To Arco

To Blackfoot

1 Test Area North
2 Naval Reactors Facility
3 Reactor Technology Complex
4 Radioactive Waste Management Complex
5 Central Facilities Area
6 Critical Infrastructure Test Range Complex
7 Materials and Fuels Complex
8 Science and Technology Complex (Idaho Falls)

www.nuclear.energy.gov
February 2010

This designation has enabled ATR to become a cornerstone of nuclear energy R&D in the United States and allows a broader use of ATR capabilities. The extensive capabilities of ATR allow a wide range of advanced nuclear energy irradiation testing to be conducted simultaneously by universities, commercial industry, international organizations, and other national laboratories without interfering with its primary missions. Increasing accessibility to ATR through the NSUF is an important step for INL in building strong ties with the nuclear industry and universities conducting nuclear energy R&D.

NE, through its IFM program, funds the ATR Life Extension Program to ensure the long-term availability of this essential nuclear power research capability.

Materials and Fuels Complex (MFC) —

The facilities at MFC are used to conduct advanced nuclear energy technology R&D. The facilities, personnel, and infrastructure at MFC support several important DOE nuclear energy, defense, and environmental management programs, most notably the development of alternative nuclear fuel-cycle technologies. MFC includes the following major facilities:

- Fuel Conditioning Facility;
- Fuel Manufacturing Facility;
- Hot Fuels Examination Facility;
- Analytical Laboratory;
- Electron Microscopy Laboratory; and
- Radioactive Scrap and Waste Facility.

Research and Education Campus (REC) —

Located in Idaho Falls, Idaho, REC includes more than 30 DOE-owned and leased buildings that house office space, CAES, and extensive laboratory facilities. The laboratories support NE's research and development activities, national security programs, and a wide range of research for other disciplines.

PLANNED PROGRAM ACCOMPLISHMENTS

FY 2010

- Enable INL facility operations supporting nuclear science, engineering, and energy-related R&D programs for DOE, NNSA, and U.S. universities.
- Conduct ATR base operations that enable 275 days of safe, compliant reactor operations per year, serving national security and civilian nuclear power R&D programs at this National Scientific User Facility.
- Continue the ATR Life Extension Program (LEP) to restore hardware, analyze system performance, and complete safety analysis required to reliably operate this $1.2B nuclear research reactor for at least another 20 years.
- Perform almost 1,000 single and recurring preventive equipment/system maintenance activities to maintain more than 150 laboratories, hot cells, and shops at the MFC and ATR complex.
- Prepare surplus nuclear material at INL for off-site shipment to improve nuclear material management.
- Establish the cost and schedule range for the new Remote-Handled Low-Level Waste Disposal Project to sustain critical laboratory capability.

ADVANCED TEST REACTOR

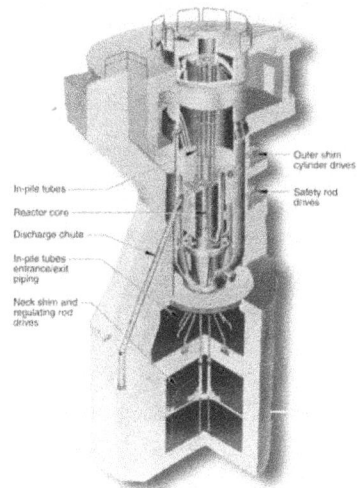

Outer shim cylinder drives
Safety rod drives
In-pile tubes
Reactor core
Discharge chute
In-pile tubes entrance/exit piping
Neck shim and regulating rod drives

www.nuclear.energy.gov
February 2010

IDAHO NUCLEAR INFRASTRUCTURE

The U.S. Department of Energy's Office of Nuclear Energy

Program Budget

Idaho Nuclear Infrastructure
($ in Millions)

Idaho Facilities Management

	FY 2010 Actual	FY 2011 Request
	$173.0	$162.5

Idaho Site-wide Safeguards & Security

	FY 2010 Actual	FY 2011 Request
	$83.4	$88.2

FY 2011

- Enable INL facility operations supporting nuclear science, engineering, and energy-related R&D programs for DOE, NNSA, and U.S. universities.

- Conduct ATR base operations to support more than 40 irradiation campaigns as scheduled while maintaining an operating efficiency greater than 80 percent.

- Award three to five university experiments using the ATR and other INL research facilities and support six university partnerships to increase available capabilities for NSUF experiments.

- Perform almost 1,000 single and recurring preventive equipment/system maintenance activities to maintain greater than 150 laboratories, hot cells, and shops at the MFC and ATR complex.

- Process approximately 400 kilograms of Experimental Breeder Reactor (EBR)-II sodium-bonded fuel, consistent with the settlement agreement.

- Approve the performance baseline for the new Remote-Handled Low-Level Waste Disposal Project to sustain critical laboratory capability.

www.nuclear.energy.gov
February 2010

Radioactive Waste: Production, Storage, Disposal

U.S. Nuclear Regulatory Commission

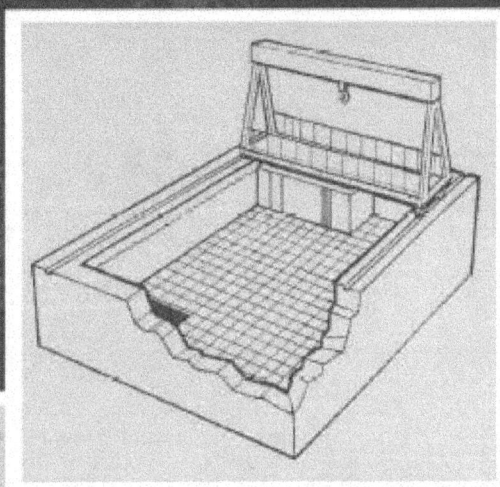

Contents

NRC headquarters offices are located in Rockville, Maryland.

Contents

(continued)

Contents

Radioactive Waste: An Introduction

Radioactive wastes are the leftovers from the use of nuclear materials for the production of electricity, diagnosis and treatment of disease, and other purposes.

The materials are either naturally occurring or man-made. Certain kinds of radioactive materials, and the wastes produced from using these materials, are subject to regulatory control by the federal government or the states.

The Department of Energy (DOE) is responsible for radioactive waste related to nuclear weapons production and certain research activities. The Nuclear Regulatory Commission (NRC) and some states regulate commercial radioactive waste that results from the production of electricity and other non-military uses of nuclear material.

Various other federal agencies, such as the Environmental Protection Agency, the Department of Transportation, and the Department of Health and Human Services, also have a role in the regulation of radioactive material.

The NRC regulates the management, storage and disposal of radioactive waste produced as a result of NRC-licensed activities. The agency has entered into agreements with 32 states, called Agreement States, to allow these states to regulate the management, storage and disposal of certain nuclear waste.

1

Radioactive Waste: Production, Storage, Disposal

Nuclear power plants, such as this Calvert Cliffs plant near Lusby, Maryland, produce electricity and, as a byproduct, produce radioactive waste.

The commercial radioactive waste that is regulated by the NRC or the Agreement States and that is the subject of this brochure is of three basic types: high-level waste, mill tailings, and low-level waste.

High-level radioactive waste consists of "irradiated" or used nuclear reactor fuel (i.e., fuel that has been used in a reactor to produce electricity). The used reactor fuel is in a solid form consisting of small fuel pellets in long metal tubes.

2

Radioactive Waste: Production, Storage, Disposal

Mill tailings wastes are the residues remaining after the processing of natural ore to extract uranium and thorium.

Commercial radioactive wastes that are not high-level wastes or uranium and thorium milling wastes are classified as low-level radioactive waste. The low-level wastes can include radioactively contaminated protective clothing, tools, filters, rags, medical tubes, and many other items.

NRC licensees are encouraged to manage their activities so as to limit the amount of radioactive waste they produce. Techniques include avoiding the spread of radioactive contamination, surveying items to ensure that they are radioactive before placing them in a radioactive waste container, using care to avoid mixing contaminated waste with other trash, using radioactive materials whose radioactivity diminishes quickly and limiting radioactive material usage to the minimum necessary to establish the objective.

Licensees take steps to reduce the volume of radioactive waste after it has been produced. Common means are compaction and incineration. Approximately 59 NRC licensees are authorized to incinerate certain low-level wastes, although most incineration is performed by a small number of commercial incinerators.

The radioactivity of nuclear waste decreases with the passage of time, through a process called radioactive decay. ("Radioactivity" refers to the spontaneous disintegration of an unstable atomic nucleus, usually accompanied by the emission of ionizing radiation.) The amount of time

3

362

Radioactive Waste: Production, Storage, Disposal

necessary to decrease the radioactivity of radioactive material to one-half the original amount is called the radioactive half-life of the radioactive material. Radioactive waste with a short half-life is often stored temporarily before disposal in order to reduce potential radiation doses to workers who handle and transport the waste, as well as to reduce the radiation levels at disposal sites.

In addition, NRC authorizes some licensees to store short-half-lived material until the radioactivity is indistinguishable from ambient radiation levels, and then dispose of the material as non-radioactive waste.

Currently, there are no permanent disposal facilities in the United States for high-level nuclear waste; therefore commercial high-level waste (spent fuel) is in temporary storage, mainly at nuclear power plants.

Most uranium mill tailings are disposed of in place or near the mill, after constructing a barrier of a material such as clay on top of the pile to prevent radon from escaping into the atmosphere and covering the mill tailings pile with soil, rocks or other materials to prevent erosion.

For low-level waste, three commercial land disposal facilities are available, but they accept waste only from certain states or accept only limited types of low-level wastes. The remainder of the low-level waste is stored primarily at the site where

4

363

Radioactive Waste: Production, Storage, Disposal

This low-level radioactive waste disposal site in Richland, Washington, accepts wastes from the Northwest and Rocky Mountain states.

it was produced, such as at hospitals, research facilities, clinics and nuclear power plants.

The following sections of this pamphlet provide separate discussions on high-level and low-level radioactive waste and mill tailings.

5

High-Level Radioactive Waste

High-Level Radioactive Waste

What is high-level waste?

After uranium fuel has been used in a reactor for a while, it is no longer as efficient in splitting its atoms and producing heat to make electricity. It is then called "spent" nuclear fuel. About one-fourth to one-third of the total fuel load is spent and is removed from the reactor every 12 to 18 months and replaced with fresh fuel. The spent nuclear fuel is high-level radioactive waste.

What is the role of NRC?

The NRC regulates all commercial reactors in the United States, including nuclear power plants that produce electricity, and university research reactors. The agency regulates the possession, transportation, storage and disposal of spent fuel produced by the nuclear reactors.

How hazardous is high-level waste?

Spent nuclear fuel is highly radioactive and potentially very harmful. Standing near unshielded spent fuel could be fatal due to the high radiation levels. Ten years after removal of spent fuel from a reactor, the radiation dose 1 meter away from a typical spent fuel assembly exceeds 20,000 rems per hour. A dose of 5,000 rems would be expected to cause immediate incapacitation and death within one week.

7

High-Level Radioactive Waste

Some of the radioactive elements in spent fuel have short half-lives (for example, iodine-131 has an 8-day half-life) and therefore their radioactivity decreases rapidly. However, many of the radioactive elements in spent fuel have long half-lives. For example, plutonium-239 has a half-life of 24,000 years, and plutonium-240 has a half-life of 6,800 years. Because it contains these long half-lived radioactive elements, spent fuel must be isolated and controlled for thousands of years.

A second hazard of spent fuel, in addition to high radiation levels, is the extremely remote possibility of an accidental "criticality," or self-sustained fissioning and splitting of the atoms of uranium and plutonium.

NRC regulations therefore require stringent design, testing and monitoring in the handling and storage of spent fuel to ensure that the risk of this type of accident is extremely unlikely. For example, special control materials (usually boron) are placed in spent fuel containers to prevent a criticality from occurring. Nuclear engineers and physicists carefully analyze and monitor the conditions of handling and storage of spent fuel to guard further against an accident.

A barrier or radiation protection shield must always be placed between spent nuclear fuel and human beings.

Water, concrete, lead, steel, depleted uranium or other suitable materials calculated to be sufficiently protective by trained engineers and health physicists, and verified by radiation measurements, are typically used as radiation shielding for spent nuclear fuel.

8

366 703-739-3790 TCNNaturalGas.com

High-Level Radioactive Waste

Most spent fuel from nuclear power plants is stored under water, as shown at the Diablo Canyon plant in California.

How and where is the waste stored?

Spent fuel may be stored in either a wet or dry environment. In addition, it may be stored either at the reactor where it was used or away from the reactor at another site.

The various techniques are as follows:

Wet Storage

Currently most spent nuclear fuel is safely stored in specially designed pools at individual reactor sites around the country. The water-pool option involves storing spent fuel in rods under at least 20 feet of water, which provides adequate

9

High-Level Radioactive Waste

shielding from the radiation for anyone near the pool. The rods are moved into the water pools from the reactor along the bottom of water canals, so that the spent fuel always is shielded to protect workers.

A typical spent fuel rod is about 12 feet long and 3/4 inch in diameter. The rods are arranged in somewhat square arrays, known as fuel assemblies, that range in size from an array of 6 rods by 6 rods to an array of 17 rods by 17 rods. The fuel pools vary in size from a capacity of 216 to 8,083 fuel assemblies.

Most pools were originally designed to store several years worth of spent fuel. Due to delays in developing disposal facilities for the spent fuel, licensees have redesigned and rebuilt equipment in the pools over the years to allow a greater number of spent fuel rods to be stored. However, this storage option is limited by the size of the spent fuel pool and the need to keep individual fuel rods from getting too close to other rods and initiating a criticality or nuclear reaction.

Dry Storage

If pool capacity is reached, licensees may move toward use of above-ground dry storage casks. The first dry storage installation was licensed by the NRC in 1986. In this method, spent fuel is surrounded by inert gas inside a container called a cask. The casks can be made of metal or concrete, and some can be used for both storage and transportation. They are either placed horizontally or stand vertically on a concrete pad.

Seventeen nuclear power plants are currently storing spent fuel under the dry storage option.

10

368 703-739-3790 TCNNaturalGas.com

High-Level Radioactive Waste

Spent fuel may be stored in dry casks either horizontally, as shown at the H.B. Robinson nuclear power plant in South Carolina, or vertically, as shown at the Surry nuclear power plant in Virginia.

11

High-Level Radioactive Waste

Away-from-Reactor Storage

General Electric Company has a facility to store spent fuel away from reactors, using the wet storage pool technology, at Morris, Illinois. GE received a license to receive and store nuclear material at this facility in 1971. The facility is essentially full, and the company has completed contracts with specific utilities (under which it had agreed to accept their used fuel) and has no plans to accept additional spent fuel.

Storage Differences

Both pool storage and dry storage are safe methods, but there are significant differences. Pool storage requires a greater and more consistent operational vigilance on the part of utilities or other licensees and the satisfactory performance of many mechanical systems using pumps, piping and instrumentation.

Dry storage, which is almost completely passive, is simpler, uses fewer support systems and offers fewer opportunities for things to go wrong through human or mechanical error. Dry storage is not suitable for fuel until the fuel has been out of the reactor for a few years and the amount of heat generated by radioactive decay has been reduced.

Monitored Retrievable Storage

The Nuclear Waste Policy Act (NWPA) of 1982 authorized the Department of Energy (DOE) to construct a monitored retrievable storage (MRS) facility for storage of high-level waste, with certain restrictions.

12

370 703-739-3790 TCNNaturalGas.com

> ## High-Level Radioactive Waste
>
> Representatives of state and local governments and Indian tribes and members of the public would be invited to participate in meetings on an MRS facility.
>
> NRC would publish notice of receipt of DOE's application to build an MRS facility and hold a public hearing, if requested, before issuance of the license.
>
> ### How much high-level waste is there?
>
> About 160,000 spent fuel assemblies, containing 45,000 tons of spent fuel from nuclear power plants, are currently in storage in the United States. Of these, about 156,500 assemblies are stored at nuclear power plants, and approximately 3,500 assemblies are stored at away-from-reactor storage facilities, such as the General Electric plant at Morris, Illinois. The vast majority of the assemblies are stored in water pools, and less than 5% are stored in dry casks.
>
> About 7,800 used fuel assemblies are taken out of reactors each year and are stored until a disposal facility becomes available.
>
> If all the 160,000 spent fuel assemblies currently in storage were assembled in one place, they would only cover a football field about 5 1/2 yards high.
>
> 13

High-Level Radioactive Waste

How and where will the high-level waste be disposed of?

DOE is developing plans for a permanent disposal facility for spent fuel from nuclear power plants (as well as for the high-level waste that has been produced by the nation's nuclear weapons production activities).

Congress has directed DOE to focus on a proposed site at Yucca Mountain, Nevada, for the disposal facility. This has aroused some controversy, particularly with state and local authorities.

Studies are still underway to determine if the site is adequate for permanent disposal of the high-level waste. NRC has a rigorous regulatory program for review of these ongoing DOE site investigations.

DOE would design, build and operate the facility, subject to federal regulations and oversight by the NRC. The NRC must approve the site and design for the disposal facility, as well as inspect it during construction and operation.

14

372

High-Level Radioactive Waste

The Nuclear Waste Policy Act directed the Department of Energy to study Yucca Mountain, Nevada, to determine whether it would be suitable for disposal of high-level radioactive waste.

Once DOE submits an application to construct a repository, the NWPA calls for NRC to complete its review within three years.

If the NRC authorizes construction, DOE will proceed with constructing the repository and would submit a license application update (containing additional details on design and construction of the facility) to the NRC. This would be followed by an NRC decision on whether to license operation of the repository.

15

High-Level Radioactive Waste

NRC Safety Requirements

As required by the NWPA, the NRC has issued technical requirements and criteria for approving or disapproving DOE's application. These are contained in Part 60 of the NRC's regulations. Examples include:

■ Radiation doses during repository operations must be kept below regulatory limits. These limits are 100 millirems per year for members of the general public (which is about a third of the average American's annual dose from nature) and 5,000 millirems per year for workers.

■ Waste must be retrievable for 50 years after waste emplacement begins.

■ The container in which the high-level waste will be placed must maintain its integrity for 300 to 1,000 years.

■ The waste packages must not contain explosive or flammable materials or liquids that could endanger the repository.

Public Involvement

Representatives of state and local governments and Indian tribes are invited to participate in meetings on the high-level waste repository. Members of the public may attend as observers.

NRC will publish notice of receipt of DOE's application to build a repository and hold a public hearing before issuance of the construction authorization. When DOE submits an application to receive and possess high-level waste at the

16

High-Level Radioactive Waste

LM-300 drill rig at Yucca Mountain, Nevada, obtained underground rock and soil samples that scientists examined to help determine site suitability for high-level waste disposal.

17

High-Level Radioactive Waste

Tunnel boring machine excavated Yucca Mountain to allow analysis of underground conditions and suitability of site for high-level waste disposal.

facility, NRC will again announce receipt of the application and will publish notice of the opportunity for an optional additional public hearing.

The NRC has established an Internet web site to inform interested parties of upcoming meetings, including those on radioactive waste. The address is http://www.nrc.gov/public-involve/public-meetings.html on the Internet. Members of the public who do not have access to the Internet may obtain information on public meetings by calling 800-397-4209.

18

Low-Level Radioactive Waste

Low-Level Radioactive Waste

What is low-level waste?

Low-level radioactive waste includes items that have become contaminated with radioactive material or have become radioactive through exposure to neutron radiation. This waste is typically contaminated protective shoe covers and clothing, wiping rags, mops, filters, reactor water treatment residues, equipment and tools, luminous dials, medical tubes, swabs, injection needles, syringes, and laboratory animal carcasses and tissues. The most intensely radioactive wastes are typically found in the water treatment residues, discarded parts from nuclear reactors and small gauges containing radioactive material.

VOLUME (Thousands of Cubic Feet)

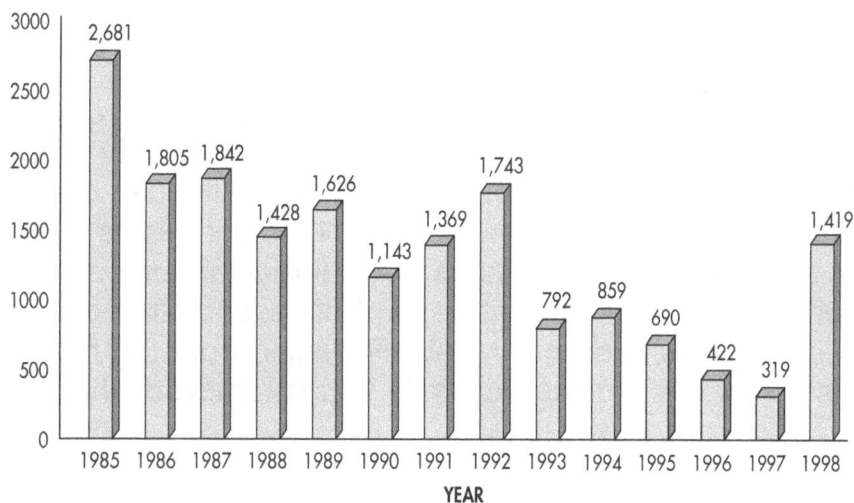

Year	Volume
1985	2,681
1986	1,805
1987	1,842
1988	1,428
1989	1,626
1990	1,143
1991	1,369
1992	1,743
1993	792
1994	859
1995	690
1996	422
1997	319
1998	1,419

YEAR

This chart shows the volume of low-level waste received at U.S. disposal facilities from 1985 to 1998.

19

Low-Level Radioactive Waste

The NRC has adopted a waste classification system for low-level radioactive waste based on its potential hazards, and has specified disposal and waste form requirements for each of the general classes of waste: Class A, Class B and Class C waste. Although the classification of waste can be complex, Class A waste generally contains lower concentrations of long half-lived radioactive material than Class B and C wastes.

Where does low-level waste come from?

In 1998, low-level waste disposal facilities received about 1,419 thousand cubic feet of commercially generated radioactive waste. Of this 14.8% came from nuclear reactors, 6.7% from industrial users, 2% from government sources (other than nuclear weapons sites), 0.3% from academic users, 0.1% from medical facilities, and the rest was undefined.

Nuclear Reactors

During normal operation of a nuclear reactor, some small amounts of radioactive material may be released into, or produced in, the water surrounding the fuel. Although reactor operators clean the water by using filters and resins, some of this material contaminates internal reactor components, such as pipes, pumps, valves, and filters, and other objects such as tools and equipment. Radiation from the reactor also produces radioactive waste that is removed when the reactor is decommissioned.

20

378

Low-Level Radioactive Waste

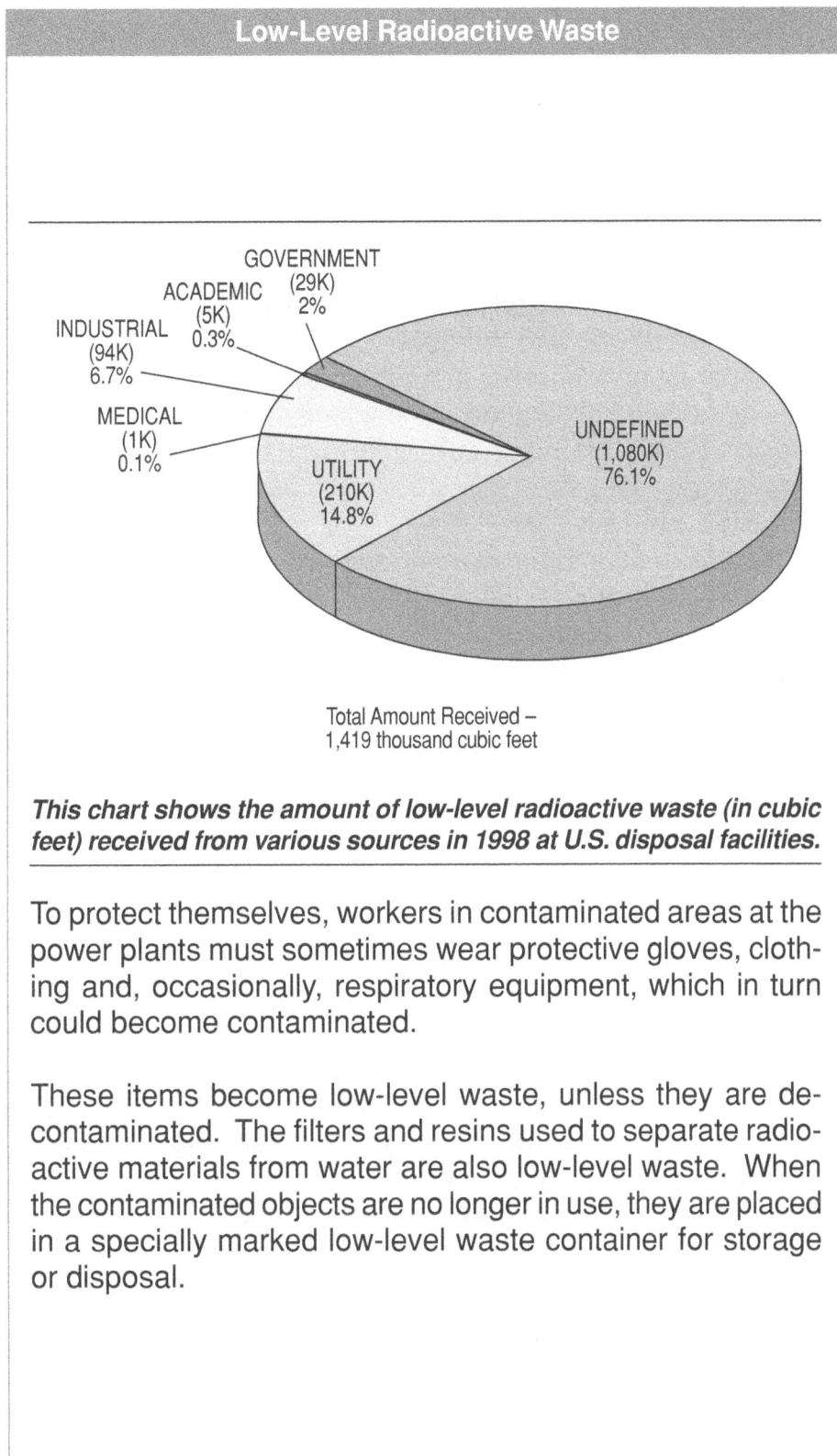

GOVERNMENT
(29K)
2%

ACADEMIC
(5K)
0.3%

INDUSTRIAL
(94K)
6.7%

MEDICAL
(1K)
0.1%

UTILITY
(210K)
14.8%

UNDEFINED
(1,080K)
76.1%

Total Amount Received –
1,419 thousand cubic feet

This chart shows the amount of low-level radioactive waste (in cubic feet) received from various sources in 1998 at U.S. disposal facilities.

To protect themselves, workers in contaminated areas at the power plants must sometimes wear protective gloves, clothing and, occasionally, respiratory equipment, which in turn could become contaminated.

These items become low-level waste, unless they are decontaminated. The filters and resins used to separate radioactive materials from water are also low-level waste. When the contaminated objects are no longer in use, they are placed in a specially marked low-level waste container for storage or disposal.

21

Low-Level Radioactive Waste

Medical Facilities

At medical facilities, radioactive materials are used in numerous diagnostic and therapeutic procedures for patients. During these procedures, test tubes, syringes, bottles, tubing and other objects come into contact with radioactive material. Some of the material remains in the objects, contaminating them.

In medical research, laboratory animals are sometimes injected with radioactive material for research purposes to combat diseases, such as AIDS and cancer. The animal carcasses containing the radioactive material become low-level radioactive waste and must be handled appropriately.

Hospitals may store waste containing radioactive material with short half lives until it decays to background radiation levels for ultimate disposal with non-radioactive medical waste. Waste containing longer-lived radioactive material is stored or sent to a low-level radioactive waste disposal facility.

Industry and Research Institutes

Commercial and industrial firms use radioactive materials to measure the thickness, density or volume of materials; to determine the age of prehistoric and geological objects; to examine welds and structures for flaws; to analyze wells for oil and gas exploration; and for various other types of research and development.

During research and chemical analysis, test tubes, bottles, tubing and process equipment come into contact with the

22

Low-Level Radioactive Waste

NUCLEAR MATERIALS USED IN SOME EVERYDAY THINGS:

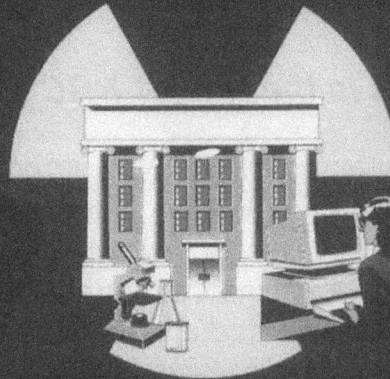

Research

Medicine

Use of nuclear materials for a variety of purposes, such as in exit signs, research, smoke detectors, and medicine, results in the production of nuclear waste.

23

Low-Level Radioactive Waste

radioactive material, become contaminated and are classified as low-level waste. Waste may also be produced during the manufacture of devices, such as certain gauges, luminous watches, exit signs and smoke detectors, that contain radioactive material.

What is the role of NRC?

The NRC regulates about 4,900 licenses for the possession and use of radioactive materials. In addition, 32 Agreement States regulate approximately 16,250 radioactive materials licenses. Agreement States are those states that have accepted responsibility, through agreement with the NRC, over the licensing of radioactive materials within their state.

The NRC and the Agreement States oversee licensees' management and disposal of radioactive waste products.

How hazardous is low-level waste?

The danger of exposure to radiation in low-level radioactive waste varies widely according to the types and concentration of radioactive material contained in the waste. Low-level waste containing some radioactive materials used in medical research, for example, is not particularly hazardous unless inhaled or consumed, and a person can stand near it without shielding. Low-level waste from processing water at a reactor, on the other hand, may be quite hazardous. For example, low-level waste could cause exposures that could lead to death or an increased risk of cancer.

24

382

Low-Level Radioactive Waste

How is low-level waste stored?

Storage of low-level radioactive waste requires an NRC or Agreement State license. NRC or Agreement State regulations require the waste to be stored in a manner that keeps radiation doses to workers and members of the public below NRC-specified levels. Licensees must further reduce these doses to levels that are as low as reasonably achievable. Actual doses, in most cases, are a small fraction of the NRC limits.

Low-level radioactive waste is packaged in containers appropriate to its level of hazard. Some low-level radioactive wastes require shielding with lead, concrete or other materials to protect workers and members of the public.

Workers are trained to maintain a safe distance from the more highly radioactive materials, to limit the amount of time they spend near the materials, and to monitor the waste to detect any releases.

Nuclear power plants may store waste in special buildings that provide an extra degree of shielding. Safe distances must be maintained between the buildings containing radioactive material and the fence restricting public access to licensee property.

Hospitals typically keep their waste stored in special containers or separate rooms.

Radioactive waste storage areas are posted to identify the radioactive waste so that workers and the public will not inadvertently enter the area.

25

Low-Level Radioactive Waste

Low-level waste may be stored to allow short-lived radio-nuclides to decay to innocuous levels and to provide safe-keeping when access to disposal sites is not available. The NRC believes storage can be safe over the short term as an interim measure, but favors disposal rather than storage over the long term.

How and where is low-level waste disposed of?

There are two low-level disposal facilities that accept a broad range of low-level wastes. They are located in Barnwell, South Carolina, and Richland, Washington.

This low-level waste disposal facility in Barnwell, South Carolina, buries waste underground.

26

Low-Level Radioactive Waste

In addition, Envirocare of Utah is licensed by the NRC to operate a facility near Clive, Utah, for disposal of uranium and thorium mill tailings. The facility also accepts certain other radioactive wastes under a State of Utah license. It primarily accepts low-level waste with small concentrations of radioactive material that are generated after a facility shuts down permanently and needs to remove a large bulk of contaminated material—such as contaminated soil or debris from demolished buildings—in preparation for license termination.

Four former low-level radioactive waste disposal sites are closed and no longer accept wastes. They are located in or near Sheffield, Illinois; Morehead, Kentucky; Beatty, Nevada; and West Valley, New York.

The low-level wastes at the Barnwell and Richland facilities and the four closed sites are or will be buried under several feet of soil in near-surface shallow trenches, usually in the containers in which they were shipped.

Laws and Regulations

The Low-Level Radioactive Waste Policy Amendments Act of 1985 made the states responsible for low-level radioactive waste disposal. It encouraged the states to enter into compacts that would allow several states to dispose of waste at a joint disposal facility. Most states have entered into compacts. However, to date no new disposal facilities have been built.

NRC and state regulations establish requirements for the siting, design and operation of disposal facilities, including buffer

27

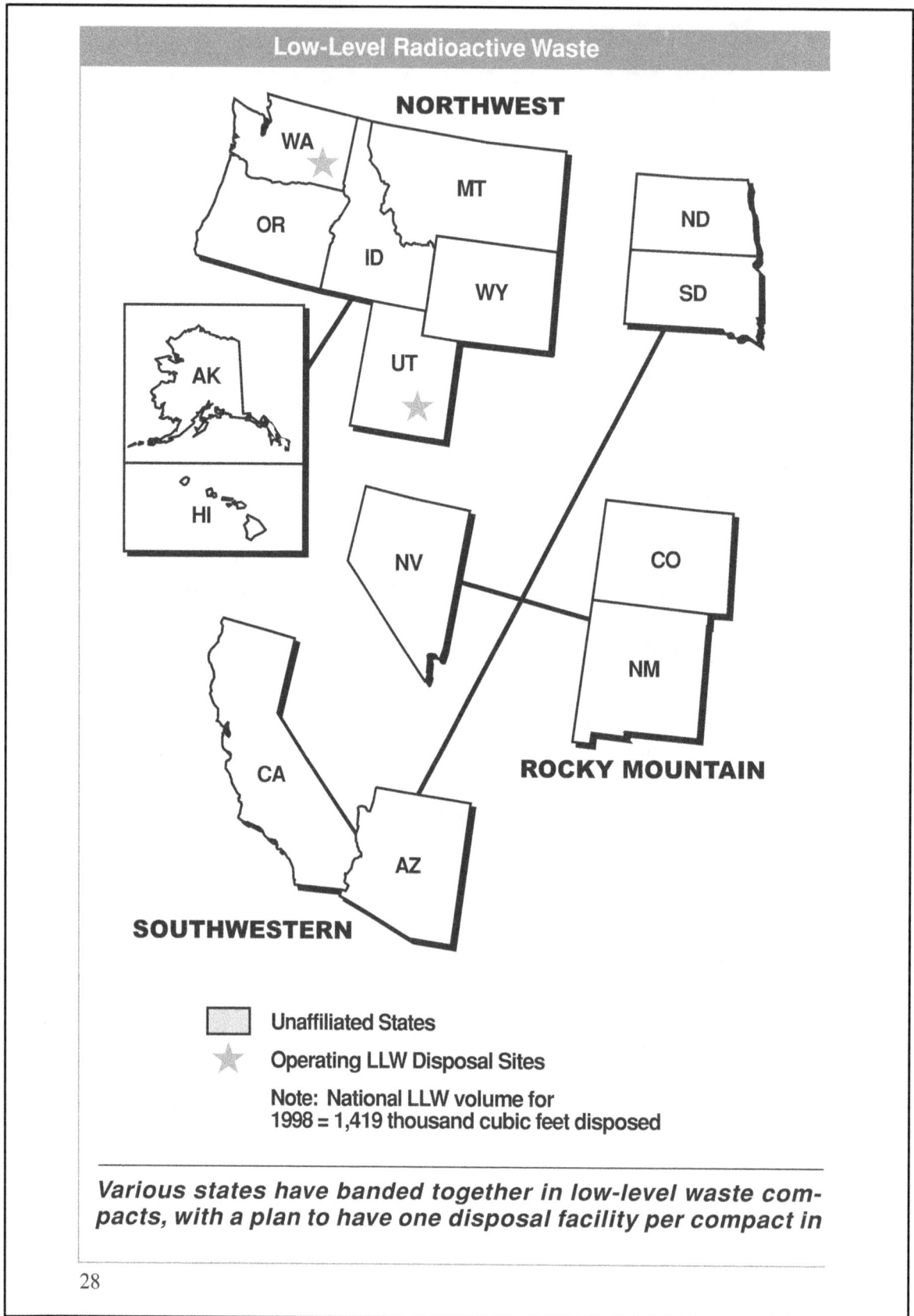

Low-Level Radioactive Waste

NORTHWEST

WA

OR

MT

ID

WY

ND

SD

AK

UT

HI

NV

CO

CA

NM

AZ

ROCKY MOUNTAIN

SOUTHWESTERN

▢ Unaffiliated States

⭐ Operating LLW Disposal Sites

Note: National LLW volume for
1998 = 1,419 thousand cubic feet disposed

*Various states have banded together in low-level waste com-
pacts, with a plan to have one disposal facility per compact in*

28

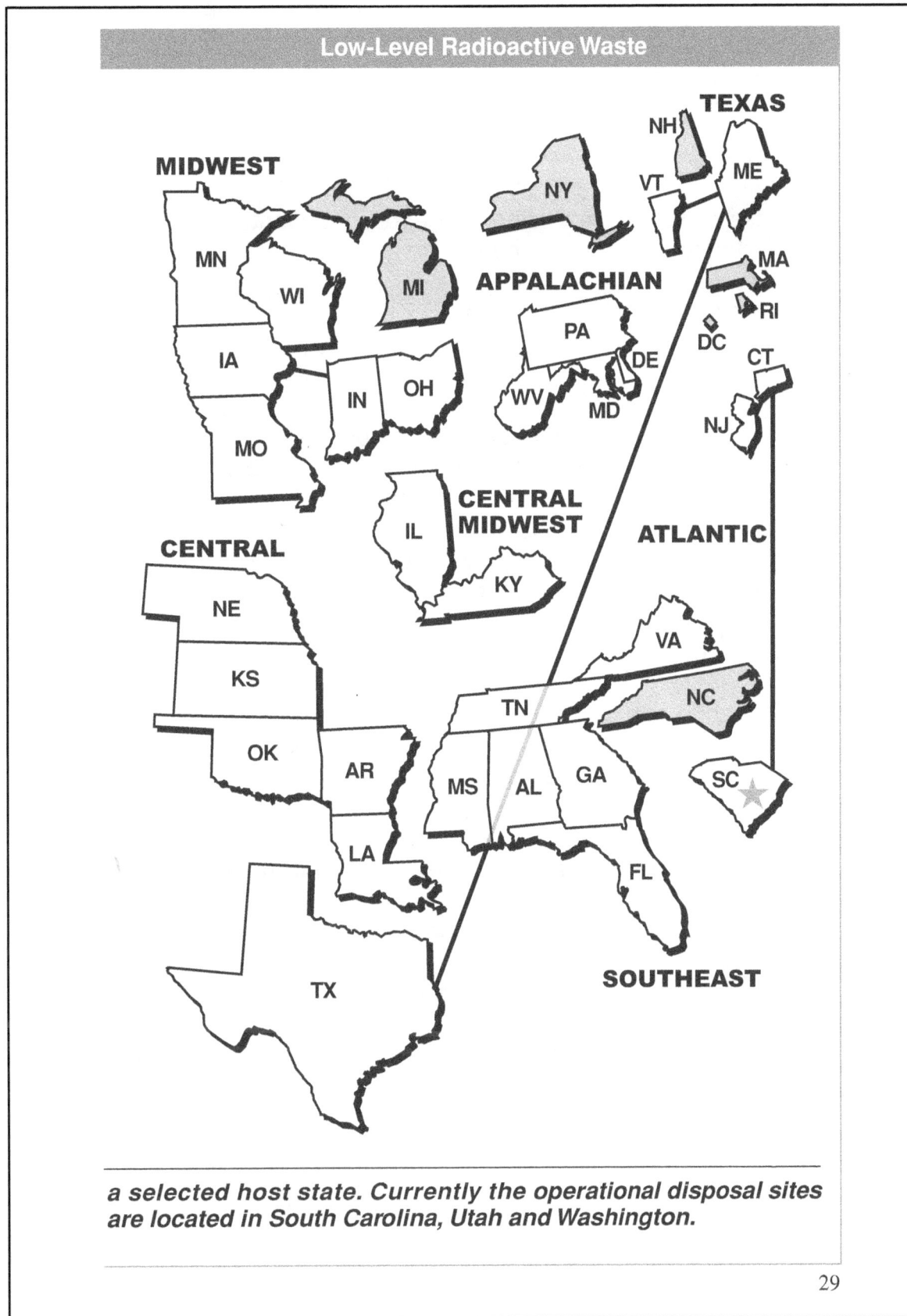

Low-Level Radioactive Waste

MIDWEST

MN

WI

MI

IA

IN   OH

MO

TEXAS

NH

VT   ME

NY

APPALACHIAN

PA

WV   DE

MD

MA

RI

DC

CT

NJ

CENTRAL MIDWEST

IL

KY

CENTRAL

NE

KS

OK

AR

LA

TX

ATLANTIC

VA

NC

SC ★

TN

MS   AL   GA

FL

SOUTHEAST

a selected host state. Currently the operational disposal sites are located in South Carolina, Utah and Washington.

29

387

Low-Level Radioactive Waste

zones of land surrounding and under the waste to permit monitoring and possible corrective actions.

When a disposal facility ceases operations, a post-closure period of maintenance and monitoring is required to confirm that the closed site is safely performing as expected before transfer to a government custodial agency for long-term control. Access to the site may be restricted for a long time, but NRC and state regulations do not allow reliance on institutional controls after 100 years following site closure. After 100 years, passive controls, such as custodial care, waste markers and land records, will be relied on to prevent disturbance of the emplaced waste.

Public Involvement

NRC and state procedures for development of a new low-level waste disposal facility provide several opportunities for public involvement, including:

- Public review and comment on a license application;
- Participation in the license review by the state or tribal governing body;
- Public review and comment on the required draft environmental impact statement;
- An opportunity for public hearings on the initial license and subsequent amendments;
- Attendance at any of the NRC's meetings with the license applicant.

30

Mill Tailings

Tailing wastes are generated during the milling of certain ores to extract uranium and thorium. These wastes have relatively low concentrations of radioactive materials with long half-lives. Tailings contain radium (which, through radioactive decay, becomes radon), thorium, and small residual amounts of uranium that were not extracted during the milling process.

The Rio Algom uranium mill and tailings site in Utah is undergoing reclamation.

31

Mill Tailings

LOCATIONS OF URANIUM MILL TAILINGS SITES

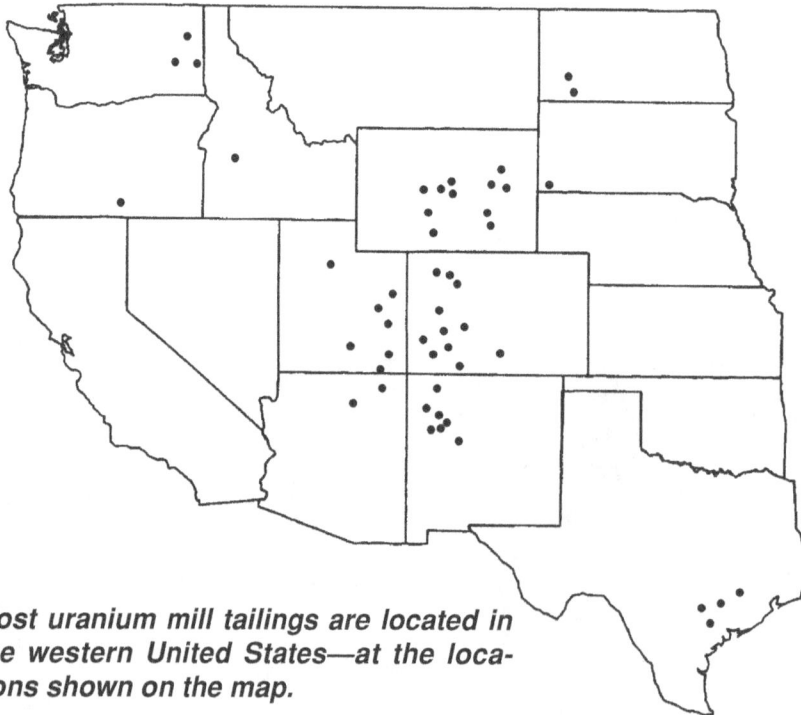

Most uranium mill tailings are located in the western United States—at the locations shown on the map.

The Office of Surface Mining, U.S. Department of Interior and individual states regulate mining. NRC regulates milling and the disposal of tailings in non-Agreement States, while State agencies regulate these activities in Agreement States when the agreement specifically includes tailings.

Mill tailings consist of fine-grained, sand-like and silty materials, usually deposited in large piles next to the mill that processed the ore. Uranium mills are located principally in the western United States, where deposits of uranium ore are more plentiful.

32

Mill Tailings

NRC requires licensees to meet Environmental Protection Agency standards for cleanup of uranium and thorium mill sites after the milling operations have permanently closed. This includes requirements for long-term stability of the mill tailings piles, radon emissions control, water quality protection and cleanup, and cleanup of lands and buildings.

NRC regulations require that a cover be placed over the mill tailings to control the release of radon gases at the end of milling operations. The cover must be effective in controlling radon releases for 1,000 years to the extent reasonably achievable and, in any case, for no less than 200 years.

The uranium mill tailings contain chemical and radiological material discarded from the mill. Radium and thorium, which are the dominant radioactive materials in mill tailings, have long half-lives (1,600 and 77,000 years respectively). Therefore Congress requires perpetual government custody of the tailings disposal sites.

33

Radioactive Waste: Production, Storage, Disposal

For Additional Information Contact:

Office of Public Affairs—Headquarters
U.S. Nuclear Regulatory Commission
Washington, DC 20555

Telephone: 301-415-8200
Fax: 301-415-2234
Internet: opa@nrc.gov
website: www.nrc.gov

Regional Public Affairs Offices

Region I
475 Allendale Road
King of Prussia, PA 19406-1415
(610) 337-5330

Region II
61 Forsyth Street
Suite 23 T85
Atlanta, GA 30303-3415
(404) 562-4416

Region III
801 Warrenville Road
Lisle, IL 60532-4351
(630) 829-9663

Region IV
611 Ryan Plaza Drive
Suite 400
Arlington, TX 76011-8064
(817) 860-8128

34

U.S. Nuclear Regulatory Commission

Washington, DC 20555-0001

Office of Public Affairs

NUREG/BR-0216, Rev. 2
May 2002

Civilian Nuclear Waste Disposal

Mark Holt
Specialist in Energy Policy

September 4, 2009

Congressional Research Service

7-5700

www.crs.gov

RL33461

CRS Report for Congress
Prepared for Members and Committees of Congress

Summary

Management of civilian radioactive waste has posed difficult issues for Congress since the beginning of the nuclear power industry in the 1950s. Federal policy is based on the premise that nuclear waste can be disposed of safely, but proposed storage and disposal facilities have frequently been challenged on safety, health, and environmental grounds. Although civilian radioactive waste encompasses a wide range of materials, most of the current debate focuses on highly radioactive spent fuel from nuclear power plants.

The Nuclear Waste Policy Act of 1982 (NWPA) calls for disposal of spent nuclear fuel in a deep geologic repository. NWPA established an office in the Department of Energy (DOE) to develop such a repository and required the program's civilian costs to be covered by a fee on nuclear-generated electricity, paid into the Nuclear Waste Fund. Amendments to NWPA in 1987 restricted DOE's repository site studies to Yucca Mountain in Nevada.

DOE submitted a license application for the proposed Yucca Mountain repository to the Nuclear Regulatory Commission (NRC) on June 3, 2008, and NRC docketed the application September 8, 2008. The NRC license is to be based on radiation exposure standards set by the Environmental Protection Agency (EPA), which issued revised standards September 30, 2008. The State of Nevada strongly opposes the Yucca Mountain project, disputing DOE's analysis that the repository would meet EPA's standards. Risks cited by repository opponents include excessive water infiltration, earthquakes, volcanoes, and human intrusion.

The Obama Administration has decided to "terminate the Yucca Mountain program while developing nuclear waste disposal alternatives," according to the DOE FY2010 budget justification. Alternatives to Yucca Mountain are to be evaluated by a "blue ribbon" panel of experts convened by the Administration. At the same time, according to the justification, the NRC licensing process for the Yucca Mountain repository is to continue, "consistent with the provisions of the Nuclear Waste Policy Act."

The FY2010 budget request of $198.6 million for DOE's Office of Civilian Radioactive Waste Management would provide only enough funding to continue the Yucca Mountain licensing process and to evaluate alternative policies, according to DOE. The request is about $90 million below the FY2009 funding level, which was nearly $100 million below the FY2008 level. All work related solely to preparing for construction and operation of the Yucca Mountain repository is being halted, according to the DOE budget justification.

The House-passed version of the FY2010 Energy and Water Development Appropriations Bill (H.R. 3183) approves the Administration's funding cuts but includes a requirement that Yucca Mountain be one of the options considered by the "blue ribbon" nuclear waste panel. The Senate version of the bill also approves the DOE nuclear waste funding cut but does not include the House requirement on the blue-ribbon panel; in addition, the Senate bill would reduce funding for NRC's Yucca Mountain licensing activities. Senator Reid of Nevada, a long-time opponent of the proposed Yucca Mountain repository, announced on July 29, 2009, that the Administration had agreed to terminate the Yucca Mountain licensing effort in the FY2011 budget request.

Congressional Research Service

396 703-739-3790 TCNNaturalGas.com

Contents

Figures

Tables

Contacts

Most Recent Developments

The Obama Administration has decided to "terminate the Yucca Mountain program while developing nuclear waste disposal alternatives," according to the Department of Energy (DOE) FY2010 budget justification, submitted to Congress May 7, 2009. Under the Nuclear Waste Policy Act, the Yucca Mountain site in Nevada is the only location under consideration by DOE for construction of a national high-level radioactive waste repository. DOE had submitted a license application for the proposed repository to the Nuclear Regulatory Commission (NRC) on June 3, 2008.

President Obama's FY2010 budget calls for a "blue ribbon" panel of experts to evaluate alternatives to the Yucca Mountain repository. At the same time, according to the DOE budget justification, the NRC licensing process for the Yucca Mountain repository is to continue, "consistent with the provisions of the Nuclear Waste Policy Act."

The FY2010 budget request of $198.6 million for DOE's Office of Civilian Radioactive Waste Management, which runs the nuclear waste program, would provide only enough funding to continue the Yucca Mountain licensing process and to evaluate alternative policies, according to DOE. The request is about $90 million below the FY2009 funding level, which was nearly $100 million below the FY2008 level. All work related solely to preparing for construction and operation of the Yucca Mountain repository is being halted, according to the DOE budget justification.

The House version of the FY2010 Energy and Water Development Appropriations Bill (H.R. 3183, H.Rept. 111-203), passed July 17, 2009, approves the Administration's funding cuts but includes a requirement that Yucca Mountain be one of the options considered by the "blue ribbon" nuclear waste panel. The Senate version of the bill, passed July 29, 2009 (S.Rept. 111-45), also approves the DOE nuclear waste funding cut but does not include the House requirement on the blue-ribbon panel; in addition, the Senate bill would reduce funding for NRC's Yucca Mountain licensing activities. Senator Reid of Nevada, a long-time opponent of the proposed Yucca Mountain repository, announced on the same day as Senate passage that the Administration had agreed to terminate the Yucca Mountain licensing effort in the FY2011 budget request.

Introduction

Nuclear waste has sometimes been called the Achilles' heel of the nuclear power industry; much of the controversy over nuclear power centers on the lack of a disposal system for the highly radioactive spent fuel that must be regularly removed from operating reactors. Low-level radioactive waste generated by nuclear power plants, industry, hospitals, and other activities is also a longstanding issue.

Spent Nuclear Fuel Program

Under the Nuclear Waste Policy Act of 1982 (NWPA) and 1987 amendments, the Department of Energy (DOE) is focusing on Yucca Mountain, Nevada, to house a deep underground repository for spent nuclear fuel and other highly radioactive waste. The State of Nevada has strongly opposed DOE's efforts on the grounds that the site is unsafe, pointing to potential volcanic activity, earthquakes, water infiltration, underground flooding, nuclear chain reactions, and fossil fuel and mineral deposits that might encourage future human intrusion.

Under the Bush Administration, DOE determined that Yucca Mountain was suitable for a repository and that licensing of the site by the Nuclear Regulatory Commission (NRC) should proceed. DOE submitted a license application for the repository to NRC on June 3, 2008, and projected that the repository could begin receiving waste in 2020, about 22 years later than the 1998 goal specified by NWPA.[1]

However, the Obama Administration decided that the Yucca Mountain repository should not be opened, although it requested FY2010 funding to continue the NRC licensing process. The Administration announced plans to convene a "blue ribbon" panel to develop alternative waste disposal strategies. (For a discussion of policy options, see CRS Report R40202, *Nuclear Waste Disposal: Alternatives to Yucca Mountain*, by Mark Holt.)

The safety of geologic disposal of spent nuclear fuel and high-level waste (HLW), as planned in the United States, depends largely on the characteristics of the rock formations from which a repository would be excavated. Because many geologic formations are believed to have remained undisturbed for millions of years, it appeared technically feasible to isolate radioactive materials from the environment until they decayed to safe levels. "There is strong worldwide consensus that the best, safest long-term option for dealing with HLW is geologic isolation," according to the National Research Council.[2]

But, as the Yucca Mountain controversy indicates, scientific confidence about the concept of deep geologic disposal has turned out to be difficult to apply to specific sites. Every high-level waste site that has been proposed by DOE and its predecessor agencies has faced allegations or discovery of unacceptable flaws, such as water intrusion or earthquake vulnerability, that could release radioactivity into the environment. Much of the problem results from the inherent uncertainty involved in predicting waste site performance for the one million years that nuclear waste is to be isolated.

President Obama's FY2010 budget calls for long-term research on technologies that could reduce the volume and toxicity of nuclear waste. The Bush Administration had proposed to develop large-scale facilities to reprocess and recycle spent nuclear fuel by separating long-lived elements, such as plutonium, that could be made into new fuel and "transmuted" into shorter-lived radioactive isotopes. Spent fuel reprocessing, however, has long been controversial because of the potential weapons use of separated plutonium and cost concerns. The Obama Administration proposes to refocus DOE's nuclear waste research on fundamental science and away from the design and development of reprocessing facilities.

---

[1] Nuclear Energy Institute, Key Issues, Yucca Mountain, http://www.nei.org/keyissues/nuclearwastedisposal/yuccamountain/, viewed April 11, 2008.

[2] National Research Council, Board on Radioactive Waste Management, *Rethinking High-Level Radioactive Waste Disposal: A Position Statement of the Board on Radioactive Waste Management* (1990), p. 2.

Other Programs

Other types of civilian radioactive waste have also generated public controversy, particularly low-level waste, which is produced by nuclear power plants, medical institutions, industrial operations, and research activities. Civilian low-level waste currently is disposed of in large trenches at sites in South Carolina and Washington state. However, the Washington facility does not accept waste from outside its region, and the South Carolina site is available only to the three members of the Atlantic disposal compact (Connecticut, New Jersey, and South Carolina) as of June 30, 2008. The lowest-concentration class of low-level radioactive waste is accepted from any waste generator by a Utah commercial disposal facility.

Threats by states to close their disposal facilities led to congressional authorization of regional compacts for low-level waste disposal in 1985. No new sites have been opened by any of the 10 approved disposal compacts, although a site in Texas received conditional approval in January 2009 and may open in 2010.

Nuclear Utility Lawsuits

NWPA section 302 authorized DOE to enter into contracts with U.S. generators of spent nuclear fuel and other highly radioactive waste; under the contracts, DOE was to dispose of the waste in return for a fee on nuclear power generation. The act prohibited nuclear reactors from being licensed to operate without a nuclear waste disposal contract with DOE, and all reactor operators subsequently signed them.[3] As required by NWPA, the contracts specified that DOE would begin disposing of nuclear waste no later than January 31, 1998.

After DOE missed the contractual deadline, nuclear utilities began filing lawsuits to recover their additional storage costs—costs they would not have incurred had DOE begun accepting waste in 1998 as scheduled. DOE reached its first settlement with a nuclear utility, PECO Energy Company (now part of Exelon), on July 19, 2000. The agreement allowed PECO to keep up to $80 million in nuclear waste fee revenues during the subsequent 10 years. However, other utilities sued DOE to block the settlement, contending that nuclear waste fees may be used only for the DOE waste program and not as compensation for missing the disposal deadline. The U.S. Court of Appeals for the 11th Circuit agreed, ruling September 24, 2002, that any compensation would have to come from general revenues or other sources than the waste fund.

The Department of Justice has since negotiated settlements with four utilities: Exelon, Scana, Duke, and the Omaha Public Power District, plus an additional tentative settlement. All five settlements would involve 36 of the 118 reactors (operating and closed) that are covered by DOE waste disposal contracts under NWPA. Under the settlements, utilities submit annual reimbursement claims to DOE for any delay-related nuclear waste storage costs they incurred during that year. Any disagreements over reimbursable claims between DOE and a utility would go to arbitration. Through the end of calendar year 2008, $406 million had been paid under the

---

[3] The Standard Contract for Disposal of Spent Nuclear Fuel and/or High-Level Radioactive Waste can be found at 10 CFR 961.11.

settlements. The payments are made from the U.S. Treasury's Judgment Fund, a permanent account that is used to cover damage claims against the U.S. government.[4]

Other nuclear utilities have not reached settlements, but have continued pursuing their damage claims through the U.S. Court of Federal Claims. Unlike the settlements, which cover all past and future damages resulting from DOE's nuclear waste delays, awards by the Court of Claims can cover only damages that have already been incurred; therefore, utilities must continue filing claims as they accrue additional delay-related costs. About 20 cases involving initial damage claims have been tried in the Court of Claims so far, and about 30 more are pending. In addition, about half a dozen second-round suits have been filed by utilities that had already filed initial claims.

In the cases that have been tried so far, the Court of Claims has awarded judgments to the plaintiffs totaling $790 million. Of that amount, only $34.9 million has been paid—for one case filed by the Tennessee Valley Authority (TVA). The remaining cases are under appeal. Added to the $406 million in settlement payments, the $34.9 million TVA award brings the federal government's total nuclear waste damage payments so far to $440.9 million.

Future Liability Estimates

DOE estimates that its potential liabilities for waste program delays will total $11 billion through 2056 (in current dollars) if the Department is able to begin taking spent nuclear fuel from plant sites by 2020, which had been the Bush Administration's most recent goal. DOE's methodology for this estimate is shown in **Figure 1**. The yellow line shows DOE's estimate of how much spent fuel would have been removed from nuclear plant sites had shipments begun on the NWPA deadline of January 1998. The rate of waste acceptance under that scenario is 900 metric tons per year from 1998 through 2015 and 2,100 tons/year thereafter. That assumed acceptance rate was negotiated by DOE as part of the settlements discussed above. The annual costs reimbursed by DOE under the settlements cover utilities' expenses for storing waste that would have already been taken away under the assumed acceptance rate (the yellow line).

The green and red lines in **Figure 1** show DOE's planned waste acceptance rate if waste shipments begin by 2017 or 2020. Under those scenarios, DOE would take away 400 metric tons the first year, 600 the second year, 1,200 the third year, 2,000 the fourth year, and 3,000 per year thereafter. This is the rate assumed by DOE's Total System Life Cycle Cost Report.[5] At that higher acceptance rate, DOE would be able to eventually catch up with the amount of waste that it was assumed to take under the settlements (the yellow line). If waste acceptance began by 2017 (the green line), the backlog would be eliminated by 2046, and if acceptance began by 2020 (the red line) the backlog would be gone by 2056. Under the settlements, therefore, there would be no further annual damage payments after those years, if DOE were able to achieve the 2017 or 2020 acceptance scenario.

---

[4] Information in this section about nuclear waste settlements, court judgments, and liability estimates is based on a telephone conversation with David K. Zabransky, Nuclear Utility Specialist, Office of Civilian Radioactive Waste Management, U.S. Department of Energy, March 25, 2009.

[5] U.S. Department of Energy, Office of Civilian Radioactive Waste Management, *Analysis of the Total System Life Cycle Cost of the Civilian Radioactive Waste Management Program, Fiscal year 2007*, DOE/RW-0591, Washington, DC, July 2008, p. 20, http://ocrwm.doe.gov/about/budget/pdf/TSLCC_2007_8_05_08.pdf.

Figure 1. DOE Estimate of Future Liabilities for Nuclear Waste Delays

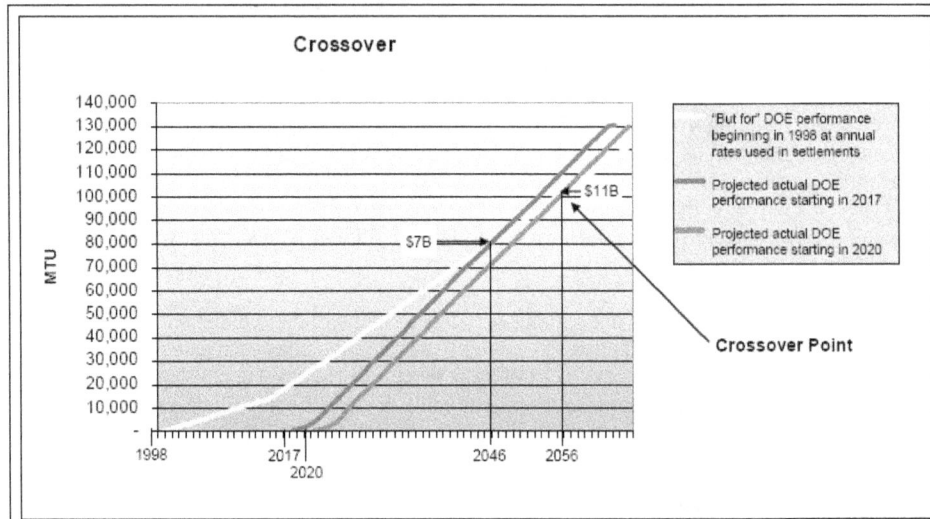

Source: Christopher A. Kouts, Principal Deputy Director, Office of Civilian Radioactive Waste Management, U.S. Department of Energy, "Yucca Mountain Program Status Update," July 22, 2008, p. 18.

DOE bases its estimate of the total damage payments that would be paid through 2046 or 2056 on the amounts paid to date under the settlement claims. If damage awards by the Court of Claims (currently involving about two-thirds of U.S. reactors) exceed those assumed levels, then future payments would be higher than the DOE estimate in **Figure 1**.

Further delays in the start of waste acceptance would delay the point at which DOE would catch up to the cumulative waste shipments assumed under the settlement scenario (yellow line) and would no longer have to make annual damage payments. DOE estimates that each year's delay in the startup date would increase the total eventual damage payments by as much as $500 million.

DOE filed a license application with the Nuclear Regulatory Commission (NRC) for the proposed Yucca Mountain repository in June 2008, and has estimated that annual program spending would have to increase to nearly $2 billion (from around $300 million in FY2009) to allow waste shipments to begin by 2020 if the license were approved.[6] However, President Obama's FY2010 budget request would sharply reduce Yucca Mountain funding, as noted above.

The House and Senate have both accepted the Administration's FY2010 nuclear waste budget cuts, so it appears unlikely that spent nuclear fuel shipments to Yucca Mountain could begin by 2020, even if full funding for the project were to be restored in the future. Waste acceptance by 2020 might be possible if Congress authorizes one or more temporary storage sites within the next few years, although previous efforts to develop such facilities have been blocked by state and local opposition.

---

[6] *Ibid.*, p. B-2.

Delays in the federal waste disposal program could also lead to future environmental enforcement action over DOE's own high-level waste and spent fuel, mostly resulting from defense and research activities. Some of the DOE-owned waste is currently being stored in non-compliance with state and federal environmental laws, making DOE potentially subject to fines and penalties if the waste is not removed according to previously negotiated compliance schedules.

Congressional Action

President Obama's proposal to terminate the Yucca Mountain project and search for disposal alternatives has prompted considerable congressional response. The Senate Energy and Natural Resources Committee approved a comprehensive energy bill on June 17, 2009, that would establish a federal advisory commission to study nuclear waste options and submit recommendations to Congress (S. 1492, S.Rept. 111-48). President Obama's FY2010 budget request for $5 million for a blue-ribbon panel to study nuclear waste alternatives has been approved by the House and Senate and is awaiting conference action, with the primary difference being the House's requirement that Yucca Mountain be one of the alternatives considered.

Senator Graham introduced a bill (S. 861) that would require the President to certify that the Yucca Mountain site continues to be the designated location for a nuclear waste repository under the Nuclear Waste Policy Act. If such a certification is not made within 30 days after enactment or is subsequently revoked, the Treasury is to refund all payments, plus interest, made by nuclear reactor owners to the Nuclear Waste Fund. DOE is to begin shipping defense-related high-level radioactive waste to Yucca Mountain by 2017 or pay $1 million per day to each state in which such waste is located.

Delays in nuclear waste disposal could affect the approximately 30 new U.S. reactors currently being proposed, because new reactors cannot be licensed without an NRC determination that sufficient waste disposal capacity will be available. Several bills have been introduced (see Legislation section) to prohibit NRC from denying a reactor license because of a lack of disposal capacity. Several recent bills would also encourage nuclear waste reprocessing and recycling, and would place the Nuclear Waste Fund "off budget" so that appropriations from the Waste Fund would no longer be subject to budget caps.

The Bush Administration proposed legislation in the 109th Congress (H.R. 5360, S. 2589) that was designed to speed the development of the Yucca Mountain repository. The bill would have reduced the scope of environmental reviews for the repository, changed the budget scoring of waste fee receipts so that program funding could be increased more easily, exempted nuclear waste sent to Yucca Mountain from disposal requirements under the Resource Conservation and Recovery Act (RCRA), allowed preemption of state and local transportation requirements, and permanently withdrawn the site from public lands use.

In addition, the Bush Administration bill would have required NRC to assume that sufficient disposal capacity would be available for waste produced by new reactors (known as the "waste confidence" determination). It also would have repealed the 70,000 metric ton limit on the amount of waste that could be emplaced at Yucca Mountain, a limit that is expected to be exceeded by currently operating reactors during their lifetimes.

The State of Nevada strongly opposed the Administration's Yucca Mountain legislation. As an alternative approach, Senator Reid introduced legislation on March 6, 2007, to require commercial nuclear reactor operators to place their spent nuclear fuel into on-site dry storage casks, which would then become the permanent responsibility of DOE (S. 784). Opponents of the proposal contend that it would leave spent fuel at reactor sites indefinitely and undermine the Nuclear Waste Policy Act. However, supporters argue that the waste would be safer in dry storage at reactor sites than if it were shipped across the country to Yucca Mountain.

Because of delays in the Yucca Mountain project, the Senate Appropriations Committee included statutory authorization for the Secretary of Energy to designate interim storage sites for spent nuclear fuel as part of the FY2007 Energy and Water Development Appropriations bill (H.R. 5427, Sec. 313). However, the 109th Congress adjourned without enacting the measure. The Senate Committee's provisions would have required the Secretary, after consultation with the governor, to designate a storage site in each state with a nuclear power plant, if feasible, or to designate regional storage facilities.

President Bush recommended the Yucca Mountain site to Congress on February 15, 2002, and Nevada Governor Guinn submitted a notice of disapproval, or "state veto," April 8, 2002, as allowed by NWPA. The state veto would have blocked further repository development at Yucca Mountain if a resolution approving the site had not been passed by Congress and signed into law within 90 days of continuous session. An approval resolution was signed by President Bush July 23, 2002 (P.L. 107-200).[7]

Characteristics of Nuclear Waste

Radioactive waste is a term that encompasses a broad range of material with widely varying characteristics. Some waste has relatively slight radioactivity and is safe to handle, while other types are intensely hot in both temperature and radioactivity. Some decays to safe levels of radioactivity in a matter of days or weeks, while other types will remain dangerous for thousands of years. Major types of radioactive waste are described below:[8]

Spent nuclear fuel. Fuel rods that have been permanently withdrawn from a nuclear reactor because they can no longer efficiently sustain a nuclear chain reaction (although they contain uranium and plutonium that could be extracted through reprocessing to make new fuel). By far the most radioactive type of civilian nuclear waste, spent fuel contains extremely hot but relatively short-lived fission products (fragments of the nuclei of uranium and other fissile elements) as well as long-lived radionuclides (radioactive atoms) such as plutonium, which remains dangerously radioactive for tens of thousands of years or more.

---

[7] Senator Bingaman introduced the approval resolution in the Senate April 9, 2002 (S.J.Res. 34), and Representative Barton introduced it in the House April 11, 2002 (H.J.Res. 87). The Subcommittee on Energy and Air Quality of the House Committee on Energy and Commerce approved H.J.Res. 87 on April 23 by a 24-2 vote, and the full Committee approved the measure two days later, 41-6 (H.Rept. 107-425). The resolution was passed by the House May 8, 2002, by a vote of 306-117. The Senate Committee on Energy and Natural Resources approved S.J.Res. 34 by a 13-10 vote June 5, 2002 (S.Rept. 107-159). Following a 60-39 vote to consider S.J.Res. 34, the Senate passed H.J.Res. 87 by voice vote July 9, 2002.

[8] Statutory definitions for "spent nuclear fuel," "high-level radioactive waste," and "low-level radioactive waste" can be found in Section 2 of the Nuclear Waste Policy Act of 1982 (42 U.S.C. 10101). "Transuranic waste" is defined in Section 11ee. of the Atomic Energy Act (42 U.S.C. 2014); Section 11e.(2) of the Act includes uranium mill tailings in the definition of "byproduct material." "Mixed waste" consists of chemically hazardous waste as defined by EPA regulations (40 CFR Part 261, Subparts C and D) that contains radioactive materials as defined by the Atomic Energy Act.

High-level waste. Highly radioactive residue created by spent fuel reprocessing (almost entirely for defense purposes in the United States). High-level waste contains most of the radioactive fission products of spent fuel, but most of the uranium and plutonium usually has been removed for re-use. Enough long-lived radioactive elements typically remain, however, to require isolation for 10,000 years or more.

Transuranic (TRU) waste. Relatively low-activity waste that contains more than a certain level of long-lived elements heavier than uranium (primarily plutonium). Shielding may be required for handling of some types of TRU waste. In the United States, transuranic waste is generated almost entirely by nuclear weapons production processes. Because of the plutonium, long-term isolation is required. TRU waste is being sent to a deep underground repository, the Waste Isolation Pilot Plant (WIPP), near Carlsbad, New Mexico.

Low-level waste. Radioactive waste not classified as spent fuel, high-level waste, TRU waste, or byproduct material such as uranium mill tailings (below). Four classes of low-level waste have been established by NRC, ranging from least radioactive and shortest-lived to the longest-lived and most radioactive. Although some types of low-level waste can be more radioactive than some types of high-level waste, in general low-level waste contains relatively low amounts of radioactivity that decays relatively quickly. Low-level waste disposal facilities cannot accept material that exceeds NRC concentration limits.

Uranium mill tailings. Sand-like residues remaining from the processing of uranium ore. Such tailings have very low radioactivity but extremely large volumes that can pose a hazard, particularly from radon emissions or groundwater contamination.

Mixed waste. Chemically hazardous waste that includes radioactive material. High-level, low-level, and TRU waste, and radioactive byproduct material, often falls under the designation of mixed waste. Such waste poses complicated institutional problems, because the radioactive portion is regulated by DOE or NRC under the Atomic Energy Act, while the Environmental Protection Agency (EPA) regulates the non-radioactive elements under the Resource Conservation and Recovery Act (RCRA).

Spent Nuclear Fuel

When spent nuclear fuel is removed from a reactor, usually after several years of power production, it is thermally hot and highly radioactive. The spent fuel is in the form of fuel assemblies, which consist of arrays of metal-clad fuel rods 12-15 feet long.

A fresh fuel rod, which emits relatively little radioactivity, contains uranium that has been enriched in the isotope U-235 (usually 3%-5%). But after nuclear fission has taken place in the reactor, many of the uranium nuclei in the fuel rods have been split into a variety of highly radioactive fission products; others have absorbed neutrons to become radioactive plutonium, some of which has also split into fission products. Radioactive gases are also contained in the spent fuel rods. Newly withdrawn spent fuel assemblies are stored in deep pools of water adjacent to the reactors to keep them from overheating and to protect workers from radiation.

Spent fuel discharged from U.S. commercial nuclear reactors is currently stored at 72 power plant sites around the nation, plus two small central storage facilities. A typical large commercial nuclear reactor discharges an average of 20-30 metric tons of spent fuel per year—an average of about 2,150 metric tons annually for the entire U.S. nuclear power industry. The nuclear industry estimated that the total amount of commercial spent fuel was 56,586 metric tons by January 2008,[9] an amount projected to reach 62,000 metric tons by 2010. Including 7,000 metric tons of DOE spent fuel and high-level waste that is also planned for disposal at Yucca Mountain, the total amount would nearly reach NWPA's 70,000-metric-ton limit by 2010.

As long as nuclear power continues to be generated, the amounts stored at plant sites will continue to grow until an interim storage facility or a permanent repository can be opened—or until alternative treatment and disposal technology is developed. DOE recently updated its estimate of the total amount of U.S. commercial spent fuel that may eventually require disposal from 105,000 metric tons[10] to 130,000 metric tons.[11]

New storage capacity at operating nuclear plant sites or other locations will be required if DOE is unable to begin accepting waste into its disposal system until 2020 or later. Most utilities are expected to construct new dry storage capacity for their older, cooler fuel. On-site dry storage facilities currently in operation or planned typically consist of metal casks or concrete modules. Forty-seven licensed dry storage facilities are currently operating in the United States.[12] NRC has determined that spent fuel could be stored safely at reactor sites for up to 100 years.[13]

The terrorist attacks of September 11, 2001, heightened concerns about the vulnerability of stored spent fuel. Concerns have been raised that an aircraft crash into a reactor's pool area or sabotage could drain the pool and cause the spent fuel inside to overheat. A report released by NRC January 17, 2001, found that overheating could cause the zirconium alloy cladding of spent fuel to catch fire and release hazardous amounts of radioactivity, although it characterized the probability of such a fire as low.

In a report released April 6, 2005, the National Academy of Sciences (NAS) found that "successful terrorist attacks on spent fuel pools, though difficult, are possible." To reduce the likelihood of spent fuel cladding fires, the NAS study recommended that hotter and cooler spent fuel assemblies be interspersed throughout spent fuel pools, that spray systems be installed above the pools, and that more fuel be transferred from pools to dry cask storage.[14] NRC has agreed to consider some of the recommendations, although it contends that current security measures would prevent successful attacks. The nuclear industry contends that the several hours required for uncovered spent fuel to heat up enough to catch fire would allow ample time for alternative measures to cool the fuel. The FY2006 Energy and Water appropriations bill (P.L. 109-103) gave NRC an additional $21 million to implement the NAS recommendations.

[9] "Spent Fuel Inventory at 56,586 Metric Tons," NuclearFuel, January 28, 2008, p. 10.

[10] DOE Office of Civilian Radioactive Waste Management, *OCRWM Annual Report to Congress, Fiscal Year 2002*, DOE/RW-0560, October 2003, Appendix C.

[11] DOE Office of Civilian Radioactive Waste Management, *Draft Supplemental Environmental Impact Statement for a Geologic Repository for the Disposal of Spent Nuclear Fuel and High-Level Radioactive Waste at Yucca Mountain, Nye County, Nevada*, Summary, DOE/EIS-0250F-S1D, October 2007, p. S-47.

[12] ACI Nuclear Energy Solutions, *U.S. Commercial Reactor Dry Storage Summary*, July 20, 2009.

[13] Nuclear Regulatory Commission, *Waste Confidence Decision Review*, 55 *Federal Register* 38474, September 18, 1990.

[14] National Academy of Sciences, *Safety and Security of Commercial Spent Nuclear Fuel Storage: Public Report*, released April 6, 2005, p. 2.

Commercial Low-Level Waste

Nearly 2.1 million cubic feet of low-level waste with about 800,000 curies of radioactivity was shipped to commercial disposal sites in 2008, according to DOE.[15] Volumes and radioactivity can vary widely from year to year, based on the status of nuclear decommissioning projects and cleanup activities that can generate especially large quantities.

Low-level radioactive waste is divided into three major categories for handling and disposal: Class A, B, and C. Classes B and C have constituted less than 1% of the volume of U.S. low-level waste disposal during the past five years but contain most of its radioactivity. For more background on radioactive waste characteristics, see CRS Report RL32163, *Radioactive Waste Streams: Waste Classification for Disposal*, by Anthony Andrews.

Current Policy and Regulation

Spent fuel and high-level waste are a federal responsibility, while states are authorized to develop disposal facilities for commercial low-level waste. In general, disposal requirements have grown more stringent over the years, in line with overall national environmental policy and heightened concerns about the hazards of radioactivity.

Spent Nuclear Fuel

Current Program

The Nuclear Waste Policy Act of 1982 (NWPA, P.L. 97-425) established a system for selecting a geologic repository for the permanent disposal of up to 70,000 metric tons (77,000 tons) of spent nuclear fuel and high-level waste. DOE's Office of Civilian Radioactive Waste Management (OCRWM) was created to carry out the program. The Nuclear Waste Fund, holding receipts from a fee on commercial nuclear power and federal contributions for emplacement of high-level defense waste, was established to pay for the program. DOE was required to select three candidate sites for the first national high-level waste repository.

After much controversy over DOE's implementation of NWPA, the act was substantially modified by the Nuclear Waste Policy Amendments Act of 1987 (Title IV, Subtitle A of P.L. 100-203, the Omnibus Budget Reconciliation Act of 1987). Under the amendments, the only candidate site DOE may consider for a permanent high-level waste repository is at Yucca Mountain, Nevada. If that site cannot be licensed, DOE must return to Congress for further instructions.

The 1987 amendments also authorized construction of a monitored retrievable storage (MRS) facility to store spent fuel and prepare it for delivery to the repository. But because of fears that the MRS would reduce the need to open the permanent repository and become a de facto repository itself, the law forbids DOE from selecting an MRS site until recommending to the President that a permanent repository be constructed. The repository recommendation occurred in February 2002, but DOE has not announced any plans for an MRS.

---

[15] U.S. Department of Energy, Management Information Manifest System, http://mims.apps.em.doe.gov/mims.asp#.

President Obama has proposed that OCRWM and the DOE Office of Nuclear Energy (NE) be headed by the same person, leading to speculation that the two offices will be functionally combined. Much of NE's FY2010 funding request is focused on nuclear waste research. The President's nominee, Warren F. "Pete" Miller Jr., has been confirmed by the Senate for the NE post but not for OCRWM.

Waste Facility Schedules

In proposing to terminate the Yucca Mountain project, the Obama Administration has not announced a new schedule for the DOE nuclear waste program. The Administration has proposed that a blue-ribbon panel be convened in FY2010 to study waste disposal options, but no other details have been announced. The Administration currently is continuing the Yucca Mountain licensing process, which, under NWPA, is supposed to take no more than four years after the license submission, which was June 3, 2008. However, Senator Reid has announced that the Administration will seek no further repository licensing funding in FY2011.

Over the objections of the State of Nevada, NRC formally accepted the Yucca Mountain license application for docketing and review on September 8, 2008. At the same time, NRC staff recommended that the Commission adopt DOE's environmental impact statement for the project, but with a stipulation that supplemental groundwater analysis be conducted.[16]

The major activity at the Yucca Mountain site so far has been the construction and operation of an "exploratory studies facility" (ESF) with a 25-foot-diameter tunnel boring machine. The ESF consists primarily of a five-mile tunnel with ramps leading to the surface at its north and south ends. The tunnel boring machine began excavating the north ramp in October 1994 and broke through to the surface at the south entrance April 25, 1997. Underground studies have been conducted at several side alcoves that were excavated off the main tunnel. Budget cuts in FY2009 have halted most activity at the Yucca Mountain site.

Private Interim Storage

In response to delays in the federal nuclear waste program, a utility consortium signed an agreement with the Skull Valley Band of the Goshute Indians in Utah on December 27, 1996, to develop a private spent fuel storage facility on tribal land. The Private Fuel Storage (PFS) consortium submitted a license application to NRC on June 25, 1997, and an NRC licensing board recommended approval on February 24, 2005. On September 9, 2005, NRC denied the State of Utah's final appeals and authorized the NRC staff to issue the license. The 20-year license for storing up to 44,000 tons of spent fuel in dry casks was issued on February 21, 2006, although NRC noted that Interior Department approval would also be required.

On September 7, 2006, the Department of the Interior issued two decisions against the PFS project. The Bureau of Indian Affairs disapproved a proposed lease of tribal trust lands to PFS, concluding there was too much risk that the waste could remain at the site indefinitely.[17] The

---

[16] Licensing documents are posted at http://www.rw.doe.gov/.

[17] Bureau of Indian Affairs, *Record of Decision for the Construction and Operation of an Independent Spent Fuel Storage Installation (ISFSI) on the Reservation of the Skull Valley Band of Goshute Indians (Band) in Tooele County, Utah*, September 7, 2006.

Bureau of Land Management rejected the necessary rights-of-way to transport waste to the facility, concluding that a proposed rail line would be incompatible with the Cedar Mountain Wilderness Area and that existing roads would be inadequate.[18]

In reaction to the Interior Department decisions, Senator Hatch, a staunch opponent of the PFS proposal, declared the project "stone cold dead."[19] However, the Skull Valley Band of Goshutes and PFS filed a federal lawsuit July 17, 2007, to overturn the Interior decisions on the grounds that they were politically motivated.[20]

Regulatory Requirements

NWPA requires that high-level waste facilities be licensed by the NRC in accordance with general standards issued by EPA. Under the Energy Policy Act of 1992 (P.L. 102-486), EPA was required to write new standards specifically for Yucca Mountain. NWPA also requires the repository to meet general siting guidelines prepared by DOE and approved by NRC. Transportation of waste to storage and disposal sites is regulated by NRC and the Department of Transportation (DOT). Under NWPA, DOE shipments to Yucca Mountain must use NRC-certified casks and comply with NRC requirements for notifying state and local governments. Yucca Mountain shipments must also follow DOT regulations on routing, placarding, and safety.

NRC's licensing requirements for Yucca Mountain, at 10 C.F.R. 63, require compliance with EPA's standards (described below) and establish procedures that DOE must follow in seeking a repository license. For example, DOE must conduct a repository performance confirmation program that would indicate whether natural and man-made systems were functioning as intended and assure that other assumptions about repository conditions were accurate.

The Energy Policy Act of 1992 (P.L. 102-486) made a number of changes in the nuclear waste regulatory system, particularly that EPA must issue new environmental standards specifically for the Yucca Mountain repository site. General EPA repository standards previously issued and subsequently revised no longer apply to Yucca Mountain. DOE and NRC had raised concern that some of EPA's general standards might be impossible or impractical to meet at Yucca Mountain.[21]

The new standards, which limit the radiation dose that the repository could impose on individual members of the public, were required to be consistent with the findings of a study by the National Academy of Sciences (NAS), which was issued August 1, 1995.[22] The NAS study recommended that the Yucca Mountain environmental standards establish a limit on risk to individuals near the repository, rather than setting specific limits for the releases of radioactive material or on radioactive doses, as under previous EPA standards. The NAS study also examined the potential for human intrusion into the repository and found no scientific basis for predicting human behavior thousands of years into the future.

---

[18] Bureau of Land Management, *Record of Decision Addressing Right-of-Way Applications U 76985 and U 76986 to Transport Spent Nuclear Fuel to the Reservation of the Skull Valley Band of Goshute Indians*, September 7, 2006.

[19] Senator Orrin Hatch, *Utahns Deliver Killing Blow to Skull Valley Nuke Waste Plan*, News Release, September 7, 2006.

[20] Winslow, Ben, "Gosutes, PFS Sue Interior," *Deseret Morning News*, July 18, 2007.

[21] See, for example: NRC, "Analysis of Energy Policy Act of 1992 Issues Related to High-Level Waste Disposal Standards, SECY-93-013, January 25, 1993, attachment p. 4.

[22] National Research Council. *Technical Bases for Yucca Mountain Standards*. National Academy Press. 1995.

Pursuant to the Energy Policy Act, EPA published its proposed Yucca Mountain radiation protection standards on August 27, 1999. The proposal would have limited annual radiation doses to 15 millirems for the "reasonably maximally exposed individual," and to 4 millirems from groundwater exposure, for the first 10,000 years of repository operation. EPA calculated that its standard would result in an annual risk of fatal cancer for the maximally exposed individual of seven chances in a million. The nuclear industry criticized the EPA proposal as being unnecessarily stringent, particularly the groundwater standard. On the other hand, environmental groups contended that the 10,000-year standard proposed by EPA was too short, because DOE had projected that radioactive releases from the repository would peak after about 400,000 years.

EPA issued its final Yucca Mountain standards on June 6, 2001. The final standards included most of the major provisions of the proposed version, including the 15 millirem overall exposure limit and the 4 millirem groundwater limit. Despite the Department's opposition to the EPA standards, DOE's site suitability evaluation determined that the Yucca Mountain site would be able to meet them. NRC revised its repository regulations September 7, 2001, to conform to the EPA standards.

A three-judge U.S. Court of Appeals panel on July 9, 2004, struck down the 10,000-year regulatory compliance period in the EPA and NRC Yucca Mountain standards.[23] The court ruled that the 10,000-year period was inconsistent with the NAS study on which the Energy Policy Act required the Yucca Mountain regulations to be based. In fact, the court found, the NAS study had specifically rejected a 10,000-year compliance period because of analysis that showed peak radioactive exposures from the repository would take place several hundred thousand years in the future.

In response to the court decision, EPA proposed a new version of the Yucca Mountain standards on August 9, 2005. The proposal would have retained the dose limits of the previous standard for the first 10,000 years but allowed a higher annual dose of 350 millirems for the period of 10,000 years through 1 million years. EPA also proposed to base the post-10,000-year Yucca Mountain standard on the median dose, rather than the mean, potentially making it easier to meet.[24] Nevada state officials called EPA's proposed standard far too lenient and charged that it was "unlawful and arbitrary."[25]

EPA issued its final rule to amend the Yucca Mountain standards on September 30, 2008. The final rule reduces the annual dose limit during the period of 10,000 through 1 million years from the proposed 350 millirems to 100 millirems, which the agency contended was consistent with international standards. Under the final rule, compliance with the post-10,000-year standard will be based on the arithmetic mean of projected doses, rather than the median as proposed. The 4 millirem groundwater standard will continue to apply only to the first 10,000 years.[26] NRC will have to revise its repository licensing regulations to conform to the new EPA standards. (For more

---

[23] *Nuclear Energy Institute v. Environmental Protection Agency*, U.S. Court of Appeals for the District of Columbia Circuit, No. 01-1258, July 9, 2004.

[24] Especially high doses at the upper end of the exposure range would raise the mean, or average, more than the median, or the halfway point in the data set.

[25] Office of the Governor, Agency for Nuclear Projects. *Comments by the State of Nevada on EPA's Proposed New Radiation Protection Rule for the Yucca Mountain Nuclear Waste Repository*. November 2005.

[26] Posted on the EPA website at http://www.epa.gov/radiation/yucca.

information, see CRS Report RL34698, *EPA's Final Health and Safety Standard for Yucca Mountain*, by Bonnie C. Gitlin.)

DOE estimated in its June 2008 Final Supplemental Environmental Impact Statement (FSEIS) for the Yucca Mountain repository that the maximum mean annual individual dose after 10,000 years would be 2 millirems. That is substantially below the level estimated by the 2002 Final Environmental Impact Statement, which calculated that the peak doses—occurring after 400,000 years—would be about 150 millirems (Volume 1, Chapter 5). The FSEIS attributed the reduction to changes in DOE's computer model and in the assumptions used, noting that "various elements of DOE's modeling approach may be challenged as part of the NRC licensing process."[27]

Alternative Technologies

Several alternatives to the geologic disposal of spent fuel have been studied by DOE and its predecessor agencies, as well as technologies that might make waste disposal easier. However, most of these technologies involve large technical obstacles, uncertain costs, and potential public opposition.

Among the primary long-term disposal alternatives to geologic repositories are disposal in deep ocean trenches and transport into space, neither of which is currently being studied by DOE. Other technologies have been studied that, while probably not replacing geologic disposal, might make geologic disposal safer and more predictable. Chief among these is the reprocessing or "recycling" of spent fuel so that plutonium, uranium, and other long-lived radionuclides could be converted to faster-decaying fission products in special nuclear reactors or particle accelerators.

Funding

The FY2010 OCRWM budget request of $198.6 million would provide only enough funding to continue the Yucca Mountain licensing process and to evaluate alternative policies, according to DOE. The request is about $90 million below the FY2009 funding level, which was nearly $100 million below the FY2008 level. More than 2,000 waste program contract employees will be terminated by the end of FY2009, according to the budget justification. Most of the program's remaining work is to be taken over by federal staff.

All work related solely to preparing for construction and operation of the Yucca Mountain repository is being halted, according to the DOE budget justification. Such activities include development of repository infrastructure, waste transportation preparations, and system engineering and analysis.

The House agreed with the Administration's plans to provide funding solely for Yucca Mountain licensing activities and for a blue-ribbon panel to review waste management options. The House approved the Administration budget request, including $5 million for the blue ribbon review. However, the House-passed bill specifies that the review must include Yucca Mountain as one of the alternatives, despite the Administration's contention that the site should no longer be considered. According to the House Appropriations Committee report, "It might well be the case that an alternative to Yucca Mountain better meets the requirements of the future strategy, but the

---

[27] FSEIS, p. S-42. Posted on the DOE website at http://www.rw.doe.gov/ym_repository/seis/docs/002_Summary.pdf.

review does not have scientific integrity without considering Yucca Mountain." The House panel also recommended that at least $70 million of the program's funding be devoted to maintaining expertise by the Yucca Mountain Project management contractor to support the licensing effort, rather than relying entirely on federal staff. The Senate also recommended approval of the Administration request, but without any restrictions on the blue ribbon panel.

Funding for the nuclear waste program is provided under two appropriations accounts, as shown in **Table 1**. The Administration's FY2010 request is divided evenly between an appropriation from the Nuclear Waste Fund, which holds fees paid by nuclear utilities, and the Defense Nuclear Waste Disposal account, which pays for disposal of high-level waste from the nuclear weapons program. The Senate Appropriations Committee report calls for the Secretary of Energy to suspend fee collections, "given the Administration's decision to terminate the Yucca Mountain repository program while developing disposal alternatives."

Additional funding from the Nuclear Waste Fund for the Yucca Mountain licensing process is included in the NRC budget request. The House provided the full $56 million requested, while the Senate voted to cut the request to $29 million. Senator Reid, a long-time opponent of the proposed Yucca Mountain repository, announced on July 29, 2009, that the Administration had agreed to terminate the Yucca Mountain licensing effort in the FY2011 budget request.

Although nuclear utilities pay fees to the Nuclear Waste Fund to cover the disposal costs of civilian nuclear spent fuel, DOE cannot spend the money in the fund until it is appropriated by Congress. Through June 30, 2009, utility nuclear waste fees and interest totaled $30.48 billion, of which $7.295 billion had been disbursed to the waste disposal program, according to DOE's program summary report, leaving a balance of $23.185 billion in the Nuclear Waste Fund. In addition to the disbursements from the Nuclear Waste Fund, the waste disposal program received defense waste disposal appropriations totaling $3.49 billion through FY2008, according to DOE.[28]

DOE's latest update of its *Analysis of the Total System Life Cycle Cost of the Civilian Radioactive Waste Management Program* was released on August 5, 2008.[29] According to that estimate, the Yucca Mountain program would cost $96.2 billion in 2007 dollars from the beginning of the program in 1983 to repository closure in 2133. DOE's previous estimate, issued in 2001, was $57.5 billion in 2000 dollars. Major factors in the increase are inflation and a higher estimate of spent fuel to be generated by existing reactors. Spent fuel from proposed new reactors is not included in the cost estimate.

---

[28] DOE, Office of Civilian Radioactive Waste Management, Office of Program Management, *Monthly Summary of Program Financial and Budget Information*, as of July 1, 2009, available at http://www.ocrwm.doe.gov/about/budget/money.shtml. The report notes that some figures may not add due to independent rounding.

[29] Available on the OCRWM website at http://www.rw.doe.gov/about/budget/pdf/TSLCC_2007_8_05_08.pdf.

Civilian Nuclear Waste Disposal

Table 1. DOE Civilian Spent Fuel Management Funding

(in millions of current dollars)

Program	FY2007 Approp.	FY2008 Approp.	FY2009 Approp.	FY2010 Request	FY2010 House	FY2010 Senate
Yucca Mountain	298.1	267.1	—a	—	—	—
Transportation	35.3	18.3	—	—	—	—
Management and Integration	46.7	26.4	—	—	—	—
Program Direction	64.4	74.7	74.9	70.0	70.0	70.0
Total	**445.7**	**386.4**	**288.4**	**196.8**	**196.8**	**196.8**
Source of Funding						
Nuclear Waste Fund appropriations	99.2	187.3	145.4	98.4	98.4	98.4
Defense waste appropriations	346.5	199.2	143.0	98.4	98.4	98.4

Sources: DOE FY2010 Congressional Budget Request, H.Rept. 111-203, S.Rept. 111-45.

a. Subcategories not specified.

Low-Level Radioactive Waste

Current Policy

Selecting disposal sites for low-level radioactive waste, which generally consists of low concentrations of relatively short-lived radionuclides, is a state responsibility under the 1980 Low-Level Radioactive Waste Policy Act and 1985 amendments. Most states have joined congressionally approved interstate compacts to handle low-level waste disposal. Under the 1985 amendments, the nation's three (at that time) operating commercial low-level waste disposal facilities could start refusing to accept waste from outside their regional interstate compacts after the end of 1992. One of the three sites closed, and the remaining two are using their congressionally granted authority to prohibit waste from outside their regional compacts. Another site, in Utah, has since become available nationwide for most Class A low-level waste, but no site is currently open to nationwide disposal of all major types of low-level waste.

Despite the 1992 deadline, no new disposal sites have been opened under the Low-Level Waste Act. Legislation providing congressional consent to a disposal compact among Texas, Maine, and Vermont was signed by President Clinton September 20, 1998 (P.L. 105-236). However, on October 22, 1998, a proposed disposal site near Sierra Blanca, Texas, was rejected by the Texas Natural Resource Conservation Commission, and Maine has since withdrawn. Texas Governor Perry signed legislation June 20, 2003, authorizing the Texas Commission on Environmental Quality (TCEQ) to license adjoining disposal facilities for commercial and federally generated low-level waste. Pursuant to that statute, an application to build a disposal facility for commercial and federal low-level waste in Andrews County, Texas, was filed August 2, 2004, by Waste Control Specialists LLC. TCEQ voted January 14, 2009, to issue the license after the necessary land and mineral rights have been acquired.[30] Waste Control Specialists has predicted that the facility could start receiving waste by mid-2010.[31]

The Midwestern Compact voted June 26, 1997, to halt development of a disposal facility in Ohio. Nebraska regulators rejected a proposed waste site for the Central Compact December 21, 1998, drawing a lawsuit from five utilities in the region. A U.S. district court judge ruled September 30, 2002, that Nebraska had exercised bad faith in disapproving the site and ordered the state to pay $151 million to the compact. A settlement was reached August 9, 2004, resulting in a payment of $145.8 million,[32] and the compact is seeking access to the planned Texas disposal facility. Most other regional disposal compacts and individual states that have not joined compacts are making little progress toward finding disposal sites.

The disposal facility at Barnwell, South Carolina, is currently accepting all Class A, B, and C low-level waste from the Atlantic Compact (formerly the Northeast Compact), in which South Carolina joined original members Connecticut and New Jersey on July 1, 2000. Under the compact, South Carolina can limit the use of the Barnwell facility to the three compact members, and a state law enacted in June 2000 phased out acceptance of non-compact waste through June 30, 2008. The Barnwell facility previously had stopped accepting waste from outside the Southeast Compact at the end of June 1994. The Southeast Compact Commission in May 1995 twice rejected a South Carolina proposal to open the Barnwell site to waste generators outside the Southeast and to bar access to North Carolina until that state opened a new regional disposal facility, as required by the compact. The rejection of those proposals led the South Carolina General Assembly to vote in 1995 to withdraw from the Southeast Compact and begin accepting waste at Barnwell from all states but North Carolina. North Carolina withdrew from the Southeast Compact July 26, 1999.

The only other existing disposal facility for all three major classes of low-level waste is at Hanford, Washington. Controlled by the Northwest Compact, the Hanford site will continue taking waste from the neighboring Rocky Mountain Compact under a contract. Since the South Carolina facility closed to out-of-region waste, the 36 states and the District of Columbia that are outside the Northwest, Rocky Mountain, and Atlantic Compacts have had no disposal site for Class B and C low-level waste. Waste generators in those states must store their Class B and C waste on site until new disposal sites are available.

Regulatory Requirements

Licensing of commercial low-level waste facilities is carried out under the Atomic Energy Act by NRC or by "agreement states" with regulatory programs approved by NRC. NRC regulations governing low-level waste licenses must conform to general environmental protection standards and radiation protection guidelines issued by EPA. Transportation of low-level waste is jointly regulated by NRC and the Department of Transportation.

Most states considering new or expanded low-level waste disposal facilities, including Texas and Utah, are agreement states. Most states, both agreement and non-agreement, have established substantially stricter technical requirements for low-level waste disposal than NRC's, such as

---

[30] TCEQ website: http://www.tceq.state.tx.us/permitting/radmat/licensing/wcs_license_app.html#wcs_status.

[31] Weil, Jenny, "Texas Regulators Approve License for LLW Disposal," *Inside NRC*, January 19, 2009, p. 3.

[32] USAToday.com, August 1, 2005, http://www.usatoday.com/news/nation/2005-08-01-nukewaste_x.htm.

banning shallow land burial and requiring concrete bunkers and other engineered barriers. NRC would issue the licenses in non-agreement states.

Concluding Discussion

Disposal of radioactive waste will be a key issue in the continuing nuclear power debate. Without a national disposal system, spent fuel from nuclear power plants must be stored on-site indefinitely. This situation may raise public concern near proposed reactor sites, particularly at sites without existing reactors where spent nuclear fuel is already stored.

Under current law, the federal government's nuclear waste disposal policy is focused on the planned Yucca Mountain repository. However, deep funding reductions and uncertainty that have accompanied President Obama's plan to terminate the Yucca Mountain project have brought most activities in the DOE waste program to a halt. Although licensing of the Yucca Mountain repository is to continue through FY2010, DOE will face relentless opposition from the State of Nevada and a possible funding cutoff in FY2011.

Because of their waste-disposal contracts with DOE, owners of existing reactors are likely to continue seeking damages from the federal government if disposal delays continue. DOE's 2004 settlement with the nation's largest nuclear operator, Exelon, could require payments of up to $600 million from the federal judgment fund, and DOE estimates that payments could rise to $11 billion if Yucca Mountain does not open before 2020. The nuclear industry has predicted that future damages could reach tens of billions of dollars if the federal disposal program fails altogether.

Lack of a nuclear waste disposal system could also affect the licensing of proposed new nuclear plants, both because of NRC licensing guidelines and various state laws.[33] In addition, further repository delays could force DOE to miss compliance deadlines for defense waste disposal.

Problems being created by nuclear waste disposal delays would presumably be addressed by the President's proposed "blue ribbon" panel of experts on nuclear waste management alternatives. Major options include centralized interim storage, continued storage at existing nuclear sites, reprocessing and waste treatment technology, alternative repository sites, or a combination. Given the delays resulting from the current budget cuts, longer on-site storage is almost a certainty under any option. Any of the options would also face intense controversy, especially among states and regions that might be potential hosts for future waste facilities. As a result, substantial debate would be expected over any proposals to change the Nuclear Waste Policy Act.

---

[33] Lovell, David L., Wisconsin Legislative Council Staff, *State Statutes Limiting the Construction of Nuclear Power Plants*, October 5, 2006.

Legislation

H.R. 513 (Forbes)

New Manhattan Project for Energy Independence. Includes grants and prizes for nuclear waste treatment technology. Introduced January 14, 2009; referred to Committee on Science and Technology.

H.R. 2250 (Burton)

Energy Independence Now Act of 2009. Includes a provision prohibiting the Nuclear Regulatory Commission from denying a nuclear reactor license because of a lack of nuclear waste disposal capacity. Introduced May 5, 2009; referred to multiple committees.

H.R. 2300 (Bishop)

Among other provisions, would authorize DOE to enter into temporary spent nuclear fuel storage agreements with volunteer sites, establish payments to settle nuclear utility breach-of-contract claims for DOE waste disposal delays, and prohibit NRC from considering nuclear waste storage when licensing new nuclear facilities. Introduced May 7, 2009; referred to multiple committees.

H.R. 2539 (Thornberry)

No More Excuses Energy Act of 2009. Includes a provision prohibiting the Nuclear Regulatory Commission from denying a nuclear reactor license because of a lack of nuclear waste disposal capacity. Introduced May 5, 2009; referred to multiple committees.

H.R. 3183 (Pastor)

Energy and Water Development and Related Agencies Appropriations Act, 2010. Includes funding for nuclear waste programs. Reported as an original measure by the House Appropriations Committee July 13, 2009. Passed House July 17, 2009, by vote of 320-97 (H.Rept. 111-203); passed Senate July 29, 2009, by vote of 85-9 (S.Rept. 111-45).

H.R. 3385 (Barton)

Would authorize DOE to use the Nuclear Waste Fund to pay for grants or long-term contracts for spent nuclear fuel recycling or reprocessing and place the Waste Fund off-budget. Introduced July 29, 2009; referred to committees on Energy and Commerce and the Budget.

S. 591 (Reid)

National Commission on High-Level Radioactive Waste and Spent Nuclear Fuel Establishment Act of 2009. Would establish a national commission to study nuclear waste management improvements. Introduced March 12, 2009; referred to Committee on Environment and Public Works.

S. 807 (Nelson of Nebraska)

SMART Energy Act. Includes provision authorizing DOE to begin construction of a spent fuel recycling research and development facility. Introduced April 2, 2009; referred to Committee on Finance.

S. 861 (Graham)

Rebating America's Deposits Act. Requires the President to certify that the Yucca Mountain site continues to be the designated location for a nuclear waste repository under the Nuclear Waste Policy Act. If such a certification is not made within 30 days after enactment or is subsequently revoked, the Treasury is to refund all payments, plus interest, made by nuclear reactor owners to the Nuclear Waste Fund. DOE is to begin shipping defense-related high-level radioactive waste to Yucca Mountain by 2017 or pay $1 million per day to each state in which such waste is located. Introduced April 22, 2009; referred to Committee on Energy and Natural Resources.

S. 1333 (Barrasso)

Clean, Affordable, and Reliable Energy Act of 2009. Includes provisions to take the Nuclear Waste Fund off-budget, authorize DOE to use the Nuclear Waste Fund to pay for grants or long-term contracts for spent nuclear fuel recycling or reprocessing, and prohibit NRC from denying licenses for new nuclear facilities because of a lack of waste disposal capacity. Introduced June 24, 2009; referred to Committee on Finance.

S. 1462 (Bingaman)

American Clean Energy Leadership Act of 2009. Includes provision to establish a federal commission to study nuclear waste management alternatives and make recommendations to Congress. Approved by Energy and Natural Resources Committee June 17, 2009, and reported as an original bill July 16, 2009 (S.Rept. 111-48).

Congressional Hearings, Reports, and Documents

U.S. Congress. House. Committee on Energy and Commerce. Subcommittee on Energy and Air Quality. *A Review of the President's Recommendation to Develop a Nuclear Waste Repository at Yucca Mountain, Nevada.* Hearing, 107th Congress, 2nd session. April 18, 2002. Washington: GPO, 2002. 294 p. Serial no. 107-99.

U.S. Congress. Senate. Committee on Energy and Natural Resources. *Low-Level Radioactive Waste.* Hearing, 108th Congress, 2nd session. September 30, 2004. Washington: GPO, 2005. 62 p.

U.S. Congress. Senate. *Yucca Mountain Repository Development.* Hearings, 107th Congress, 1st session. May 16, 22, and 23, 2002. Washington: GPO, 2002. 240 p. S.Hrg. 107-483.

For Additional Reading

Harvard University. John F. Kennedy School of Government. Belfer Center for Science and International Affairs. *The Economics of Reprocessing vs. Direct Disposal of Spent Nuclear Fuel.* DE-FG26-99FT4028. December 2003.

Nuclear Waste Technical Review Board. *Report to the U.S. Congress and the U.S. Secretary of Energy.* September 2008.
http://www.nwtrb.gov/reports/reports.html

University of Illinois. Program in Arms Control, Disarmament, and International Security. *'Plan D' for Spent Nuclear Fuel.* 2009. http://acdis.illinois.edu/publications/207/publication-PlanDforSpentNuclearFuel.html.

U.S. Department of Energy. *Office of Civilian Radioactive Waste Management home page*; covers DOE activities for disposal, transportation, and other management of civilian nuclear waste. http://www.ocrwm.doe.gov.

U.S. General Accounting Office. *Low-Level Radioactive Waste: Disposal Availability Adequate in the Short Term, but Oversight Needed to Identify Any Future Shortfalls.* GAO-04-604. June 2004. 53 p.

Walker, J. Samuel. *The Road to Yucca Mountain: The Development of Radioactive Waste Policy in the United States.* University of California Press. 2009. 228 p.

Author Contact Information

Mark Holt
Specialist in Energy Policy
mholt@crs.loc.gov, 7-1704

Congressional Research Service

Nuclear Waste Disposal: Alternatives to Yucca Mountain

Mark Holt
Specialist in Energy Policy

February 6, 2009

Congressional Research Service

7-5700

www.crs.gov

R40202

CRS Report for Congress

Prepared for Members and Committees of Congress

Summary

Congress designated Yucca Mountain, NV, as the nation's sole candidate site for a permanent high-level nuclear waste repository in 1987, following years of controversy over the site-selection process. Over the strenuous objections of the State of Nevada, the Department of Energy (DOE) submitted a license application for the proposed Yucca Mountain repository in June 2008 to the Nuclear Regulatory Commission (NRC). During the 2008 election campaign, now-President Obama lent support to Nevada's fight against the repository, contending in an issue statement that he and now-Vice President Biden "do not believe that Yucca Mountain is a suitable site."

Under the current nuclear waste program, DOE hopes to begin transporting spent nuclear fuel and other highly radioactive waste to Yucca Mountain by 2020. That schedule is 22 years beyond the 1998 deadline established by the Nuclear Waste Policy Act (NWPA). Because U.S. nuclear power plants will continue to generate nuclear waste after a repository opens, DOE estimates that all waste could not be removed from existing reactors until about 2066 even under the current Yucca Mountain schedule. Not all the projected waste could be disposed of at Yucca Mountain, however, unless NWPA's current limit on the repository's capacity is increased.

If the Obama Administration decides to halt the Yucca Mountain project, it has a variety of tools available to implement that policy. Although the President cannot directly affect NRC proceedings, the Secretary of Energy could withdraw the Yucca Mountain license application under NRC rules. The President could also urge Congress to cut or eliminate funding for the Yucca Mountain project, and propose legislation to restructure the nuclear waste program.

Abandonment of Yucca Mountain would probably further delay the federal government's removal of nuclear waste from reactor sites and therefore increase the government's liabilities for missing the NWPA deadline. DOE estimates that such liabilities will reach $11 billion even if Yucca Mountain opens as currently planned. DOE's agreements with states to remove defense-related high-level waste could also be affected. If the Yucca Mountain project were halted without a clear alternative path for waste management, the licensing of proposed new nuclear power plants could be affected as well. NRC has determined that waste can be safely stored at reactor sites for at least 30 years after a reactor shuts down and is proposing to extend that period to 60 years. While that proposal would allow at least 100 years for waste to remain at reactor sites (including a 40-year reactor operating period), NRC's policy is that new reactors should not be licensed without "reasonable confidence that the wastes can and will in due course be disposed of safely."

Current law provides no alternative repository site to Yucca Mountain, and it does not authorize DOE to open temporary storage facilities without a permanent repository in operation. Without congressional action, therefore, the default alternative to Yucca Mountain would be indefinite on-site storage of nuclear waste at reactor sites and other nuclear facilities. Private central storage facilities can also be licensed under current law; such a facility has been licensed in Utah but its operation has been blocked by the Department of the Interior.

Congress has considered legislation repeatedly since the mid-1990s to authorize a federal interim storage facility for nuclear waste but none has been enacted. Reprocessing of spent fuel could reduce waste volumes and long-term toxicity, but such facilities are costly and raise concerns about the separation of plutonium that could be used in nuclear weapons. Storage and reprocessing would still eventually require a permanent repository, and a search for a new repository site would need to avoid the obstacles that have hampered previous U.S. efforts.

Congressional Research Service

Contents

Contacts

Proposals for a New Direction

Nevada's Yucca Mountain has been the sole candidate site for the nation's first permanent high-level nuclear waste repository since Congress singled it out in 1987 and halted consideration of any other location. After numerous delays, the Department of Energy (DOE), which was supposed to open a waste repository by 1998, submitted a repository license application for Yucca Mountain to the Nuclear Regulatory Commission (NRC) in June 2008. If NRC approves the license, DOE hopes to begin shipping nuclear waste to the repository by 2020.

The congressional decision to focus solely on Yucca Mountain was highly controversial and continues to face harsh criticism, particularly from the State of Nevada. Yucca Mountain opponents dispute DOE's determination that the site is suitable for long-term disposal of nuclear waste and call for fundamental change in the program. Nonetheless, the proposed Yucca Mountain repository has consistently maintained sufficient congressional support to continue moving forward. Congress explicitly rejected a Nevada "state veto" of the site in 2002, has blocked repeated efforts to halt the program's funding, and has not taken up any of numerous legislative proposals to delay the program or find a new site. President George W. Bush also steadfastly supported the Yucca Mountain project.

But Administration support for Yucca Mountain will apparently change under President Obama. In their campaign statement on nuclear energy policy, Obama and Vice President Biden laid out the following position:

> In terms of waste storage, Barack Obama and Joe Biden do not believe that Yucca Mountain is a suitable site. They will lead federal efforts to look for safe, long-term disposal solutions based on objective, scientific analysis. In the meantime, they will develop requirements to ensure that the waste stored at current reactor sites is contained using the most advanced dry-cask storage technology available.[1]

The Obama-Biden campaign statement on Yucca Mountain raises numerous questions about the future direction of U.S. nuclear waste policy. In particular, what type of long-term disposal solutions could be considered? Every option for handling nuclear waste ultimately requires a method of long-term isolation from the environment. If Yucca Mountain were rejected, it would appear that a new repository site search would need to be undertaken at some point. Given the criticism that DOE has drawn over its handling of the waste program, pressure may intensify for such a search to be handed over to a new organization entirely.

The current effort to develop a repository at Yucca Mountain began with the enactment of the Nuclear Waste Policy Act of 1982 (NWPA, P.L. 97-425), and opening a repository at a different location could take a long time as well, even if the process were started right away. During such an indefinite time period, how would the licensing of new nuclear power plants be affected? Would spent nuclear fuel and other highly radioactive waste remain at commercial reactors and other existing nuclear facilities, or would it be moved to centralized interim storage? Previous U.S. efforts to develop interim central nuclear waste storage facilities have drawn fierce opposition.

---

[1] Obama for America, "Barack Obama and Joe Biden: New Energy for America," campaign issue statement, 2008, http://www.barackobama.com/pdf/factsheet_energy_speech_080308.pdf.

Since the 1970s, U.S. nuclear waste policy has been based on the "once through" fuel cycle, in which nuclear fuel is to be used once in a reactor and then permanently disposed of. The major alternative is the "closed" fuel cycle, in which spent nuclear fuel would be reprocessed into new fuel for advanced reactors or particle accelerators. Fast reactors or accelerators would destroy the longest-lived radioactive components of the fuel, leaving only relatively short-lived radioactive isotopes, which would decay to background levels within 1,000 years, for permanent disposal. Under that scenario, spent fuel could be stored at reprocessing facilities while awaiting its turn to be made into new fuel, and the relatively short life of the resulting waste could make it easier to site a permanent repository. However, the material for nuclear fuel that results from reprocessing (primarily plutonium) can also be used for nuclear explosives, raising concerns about nuclear weapons proliferation.

Because current law specifies that Yucca Mountain is the sole candidate site for a high-level waste repository, legislation would probably be needed if a major change in direction in the nuclear waste program is sought. But the Obama Administration does have authority under current law to withdraw the Yucca Mountain license application, propose reductions in the program's funding, and take other administrative actions to delay or halt the development of a repository at the Yucca Mountain site. This report discusses those options and the likely impact of indefinite delays in the waste program. It then discusses the mid- and long-term alternatives to the existing waste program, and finally reviews the history of U.S. efforts to site nuclear waste facilities.

Baseline: Current Waste Program Projections

DOE's latest schedules for nuclear waste shipments and projected costs under the existing program provide a baseline for analyzing Yucca Mountain alternatives. Although the planned opening of the Yucca Mountain repository is at least 22 years later than NWPA's 1998 deadline, and removing waste from existing storage sites would require many decades, a major redirection of the waste program would probably involve even longer time frames. Of course, there is no certainty that DOE will be able to meet its current schedules, or that the Yucca Mountain repository will receive a license from NRC under the current program. Moreover, policymakers could conclude that the benefits of redirecting the nuclear waste program now would outweigh the almost certain delays in developing a permanent repository and the increased costs of interim storage.

Under DOE's current schedule, about 400 metric tons of spent nuclear fuel would be shipped to Yucca Mountain from reactor sites in 2020. Shipments would rise to 600 metric tons in 2021, 1,200 metric tons in 2022, and 2,000 metric tons in 2023, and reach the planned maximum annual capacity of 3,000 metric tons in 2024.[2] Because the total U.S. commercial reactor fleet discharges an average of about 2,000 metric tons of spent fuel per year, the above shipment schedule would not begin reducing the backlog of spent fuel stored at reactor sites until 2024.

---

[2] U.S. Department of Energy, Office of Civilian Radioactive Waste Management, *Total System Life Cycle Cost Report*, DOE/RW-0591, Washington, DC, July 2008, p. 20, http://www.ocrwm.doe.gov/about/budget/pdf/ TSLCC_2007_8_05_08.pdf.

DOE estimates that the amount of commercial spent fuel stored in pools of water and dry casks at reactor sites and other facilities was 57,700 metric tons at the end of 2007.[3] If commercial spent fuel continues to accumulate at the rate of 2,000 metric tons per year, then inventories would reach 81,000 metric tons before shipments are to begin in 2020 and peak at nearly 85,000 in 2023, after which shipments to Yucca Mountain would exceed reactor discharges by 1,000 tons per year. DOE projections indicate that shipments of all spent fuel from previous and existing U.S. nuclear power plants would continue until about 2066, totaling 109,300 metric tons (under existing reactor license periods and extensions). In addition, the equivalent of 12,800 metric tons of defense-related spent nuclear fuel and high-level radioactive waste would be received at Yucca Mountain during the same period, for a total of 122,100 metric tons.[4]

Although DOE's cost projections assume that all spent fuel from existing reactors, plus defense waste, will be shipped to the planned Yucca Mountain repository, NWPA section 114(d) caps Yucca Mountain's capacity at the equivalent of 70,000 metric tons of spent fuel until a second repository begins operating. No such repository is currently authorized. A recent DOE report on the need for a second repository concludes that all existing and anticipated spent fuel and high-level waste could be physically accommodated at Yucca Mountain.[5] Legislation to lift the 70,000-ton limit proposed by the Bush Administration was introduced during the 109th (H.R. 5360) and 110th (S. 37) Congresses but not acted upon.

As amended in 1987, NWPA provides no backup plan for spent fuel management if the Yucca Mountain repository were to be halted. A "monitored retrievable storage" (MRS) facility is authorized by NWPA section 142, but construction is prohibited until NRC has authorized the construction of the Yucca Mountain repository.

Section 302 of NWPA requires nuclear power plant operators to sign contracts with DOE under which the nuclear plants must pay fees to the federal government in return for DOE's spent fuel disposal services. The nuclear power plant fees are deposited in a Treasury account called the Nuclear Waste Fund to pay for the DOE waste program but cannot be spent without congressional appropriation. The Fund's balance was about $20 billion at the end of FY2008.[6] Because the DOE waste program will also handle defense-related waste, Congress typically supplements annual appropriations from the Nuclear Waste Fund with appropriations from general revenues.

Annual spending for the nuclear waste program, focusing on Yucca Mountain site studies and the license application, has averaged about $400 million in recent years. However, DOE projects that to build the repository and develop a transportation system within the next 12 years, annual funding would need to increase to nearly $2 billion during the peak of construction. Disposal of all 122,100 metric tons of currently anticipated waste at Yucca Mountain is projected to cost $96.18 billion (in 2007 dollars) through 2133.[7]

---

[3] U.S. Department of Energy, Office of Civilian Radioactive Waste Management, *Report to the President and the Congress by the Secretary of Energy on the Need for a Second Repository*, DOE/RW-0595, Washington, DC, December 2008, p. 6, http://www.rw.doe.gov/info_library/program_docs/Second_Repository_Rpt_120908.pdf.

[4] DOE, *Life Cycle Cost Report,* op. cit., p. A-3.

[5] DOE, *Need for a Second Repository,* op. cit., p. 1.

[6] Department of Energy, Office of Civilian Radioactive Waste Management, *Monthly Summary of Program Financial & Budget Information,* as of September 1, 2008, p. 7.

[7] U.S. Department of Energy, Office of Civilian Radioactive Waste Management, *Total System Life Cycle Cost Report,* DOE/RW-0591, Washington, DC, July 2008, p. B-2.

NWPA section 302(d) restricts the use of the Nuclear Waste Fund to disposal activities and research authorized by the act. The section specifically prohibits DOE from expending the funds for any facility besides those expressly authorized by NWPA or subsequent act of Congress. As amended in 1987, NWPA currently authorizes only a repository at Yucca Mountain and a monitored retrievable storage facility tied to the operation of a Yucca Mountain repository.

The contracts that DOE signed with nuclear utilities required DOE to begin taking waste from reactor sites by January 31, 1998. Because that deadline was missed, DOE has been ruled liable for all waste storage costs that nuclear utilities would not have incurred had shipments to the planned repository begun on time. The U.S. Court of Federal Claims has already issued several judgments against DOE. Claims are paid from the federal judgment fund, rather than the Nuclear Waste Fund, and require no congressional appropriations. DOE calculates that its nuclear waste liabilities to nuclear reactor operators under current law will ultimately total $11 billion if shipments begin by 2020 as currently planned and potentially much more if waste operations are further delayed.[8]

Options for Halting or Delaying Yucca Mountain

The Yucca Mountain repository is now in the final stages of the lengthy approval process established by NWPA. Therefore, if the Obama Administration does "not believe that Yucca Mountain is a suitable site," what options are available at this point to stop the project?

NWPA sections 113-116 prescribed the following actions by the Secretary of Energy, the President, and the Nuclear Regulatory Commission toward developing a nuclear waste repository at the Yucca Mountain site:

- The Secretary was to determine whether the site would be suitable for a repository, and, if so, notify the State of Nevada and recommend that the President approve the project.

- If the President agreed with the Secretary's recommendation, the President was to submit an approval recommendation to Congress.

- After the presidential recommendation, the Nevada Legislature or Governor was allowed to submit a notice of disapproval to Congress.

- State disapproval would block the repository unless Congress voted within 90 days for an approval resolution that was signed by the President.

- Once the presidential site designation took effect, the Secretary was required to submit a repository license application to NRC.

All these steps have now taken place, and the Yucca Mountain license application has been docketed by NRC for consideration. NRC is an independent regulatory body not directly under the President's control. But the President has a variety of tools at his disposal that could dramatically affect the nuclear waste program.

---

[8] Christopher A. Kouts, Office of Civilian Radioactive Waste Management, "Yucca Mountain Program Status Update," Presentation to Environmental Protection Agency Workshop on Energy and Environmental Sustainability in a Carbon Constrained Future, New York, NY, September 11, 2008, p. 9, http://www.epa.gov/region2/energyworkshop/workshop_presentations/session2/nuclear_session/panel1_nuclear_waste_disposal.pdf.

Withdraw License Application

To stop further action on Yucca Mountain, perhaps the most dramatic step would be for DOE to withdraw the repository license application, as allowed by NRC procedures.[9] Such a withdrawal could be temporary, pending completion of some of the study options described below, or it could be permanent, with the intention of completely ending the Yucca Mountain project. If the license application were permanently withdrawn, the previous presidential site designation may have to be reversed as well, because NWPA section 114 requires that the license application be submitted 90 days after the presidential designation takes effect (a deadline that was missed by more than five years). The fact that, as noted above, the presidential designation of the Yucca Mountain site took effect pursuant to a congressional override of Nevada's "state veto" could be a complicating factor.

Reduce Appropriations

Restricting funding for the Yucca Mountain Project could be another approach. Congressional opponents of the waste program have succeeded in cutting its funds in recent years, although not enough to prevent DOE from submitting the license application. The Bush Administration requested an appropriation of $37.3 million from the Nuclear Waste Fund for NRC's license review activities during FY2009,[10] and substantial reductions could force significant delays in the planned four-year licensing schedule.

As noted above, DOE contends that it will need large funding increases to design and build the Yucca Mountain repository by 2020, so even steady funding would probably push back that schedule by many years. Of course, eliminating funding altogether for NRC licensing and DOE repository construction would halt further work on the project. The Obama Administration's ability to reduce or eliminate Yucca Mountain funding, if it so desired, would depend on its influence with Congress.

Key Policy Appointments

Although NRC is independent of the Administration, and Commissioners cannot be removed by the President without cause, the President can change the makeup of the Commission over time. Significant changes in the Commission could affect the Yucca Mountain licensing process and its ultimate outcome. The five NRC Commissioners serve five-year terms that are staggered so that one expires each year, on June 30. No more than three Commissioners may be from the same political party. One of the five slots is currently vacant. The President can also redesignate the Chairmanship of the NRC to a different Commissioner at any time.

Other presidential appointments could also have a strong effect on the Yucca Mountain project, the Secretary of Energy in particular. Senate Majority Leader Harry Reid had promised to block any nominee for that post who supported the Yucca Mountain site.[11] The Energy Secretary could

---

[9] 10 C.F.R. § 2.107.

[10] Nuclear Regulatory Commission, *Performance Budget Fiscal Year 2009*, NUREG-1100 Volume 24, February 2008, p. 2, http://www.nrc.gov/reading-rm/doc-collections/nuregs/staff/sr1100/v24/

[11] Erica Werner, "Reid Won't Allow Energy Secretary Who Supports Yucca Waste Dump," *Associated Press*, December 4, 2008, http://www.rgj.com.

---

initiate a redirection of the nuclear waste program, revisit DOE's complex computer models that predict low radioactive releases from Yucca Mountain, or, as noted above, withdraw the license application.

Senator Reid expressed support for President Obama's Energy Secretary, former Lawrence Berkeley National Laboratory Director Steven Chu, contending that "Dr. Chu also knows, like most Nevadans, that Yucca Mountain is not a viable solution for dumping and dealing with nuclear waste."[12] In a 2005 interview posted on the Lawrence Berkeley National Laboratory website, Chu noted projections from his lab that waste canisters in Yucca Mountain would begin to fail after about 5,000 years, which would require the underlying rock formations to prevent unacceptable migration of radioactive material into the groundwater.[13] However, Chu also signed a nuclear policy statement with other national laboratory directors in August 2008 that called for "licensing of the Yucca Mountain Repository as a long-term resource."[14]

Another important appointment for Yucca Mountain is the Administrator of the Environmental Protection Agency (EPA), Lisa Jackson, who was confirmed January 22, 2009. Under NWPA, EPA sets the radiation protection standard that NRC must use in licensing Yucca Mountain. After an earlier version of the standard was struck down by a federal court, EPA published a final standard October 15, 2008, which sets individual radiation exposure limits of 15 millirems for the first 10,000 years after disposal and 100 millirems after 10,000 through one million years.[15] The State of Nevada has sued to overturn the EPA regulations, contending that there is no justification for a higher limit after 10,000 years.[16] If the EPA standard is overturned, licensing of the Yucca Mountain site could become more difficult.[17]

Waste Program Review

Rather than moving immediately to halt the Yucca Mountain repository, the Obama Administration could delay or suspend the project through the approaches described above and then initiate a major program review – a step implied by the Obama-Biden campaign policy statement. The scope of the review could include such topics as management issues, research and development needs, foreign waste management experience, and broad policy options.

Such a review could be conducted by an interagency task force within the new Administration or by an outside entity such as the National Academy of Sciences (NAS) or a presidential commission. Independent scientific reviews of the waste program are currently provided by the

---

[12] Senator Harry Reid, "Statement on the nomination of Dr. Steven Chu to be Secretary of Energy," press release, December 15, 2008, http://reid.senate.gov/newsroom/pr_121508_energysecnom.cfm.

[13] Lawrence Berkeley National Laboratory, "Growing energy: Berkeley Lab's Steve Chu on what termite guts have to do with global warming," press release, September 30, 2005, http://berkeley.edu/news/media/releases/2005/10/03_chu.shtml.

[14] DOE National Laboratory Directors, *A Sustainable Energy Future: The Essential Role of Nuclear Energy*, August 2008, p. 1, http://www.ne.doe.gov/pdfFiles/rpt_SustainableEnergyFuture_Aug2008.pdf.

[15] Nuclear Regulatory Commission, "Public Health and Environmental Radiation Protection Standards for Yucca Mountain, Nevada," 73 *Federal Register* 61256, October 15, 2008.

[16] *State of Nevada v. U.S. Environmental Protection Agency* (U.S. Court of Appeals for the District of Columbia Circuit 2008).

[17] For more on this issue, see CRS Report RL34698, *EPA's Final Health and Safety Standard for Yucca Mountain*, by Bonnie C. Gitlin.

Nuclear Waste Technical Review Board (NWTRB), which, under NWPA Title V, the President appoints from nominees provided by NAS. However, the President does not set the agenda for independent agencies such as NWTRB, so it is not clear what role the Board might play in a major policy review.

Consequences of a Yucca Mountain Policy Shift

A decision by the incoming Administration to halt or delay development of the Yucca Mountain repository could have significant impact on the federal budget, proposed new U.S. nuclear power plants, waste storage at existing reactor sites, and disposal of defense-related nuclear waste. Such consequences could be most pronounced if such a policy shift involved simply a halt of Yucca Mountain without legislation to forge a new direction – legislation that presumably would address the issues discussed below.

No matter what decision is made on Yucca Mountain, there is a broad scientific consensus that long-term isolation of nuclear waste from the environment – for at least 1,000 years – will still be required. In other words, if Yucca Mountain were abandoned, another repository site in the United States would almost certainly have to be found eventually. Reprocessing and recycling of nuclear spent fuel can reduce the amount of long-lived radioactive waste requiring isolation but cannot entirely eliminate the need for such isolation. Alternatives to deep geologic waste isolation have been studied, such as space and subseabed disposal, but they face daunting technical obstacles, and none has ever been developed beyond the conceptual stage.

If development of the Yucca Mountain repository were delayed or halted, commercial spent fuel and defense-related nuclear waste would almost certainly remain at numerous on-site storage facilities longer than currently planned. A new repository to replace Yucca Mountain would be unlikely to open by 2020 to prevent delays in DOE's current shipping schedule. Federal centralized interim storage has been proposed repeatedly as a solution, including the MRS facility authorized by NWPA, but no such facility has been developed. If new legislation were to authorize a central interim storage facility, it possibly could begin receiving waste by 2020 and prevent further delays. Another possibility is a private spent fuel storage facility in Utah that has already received an NRC license and might be opened relatively quickly if other administrative approvals were granted (as discussed in a subsequent section). However, the Utah facility's licensed capacity is limited to 40,000 metric tons of spent fuel, which could be stored at the site for no longer than 40 years.

Federal Liabilities for Disposal Delays

As noted above, DOE is liable for utilities' nuclear waste storage costs resulting from the missed NWPA disposal deadline. According to DOE, "for each additional year of delay, the Department estimates that there may be hundreds of millions of dollars of additional damages."[18] These mandatory payments would be a direct cost to the federal government, but they would stretch over several decades because utilities cannot recover damages until their extra storage costs are

[18] Department of Energy, Office of Civilian Radioactive Waste Management, *Report to Congress on the Demonstration of the Interim Storage of Spent Nuclear Fuel from Decommissioned Nuclear Power Reactor Sites*, DOE/RW-0596, Washington, DC, December 2008, p. 6, http://www.rw.doe.gov/info_library/program_docs/ES_Interim_Storage_Report_120108.pdf.

actually incurred. DOE projects that if disposal begins by 2020, $11 billion in liabilities will be incurred by 2056.[19]

If the Yucca Mountain site were abandoned without an alternative storage or disposal process in place, court judgments against DOE could rise far higher. The nuclear industry has raised the possibility that DOE could be found in complete default on its NWPA contracts and be ordered to refund all the nuclear waste fees that had been collected, in addition to paying utilities' extra at-reactor storage costs.[20] Through the end of FY2008, DOE had collected more than $28 billion in fees and interest payments – an amount that has been growing at about $1.5 billion per year.[21] In at least one of the nuclear utility cases before the Federal Court of Claims, a judge issued a show-cause order for why the DOE nuclear waste contracts should not be voided and all payments returned to utilities,[22] although that step was not included in the court's final decision.[23]

Licensing Complications for New Power Reactors

No new commercial reactors have been ordered in the United States since the 1970s, but concerns over potential carbon dioxide controls and high natural gas prices have prompted U.S. electric utilities to again consider the nuclear power option. License applications for 26 new reactors have been filed with NRC, and more are anticipated.[24] Further delays in the DOE nuclear waste program could pose an obstacle to licensing the proposed new reactors.

NRC established a policy in 1977 that it "would not continue to license reactors if it did not have reasonable confidence that the wastes can and will in due course be disposed of safely."[25] NRC then began a Waste Confidence proceeding that resulted in 1984 findings that there was "reasonable assurance" that a nuclear waste repository would be available by 2007-2009 and that waste could be safely stored at reactor sites for at least 30 years after reactors have shut down.[26]

After DOE's schedule for opening the Yucca Mountain repository slipped to 2010, NRC revised its Waste Confidence Decision in 1990 to find reasonable assurance that a repository "will be available within the first quarter of the twenty-first century."[27] With DOE now planning to open Yucca Mountain by 2020 at the earliest, NRC is proposing a further revision to find reasonable assurance that a repository will be available within 50-60 years after a reactor's licensed operating life and that spent fuel can be stored safely for at least 60 years after a reactor's licensed

---

[19] Kouts, op.cit., p. 9.

[20] U.S. Congress, Senate Committee on Energy and Natural Resources, *Nuclear Waste Litigation*, To examine the impacts of federal court decisions on breach of federal nuclear waste contracts, 106th Cong., 2nd sess., September 28, 2000, S.Hrg. 106-918 (Washington: GPO, 2001), p. 46.

[21] Department of Energy, Office of Civilian Radioactive Waste Management, *Monthly Summary of Program Financial & Budget Information*, as of September 1, 2008, p. 7.

[22] Jeff Beattie, "Federal Judge Suggests Voiding Utilities' Yucca Mountain Contracts," *Energy Daily*, April 29, 2005, p. 1.

[23] *Sacramento Municipal Utility District v. United States*, (Court of Federal Claims 2006).

[24] Nuclear Regulatory Commission, "Combined License Applications for New Reactors," http://www.nrc.gov/reactors/new-reactors/col.html.

[25] Nuclear Regulatory Commission, 42 *Federal Register* 34391, July 5, 1977.

[26] Nuclear Regulatory Commission, 49 *Federal Register* 34658, August 31, 1984.

[27] Nuclear Regulatory Commission, 55 *Federal Register* 38472, September 18, 1990.

life.[28] Although the NRC's latest proposed revision would allow for decades of further slippage in the Yucca Mountain schedule, it is not clear that NRC's waste-related criteria for licensing new reactors would be satisfied if the Yucca Mountain project were canceled without an alternative plan in place.

Six states – California, Connecticut, Kentucky, New Jersey, West Virginia, and Wisconsin – have specific laws that link approval for new nuclear power plants to adequate waste disposal capacity. Kansas forbids cost recovery for "excess" nuclear power capacity if no "technology or means for disposal of high-level nuclear waste" is available.[29] The U.S. Supreme Court has held that state authority over nuclear power plant construction is limited to economic considerations rather than safety, which is solely under NRC jurisdiction.[30] No nuclear plants have been ordered since the various state restrictions were enacted, so their ability to meet the Supreme Court's criteria has yet to be tested.

The nuclear waste issue has also historically been a focal point for public opposition to nuclear power. Proposed new reactors that have no clear path for removing waste from their sites could face intensified public scrutiny, particularly at proposed sites that do not already have operating reactors.

Environmental Cleanup Penalties

For defense-related nuclear waste, which resulted from production of nuclear weapons and naval reactor fuel by DOE and its predecessor agencies, indefinite delays in developing a repository could also have legal consequences for DOE. Defense-related high-level radioactive waste resulted from decades of reprocessing spent fuel to extract plutonium for nuclear warheads or highly enriched uranium from spent naval reactor fuel. As noted above, DOE's inventory of defense high-level waste and unreprocessed spent fuel, plus waste from other DOE nuclear programs, totals the equivalent of 12,800 metric tons (the mass of spent fuel before reprocessing). This material is located primarily at Hanford, WA, Savannah River, SC, and the Idaho National Laboratory.

Congress has given states the authority to enforce waste management laws against federal agencies, including DOE. Without Yucca Mountain or an alternative repository plan, DOE would not have a permanent disposal site for waste now stored at its defense-related facilities. A lack of repository capacity, according to DOE, "could threaten the Department's ability to fulfill [regulatory] agreements with the states hosting those sites to remove the waste for permanent disposal."[31]

---

[28] Nuclear Regulatory Commission, "Waste Confidence Decision Update," 73 *Federal Register* 59551, October 9, 2008.

[29] David L. Lovell, Wisconsin Legislative Council Staff, *State Statutes Limiting the Construction of Nuclear Power Plants*, October 5, 2006.

[30] *Pacific Gas & Electric Co. v. State Energy Resources Conservation and Development Commission*, 461 U.S. (190 1983).

[31] DOE, *Need for a Second Repository*, op. cit., p. 13.

Long-Term Risk

The near-term environmental impact of further Yucca Mountain delays or abandonment would be minimal for at least 100 years, according to DOE's supplemental environmental impact statement (SEIS) for the repository program.[32] That assessment is consistent with the NRC Waste Confidence Decision cited above. As long as storage facilities at reactors and other sites are maintained and guarded through institutional controls, radioactive releases to the environment are expected to be small.

On-site storage of spent fuel will continue for many decades even if Yucca Mountain begins receiving waste shipments by 2020, as discussed in the baseline program section above. Moreover, some freshly discharged spent fuel will be stored on site as long as reactors are operating. Some environmental groups have argued that it would be safer to leave all nuclear waste for an extended period in hardened on-site storage facilities rather than begin sending it to a central facility as soon as possible, because allowing the waste's radioactivity to decay would reduce the consequences of transportation accidents or sabotage when the waste is ultimately moved.[33] Moreover, waste placed at a central interim storage facility would probably have to be moved a second time to a permanent repository, potentially further increasing transportation risks.

Beyond 100 years or so, the environmental risks of surface storage resulting from the lack of an underground repository become more uncertain. At some point in the future, maintenance and security of surface storage facilities would be expected to drop below adequate levels because of unforeseen circumstances. Whether that would occur after 500 years, 1,000 years, or 10,000 years is open to speculation. Federal nuclear waste policy would presumably continue to envision a permanent disposal method well before such a breakdown, but the risk would be expected to rise if delays continued indefinitely. The SEIS predicts that substantial amounts of radioactivity would reach the accessible environment within 10,000 years after the end of institutional controls on surface storage facilities, "with eventual catastrophic consequences for human health."[34]

Nuclear Waste Policy Options

Because NWPA specifies that only Yucca Mountain may be considered for a repository site and that a federal storage facility cannot open before the repository is licensed, the government's waste management options are sharply limited under current law. Without congressional action, alternatives to Yucca Mountain would consist primarily of indefinite on-site storage or licensing of new private storage sites.

New legislation would open up much broader possibilities, ranging from a search for a new repository site and federal interim storage to reprocessing and alternative disposal technologies.

---

[32] Department of Energy, Office of Civilian Radioactive Waste Management, *Final Supplemental Environmental Impact Statement for a Geologic Repository for the Disposal of Spent Nuclear Fuel and High-Level Radioactive Waste at Yucca Mountain, Nye County, Nevada,* DOE/EIS-0250F-S1, Washington, DC, June 2008, p. S-50, http://www.rw.doe.gov/ym_repository/seis/index.shtml.

[33] U.S. Congress, House Committee on Commerce, Subcommittee on Energy and Power, *The Nuclear Waste Policy Act of 1997,* hearing on H.R. 1270, 105th Cong., 1st sess., April 29, 1997, Serial No. 105-27 (Washington: GPO, 1997), p. 125.

[34] DOE, *Final Supplemental Environmental Impact Statement,* op. cit., p. S-51.

Some – but probably not all – of the consequences of changing the current waste policy could also be mitigated through legislation. Any legislation dealing with nuclear waste siting is almost certain to prove extraordinarily controversial.

Institutional Changes

Almost since the beginning of the current nuclear waste program in 1982, DOE has regularly been accused of mismanagement and allowing political considerations to affect scientific decisions. Many proposals have been made by the nuclear industry and its critics alike to transfer the nuclear waste program to an independent organization that might be more efficient and less affected by politics. For example, a proposal in the 110th Congress (H.R. 6001, section 186) would have handed the DOE waste program to an independent High Level Waste Authority, headed by a presidentially appointed seven-member board. In implementing the nuclear waste program, the Waste Authority would have been authorized to consider all reasonable options, including alternative repository locations.

An independent waste agency could be a government agency, as in H.R. 6001, a government corporation, or a private-sector entity. Two studies conducted for DOE's Global Nuclear Energy Partnership recommended the establishment of a government corporation to manage nuclear waste, including spent fuel reprocessing. One of the studies, by a consortium led by the French firm Areva, called a government corporation the best option "because it allows the utilities to have some level of oversight while full government ownership keeps the cost of capital low."[35] A team led by EnergySolutions recommended that the government corporation have a board of directors drawn from the nuclear industry, with an independent oversight board "to assure that it meets its charter obligations."[36]

Licensing and regulation by NRC could be continued unchanged under any option. The existing funding system could also be transferred largely unchanged to a new government agency, but a new funding system would probably be needed for a private entity. Whether a private entity should take permanent title to all nuclear waste could also be an issue. The nuclear industry has long contended that nuclear waste payments by reactor owners should be available directly for nuclear waste disposal activities without the need for congressional appropriation. But even if the program's management were improved, it is far from clear whether any new waste management organization could avoid the political controversy that has persistently accompanied the DOE program.

Short of establishing an entirely new organization to run the waste program, it has been suggested that additional independent oversight could improve public confidence in DOE's decisions. This was a major reason cited for creating the Nuclear Waste Technical Review Board in 1987. Some have proposed that technical oversight by a non-federal agency would be more credible, such as the Environmental Evaluation Group (EEG) in New Mexico.[37] Congress in 1988 required DOE to sign a contract with a New Mexico university, the New Mexico Institute of Mining and Technology, to administer the EEG to provide independent review and evaluation of the Waste

---

[35] International Nuclear Recycling Alliance, *Integrated U.S. Used Fuel Strategy*, May 1, 2008, p. 10.

[36] EnergySolutions, *GNEP Deployment Studies: Overall Summary Report*, Richland, WA, May 19, 2008, p. 2-2.

[37] Allison M. Macfarlane, Rodney C. Ewing, et al., *Uncertainty Underground: Yucca Mountain and the Nation's High-Level Nuclear Waste* (Cambridge, MA: MIT Press, 2006), p. 407.

Isolation Pilot Plant (WIPP), a DOE repository for relatively low-radioactivity defense waste (P.L. 100-456, section 1433). The EEG closed in 2004 after DOE halted its funding,[38] and oversight activities are now carried out by the New Mexico Environment Department.[39]

Extended On-Site Storage

It appears unlikely, based on the history of the nuclear waste program, that any alternative storage or disposal sites could become operational earlier than the planned opening of Yucca Mountain in 2020. Therefore, any alternative to the Yucca Mountain repository would almost certainly result in longer on-site storage of nuclear waste than under the current baseline program. Essentially, extended on-site storage is the default option, with the only question being how long. On-site storage could be extended for decades under some policy changes, such as a restart of the repository site search or the pursuit of alternative disposal technologies. Waste might be moved more quickly if central interim surface storage facilities were developed (as discussed below), but designating such sites may be nearly as controversial as siting a repository.

On-site nuclear waste storage at reactor facilities takes place primarily in deep pools of water that are built into the reactor building. The water is necessary to provide cooling and radiation shielding for extremely radioactive spent fuel that is freshly discharged from the reactor. After its radioactivity has sufficiently decayed, usually after several years, spent fuel can be transferred to dry storage casks and stored outside the pools. Most spent fuel pools were not designed to hold all the spent fuel generated during a reactor's operating life. Therefore, as DOE's target date for taking spent fuel from reactor sites has slipped, nuclear reactor operators have had to expand their on-site dry storage capacity (and have sued DOE for compensation).

NRC considers extended on-site storage to be safe as long as storage facilities are adequately maintained and guarded, as discussed above. However, the National Academy of Sciences (NAS) determined in 2005 that spent fuel pools could be vulnerable to terrorist attacks, particularly as spent fuel has been stored more densely in the pools to increase their capacity. NAS found that an attack could drain the cooling water from a spent fuel pool and cause the spent fuel's zirconium cladding to overheat and catch fire, releasing "large quantities of radioactive materials to the environment."[40] The Energy and Water Development Appropriations Act for FY2006 (P.L. 109-103) included $21 million for NRC to assess the vulnerabilities found by NAS at each reactor site.

Keeping nuclear waste at reactor sites and other existing facilities has long been a major goal of Yucca Mountain opponents. Numerous bills have been introduced over the years to require DOE to take over all responsibility for storing spent fuel at reactor sites, including ownership of the waste and on-site storage facilities (such as S. 784 in the 110th Congress). Such an on-site DOE takeover is intended to reduce utilities' costs, and resulting federal liabilities, for DOE's failure to remove the spent fuel. However, states with nuclear reactors oppose indefinite on-site storage, and utilities and state regulators have opposed using the Nuclear Waste Fund to pay for on-site storage (as proposed by S. 784) rather than permanent disposal or central interim storage.

---

[38] John Fleck, "WIPP Oversight Bureau Planned," *Albuquerque Journal*, June 19, 2004.

[39] http://www.nmenv.state.nm.us/doe_oversight/wipp.htm

[40] National Academy of Sciences Board on Radioactive Waste Management, *Safety and Security of Commercial Spent Nuclear Fuel Storage: Public Report*, Washington, DC, 2005, p. 6.

Federal Central Interim Storage

DOE does not believe it has the authority under current law to develop a central interim nuclear waste storage facility other than the "monitored retrievable storage" facility authorized by NWPA,[41] and the 1987 NWPA Amendments prohibit such an MRS facility from opening until Yucca Mountain is licensed. Moreover, construction of an MRS cannot begin until NRC grants a construction permit for the repository; the MRS is limited to 15,000 metric tons of spent fuel; and the MRS cannot be located in Nevada (NWPA sections 145 and 148). Numerous legislative efforts have been mounted since the mid-1990s to establish central interim storage capacity without the MRS restrictions, but without success. No matter how they have been structured, such proposals have consistently faced overwhelming concerns that any "interim" storage facility would undercut political support for a permanent repository and therefore become a "de facto" permanent disposal site.

Central interim nuclear waste storage facilities would use dry cask technology that is similar to that currently in place at reactor sites. Sealed waste canisters would be placed in individual above-ground concrete casks or bunkers for radiation shielding and for cooling by natural air circulation. Because the systems are modular, they can be constructed relatively quickly. And because the waste is not expected to remain in storage after active maintenance of the facility has ceased, a wide variety of sites are likely to be considered geologically suitable. However, public concern about large quantities of highly radioactive waste stored for an extended period of time, and perhaps indefinitely, has proven to be a major obstacle to central storage proposals.

DOE had proposed to build an MRS facility near Oak Ridge, TN, as a central receiving point for small waste shipments from individual plant sites east of the Rocky Mountains. Spent fuel was to be repackaged if necessary at the MRS facility for consolidated long-distance shipments to the planned Western repository.[42] The Oak Ridge selection was specifically nullified by the 1987 NWPA amendments, which established a new siting procedure along with the restrictions listed above. The 1987 Amendments also established an alternative, voluntary siting procedure for the MRS and other nuclear waste facilities. Under Title IV, a presidentially appointed "nuclear waste negotiator" was authorized to reach agreements with any states or Indian tribes to host nuclear waste facilities under any "reasonable and appropriate" terms. Such agreements could not take effect without being enacted into law, however.

By the early 1990s, finding a voluntary site for a central storage facility appeared to be DOE's best chance for meeting NWPA's 1998 waste acceptance deadline. DOE began providing feasibility study grants to potential volunteers, mostly Indian tribes. Potential agreements with Indian tribes proved highly objectionable to the states in which the tribes were located, and Congress blocked the grant funding in October 1993 (P.L. 103-126). The authority for the nuclear waste negotiator expired on January 21, 1995, without any proposed siting agreements having been reached.

The next major legislative push for an alternative to the MRS took place after it became clear that DOE would be unable to meet the NWPA disposal deadline. In the 104th Congress, nuclear power supporters developed legislation to authorize DOE to open an interim surface storage facility at

---

[41] DOE, *Report to Congress on Interim Storage*, op. cit., p. 14.

[42] Department of Energy, Office of Civilian Radioactive Waste Management, *Monitored Retrievable Storage Submission to Congress*, DOE/RW-0035, Washington, DC, February 1986.

the Yucca Mountain site by 1998, well before the then-anticipated opening of the underground repository in 2010 (H.R. 1020). The State of Nevada and nuclear power opponents contended that waste should not be transported to Yucca Mountain before the repository was licensed, because if the repository ultimately did not receive a license, the waste would have to be moved again, increasing potential transportation risks and costs. The bill was approved by the Commerce Committee (H.Rept. 104-254) but was not enacted. Similar bills were passed by the House and Senate in the 105th Congress (H.R. 1270, S. 104), but President Clinton threatened a veto and a conference was not held. A final try in the 106th Congress (S. 1287) drew a presidential veto that was narrowly sustained in the Senate.

After the proposals to develop interim storage capacity at Yucca Mountain were rejected, language was included in several appropriations bills and reports to require DOE to store commercial spent fuel at unspecified federal sites. The House Appropriations Committee included language in its report on the FY2006 Energy and Water Development Appropriations Bill to require DOE "to begin the movement of spent fuel to centralized interim storage at one or more DOE sites within fiscal year 2006" (H.Rept. 109-86), although the Senate did not go along with the idea. The Senate Appropriations Committee included an extensive provision in its version of the FY2007 Energy and Water bill (H.R. 5427, section 313) to authorize the Secretary of Energy to designate interim storage sites for spent nuclear fuel. The proposal, which was not enacted, would have required the Secretary to designate a storage site in each state with a nuclear power plant, after consultation with the governor, or to designate regional storage facilities.

A more limited central storage proposal was aimed solely at nine decommissioned reactor sites. Because the decommissioned sites have no ongoing nuclear activities except spent fuel storage, the removal of spent fuel would allow all nuclear-related maintenance and security at those locations to cease, producing significant operational cost savings, according to the nuclear industry. The House Appropriations Committee included report language with the FY2008 Energy and Water bill requiring DOE to "develop a plan to take custody of spent fuel currently stored at decommissioned reactor sites" (H.Rept. 110-185). The resulting DOE report concluded that all 2,800 metric tons of spent fuel at the nine decommissioned sites could be shipped to a federal central storage facility by 2018, but that DOE had no statutory authority to implement such a plan.[43]

DOE has taken spent fuel for storage at its facilities in the past in special cases, such as the damaged core from the 1979 Three Mile Island accident and from the unique Fort Saint Vrain gas-cooled reactor in Colorado. The Three Mile Island core material was shipped to DOE's Idaho National Laboratory for research. DOE is storing the Fort Saint Vrain spent fuel pursuant to a cooperative agreement signed with the reactor supplier and local utility before the demonstration reactor was built.[44] In addition, DOE stores highly enriched, U.S.-origin spent fuel from foreign research reactors because of its potential use in nuclear weapons. Some have contended that those precedents indicate that DOE has sufficient general authority under the Atomic Energy Act to store larger amounts of commercial spent fuel. However, DOE contends that its broad authority under the Atomic Energy Act is restricted to narrow circumstances under the more recently enacted and specific waste management provisions of NWPA.[45]

---

[43] DOE, *Report to Congress on Interim Storage*, op. cit., p. 14.

[44] Department of Energy, "DOE Holds License for Colorado Spent Fuel Facility," press release, June 28, 1999, http://newsdesk.inl.gov/press_releases/1999/DOE_Holds_Licen.htm.

[45] DOE, *Report to Congress on Interim Storage*, op. cit., p. 6.

Private Central Storage

Although DOE does not believe it has authority under current law to construct a federal central interim storage facility for commercial nuclear waste, NRC regularly licenses private-sector interim storage facilities under the Atomic Energy Act.[46] Such "independent spent fuel storage installations" typically are licensed for on-site storage at reactor sites, but they can also include central storage facilities.

After a nearly nine-year licensing process, NRC issued a license for a private central storage facility on February 21, 2006, that was intended to receive waste from commercial reactor sites.[47] The facility was to be developed by a nuclear utility consortium called Private Fuel Storage (PFS) on the reservation of the Skull Valley Band of the Goshute Indians in Utah. The 20-year license, renewable for an additional 20 years, allows up to 40,000 metric tons of spent fuel to be stored in 4,000 dry casks pending shipment by DOE to a permanent repository. PFS will not take title to the spent fuel, so waste is to be returned to the utilities that own it if DOE cannot take it away before the PFS license expires.[48]

On September 7, 2006, the Department of the Interior issued two decisions blocking the PFS project. The Bureau of Indian Affairs disapproved a proposed lease of tribal trust lands to PFS, concluding there was too much risk that the waste could remain at the site indefinitely, among other objections.[49] The Bureau of Land Management rejected the necessary rights-of-way to transport waste to the facility, concluding that a proposed rail line would be incompatible with the Cedar Mountain Wilderness Area and that existing roads would be inadequate.[50] Contending that the Interior Department was motivated by political pressure from the State of Utah, which strongly opposed the facility, the Skull Valley Band of Goshutes and PFS filed a federal lawsuit July 17, 2007, to overturn the decisions.[51]

The PFS project was intended to provide a waste storage option for nuclear plants that might have trouble gaining approval for on-site storage facilities or decommissioned reactors that want to remove remaining spent fuel from their sites. If the PFS facility were considered potentially useful as part of a revised spent fuel strategy, the new Administration could revisit the Interior Department's administrative decisions that are blocking the project. However, if those decisions were reversed, the project would still need to overcome a challenge to the NRC license filed by the State of Utah.[52] Another consideration for this option is that, because the waste would have to be returned after 40 years, and utilities would be paying for the service, the PFS facility might not significantly reduce DOE's liabilities for delays in spent fuel acceptance.

---

[46] 42 U.S.C. § 2011 et seq., 10 CFR Part 72.

[47] Nuclear Regulatory Commission, *License for Independent Storage of Spent Nuclear Fuel and High-Level Radioactive Waste SNM-2513*, February 21, 2006.

[48] Private Fuel Storage, LLC, *Frequently Asked Questions: Financial Accountability*, http://www.privatefuelstorage.com/faqs/faqs.html.

[49] Bureau of Indian Affairs, *Record of Decision for the Construction and Operation of an Independent Spent Fuel Storage Installation (ISFSI) on the Reservation of the Skull Valley Band of Goshute Indians (Band) in Tooele County, Utah*, September 7, 2006.

[50] Bureau of Land Management, *Record of Decision Addressing Right-of-Way Applications U 76985 and U 76986 to Transport Spent Nuclear Fuel to the Reservation of the Skull Valley Band of Goshute Indians*, September 7, 2006.

[51] *Skull Valley Band of Goshute Indians and Private Fuel Storage, LLC, v. James E. Cason et al.* (U.S. District Court for the District of Utah, Central Division 2007).

[52] Todd D. Lovinger, "Utah Challenges License Issuance to PFS," *LLW Forum News Flash*, March 12, 2006.

Spent Fuel Reprocessing and Recycling

The major alternative to direct disposal of spent fuel (the "once through" fuel cycle) is the "closed" fuel cycle, in which spent fuel is reprocessed into new fuel. The closed fuel cycle could reduce the volume and long-term radioactivity of nuclear waste and potentially postpone the need for permanent disposal. However, a National Academy of Sciences study of reprocessing technologies found that "none of the S&T [separations and transmutation] system concepts reviewed eliminates the need for a geologic repository."[53] Recycling spent fuel could also greatly increase the amount of energy extracted from a given supply of uranium. However, the closed fuel cycle is generally considered to be substantially more expensive than the once-through cycle.[54] Moreover, the separation of plutonium from spent fuel has long been a subject of national policy debates because of its potential role in nuclear weapons proliferation.

Fuel for U.S. nuclear reactors currently consists of uranium in which the fissile isotope U-235 has been increased (enriched) to 3-5%, with the remainder being the non-fissile isotope U-238. During the fuel's several-year irradiation period in the reactor, most of the U-235 splits, or fissions, releasing energy. Some of the U-238 is transmuted into fissile isotopes of plutonium, some of which also fissions. In reprocessing, the uranium and plutonium are chemically separated to be made into new fuel, while the lighter elements resulting from the fission process, called fission products, are stored for disposal.

New fuel made from reprocessed uranium and plutonium can be recycled in existing commercial light water reactors, which is being done in other countries, primarily France. After being recycled once, however, the buildup of undesirable plutonium isotopes makes further recycling in today's commercial reactors problematic. Without multiple recycling, the plutonium and other long-lived isotopes cannot be fully fissioned or transmuted into shorter-lived radioactive isotopes, and the benefits for waste disposal would therefore be modest.

For multiple recycling of spent fuel, advanced reactors would be necessary. DOE has evaluated a wide variety of options as part of its Global Nuclear Energy Partnership (GNEP) program. These include initial recycling in existing light and heavy water reactors, and subsequent recycling in high-burnup gas-cooled reactors, reactors fueled by thorium and plutonium, and "fast" reactors (in which neutrons are not slowed by water or other materials).[55] A reprocessing and recycling system with sufficient capacity could eventually treat existing spent fuel inventories along with newly generated spent fuel.

For waste disposal, the goal of such a recycling system would be to send only the fission products and other short-lived radioisotopes to a permanent repository and feed all the uranium, plutonium, and other long-lived radioisotopes back into a reactor after each cycle. If that could be accomplished, the nuclear waste in a repository would decay to insignificant levels within about 1,000 years and eliminate longer-term uncertainty about the repository's performance. Spent fuel

---

[53] National Academy of Sciences, National Research Council, *Nuclear Wastes: Technologies for Separations and Transmutation*, Washington, DC, August 1995, p. 17.

[54] Peter R. Orszag, *Costs of Reprocessing Versus Directly Disposing of Spent Nuclear Fuel*, Congressional Budget Office, Statement Before the Senate Committee on Energy and Natural Resources, Washington, DC, November 17, 2007, http://www.cbo.gov/ftpdocs/88xx/doc8808/11-14-NuclearFuel.pdf.

[55] Department of Energy, Office of Nuclear Energy, *Draft Global Nuclear Energy Partnership Programmatic Environmental Impact Statement*, DOE/EIS-0396, Washington, DC, October 2008, http://www.gnep.energy.gov/peis.html.

recycling could also save space in an underground repository by reducing the near-term heat load, which is the primary limit on repository capacity.[56] To address nuclear nonproliferation concerns, the GNEP program is conducting research on reprocessing technology that would not separate plutonium in a pure enough form for direct use in nuclear weapons.

A potential nearer-term benefit of a reprocessing strategy would be to provide an alternative destination for spent fuel currently stored at reactor sites if Yucca Mountain were to be abandoned. Because DOE is still conducting R&D on a variety of possible reprocessing technologies, however, a U.S. reprocessing facility would probably not open earlier than the current Yucca Mountain target date.

Earlier waste shipments might be possible under proposals that have been made for foreign reprocessing of U.S. spent fuel. Several foreign reprocessing plants are currently in operation. The Senate Energy and Natural Resources Committee included a provision in a nuclear waste bill in the 104th Congress (S. 1271) that would have authorized DOE to take title to spent fuel and ship it to a reprocessing plant in England. However, the provision proved highly contentious and was dropped from the final bill passed by the Senate (S. 1936). In a 2008 report for GNEP, a consortium led by the French nuclear firm Areva recommended that U.S. spent fuel be reprocessed overseas from 2010 to 2019 before the startup of a U.S. reprocessing plant after 2020.[57] The use of inactive defense-related reprocessing facilities at DOE's Savannah River Site in South Carolina has also been suggested for U.S. commercial spent fuel.[58]

The amount of spent fuel that could be shipped to reprocessing plants would be another consideration. Existing reprocessing plants in France and England are designed to handle about 800 metric tons of spent fuel per year. Therefore, at least three plants of that size would need to be constructed in the United States (assuming minimal foreign reprocessing) to handle the 2,000 metric tons of spent fuel discharged annually from U.S. reactors. About four plants would be needed to exceed the planned shipment rate to Yucca Mountain.

Many decades would be required to implement a reprocessing and recycling strategy. For example, the Areva consortium projected that a steady-state recycling system would not be fully in place until about 2070, even if the currently planned 63,000 metric tons of spent fuel were emplaced in Yucca Mountain rather than being reprocessed. The first U.S. reprocessing plant would become operational after 2020.[59] Proposals by three other GNEP consortia included similar time frames. (For more discussion of reprocessing policy, see CRS Report CRS Report RL34579, *Advanced Nuclear Power and Fuel Cycle Technologies: Outlook and Policy Options*, by Mark Holt).

---

[56] Electric Power Research Institute, *Projected Waste Packages Resulting From Alternative Spent-Fuel Separation Processes*, EPRI NP-7262 Project 3030 Final Report, Palo Alto, CA, April 1991, pp. 5-3.

[57] International Nuclear Recycling Alliance, *Presentation for GNEP Deployment Studies*, April 2008, p. 6, http://www.gnep.energy.gov/pdfs/INRA%20Presentation.pdf.

[58] David Kramer, "Report by SRS Contractor Appears to Advocate Reprocessing at Site," *Inside Energy/with Federal Lands*, January 8, 1996, p. 8.

[59] International Nuclear Recycling Alliance, *op. cit.*, p. 43.

Non-Repository Options

The inherent difficulty of siting a permanent geologic repository for high-level nuclear waste has led to a variety of proposals over the past few decades for non-repository disposal options. NWPA section 222 authorizes DOE to conduct research on such disposal alternatives. The most seriously analyzed ideas involve launching waste into space or burying it in the deep seabed. Some plausible concepts for implementing these ideas have been developed, but a great deal of development work would still be required to determine their likely feasibility.

Congress established a DOE Office of Subseabed Disposal Research in the Nuclear Waste Policy Amendments Act of 1987 (P.L. 100-203). The office was required to organize a Subseabed Consortium among leading research institutions to develop a research plan for identifying subseabed disposal sites, developing conceptual designs for subseabed disposal systems, and assessing potential environmental impacts. However, few resources were provided for the subseabed office before it was abolished in 1996 by P.L. 104-66.[60]

Previous research on subseabed disposal was conducted by the Nuclear Energy Agency (NEA) of the Organization for Economic Cooperation and Development. The United States participated in the effort through the DOE Subseabed Disposal Project, on which about $125 million was spent from 1974-1986.[61] The NEA program studied the emplacement of nuclear waste canisters in ocean sediments with gravity-driven penetrators or in drilled holes. NEA concluded in 1988 that the sediments would probably contain the waste well enough to keep the maximum dose to humans – occurring after about 100,000 years – "many orders of magnitude below present standards" and pose "insignificant risk to the deep sea environment." However, NEA also concluded that more research would be needed to confirm the safety of the subseabed disposal concepts.[62]

Subseabed disposal is currently prohibited under the 1996 Protocol to the 1972 London Dumping Convention, which was signed by the United States on March 31, 1998, and entered into force March 24, 2006, but has not been ratified by the Senate. The Protocol amended the definition of "dumping" to include "any storage of wastes or other matter in the seabed and the subsoil thereof from vessels, aircraft, platforms or other man-made structures at sea."[63] Previously it had been unclear whether the Convention prohibited subseabed disposal. Annex 1 of the Protocol requires parties to the agreement to complete a scientific study of sea disposal of radioactive material other than high-level waste by 2019 and every 25 years thereafter.

Disposal of nuclear waste in outer space has also been studied by DOE and its predecessor agencies. In a 1974 draft environmental statement on nuclear waste management, the Atomic Energy Commission (AEC) reviewed government studies of such concepts as "solar system escape, solar impact, high-earth orbit, and a solar orbit other than that of the planets." The report concluded that space disposal "does not seem an attractive alternative to the geological

[60] Steven Nadis, "The Sub-Seabed Solution," *The Atlantic Monthly Digital Edition*, October 1996.

[61] Office of Technology Assessment, *Staff Paper on the Subseabed Disposal of High-Level Radioactive Waste*, Washington, DC, May 1986, p. 3.

[62] Organization for Economic Cooperation and Development, Nuclear Energy Agency, *Feasibility of Disposal of High-Level Radioactive Waste into the Seabed, Overview of Research and Conclusions*, Volume 1, Paris, 1988, p. 60.

[63] *1996 Protocol to Convention on Prevention of Marine Pollution by Dumping of Wastes*, Treaty Doc. 110-5, September 4, 2007.

development program."[64] Major concerns include launch costs, launch safety, and the potential for future waste re-entry into Earth's atmosphere. Proposed alternatives to conventional rocket-based launch systems, such as laser propulsion and electromagnetic rail guns, might have safety and cost advantages, but a major federal commitment would be needed to determine their feasibility.[65]

Other disposal concepts studied by DOE and its predecessors include waste emplacement in polar ice sheets, deep boreholes, and deep well injection of liquid waste. AEC's draft environmental statement dismissed those alternatives as not viable,[66] and they have since received relatively little attention.

New Repository Site

Even if nuclear waste is placed in extended surface storage and is reprocessed to remove the longest-lived radioactive isotopes, a permanent disposal method almost certainly would still be required. Barring the non-repository options discussed above, that would mean that the abandonment of Yucca Mountain for any reason would eventually require a search for another repository site.

The history of site selection efforts under NWPA indicates that a new repository site search would be slow-moving and extremely controversial. Vast areas of the United States would again be under consideration, after having been eliminated by the 1987 congressional designation of Yucca Mountain as the sole candidate site. Every decision made by whatever entity were to be placed in charge of the site search would probably face intense opposition, especially as the search began to narrow. Designing a selection process that could overcome such pressures would be a major challenge.

NWPA was intended to set up a fair and technically sound process for selecting among numerous potential repository sites that DOE and its predecessor agencies had been considering. Without such a legislative mandate, DOE's previous efforts to find a waste site appeared unlikely to overcome the controversy that had arisen at every potential location.[67] However, the explicit waste siting process created by NWPA lasted only about five years before being paralyzed by renewed controversy.

Under NWPA as originally enacted, the Secretary of Energy was required to establish guidelines that DOE would follow in nominating at least five suitable sites, of which three were to be recommended to the President for detailed study, or "characterization" by January 1, 1985. Sites that DOE had been considering included salt domes along the Gulf Coast, bedded salt in the Great Plains and Midwest, volcanic tuff in the West, and basalt in the Pacific Northwest. Energy Secretary John S. Herrington recommended Hanford, WA; Deaf Smith County, TX; and Yucca

---

[64] Atomic Energy Commission, *Draft Environmental Statement on Management of Commercial High Level and Transuranium-Contaminated Radioactive Waste*, WASH-1539, September 1974, pp. 5.3-6.

[65] Jonathan Coopersmith, "Nuclear Waste in Space?," *The Space Review*, August 22, 2005, http://www.thespacereview.com/article/437/1.

[66] Atomic Energy Commission, op. cit., p. 5.3.

[67] Luther J. Carter, *Nuclear Imperatives and Public Trust: Dealing with Radioactive Waste* (Washington, DC: Resources for the Future, 1987), p. 198.

Mountain for site characterization on May 27, 1986.[68] After completing the characterization of the three sites, the Secretary was to recommend one of them to the President for the nation's first permanent nuclear waste repository. The President was required to submit his choice to Congress by March 31, 1987.

To address concerns about whether a single site or region should take all of the nation's high-level nuclear waste, NWPA limited the first repository to 70,000 metric tons until a second repository was opened. A separate track was established for locating a second repository site. By July 1, 1989, the Secretary of Energy was to nominate five sites for a second repository, including at least three sites that had not been among the five sites nominated for the first repository, and recommend three of them to the President. The recommended sites were to be located, "to the extent practicable," in different geologic media. The President was to recommend a second repository site to Congress by March 31, 1990, from any of the sites previously characterized.

Unlike the process for the first repository, which started with specific candidate sites that were already under consideration, the site search for the second repository was conducted more systematically. DOE began by focusing on major formations of granite and other crystalline rock, which had not been included in the first repository effort, in 17 states in the upper Midwest and Atlantic coast. In consultation with states, DOE developed a screening methodology to rank candidate bodies of rock for their potential suitability as a repository.[69] DOE released preliminary rankings that identified 12 promising rock bodies in seven states in January 1986.[70]

DOE's identification of potential sites for the second repository drew intense opposition from the affected states. The three potential host states for the first repository also raised strong objections, which intensified when Secretary Herrington announced on May 28, 1986, that work on the second repository would be indefinitely postponed. Herrington said the decision was based on lower growth projections for nuclear power that delayed the need for a second repository, but officials from the first repository candidate states in the West contended that the Reagan Administration had responded to political pressure from the Eastern candidate states and had unraveled a key regional compromise in NWPA.[71] Opposition from Tennessee to DOE's proposed MRS site near Oak Ridge added to the controversy. The Senate Appropriations Committee made note of the deteriorating situation:

> Intense and widespread criticism, controversial programmatic decisions by the Secretary of Energy, and a proliferation of substantial litigation have taken a toll on progress toward the goals of the program.[72]

---

[68] Department of Energy, *Recommendation by the Secretary of Energy Regarding the Suitability of the Yucca Mountain Site*, February 2002, p. 4, http://www.ocrwm.doe.gov/ym_repository/sr/sar.pdf.

[69] Department of Energy, Office of Civilian Radioactive Waste Management, *Draft Mission Plan for the Civilian Radioactive Waste Management Program*, DOE/RW-0005 DRAFT Volume 1, Washington, DC, April 1984, pp. 3-A-23.

[70] Carter, op. cit., p. 410.

[71] Mary Louise Wagner, "DOE Decision to Halt Second Repository Program Could Derail Entire Waste Act," *NuclearFuel*, June 2, 1986, p. 7.

[72] U.S. Congress, Senate Committee on Appropriations, *Energy and Water Development Appropriation Bill, 1987*, Report to accompany H.R. 5162, 99th Cong., 2nd sess., September 15, 1986, S.Rept. 99-441 (Washington: GPO, 1986), p. 157.

In addition to the controversy over site selection, it had become apparent that NWPA's timelines for characterizing the candidate sites and the anticipated cost of the characterization effort were unrealistic. With the future of the nuclear waste program in doubt, the 100th Congress decided to reopen NWPA for fundamental revision. The resulting NWPA Amendments Act of 1987 cancelled the second repository program, nullified DOE's selection of Oak Ridge for an MRS facility, and statutorily designated Yucca Mountain as the sole candidate site for a repository. Supporters of the legislation contended that characterizing only one site rather than three would be faster and save money, and noted that Yucca Mountain had been the most highly rated of the three candidates by DOE.[73] Some lawmakers, however, contended that the statutory designation of Yucca Mountain was made primarily for political reasons.[74]

The NWPA Amendments Act also provided for annual payments as an inducement to states for hosting nuclear waste facilities. States could receive up to $20 million per year for hosting a repository and $10 million for an MRS site if they agreed not to exercise their right under the law to disapprove those facilities. However, Nevada expressed no interest in the payments and, as noted previously, exercised its "state veto" in 2002. DOE did not conduct an MRS site search under the NWPA Amendments, relying instead on the Nuclear Waste Negotiator to find a voluntary site, as discussed earlier.

Although naming a single site for characterization was intended to speed up the development of a nuclear waste repository, the process actually took another 15 years after the 1987 Amendments, plus another five years to complete the license application to NRC. Supporters of the waste program contend that chronic underfunding by Congress was a major reason for the slow progress, while opponents primarily blamed DOE management problems. The State of Nevada was also able to slow the repository by denying state permits for various characterization activities and through successful lawsuits, such as the challenge to EPA's environmental standards. Nevada filed a 1,500-page petition with NRC in December 2008 to intervene in the Yucca Mountain license proceeding, raising dozens of safety, environmental, and other contentions.[75]

The history of U.S. efforts to site a nuclear waste repository illustrates the difficulty in successfully addressing local, state, and regional objections to such facilities. The United States did not succeed with the administrative process started by the Atomic Energy Commission, with the site-ranking system used for the NWPA first repository selection, the broad screening process used for the second repository, the benefits offered under the NWPA Amendments, or the voluntary selection process by the Nuclear Waste Negotiator. If Yucca Mountain is abandoned, that would arguably spell failure for the statutory designation method as well.

Just because those approaches were unsuccessful in the past does not mean they could not work in the future with program design modifications, better management, and changed circumstances. For example, it has been recently suggested that a negotiated benefits agreement with Nevada

[73] Department of Energy, Office of Civilian Radioactive Waste Management, *A Multiattribute Utility Analysis of Sites Nominated for Characterization for the First Radioactive-Waste Repository—A Decision-Aiding Methodology*, DOE/RW-0074, Washington, DC, May 1986, Chapter 5.

[74] Sen. Quentin Burdick, "Nuclear Waste Provisions," Remarks in the Senate, *Congressional Record*, vol. 133, part 26 (December 21, 1987), p. 37697.

[75] State of Nevada's Petition to Intervene as a Full Party in the Matter of Docket No. 63-001 (High Level Waste Repository), before the Nuclear Regulatory Commission, December 19, 2008. http://www.state.nv.us/nucwaste/news2008/pdf/nv081219nrc.pdf

might now be feasible, given the current economic downturn.[76] During the debate on the 1987 Amendments, Representative Morris Udall, Chairman of the House Interior Committee, contended that the original NWPA selection process would have worked had it been implemented properly:

> We created a principled process for finding the safest, most sensible places to bury these dangerous wastes. We were confident that while no State wanted a nuclear waste repository, the States ultimately chosen would accept the outcome because the selection process would have been fair and technically credible.

> Today, just 5 years later, this great program is in ruins. To help a few office seekers in the last election, the administration killed the eastern repository program, shattering the delicate regional balance at the heart of the 1982 act. Since then the Western States have felt they are being treated unfairly, and they no longer trust the technical integrity of the Department of Energy's siting decisions.[77]

Others have expressed doubt that a purely scientific and objective selection process is possible, given the inherent difficulties in making extremely long-term projections of repository behavior. The Director of the Office of Civilian Radioactive Waste Management recently described the siting of nuclear waste facilities as a "technically informed political decision."[78]

DOE's long but ultimately successful struggle to open a deep geologic repository for mid-level waste – the WIPP facility near Carlsbad, NM – indicates that siting of nuclear waste facilities is not necessarily impossible. State and local officials had invited AEC to consider the deep salt beds in the economically depressed area in the early 1970s for a high-level waste repository. After a great deal of statewide controversy, although with consistent local support, Congress authorized WIPP in 1979 to hold defense-related transuranic waste (P.L. 96-164).[79]

Transuranic (TRU) waste is not considered to be as hazardous as spent fuel and high-level waste, but it nevertheless requires long-term isolation in a geologic repository. TRU waste consists of relatively low-radioactivity material contaminated with more than a minimum concentration of long-lived plutonium.

DOE's efforts to implement the 1979 WIPP authorization were hampered by concerns by state officials that spent fuel and high-level waste would eventually be disposed of along with the transuranic waste.[80] After a dozen years of controversy over the project's implementation, Congress in 1992 enacted the Waste Isolation Pilot Plant Land Withdrawal Act (P.L. 102-579), detailing the regulations and procedures that DOE would have to follow to open the facility and banning high-level waste and spent fuel. Slow progress prompted Congress to amend the WIPP Land Withdrawal Act in 1996 to exempt WIPP waste from some land disposal restrictions and provide $20 million for New Mexico bypass roads for waste shipments (P.L. 104-201). The first

---

[76] Elaine Hiruo, "Funding Prospects Look Bleak for Yucca Project," *NuclearFuel*, January 12, 2009, p. 11.

[77] Rep. Morris Udall, House Debate, *Congressional Record*, vol. 133, part 26 (December 21, 1987), p. 37068.

[78] Edward F. Sproat III, Director, DOE Office of Civilian Radioactive Waste Management, speech to the Center for Strategic and International Studies, November 6, 2008.

[79] Carter, op. cit., p. 177.

[80] Carter, op. cit., p. 188.

waste was shipped to the repository in March 1999, nearly 20 years after the facility was authorized.[81]

It has recently been suggested that WIPP again be considered as a site for high-level waste disposal. Spent nuclear fuel could be more technically problematic, because the physical flow of the salt within a period of years will close in on stored waste and eliminate the option to retrieve the waste after 100 years or so, as could be done at Yucca Mountain.[82] Such "salt creep" occurs more quickly at higher temperatures, which could result from the disposal of high-level waste and spent fuel. A potential advantage of salt creep is that it can provide a natural seal around the waste.[83] Nevertheless, the State of New Mexico continues to strongly oppose any disposal of high-level waste at WIPP.[84]

Concluding Discussion

Significant scientific uncertainty – if not clear technical unsuitability – has arisen at every potential high-level nuclear waste repository site evaluated by the federal government. Such doubts have fed the public controversy that inevitably accompanies the announcement of such sites. As a result, the federal government has not succeeded in opening any central facilities for permanent disposal or interim storage of spent nuclear fuel and high-level waste.

The controversial nature of siting nuclear waste facilities increases the likelihood that alternatives to the proposed Yucca Mountain repository would leave waste at existing storage sites longer than under the current program schedule. Major consequences under current law could include increased liability by the federal government for utility storage costs, and fines and penalties for missing cleanup deadlines at defense-related nuclear facilities. Although NRC has determined that waste can be stored safely at reactor sites for many decades, the licensing of new plants could be affected by the lack of a definite disposal plan. Extremely long disposal delays would also increase the risk that adequate maintenance and security at storage sites would end before the waste could be removed.

Central interim storage of nuclear waste has regularly been suggested as the quickest way to begin moving waste from existing storage sites. However, without a plan for permanent disposal, the development of interim sites could be especially controversial. Reprocessing of spent fuel has long been proposed as a way to reduce the hazards of nuclear waste by removing plutonium and other long-lived radioactive material. While such a technological approach could make it easier to site a permanent repository, the separation of plutonium raises significant opposition because of its potential use in nuclear weapons and effects on U.S. nonproliferation policy. DOE is researching reprocessing techniques that could reduce the separation of pure plutonium, but their effectiveness and potential high cost continues to be a subject of controversy.

---

[81] Shawn Terry, "Waste Isolation Pilot Plant Opens Doors," *Inside Energy/with Federal Lands*, March 29, 1999, p. 1.

[82] Rick Michal, "James Conca: On WIPP and Other Things Nuclear," *Nuclear News*, February 2008, p. 44.

[83] D.J. Clayton, "Effects of Heat Generation on Nuclear Waste Disposal in Salt," American Geophysical Union Fall Meeting, abstract #H53A-1010, 2008, http://adsabs.harvard.edu/abs/2008AGUFM.H53A1010C.

[84] "New Mexico Bars High-Level Waste From Carlsbad Salt Caverns," *Environment News Service*, November 4, 2004.

The 1987 designation of Yucca Mountain as the nation's sole candidate site for a national high-level nuclear waste repository was a calculated risk that the site could be developed successfully. There is no backup plan in place. Yucca Mountain opponents contend that, as a result, the federal government has stuck with the site no matter what technical problems have been discovered. But if Yucca Mountain is determined to have significant problems, an alternative course will have little existing policy framework to build upon.

Author Contact Information

Mark Holt
Specialist in Energy Policy
mholt@crs.loc.gov, 7-1704

The Yucca Mountain Litigation: Breach of Contract Under the Nuclear Waste Policy Act of 1982

Todd Garvey
Legislative Attorney

December 22, 2009

Congressional Research Service

7-5700

www.crs.gov

R40996

CRS Report for Congress

Prepared for Members and Committees of Congress

The Yucca Mountain Litigation: Breach of Contract Under the Nuclear Waste Policy Act

Summary

Over 25 years ago, Congress addressed growing concerns regarding nuclear waste management by calling for federal collection of spent nuclear fuel (SNF) for safe, permanent disposal. To this end, the Department of Energy (DOE) was authorized by statute to enter into contracts with nuclear power providers that required the DOE to gather and dispose of spent nuclear fuel in exchange for payments by the providers into the newly established Nuclear Waste Fund (NWF). Congress subsequently named Yucca Mountain in the State of Nevada as the sole candidate site for the underground geological storage of collected SNF. Congress also mandated that federal disposal begin no later than January 31, 1998. Over 10 years ago, DOE breached these contracts by failing to begin the acceptance and disposal of SNF by the statutory deadline established in the Nuclear Waste Policy Act (NWPA). As a result, nuclear utilities have been forced to spend hundreds of millions of dollars for on-site temporary storage of toxic SNF that was expected to be transferred to the federal government for storage and disposal.

Seventy-one breach of contract claims have been filed against the DOE since 1998, resulting in approximately $1.2 billion in damages awarded thus far. Most of these awards, however, remain in appeals as the U.S. Court of Appeals for the Federal Circuit and the U.S. Court of Appeals for the District of Columbia Circuit engage in a dispute over each other's jurisdictional authority. Estimates for the total potential liability incurred by the DOE as a result of the Yucca Mountain litigation range as high as $50 billion. Moreover, after decades of political, legal, administrative, and environmental delays, the Obama Administration, with the support of Congress, defunded the Yucca Mountain project for FY2010, and announced an intention to pursue other alternatives for the disposal of SNF. Accordingly, contract damages will continue to build as there seems to be no prospect for a completed facility capable of storing SNF anywhere on the horizon.

DOE's liability for breach of contract was first established in 1996 by the U.S. Court of Appeals for the District of Columbia in *Indiana Michigan Power Co. v. U.S.* After DOE hesitated to act on its legal obligations, citing the absence of a completed SNF storage facility, the court issued a writ of mandamus mandating that DOE "proceed with contractual remedies in a manner consistent with NWPA's command that it undertake an unconditional obligation to begin disposal of SNF by January 31, 1998." The mandamus, issued in *Northern States Power Co. v. U.S.*, essentially prohibited the DOE from deflecting liability by arguing that the lack of an existing storage facility constituted an "unavoidable delay."

In 2006, the U.S. Court of Federal Claims (CFC) held that the D.C. Circuit mandamus order in *Northern States* was void for lack of jurisdiction and could not preclude the DOE from raising the "unavoidable delay" defense in the former's court. The case was appealed to the Federal Circuit, where the court recently granted an *en banc* hearing to resolve the jurisdictional question. If the Federal Circuit affirms the CFC, and voids the D.C. Circuit mandamus, DOE liability under the NWPA could be drastically reduced if the department can successfully show that the lack of an operational storage facility constitutes an "unavoidable delay."

This report will present a brief overview of the NWPA and its subsequent amendments, provide a survey of key issues that have emerged from the protracted waste storage litigation, describe the ongoing jurisdictional conflict between the D.C. Circuit and the U.S. Court of Federal Claims, and consider the potential for future liability arising from delays relating to the storage and disposal of nuclear waste.

Congressional Research Service

The Yucca Mountain Litigation: Breach of Contract Under the Nuclear Waste Policy Act

Contents

Figures

Contacts

Congressional Research Service

The Yucca Mountain Litigation: Breach of Contract Under the Nuclear Waste Policy Act

Introduction

Over 25 years ago, Congress addressed growing concerns regarding nuclear waste management by calling for federal collection of spent nuclear fuel (SNF) for safe, permanent disposal. To this end, the Department of Energy (DOE) was authorized by statute to enter into contracts with nuclear power providers that required the DOE to gather and dispose of spent nuclear fuel[1] in exchange for payments by the providers into the statutorily established Nuclear Waste Fund (NWF). Congress subsequently named Yucca Mountain in the State of Nevada as the sole candidate site for the underground geological storage of collected SNF. Congress also mandated that federal disposal begin no later than January 31, 1998. Over 10 years ago, DOE breached these contracts by failing to begin the acceptance and disposal of SNF by the statutory deadline established in the Nuclear Waste Policy Act (NWPA). As a result, nuclear utilities have been forced to spend hundreds of millions of dollars on temporary storage for toxic SNF that was expected to be transferred to the federal government for storage and disposal.[2] The breach has triggered a prolonged series of suits by nuclear power providers, most of which continue unresolved to this day.

Seventy-one breach of contract[3] claims have been filed against the DOE since 1998, resulting in approximately $1.2 billion in damages and settlements thus far.[4] Most of these awards, however, remain in appeals as the U.S. Court of Appeals for the Federal Circuit and the U.S. Court of Appeals for the District of Columbia Circuit engage in a dispute over each other's jurisdictional authority. Estimates for the total potential liability incurred by the DOE as a result of the Yucca Mountain litigation range as high as $50 billion.[5] Moreover, after decades of political, legal, administrative, and environmental delays, the Obama Administration, with the support of Congress, has defunded the further development of the Yucca Mountain project for FY2010, and announced an intention to pursue other alternatives for the disposal of SNF. Accordingly, contract damages will continue to build as there seems to be no prospect for a completed facility capable of storing SNF anywhere on the horizon.[6]

---

[1] Spent nuclear fuel consists of radioactive fuel rods, containing uranium and plutonium, that have been permanently withdrawn from a nuclear reactor because they can no longer efficiently sustain a nuclear chain reaction. See, CRS Report RL 33461, *Civilian Nuclear Waste Disposal*, by Mark Holt, at 8.

[2] U.S. nuclear power plants spend hundreds of millions of dollars a year to store radioactive SNF at the bottom of 40-feet deep pools or in "dry casks" located outside of the facility. Steve Hargreaves, *Nuclear Waste: Coming to a Town Near You?*, CNNMoney.com, November 4, 2009, *available at*: http://www.money.cnn.com.

[3] Each one of these claims included a Fifth Amendment takings claim in addition to the breach of contract claim. The takings claims, however, were dismissed early in the litigation. *See e.g.*, Consumers Energy Co. v. U.S., 84 Fed. Cl. 152 (2008).

[4] The $1.2 billion consists of approximately $400 million in out-of-court settlements and approximately $800 million in damages awarded by the Court of Federal Claims. Of the $1.2 billion, the federal government has paid only $565 million in settlements and damages. The remaining judgments are in the appeals process and are not yet final. *See*, Compilation of Office of General Counsel Materials Provided to the President-Elect's DOE Transition Team, at 46, *available at*: http://www.management.energy.gov/documents; Statement of Kim Cawley, Chief, Natural and Physical Resources Costs Estimates Unit, Congressional Budget Office Before the House Committee on the Budget, July 16, 2009 (hereinafter *CBO Testimony*).

[5] Marcia Coyle, *Nuclear Dispute Fallout*, The National Law Journal, September 14, 2009.

[6] See, *CBO Testimony*, at 1 ("The Department of Energy has not yet disposed of any civilian nuclear waste and currently has no identifiable plan for handling that responsibility.").

This report analyzes the more than 13 years of ongoing litigation over the government's obligations to collect and dispose of SNF under the Nuclear Waste Policy Act of 1982.[7] Part I will provide a brief overview of the NWPA and its subsequent amendments. Part II will provide a survey of key issues that have emerged from the protracted litigation and describe the ongoing jurisdictional conflict between the D.C. Circuit on the one hand, and the Federal Circuit and U.S. Court of Federal Claims on the other. Part III will describe the current Administration's plan to develop alternatives to nuclear waste storage at Yucca Mountain and consider the potential costs of the Yucca Mountain project.

Part I: The Road to Litigation

The Nuclear Waste Policy Act of 1982

Identifying the serious hazards of nuclear waste, Congress passed the Nuclear Waste Policy Act of 1982 (NWPA) in an effort to centralize the long-term management of nuclear waste by making the federal government responsible for collecting, transporting, storing and disposing of the nation's SNF.[8] In order to achieve this goal, the NWPA established a statutory system for selecting a site for a geologic repository for the permanent disposal of nuclear waste.[9] The DOE was authorized by the statute to carry out the disposal program[10] and develop the permanent nuclear waste repository. Commercial nuclear power owners and operators would fund a large portion of the program through significant annual contributions, or fees, to the newly established Nuclear Waste Fund (NWF).[11]

To carry out the statutory scheme created by the NWPA, the DOE was also authorized to enter into contracts with nuclear facilities to allow the department to take possession of nuclear waste and ensure its storage and disposal in the prospective permanent repository.[12] Section 302 of the NWPA sets out the critical statutory deadline established in the NWPA and forms the main basis for litigation. This provision explicitly mandates:

> (A) Following commencement of operation of a repository, the Secretary shall take title to the high-level radioactive waste or spent nuclear fuel involved as expeditiously as practicable upon the request of the generator or owner of such waste or spent fuel; and

> (B) In return for payment of fees established by this section, the Secretary, *beginning not later than January 31, 1998,* will dispose of the high-level radioactive waste or spent nuclear fuel involved as provided in this subtitle.[13]

---

[7] This report does not discuss the significant amount of environmental litigation relating to the licensing of the Yucca Mountain facility by the Nuclear Regulatory Commission.

[8] P.L. 97-425, *codified at* 42 U.S.C. §§ 10101 *et seq.*

[9] *Id.* at §§ 111-125.

[10] The NWPA created the Office of Civilian and Radioactive Waste Management to carry out the DOE's obligations under the act. *Id.* at § 304.

[11] *Id.* at § 302.

[12] *Id.* at § 302(a).

[13] *Id.* at §302(a)(5) (emphasis added).

In an effort to streamline the collection and disposal process, the DOE elected to create a single "Standard Contract for Disposal of Spent Nuclear Fuel and/or High Level Radioactive Waste" (Standard Contract) for use with all nuclear power providers. DOE chose to develop the Standard Contract through the formal notice-and-comment rulemaking process. The final contract, published in the Federal Register, somewhat modified the language of the NWPA and provides:

> The services to be provided by DOE under this contract shall begin, after commencement of facility operations, not later than January 31, 1998 and shall continue until such time as all SNF ... has been disposed of.[14]

Although the NWPA did not expressly mandate that all nuclear utility providers enter into an agreement with the DOE for the disposal of nuclear waste, the utilities were required to enter into the Standard Contract as a condition of renewing or obtaining the required operating license from the Nuclear Regulatory Commission.[15] All operating nuclear facilities, therefore, became parties to the Standard Contract.

By 1987, pursuant to its obligations under the NWPA, DOE had identified three potential sites for the permanent repository: Yucca Mountain; Hanford, Washington; and Deaf Smith County, Texas. In 1987, Congress amended the NWPA to name Yucca Mountain as the sole candidate site for the permanent repository.[16] The amendments, strongly lobbied for by the congressional delegations from Washington and Texas, did not, however, end the DOE selection and approval process which continued as outlined under the NWPA.

Breach of the Standard Contract

By 1993, the DOE had made little progress in preparing to take possession of SNF, and the Yucca Mountain facility was at least a decade or more away from completion. Concerned as to whether DOE would be able to meet its contractual obligations by the end of January 1998, the utilities, which had been paying into the NWF for 11 years,[17] requested in writing that the DOE address its responsibilities under the NWPA and update the signatories of the Standard Contract on DOE's overall preparedness. DOE initially responded to this request with an informal letter, stating that DOE's interpretation of the contract was that the department had no obligations under the contract until the permanent repository at Yucca Mountain was complete.[18]

In response to this interpretive dispute, DOE sought comments from the public on the department's statutory obligations under the NWPA and the Standard Contract. After further review, DOE issued a "Final Interpretation of Nuclear Waste Acceptance Issues" which formally pronounced the department's position that it had no "legal obligation under either the [NWPA] or the Standard Contract to begin disposal of SNF by January 31, 1998, in the absence of a repository or interim storage facility."[19] Pursuant to this interpretation, the department added that it would not begin accepting nuclear waste from nuclear utilities by the date specified in the act,

---

[14] 10 C.F.R. § 961.11.

[15] 42 U.S.C. § 10222(b)(1)(A).

[16] 42 U.S.C. § 10172.

[17] Fees by nuclear providers into the NWF have been estimated at $750 million annually. *CBO Testimony*, at 3.

[18] *See,* Indiana Michigan Power Co. v. U.S., 88 F.3d 1272, 1274 (D.C. Cir. 1996).

[19] 60 Fed. Reg. 21,793-94

nor did it have authority under the NWPA to provide interim storage for spent nuclear fuel.[20] In the alternative, were section 302 to create an unconditional obligation on the part of DOE to begin disposing of nuclear waste by January 31, 1998, redress should be governed by the "unavoidable delay" provisions of the Standard Contract which expressly states that "no party shall be liable for damages in the case of unavoidable delay ..."[21]

Nuclear utility companies, having paid billions into the NWF since 1982[22] in addition to the millions spent for on-site temporary storage, turned to the federal courts to review DOE's interpretation of its own obligations under the NWPA and the Standard Contract.

Part II: Litigation

As of May 2009, 71 lawsuits had been filed against the DOE related to the department's failure to commence the collection and disposal of SNF. Of the 71 filed lawsuits, 10 have been settled, 6 were withdrawn, 4 reached final judgment, and 51 remain pending.[23] According to the Congressional Budget Office, the government's current liability—based on settlements, final judgments, and entered judgments under appeal—stands at $1.3 billion.[24] The following section will highlight key court decisions that have emerged from the ongoing contractual dispute between DOE and the nuclear power utilities.

DOE's Statutory Obligation to Begin Accepting SNF

The first NWPA-related claim against the DOE was filed in the U.S. Court of Appeals for the District of Columbia Circuit in 1996.[25] Although the DOE had not yet breached the contract, as performance was not required before January 31, 1998, Indiana Michigan Power Company sought a preemptive judicial review of the department's determination that it had no obligation to begin accepting SNF until the completion of the Yucca Mountain facility.

In *Indiana Michigan Power Co. v. Department of Energy*, the D.C. Circuit, applying the *Chevron*[26] analysis for reviewing an agency's statutory interpretation, invalidated DOE's interpretation as contrary to the plain meaning of the NWPA.[27] The court reasoned that section 302(A) and section 302(B) represented independent statutory obligations. While the obligation to "take title to" nuclear waste in section 302(A) may have been conditioned on the construction of a repository, the obligation to "dispose" of nuclear waste under section 302(B) contained no such

---

[20] *Id.* at 21,794.

[21] *Id.* at 21,797.

[22] As of July 1, 2009, fees paid into the NWF totaled $16.3 billion. The NWF has also received $12.8 billion in intergovernmental transfers. The Congressional Budget Office predicted the NWF's balance at the end of FY2009 would be $23.8 billion. *CBO Testimony*, at 3.

[23] *Id.* at 6-7.

[24] *Id.*

[25] *Indiana Michigan Power*, 88 F.3d 1272.

[26] Under the *Chevron* doctrine, a court will defer to an agency's interpretation of an ambiguous statute where the agency's "answer is based on a permissible construction of the statute." Chevron U.S.A. Inc. v. Natural Resources Defense Council, 467 U.S. 837 (1984).

[27] *Id.* at 1274.

limitation.[28] Indeed, DOE's duty to commence disposal of nuclear waste, held the court, was to begin "not later than January 31, 1998 without qualification or condition."[29] The argument put forth by DOE, and rejected by the court, was that section 302(A) and section 302(B) "must be read together," since taking title to SNF cannot be separated from disposing of SNF.[30] In construing the DOE's "disposal" obligation broadly, the court noted that "it is not unusual, particularly in the nuclear area, to recognize a division between ownership of materials and other obligations relating to such materials."[31] The court concluded that the NWPA and Standard Contract had created a "reciprocal" and binding contractual relationship between the DOE and the nuclear utilities, whereby DOE would dispose of the utilities' nuclear waste in return for the payment of fees into the NWF.[32]

The DOE did not immediately take action in response to the D.C. Circuit's holding in *Indiana Michigan*. Instead, the department informed the nuclear utilities involved that it would be unable to comply with the January 31, 1998, deadline and was not prepared to begin accepting spent nuclear fuel for disposal.[33] DOE asserted that it was waiting for the results of the Yucca Mountain Project Viability Assessment before proceeding, but predicted that the Yucca Mountain facility could potentially be opened by 2010.[34]

Prohibiting the "Unavoidable Delay" Defense

In addition to informing the nuclear utilities that it would be unable to comply with the January 31, 1998, deadline, the DOE also asserted that the department was not responsible for any monetary damages incurred by the utilities as a result of DOE's delay.[35] The department had determined that the lack of a permanent repository at Yucca Mountain constituted an "unavoidable delay" under article IX of the Standard Contract.[36] The "unavoidable delay" provision of the Standard Contract provides:

> Neither the Government nor the purchaser shall be liable under this contract for damages caused by failure to perform its obligations hereunder, if such failure arises out of the causes beyond the control and without the fault or negligence of the party failing to perform.[37]

As such, argued DOE, the terms of the Standard Contract relieved the department from any obligation to "provide a financial remedy for the delay."[38]

---

[28] *Id*. at 1276.

[29] *Id*.

[30] *Id*. ("DOE next argues that subsections (A) and (B) of 302(a)(5) are not independent provisions, but rather must be read together.").

[31] *Id*.

[32] *Id*. at 1277 ("Thus we hold that section 302(a)(5)(B) creates an obligation in DOE, reciprocal to the utilities' obligation to pay, to start disposing of the SNF no later than January 31, 1998.").

[33] *See*, Northern States Power Co. v. U.S., 128 F.3d 754, 757 (D.C. Cir. 1997).

[34] *Id*.

[35] *Id*.

[36] 10 C.F.R. § 961.11.

[37] *Id*. The provision continues: "In the event circumstances beyond the reasonable control of the Purchaser or DOE— such as acts of God, or of the public enemy, acts of Government in either its sovereign or contractual capacity, fires, floods, epidemics, quarantine restrictions ... cause delay in scheduled delivery acceptance or transport of SNF ... the parties will readjust their schedules, as appropriate, to accommodate such delay."

The nuclear utilities responded to DOE's communications in 1997 by asking the D.C. Circuit to issue a writ of mandamus, compelling DOE to adhere to the court's earlier decision in *Indiana Michigan* and begin accepting nuclear waste for disposal. In *Northern States Power Co. v. U.S.*, the court refused to grant the "drastic" and broad relief the utilities asked for, holding that the terms of the Standard Contract provided for another "potentially adequate remedy."[39] Before the court would consider compelling DOE to act, the utilities would first have to pursue the administrative remedies available under the Standard Contract for delayed performance.[40]

However, the court was unwilling to accept DOE's interpretation of its own delays as "unavoidable" under the Standard Contract. The court reiterated, in rejecting DOE's argument that a lack of an operational repository qualified as an unavoidable delay, that DOE's obligation to begin disposal of SNF by January 31, 1998, existed regardless of the existence of an operational storage facility.[41] DOE's "unavoidable delay" defense, noted the court, represented a simple "recycling [of] the arguments [previously] rejected by this court."[42] Based on the DOE's "repeated attempts to excuse its delay on the grounds that it lacks an operational repository," the D.C. Circuit, in a significant exercise of authority, issued a writ of mandamus prohibiting the DOE from concluding that the lack of an operational permanent repository constituted an "unavoidable delay" under the Standard Contract.[43] The court ordered DOE to "proceed with contractual remedies in a manner consistent with NWPA's command that it undertake an unconditional obligation to begin disposal of the SNF by January 31, 1998."[44]

In a preview of the jurisdictional dispute that would develop a decade later, DOE filed a petition for rehearing in response to the *Northern States* mandamus. DOE challenged the D.C. Circuit's exercise of authority by asserting that the court "lacked jurisdiction to construe the unavoidable delays clause of the Standard Contract," as such an interpretation of a government contract was squarely within the jurisdiction of the Court of Federal Claims under the Tucker Act.[45] The D.C. Circuit denied the motion for rehearing, holding that the court had not adjudicated a contractual dispute, but rather issued the mandamus in an effort to enforce a statutory duty.[46] Accepting the D.C. Circuit's reasoning, DOE interpreted the *Northern States* mandamus as prohibiting the department from raising the unavoidable delay clause as a defense in any future litigation.[47]

(...continued)

[38] *Northern States*, 128 F.3d at 757.

[39] *Id.* at 758,761. The remedy that was considered "potentially adequate" in Northern States was later deemed "inadequate" in Maine Yankee Atomic Power Co. v. U.S. 225 F.3d 1336 (Fed. Cir. 2000).

[40] The remedy available under the contract allows for an equitable adjustment of charges and schedules. 10 C.F.R. 961.11.

[41] *Northern States*, 128 F.3d at 760.

[42] *Id.*

[43] *Id.*

[44] *Id.*

[45] Northern States Power Co. v. United States, 1998 U.S. App. LEXIS 12919 (D.C. Cir. May 5, 1998); Tucker Act, 28 U.S.C. § 1491(a).

[46] *Id.* ("The Tucker Act does not prevent us from exercising jurisdiction over an action to enforce compliance with the NWPA.").

[47] *See, e.g.* Yankee Atomic Electric Company v. U.S., 42 Fed. Cl. 223 (1998) ("As a result, DOE maintains that it is prohibited from arguing that its failure to begin SNF disposal services is an unavoidable, non-compensable delay under Article IX.A of the Standard Contract.").

Litigation Continues: Remedies, Offsets, and Damages

After establishing the DOE's statutory obligations under the NWPA in the D.C. Circuit, many nuclear utilities awaited the expiration of the January 31, 1998, deadline before seeking monetary damages by filing their breach of contract claims in the U.S. Court of Federal Claims (CFC).[48] Under the Tucker Act, the CFC has jurisdiction over monetary claims against the United States "founded either upon the Constitution, or any Act of Congress or any regulation of an executive department, or upon any express or implied contract with the United States."[49] Decisions of the CFC are appealed to the Federal Circuit.

In considering the cases, the CFC initially had to answer the threshold question of whether the nuclear utilities were required to exhaust available administrative remedies under the Standard Contract prior to seeking judicial relief. Generally, if administrative remedies can provide adequate relief for a breach of contract claim, the plaintiff must first exhaust those remedies before seeking redress in another court.[50] Judges on the CFC came to opposite conclusions as to whether the Standard Contract could provide adequate relief to the nuclear utilities, and the issue was left for the Federal Circuit to settle on appeal.[51]

Remedies Under Standard Contract Inadequate

In an important 2000 case, entitled *Maine Yankee Atomic Power Co. v. U.S.,* the Federal Circuit concluded that adequate relief was not available to the nuclear utilities under the Standard Contract, a conclusion that would allow breach of contract claims against the DOE to go forward in the CFC.[52] DOE, with the "unavoidable delay" clause unavailable, argued that the "avoidable delays" clause of the contract provided the plaintiffs with an avenue for adequate administrative relief.[53] The "avoidable delay" provision of the Standard Contract requires that:

> in the event of any in the delivery, acceptance, or transport … caused by circumstances within the reasonable control of either [party] … the charges and schedules specified by this contract will be equitably adjusted to reflect any estimated additional costs incurred by the party not responsible for or contributing to the delay.[54]

The court disagreed, holding that the "avoidable delay" provision applied only to routine delays occurring after the parties had begun performance of their obligations under the contract, not to breaches of a "critical and central obligation of the contract," such as a failure to begin

---

[48] The D.C. Circuit, though retaining jurisdiction over review of final agency actions, rejected the notion that the U.S. Courts of Appeals had jurisdiction over breach of contract claims under the NWPA, holding that the "Court of Federal Claims, not this court, is the proper forum for adjudicating contract disputes." Wisconsin Elec. Power v. U.S. Dep't of Energy, 211 F.3d 646, 647 (D.C. Cir. 2000).

[49] 28 U.S.C. § 1491.

[50] *See,* McKart v. U.S. 395 U.S. 185, 193 (1969) ("No one is entitled to judicial relief … until the prescribed administrative remedy has been exhausted.").

[51] *See,* Yankee Atomic Elec. Co. v. U.S., 42 Fed. Cl. 223 (1998) (holding available administrative relief was not adequate); Northern States Power Co. v. U.S. 224 F.3d 1361 (Fed. Cl. 2000) (holding available administrative relief was adequate).

[52] 225 F.3d 1336 (Fed. Cir. 2000).

[53] *Id.* at 1341-1342.

[54] Standard Contract, 10 C.F.R. § 961.11.

---

performance by the statutory deadline.[55] The court added that relief in the form of a "charge or schedule adjustment," as provided under the Standard Contract, was wholly inadequate to compensate the nuclear utilities for damages they had sustained in storing spent nuclear fuel that had been covered by the contract.[56] As a result of the *Maine Yankee* decision, signatories to the Standard Contract were now free to seek monetary damages against the DOE, by filing their breach of contract claims in the CFC, without first exhausting the DOE administrative process.

NWF Offset Invalid

Following the *Indiana Michigan Power* and *Maine Yankee* decisions, and the realization that a slew of breach of contract claims were being filed in the CFC, the DOE attempted to curtail its potential contract liability by modifying contract terms with individual nuclear utilities. Under the proposed modification, DOE was willing to return a portion of payments made by a utility into the NWF and suspend any future payments, if the utility was willing to relinquish all future claims against the DOE.[57] The department entered into one such agreement with Exelon Generation Company in 2002. Other utilities that had also contributed to the NWF, however, challenged this arrangement as an invalid use of NWF funds.

The 11th Circuit, in *Alabama Power Co. v. U.S.*, invalidated the contractual modification reached between DOE and Exelon Generation Company.[58] The agreed upon "offset," the court held, was "tantamount to an expenditure of funds" from the NWF.[59] Under the NWPA, NWF funds were to be used only for the "disposal" of nuclear waste.[60] DOE could not, therefore, allocate NWF funds to individual nuclear utilities to pay for what the court classified as on-site "interim storage." Were DOE allowed to use NWF funds to offset the costs of the department's failure to dispose of SNF, it would be analogous to allowing the DOE to "pay for its own breach out of a fund paid for by the utilities."[61] Any arrangement in which the utilities were made to "bear the costs of the [department's] breach" was invalid.[62]

Calculating Damages

Although the DOE had acknowledged its partial breach of the Standard Contract by 2005, the department was still intent on limiting its overall liability.[63] In *System Fuels Inc. v. U.S.*, DOE

---

[55] Maine Yankee Atomic Power Co. v. U.S., 225 F.3d 1336, 1341-42 (Fed. Cir. 2000). The Federal Circuit considered another case on the same day, citing the *Maine Yankee* decision, and dealing with the "avoidable delay" clause. Northern States Power Co. v. U.S., 224 F.3d 1361 (Fed. Cir. 2000). The language of the opinion however, referred to the "unavoidable delay" provision as not inapplicable. Although DOE argues the court's reference to "unavoidable delays" was "inadvertent," the CFC has stated that "pending any precedential ruling to the contrary, the court is bound by the Federal Circuit's language." S. Nuclear Operating Co. v. U.S., 77 Fed. Cl. 396 (2007).

[56] *Id.* at 1342.

[57] *See,* Alabama Power Co. v. U.S., 307 F.3d 1300, 1306 (11th Cir. 2002).

[58] *Id.* at 1315. The case was brought in the Eleventh Circuit, rather than the CFC, because the issue was a statutory question on the permissible use of NWF funds under the NWPA and not a breach of contract claim.

[59] *Id.* at 1312.

[60] *Id.* at 1313 ("An expenditure on interim storage is not an act of 'disposal.'").

[61] *Id.* at 1314.

[62] *Id.*

[63] System Fuels Inc. v. U.S., 66 Fed. Cl. 722, 730 (2005) ("The government admitted on February 10, 2005 that 'DOE's delay in beginning acceptance of SNF … constitutes a partial breach of the Standard Contract.'")

argued that the date of breach, the date from which damages would be calculated, was not January 31, 1998. Rather, the department argued the date should be determined by the scheduling provisions of the contract that were to be used in determining when pick-up and delivery of SNF was to occur.[64] Under the Standard Contract, DOE was to publish Annual Capacity Reports through which individual utilities could submit Delivery Commitment Schedules (DCS) to notify DOE of the amount of SNF that required DOE disposal and to schedule SNF pick-ups.[65] According to the DOE, most utilities had not yet submitted a DCS to schedule an initial pick-up date. The CFC dismissed the argument on summary judgment, holding that the scheduling provisions were "only created for planning purposes" and not binding on either party.[66] Additionally, DOE had consistently delayed or refused to accept DCSs submitted by nuclear utilities, an action that itself would have constituted a breach of contract were the scheduling provisions binding.[67] A party's failure to submit a DCS was therefore "irrelevant" and damages would be calculated from the statutory deadline of January 31, 1998.[68]

In August of 2008, the Federal Circuit further clarified the method for calculating damages in NWPA breach of contract suits by establishing the rate at which the DOE was expected to accept SNF under the Standard Contract.[69] The anticipated rate of acceptance was essential to calculating the total amount of SNF DOE was contractually obligated to accept from the nuclear utilities from the 1998 deadline forward. DOE argued for a lower rate established under a report issued in 1991, as opposed to the initial rate of acceptance established in a 1987 DOE scheduling report.[70] The Federal Circuit, however, rejected this argument, holding instead that damages would be calculated in relation to the higher 1987 acceptance rate, as that rate most closely reflected the intent and expectations of the parties at the time of the contract.[71] The 1991 rate, held the court, was most likely the result of a "litigation strategy," put forth to "minimize DOE's exposure for its impending breach, rather than as a realistic, good faith projection for waste acceptance."[72]

Damages for the failure to accept SNF would thus be calculated from January 31, 1998, at the initial expected rate of acceptance established in 1987.

Jurisdictional Dispute Develops

From 1998 forward, the CFC had been entertaining breach of contract suits filed by the nuclear utilities against DOE without any significant discussion of jurisdiction. Then, in 2005, the court, for the first time, dismissed an NWPA breach of contract suit for lack of subject matter jurisdiction.[73] The court reasoned that the Standard Contract, created through the formal

---

[64] 66 Fed. Cl. 722 (2005).

[65] *Id.* at 730.

[66] *Id.* at 732.

[67] *Id.* at 731.

[68] *Id.* at 732.

[69] Pacific Gas & Elec. Co. v. U.S., 536 F.3d 1282 (Fed. Cir. 2008).

[70] The initial predicted rate of acceptance was 1200 MTU/year in 1998, 2000 MTU/year by 2003, and 2650 MTU/year by 2004. The proposed 1991 rate of acceptance schedule reduced those numbers to 300 MTU/year in 1998, 875 MTU/year in 2001, and 1800 MTU/year by 2010. *Id.*

[71] *Id.* at 1291-92.

[72] *Id.*

[73] Florida Power and Light Co. v. U.S., 64 Fed. Cl. 37 (2005).

administrative process, qualified as a final agency action under the jurisdiction of the U.S. Court of Appeals as established pursuant to section 119 of the NWPA. Section 119 of the NWPA grants the U.S. courts of appeals:

> original and exclusive jurisdiction over any civil action ... for review of any final decision or action of the Secretary, the President, or the Commission under this subtitle.[74]

The dismissal was appealed to the Federal Circuit for review of the jurisdictional question.

In *PSEG Nuclear v. U.S.*, the Federal Circuit reversed the lower court's decision, holding that the NWPA had not stripped the CFC of jurisdiction over contract disputes.[75] The court based its holding on the fact that section 119 was limited to official agency action taken under Title I of the NWPA. Title I relates only to the development of a nuclear waste repository. The breach of contract suits filed by the utilities, on the other hand, were based on the expiration of the statutory deadline found in Title III of the NWPA.[76] The jurisdictional grant found in Title I of the NWPA was, therefore, determined to be inapplicable to a breach of contract claim arising under Title III of the NWPA.[77] After *PSEG*, it was clear that the CFC had the authority to exercise jurisdiction over an NWPA-related breach of contract claim. However, because the court in *PSEG* limited itself only to whether the exercise of jurisdiction by the CFC was proper,[78] the larger question of whether the D.C. Circuit's previous exercise of jurisdiction over similar contract-related claims impermissibly infringed on the CFC's jurisdiction remained unresolved.

Court of Federal Claims Overturns a Decade of Precedent

Shortly after the *PSEG* decision, which ensured the CFC's jurisdiction over contract disputes arising under the NWPA, the DOE asked the CFC to invalidate the D.C. Circuit's initial exercise of jurisdiction in *Indiana Michigan*. At oral argument in *Nebraska Public Power Dist. v. U.S.*, DOE expressed a desire to raise the "unavoidable delay" defense that the D.C. Circuit had specifically prohibited through the writ of mandamus in *Northern States*.[79] The CFC decided to entertain the question and asked the parties to brief the issue of whether the D.C. Circuit mandamus precluded DOE's assertion of the "unavoidable delay" defense in the CFC. On October 31, 2006, the court handed down a sweeping decision that voided the longstanding mandamus issued by the D.C. Circuit for lack of jurisdiction.[80]

In *Nebraska Public Power*, the CFC held that the D.C. Circuit had exceeded its jurisdiction in issuing the *Indiana Michigan* decision.[81] Since the mandamus prohibiting the DOE's use of the "unavoidable delay" defense issued in *Northern States* was issued as a means of enforcing the

---

[74] P.L. 97-425 § 119.

[75] 465 F.3d 1343 (Fed. Cir. 2006).

[76] *Id.*

[77] *Id.* at 1348 ("Therefore, section 119 of the NWPA confers jurisdiction over agency actions taken during development of a repository for SNF disposal.").

[78] *Id.* ("The difference in the parties' positions amounts to whether the courts of appeals continue to have jurisdiction to decide the propriety of agency actions ... because this issue need not be resolved in this appeal, we merely agree ... that the NWPA does not strip the court of its Tucker Act jurisdiction.").

[79] 73 Fed. Cl. 650 (2006).

[80] *Id.*

[81] *Id.*

ruling in *Indiana Michigan*, the mandamus, therefore, was also void and had no preclusive effects in the CFC. The court based its decision on the jurisdictional conclusions underlying *PSEG*, the limited scope of section 119 of the NWPA, and the absence of an effective waiver of sovereign immunity.

Defining the Jurisdiction of the CFC and U.S. Appellate Courts

Nebraska Public Power focused on whether the string of claims filed under the NWPA and the Standard Contract should be classified as a review of formal agency action within the direct purview of the U.S. Appellate Courts, or as a straightforward breach of contract claim within the exclusive jurisdiction of the CFC (subject to appeal to the Federal Circuit). The opinion made clear the CFC's position that the claims relating to the January 31, 1998, statutory deadline qualified as contract claims within the CFC's exclusive jurisdiction.[82] In considering the jurisdictional role of the two courts, the CFC adopted and applied much of the reasoning behind *PSEG*, asserting that the case had "rejected many of the key jurisdictional concepts that underlie the relevant D.C. Circuit cases."[83] Although *PSEG* focused only on whether the CFC could exercise jurisdiction over the contract claims, the court in *Nebraska Public Power* went further to establish that jurisdiction as exclusive in an attempt to resolve the two competing claims to jurisdiction over cases related to the Standard Contract.[84]

The Scope of Section 119 of the NWPA

In issuing the *Northern States* mandamus, the D.C. Circuit had invoked section 119, which granted the U.S. Appellate Courts exclusive jurisdiction over final agency action under Title I of the NWPA, as the basis for its exercise of jurisdiction. However, the Federal Circuit, reviewing the exercise of jurisdiction by the CFC, had limited the scope of section 119, based on the provision's plain language, to only those claims relating to the establishment of a permanent repository for spent nuclear fuel.[85] In *Nebraska Public Power*, the CFC adopted the reasoning in *PSEG*, and applied it to the D.C. Circuit's initial exercise of jurisdiction in *Indiana Michigan*. The resulting conclusion was that *Indiana Michigan* involved "interpretations of contract provisions that have nothing to do with the creation of repositories of spent nuclear fuel," and therefore "plainly exceeded" the grant of jurisdiction to the D.C. Circuit under section 119.[86]

Contrary to the D.C. Circuit's argument that the *Northern States* mandamus was issued pursuant to a breach of a statutory and regulatory obligation, the court added that the "essential character" of the actions brought by the nuclear utilities was contractual and therefore exclusively within the jurisdiction of the CFC.[87] The mere fact that DOE developed the Standard Contract through formal administrative rulemaking procedures was not sufficient to alter the nature of the claim

---

[82] *Id.* at 664 ("in describing where the [Federal Circuit's] jurisdiction begins, the federal circuit *sub silentio* described where the D.C. Circuit's jurisdiction ends, *to wit*, that the latter court's jurisdiction does not extend beyond reviewing agency actions under Title III that relate to the creation of the repository.").

[83] *Id.* at 662.

[84] *Id* at 665 ("The decisions in Indiana Michigan and Northern states bounded across the [jurisdictional] line, thereby intruding on this court's jurisdiction.").

[85] *Id.* at 664-666.

[86] *Id.* at 664.

[87] *Id.* at 665.

from an action based on contract to an action based on statutory or regulatory interpretation.[88] In classifying the claims in *Indiana Michigan* and *Northern States* as contractual, the court emphasized the utilities' reliance on the Standard Contract, the asserted claim for breach of contract, and the request for monetary damages.[89] As the "mandamus dispute in *Northern States* could be conceived as entirely contained within the terms of the contract" rather than a "regulation asserted to be in conflict with the NWPA," the D.C. Circuit had engaged in an interpretation of the Standard Contract that intruded on the CFC's exclusive jurisdiction.[90]

Waiver of Sovereign Immunity Under Section 702 of the APA

The CFC also held that the D.C. Circuit's decisions in *Indiana Michigan* and *Northern States* were not supported by a waiver of sovereign immunity.[91] Even if section 119 had granted the D.C. Circuit jurisdiction over the NWPA contract claims, the grant of jurisdiction was not accompanied by any waiver of sovereign immunity that would allow the case to go forward. Federal courts do not infer waivers of sovereign immunity lightly, requiring that any such waiver be "unequivocally expressed" by Congress.[92] The mere grant of jurisdiction to a court, such as the grant found in section 119, is not sufficient to constitute a waiver of sovereign immunity.[93] The required express waiver is generally characterized by a "specification of the remedy or relief that may be awarded against the U.S."[94] The court could find no express waiver anywhere in the NWPA.

With no express waiver in the NWPA, the D.C. Circuit had proceeded in *Indiana Michigan* as if the waiver derived from section 702 of the Administrative Procedure Act (APA). Section 702 acts as a general waiver of sovereign immunity for claims against the U.S. that are based on agency action.[95] The CFC determined that any reliance on section 702 was misplaced, as the APA general waiver applies only where there is "no other adequate remedy in a court."[96] Although the D.C. Circuit had taken the position that the CFC was unable to accord adequate relief to a plaintiff seeking equitable relief,[97] the Federal Circuit concluded that the section 702 waiver was inapplicable under these circumstances because the nuclear utilities had an adequate remedy in the CFC under the Tucker Act.[98] The Federal Circuit, citing the U.S. Supreme Court, rejected the

---

[88] *Id.* at 662-663 ("The fact that DOE chose to use 'administrative rulemaking' in developing the Standard Contract and in putting forth its interpretations thereof did not confer jurisdiction on the D.C. Circuit to resolve what are, in effect, contract claims.").

[89] *Id.* at 665.

[90] *Id.* at 666.

[91] Under doctrine of sovereign immunity, "the United States is immune from suit save to the extent it consents to be sued." Murray v. Hoboken Land & Improvement Co., 59 U.S. 272, 283-84 (1855).

[92] *Nebraska Power*, 73 Fed. Cl. at 666.

[93] *Id.*

[94] *Id.*

[95] 5 U.S.C. § 702 ("A person suffering legal wrong because of agency action, or adversely affected or aggrieved by agency action within the meaning of a relevant statute, is entitled to judicial review thereof ... The United States may be named as a defendant in any such action, and a judgment or decree may be entered against the United States.").

[96] *Nebraska Power*, 73 Fed. Cl. at 666 (*citing* 5 U.S.C. § 704).

[97] Courts have construed the Tucker Act as waiving sovereign immunity only for claims for damages. The CFC, therefore, cannot grant a plaintiff equitable relief in these circumstances. *See, e.g.*, Richardson v. Morris, 409 U.S. 464-65 (1973).

[98] *Nebraska Power*, at 672 ("[A]n adequate remedy was and is available in this court.").

---

notion that the limitation on the available remedies made relief in the CFC "inadequate."[99] Any other conclusion, reasoned the court, would allow plaintiffs to circumvent the jurisdiction of the CFC simply by attaching a prayer for equitable relief to what was essentially a damages suit. With an alternate and adequate remedy available in the CFC, the necessary trigger for section 702 had not been met. The court held, therefore, that absent a waiver of sovereign immunity under either section 119 of the NWPA or section 702 of the APA, the D.C. Circuit had improperly granted relief against the United States in *Indiana Michigan*.[100]

The court concluded by holding that the D.C. Circuit's decision in *Indiana Michigan* exceeded the court's jurisdiction without the support of a valid waiver of sovereign immunity and was therefore void.[101] The mandamus issued in *Northern States*, which was predicated on the decision in *Indiana Michigan*, was, therefore, also void and could not preclude the DOE from raising the unavoidable delay defense.[102] The court closed by ordering the parties to brief the issue of whether the DOE's failure to commence disposal of SNF by the established deadline was excused by the "unavoidable delay" clause of the Standard Contract.[103]

Federal Circuit Grants En Banc Hearing

Nebraska Power appealed the CFC's decision to the Federal Circuit and the case was argued in December of 2007. It was not until June 4, 2009, that the Federal Circuit answered, not with an opinion, but with an order for *en banc* re-hearing before the entire Federal Circuit.[104] The order for *en banc* hearing included a request that the parties file supplemental briefs addressing whether the mandamus issued by the D.C. Circuit in *Northern States* precludes the DOE from pleading the "unavoidable delay" defense to breach of contract claims currently pending before the CFC.[105] "If so," asked the court, "does the order exceed the jurisdiction of the District of Columbia Circuit?"[106]

---

[99] *Id.* at 669. The Federal Circuit has held that the U.S. Supreme Court did "not enunciate a broad rule that the Court of Federal Claims cannot supply an adequate remedy in any case seeking injunctive relief." Consol. Edison Co. of N.Y. v. U.S., 247 F.3d 1378, 1383 (Fed. Cir. 2001).

[100] *Id.* at 672-73.

[101] *Id.* at 673 ("[T]he court is left with the firm conviction that, in issuing the subject mandamus, the D.C. Circuit operated in excess of its jurisdiction and, specifically, without an appropriate waiver of sovereign immunity.").

[102] *Id.*

[103] *Id.* at 674.

[104] Nebraska Public Power Dist. v. U.S., 2009 U.S. App. LEXIS 12668 (Fed. Cir 2009).

[105] *Id.*

[106] *Id.*

The Yucca Mountain Litigation: Breach of Contract Under the Nuclear Waste Policy Act

Figure 1. Litigation Timeline

○ 1982 • Congress passes NWPA and authorizes the DOE to enter into contracts for the collection, storage, and disposal of spent nuclear fuel

○ 1987 • Congress amends the NWPA to identify Yucca Mountain as the sole candidate site for the nation's permanent nuclear waste repository

○ 1993 • DOE interprets NWPA in a way that creates no legal obligation on behalf of DOE to begin disposal of SNF in the absence of an existing nuclear waste repository

○ 1996 • D.C. Circuit, in *Indiana Michigan,* holds DOE has legal obligation to begin disposal of SNF no later than January 31, 1998

○ 1997 • D.C. Circuit, in *Northern States,* issues mandamus prohibiting DOE from concluding that the lack of an operational permanent repository constitutes an "unavoidable delay"

○ 1998 • Statutory deadline for DOE to commence disposal of SNF

○ 2000 • Federal Circuit, in *Maine Yankee,* concludes adequate relief not available to utilities under the provisions of the Standard Contract, opening the door for breach of contract suits in the Claims Court

○ 2002 • 11th Circuit, in *Alabama Power,* invalidates allocation of NWF funds to individual utilities to mitigate costs of interim storage

○ 2005 • Claims Court, in *System Fuels,* holds that damages will be calculated from statutory deadline of January 31, 1998

   • Federal Circuit, in *PSEG,* holds that NWPA did not strip Claims Court of jurisdiction over contract disputes arising under the act

○ 2006 • Claims Court, in *Nebraska Power,* holds that the D.C. Circuit's exercise of jurisdiction over NWPA contract related claims was improper and voids 1996 mandamus that had prohibited DOE from concluding that the lack of an operational permanent repository constituted an "unavoidable delay"

○ 2009 • Federal Circuit issues order for *en banc* re-hearing of *Nebraska Power*

   • Obama Administration budget proposal terminates funding for development of Yucca Mountain facility and seeks to expand nuclear waste disposal alternatives

Source: Congressional Research Service.

Part III: Future Prospects

Yucca Mountain and the Obama Administration

Following years of decreases in funding for the Yucca Mountain project, the Obama Administration recently decided to "terminate the Yucca Mountain program while developing nuclear waste disposal alternatives."[107] The Administration's FY2010 budget eliminates all funding for the actual Yucca Mountain facility, leaving only enough funds, approximately $197 million, to finance the ongoing Nuclear Regulatory Commission licensing process and to "explore alternatives for nuclear waste disposal."[108] The budget proposal also calls for the creation of a "blue ribbon panel" to explore, study, and evaluate alternatives to the Yucca Mountain facility for the permanent storage of SNF.[109]

Both the President and the Secretary of Energy have publicly stated that Yucca Mountain does not represent a viable option for the permanent storage of SNF.[110] Senate Majority Leader Harry Reid, who has led the fight against the Yucca Mountain facility, announced in July of 2009 that the Administration intends to terminate the remaining funding for the Yucca Mountain licensing process in the budget for FY2011.[111] Although the initial House-passed bill approving the Administration's proposed budget included language mandating that any review of nuclear waste disposal alternatives include Yucca Mountain as a potential option, the recently passed DOE appropriations bill only contains language mandating that DOE "consider all alternatives for nuclear waste disposal."[112]

Regardless of whether the Administration chooses to completely defund the Yucca Mountain project, new legislative action will be required to establish a permanent repository capable of storing the nation's SNF. Under the 1987 amendments to the NWPA, Yucca Mountain is the only authorized location for a permanent SNF repository.[113] Therefore, if Congress or the President chooses to pursue an alternative site, the NWPA will have to be amended to allow for such a selection. If, on the other hand, the choice is made to proceed with the Yucca Mountain project, Congress would need to amend the NWPA to lift the statutory limit on the facility's nuclear waste

---

[107] Statement of Steven Chu, Secretary, Department of Energy, Before the Senate Committee on Appropriations Subcommittee on Energy and Water Development, and Related Agencies, May 19, 2009.

[108] *Id.*

[109] *Id.*

[110] Statement of Steven Chu, Secretary, Department of Energy, Before the Senate Committee on the Budget, March 11, 2009 ("[B]oth the President and I have made clear that Yucca Mountain is not a workable option.").

[111] Lisa Mascaro, *Reid: White House to Cut Off Yucca Funding*, Las Vegas Sun, July 30, 2009. Although an internal DOE draft Program Decision Memorandum referencing a revised FY2011 budget request states that "[a]ll license defense activities will be terminated in December 2009," DOE recently met a December filing deadline for continuing the licensing process and spokesman Allen Benson confirmed that the DOE would continue to utilize the funds appropriated to the department to acquire a NRC license for the Yucca Mountain facility. *See*, Letter from House Committee on Energy and Commerce to Steven Chu, Secretary, Department of Energy, November 18, 2009; Keith Rogers, *Energy Department Keeping Nuclear Repository Options Open*, Las Vegas Review, December 14, 2009.

[112] P.L. 111-85 (2009).

[113] 42. U.S.C. § 10172 ("The Secretary shall terminate all site specific activities ... at all candidate sites, other than the Yucca Mountain site, within 90 days after December 22, 1987."); 42 U.S.C. § 10172(a) ("The Secretary may not conduct site-specific activities with respect to a second repository unless Congress has specifically authorized and appropriated funds for such activities.").

---

storage capacity. The current statutory cap of 70,000 metric tons is insufficient to hold existing amounts of SNF.[114]

Future Liability

The *Nebraska Power* decision has the potential to add many more years of litigation to the more than 13 years of litigation that the DOE and nuclear utilities have already undertaken on damage claims. Based on the question presented in the *en banc* order, which focuses solely on whether the D.C. Circuit properly exercised jurisdiction, even if the court rules that the *Northern States* mandamus is void, it is unlikely that the Federal Circuit will answer whether the lack of a permanent SNF repository constitutes an "unavoidable delay." At least one attorney involved in the proceedings predicts that five more years of litigation will be required to determine whether the delay was "unavoidable." Such an extension would add to the litigation costs already suffered by both sides.[115] To date, the Department of Justice, which has handled the DOE defense, has spent over $150 million on litigation costs. The nuclear utilities reportedly spend $5 million to $7 million in litigation costs on each individual case.[116]

In an attempt to curtail damages, DOE has sought to reach settlement agreements with a number of nuclear utilities. As of September 2009, the government has entered into agreements with nuclear utilities that operate 36 of the 118 nuclear facilities covered by the Standard Contract.[117] Under the settlements, contract parties submit annual reimbursement claims to DOE for any delay-related nuclear waste storage costs that they incurred during that year. As the settlement agreements cover continuing damages, the affected nuclear utilities are able to submit annual claims directly to the DOE rather than re-litigating ongoing damages in the federal courts. As of the end of 2008, DOE had paid over $400 million pursuant to these settlements.[118]

However, if the Federal Circuit affirms the CFC and "voids"[119] the D.C. Circuit mandamus, DOE liability under the Standard Contract could be drastically reduced. With the D.C. Circuit mandamus removed, DOE is free to pursue the previously barred "unavoidable delay" defense. Accordingly, if DOE is able to prove that the department's breach occurred as a result of an "unavoidable delay," then the terms of the Standard Contract direct that DOE shall not be liable for any damages.[120] In this respect, receiving a favorable decision from the Federal Circuit that acts to void the D.C. Circuit mandamus is only the first step toward limiting damages under the "unavoidable delays" provision of the Standard Contract. DOE will still have to convince the CFC that the lack of an existing facility actually qualifies as an "unavoidable delay" under the Standard Contract, an argument previously, and resoundingly, rejected by the D.C. Circuit.[121]

---

[114] 42 U.S.C. § 10134(d). *See, Report to the President and the Congress By the Secretary of Energy on the Need for a Second Repository*, Dec. 2008, *available at*: http://www.ocrwm.doe.gov.

[115] Marcia Coyle, *Nuclear Dispute Fallout*, The National Law Journal, Sept. 14, 2009 at 14.

[116] *Id.* With 71 lawsuits filed against DOE, one could estimate that the nuclear utilities as a whole have expended over $400 million in litigation costs.

[117] *See*, CRS Report RL 33461, Civilian Nuclear Waste Disposal, by Mark Holt, at 3.

[118] *Id.*

[119] Such an order may raise questions as to whether one U.S. circuit court has the authority to void the decision of a sister circuit.

[120] Standard Contract, 10 C.F.R. § 961.11.

[121] *Northern States*, 128 F.3d at 760. ("The most glaring problem with DOE's position is that it is answering the wrong (continued...)

Even with a favorable decision from the Federal Circuit, DOE may be blocked from now asserting the "unavoidable delay" defense in some pending[122] NWPA breach of contract suits. In the one opinion in a breach of contract suit released by the CFC since the *Nebraska Power* decision, the court rejected DOE's attempt to raise the unavoidable delay defense by holding that DOE had waived any such defense by not raising the issue at trial.[123] In *S. Nuclear Operating Co. v. U.S.*, the court characterized the "unavoidable delay" provision as an affirmative defense that is waived if not raised at trial and preserved for appeal.[124] The court noted that, in the instant case, DOE had not raised the "unavoidable delays" clause of the Standard Contract nor challenged the D.C. Circuit's jurisdiction in issuing the *Northern States* mandamus as it had in previous cases.[125] Accordingly, it is arguable that DOE will have waived the "unavoidable delays" defense in cases in which it did not raise the issue in some form during trial.

Conclusion

The total costs to taxpayers for delays associated with the Yucca Mountain project are difficult to project, especially given the uncertainty relating to the viability of the "unavoidable delays" defense under the Standard Contract. However, absent a significant change in the direction of NWPA-related litigation, DOE predicts that damages stemming from breach of contract claims will measure close to $12.3 billion if the department is able to begin accepting SNF by 2020—an unlikely occurrence given that all construction on the partially completed Yucca Mountain facility has ceased.[126] Approximately $500 million in additional legal damages will continue to build with each year that DOE is unable to begin accepting SNF.[127] The nuclear utilities, on the other hand, estimate DOE's total potential liability at approximately $50 billion.[128] With $150 million in litigation expenses incurred by the Department of Justice, and possibly $50 billion in liability for breach of contract, the cost of delays in the Yucca Mountain project could potentially range between $12.5 billion to $51 billion.

(...continued)

question ... DOE cannot now render its obligation contingent, and free itself of the costs caused by its delay, by advancing the same failed position that we rejected before.").

[122] DOE may be able to argue that the few final rulings that have already been paid should be re-opened under Rule 60 of the Federal Rules of Civil Procedure. Rule 60 authorizes a court to relieve a party of a final judgment where "a prior judgment upon which it is based has been reversed or otherwise vacated, or it is no longer equitable that the judgment should have prospective application." Fed. R. Civ. P. 60.

[123] S. Nuclear Operating Co., v. U.S., 77 Fed. Cl. 396 (2007).

[124] *Id.*

[125] *Id.* at 456 ("Questions of the validity or applicability of the D.C. Circuit's mandamus in this court in this breach of contract suit could have been raised but were not. Defendant raised the absence of a repository as a defense in other SNF cases.").

[126] *CBO Testimony*, at 7.

[127] *Id.*

[128] *Id.*

Author Contact Information

Todd Garvey
Legislative Attorney
tgarvey@crs.loc.gov, 7-0174

GAO

United States Government Accountability Office

Report to Congressional Requesters

November 2009

NUCLEAR WASTE MANAGEMENT

Key Attributes, Challenges, and Costs for the Yucca Mountain Repository and Two Potential Alternatives

GAO

Accountability ★ Integrity ★ Reliability

GAO-10-48

November 2009

G A O
Accountability · Integrity · Reliability

Highlights

Highlights of GAO-10-48, a report to congressional requesters

NUCLEAR WASTE MANAGEMENT

Key Attributes, Challenges, and Costs of the Yucca Mountain Repository and Two Potential Alternatives

Why GAO Did This Study

High-level nuclear waste—one of the nation's most hazardous substances—is accumulating at 80 sites in 35 states. The United States has generated 70,000 metric tons of nuclear waste and is expected to generate 153,000 metric tons by 2055. The Nuclear Waste Policy Act of 1982, as amended, requires the Department of Energy (DOE) to dispose of the waste in a geologic repository at Yucca Mountain, about 100 miles northwest of Las Vegas, Nevada. However, the repository is more than a decade behind schedule, and the nuclear waste generally remains at the commercial nuclear reactor sites and DOE sites where it was generated.

This report examines the key attributes, challenges, and costs of the Yucca Mountain repository and the two principal alternatives to a repository that nuclear waste management experts identified: storing the nuclear waste at two centralized locations and continuing to store the waste on site where it was generated. GAO developed models of total cost ranges for each alternative using component cost estimates provided by the nuclear waste management experts. However, GAO did not compare these alternatives because of significant differences in their inherent characteristics that could not be quantified.

What GAO Recommends

GAO is making no recommendations in this report. In written comments, DOE and NRC generally agreed with the report.

View GAO-10-48 or key components. For more information, contact Mark Gaffigan at 202-512-3841 or gaffiganm@gao.gov.

What GAO Found

The Yucca Mountain repository is designed to provide a permanent solution for managing nuclear waste, minimize the uncertainty of future waste safety, and enable DOE to begin fulfilling its legal obligation under the Nuclear Waste Policy Act to take custody of commercial waste, which began in 1998. However, project delays have led to utility lawsuits that DOE estimates are costing taxpayers about $12.3 billion in damages through 2020 and could cost $500 million per year after 2020, though the outcome of pending litigation may affect the government's total liability. Also, the administration has announced plans to terminate Yucca Mountain and seek alternatives. Even if DOE continues the program, it must obtain a Nuclear Regulatory Commission construction and operations license, a process likely to be delayed by budget shortfalls. GAO's analysis of DOE's cost projections found that a repository to dispose of 153,000 metric tons would cost from $41 billion to $67 billion (in 2009 present value) over a 143-year period until the repository is closed. Nuclear power rate payers would pay about 80 percent of these costs, and taxpayers would pay about 20 percent.

Centralized storage at two locations provides an alternative that could be implemented within 10 to 30 years, allowing more time to consider final disposal options, nuclear waste to be removed from decommissioned reactor sites, and the government to take custody of commercial nuclear waste, saving billions of dollars in liabilities. However, DOE's statutory authority to provide centralized storage is uncertain, and finding a state willing to host a facility could be extremely challenging. In addition, centralized storage does not provide for final waste disposal, so much of the waste would be transported twice to reach its final destination. Using cost data from experts, GAO estimated the 2009 present value cost of centralized storage of 153,000 metric tons at the end of 100 years to range from $15 billion to $29 billion but increasing to between $23 billion and $81 billion with final geologic disposal.

On-site storage would provide an alternative requiring little change from the status quo, but would face increasing challenges over time. It would also allow time for consideration of final disposal options. The additional time in on-site storage would make the waste safer to handle, reducing risks when waste is transported for final disposal. However, the government is unlikely to take custody of the waste, especially at operating nuclear reactor sites, which could result in significant financial liabilities that would increase over time. Not taking custody could also intensify public opposition to spent fuel storage site renewals and reactor license extensions, particularly with no plan in place for final waste disposition. In addition, extended on-site storage could introduce possible risks to the safety and security of the waste as the storage systems degrade and the waste decays, potentially requiring new maintenance and security measures. Using cost data from experts, GAO estimated the 2009 present value cost of on-site storage of 153,000 metric tons at the end of 100 years to range from $13 billion to $34 billion but increasing to between $20 billion to $97 billion with final geologic disposal.

_____ **United States Government Accountability Office**

Contents

Tables

Figures

Abbreviations

DOE	Department of Energy
EPA	Environmental Protection Agency
NRC	Nuclear Regulatory Commission
NWPA	Nuclear Waste Policy Act of 1982

GAO
Accountability * Integrity * Reliability

United States Government Accountability Office
Washington, DC 20548

November 4, 2009

The Honorable Barbara Boxer
Chairman
Committee on Environment and Public Works
United States Senate

The Honorable Harry Reid
United States Senate

The Honorable John Ensign
United States Senate

High-level nuclear waste consists mostly of spent nuclear fuel removed from commercial power reactors and is considered one of the most hazardous substances on earth. The U.S. national inventory of 70,000 metric tons of nuclear waste—enough to fill a football field more than 15 feet deep—has been accumulating at 80 sites in 35 states since the mid-1940s and is expected to more than double to 153,000 metric tons by 2055. The current national policy of constructing a federal repository to dispose of this waste at Yucca Mountain—which is about 100 miles northwest of Las Vegas, Nevada—has already been delayed more than a decade. As a result, nuclear waste generally remains at the sites where it was generated. Experts and regulators believe the nuclear waste, if properly stored and monitored, can be kept safe and secure on-site for decades; but communities across the country have raised concerns about the waste's lethal nature and the possibility of natural disasters or terrorism, particularly at sites near urban centers or sources of drinking water. Industry has also raised concerns that local communities will not support the expansion of the nuclear energy industry without a final waste disposition pathway. Many experts and communities view nuclear energy as a potential means of meeting future energy demands while reducing reliance on fossil fuels and cutting carbon emissions, a key contributor to climate change.

In addition to the spent nuclear fuel generated by commercial power reactors, the Department of Energy (DOE) owns and manages about 19 percent of the nuclear waste—referred to as DOE-managed spent nuclear fuel and high-level waste—which consists of spent nuclear fuel from power, research, and navy shipboard reactors, and high-level nuclear waste from the nation's nuclear weapons program. (See fig. 1 for the locations where nuclear waste is stored.)

Page 1   GAO-10-48 Nuclear Waste Management

Figure 1: Current Storage Sites and Proposed Repository for High-Level Nuclear Waste

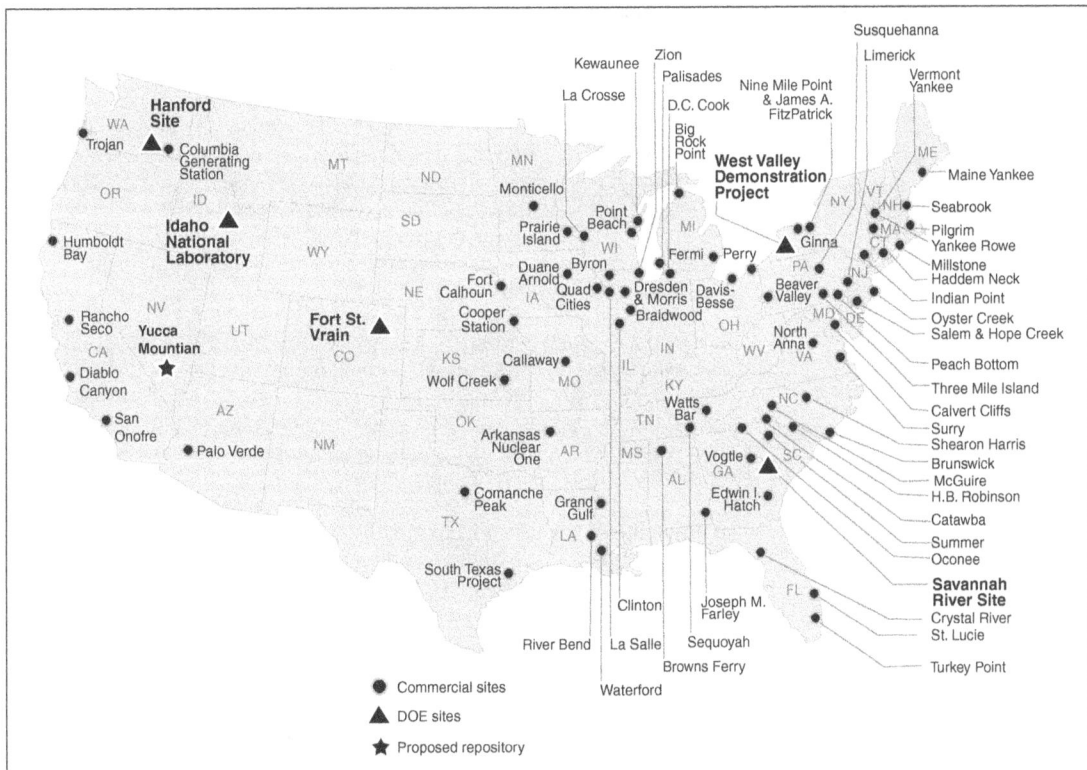

Source: DOE.

Note: Locations are approximate. DOE has reported that it is responsible for managing nuclear waste at 121 sites in 39 states, but DOE officials told us that several sites have only research reactors that generate small amounts of waste that will be consolidated at the Idaho National Laboratory for packaging prior to disposal.

Under the Nuclear Waste Policy Act of 1982 (NWPA), as amended, DOE was to evaluate one or more national geologic repositories that would be designated to permanently store commercial spent nuclear fuel and DOE-managed spent nuclear fuel and high-level waste. NWPA was amended in 1987 to direct DOE to evaluate only the Yucca Mountain site. In 2002, the president recommended and the Congress approved the Yucca Mountain site as the nation's geologic repository. The repository is intended to

GAO-10-48 Nuclear Waste Management

isolate nuclear waste from humans and the environment for thousands of years, long enough for its radioactivity to decay to near natural background levels. NWPA set January 31, 1998, as the date for DOE to start accepting nuclear waste for disposal. To meet this goal, DOE has spent more than $14 billion for design, engineering, and testing activities.[1] In June 2008, DOE submitted a license application to the Nuclear Regulatory Commission (NRC) for approval to construct the repository. In July 2008, DOE reported that its best achievable date for opening the repository, if it receives NRC approval, is in 2020. Delays in the Yucca Mountain repository have resulted in a need for continued storage of the waste onsite, leaving industry uncertain regarding the licensing of new nuclear power reactors and the nation uncertain regarding a final disposition of the waste.

In March 2009, the Secretary of Energy testified that the administration planned to terminate the Yucca Mountain repository. Since then, the administration has announced plans to study alternatives to geologic disposal at Yucca Mountain before making a decision on a future nuclear waste management strategy, which the administration said could include reprocessing or other complementary strategies.

In this context, you asked us to identify key aspects of DOE's nuclear waste management program and other possible management approaches. Specifically, you asked us to examine (1) the key attributes, challenges, and costs of the Yucca Mountain repository; (2) and identify alternative nuclear waste management approaches; (3) the key attributes, challenges, and costs of storing the nuclear waste at two centralized sites; and (4) the key attributes, challenges, and costs of continuing to store the nuclear waste at its current locations. The centralized storage and onsite storage options—both with disposal scenarios—were the two most likely alternative approaches identified by the experts we interviewed. We are also providing information on what is known about sources of funding—primarily taxpayers and nuclear power rate payers—for the Yucca Mountain repository and the two alternative approaches.

To examine the key attributes, challenges, and costs of the Yucca Mountain repository, we obtained reports and supporting documentation

---

[1] In constant fiscal year 2009 dollars. Funding comes primarily from fees collected from electric power companies operating commercial reactors and appropriations for DOE-managed spent nuclear fuel and high-level waste.

Page 3   GAO-10-48 Nuclear Waste Management

from DOE, NRC, the National Academy of Sciences, and the Nuclear Waste Technical Review Board. Specifically, we used DOE's report on the Yucca Mountain repository's total lifecycle cost to analyze the cost for disposing of either (1) 70,000 metric tons of nuclear waste, which is the statutory cap on the amount of waste that can be disposed of at Yucca Mountain, or (2) 153,000 metric tons, which is the estimated total amount of nuclear waste that has already been generated and will be generated if all currently operating commercial reactors operate for a 60-year lifespan.[2] We then discounted these costs to 2009 present value.

To identify alternative nuclear waste management approaches, we interviewed DOE officials, experts at the National Academy of Sciences and the Nuclear Waste Technical Review Board, and executives at the Nuclear Energy Institute, among others. Based on their comments, we identified two generic alternative approaches for managing this waste for at least a 100-year period before it is disposed in a repository: storing the nuclear waste at two centralized facilities—referred to as centralized storage—and continuing to store the nuclear waste on site at their current facilities—referred to as on-site storage. To examine the key attributes, challenges, and costs of each alternative, we asked nuclear waste management experts from federal agencies, industry, academic institutions, and concerned groups to comment on the attributes and challenges of each alternative, provide relevant cost data, and comment on the assumptions and cost components that we used to develop cost models for the alternatives. We then used the models to produce the total cost ranges for each alternative with and without final disposal in a geologic repository at the end of a 100-year specific time period. In addition, we analyzed onsite storage for longer periods than 100 years. We analyzed costs associated with storing 70,000 metric tons and 153,000 metric tons and discounted the costs to 2009 present value.

We did not compare the Yucca Mountain cost range to the ranges of other alternatives because of significant differences in inherent characteristics of these alternatives that our modeling work could not quantify. For example, the safety, health, and environmental risks for each are very different, which needs to be considered in the policy debate on nuclear waste management decisions. (See app. I for additional information about our scope and methodology, app. II for our methodology for soliciting

---

[2]DOE, *Analysis of the Total System Lifecycle Cost of the Civilian Radioactive Waste Management Program, Fiscal Year 2007*, DOE/RW-0591 (Washington, D.C., July 2008).

GAO-10-48 Nuclear Waste Management

comments from nuclear waste management experts, and app. III for a list of these experts.)

We conducted this performance audit from April 2008 to October 2009 in accordance with generally accepted government auditing standards. Those standards require that we plan and perform the audit to obtain sufficient, appropriate evidence to provide a reasonable basis for our findings and conclusions based on our audit objectives. We believe that the evidence obtained provides a reasonable basis for our findings and conclusions based on our audit objectives.

Background

Nuclear waste is long-lived and very hazardous—without protective shielding, the intense radioactivity of the waste can kill a person within minutes or cause cancer months or even decades after exposure.[3] Thus, careful management is required to isolate it from humans and the environment. To accomplish this, the National Academy of Sciences first endorsed the concept of nuclear waste disposal in deep geologic formations in a 1957 report to the U.S. Atomic Energy Commission, which has since been articulated by experts as the safest and most secure method of permanent disposal.[4] However, progress toward developing a geologic repository was slow until NWPA was enacted in 1983. Citing the potential risks of the accumulating amounts of nuclear waste, NWPA required the federal government to take responsibility for the disposition of nuclear waste and required DOE to develop a permanent geologic repository to protect public health and safety and the environment for

[3]For the purposes of our report, nuclear waste includes both spent nuclear fuel—fuel that has been withdrawn from a nuclear reactor following irradiation—and high-level radioactive waste—generally the material resulting from the reprocessing of spent nuclear fuel. Nuclear waste—specifically spent nuclear fuel—is also very thermally hot. As the radioactive elements in spent nuclear fuel decay, they give off heat. However, according to DOE data, a spent nuclear fuel assembly can lose nearly 80 percent of its heat 5 years after it has been removed from a reactor and about 95 percent of its heat after 100 years.

[4]National Academy of Sciences, *The Disposal of Radioactive Waste on Land*, (Washington, D.C., September 1957). This report suggested several potential alternatives for disposal of nuclear waste, stressing that although there are many potential sites for geologic disposal of waste at various depths and in various geologic formations, further research was needed regarding specific waste forms and specific geologic formations, including disposal in deep underground formations. The report stated, "the hazard related to radioactive waste is so great that no element of doubt should be allowed to exist regarding safety." Subsequent reports by the National Academy of Sciences and others have continued to endorse geologic isolation of nuclear waste and have suggested that engineered barriers, such as corrosion-resistant containers, can provide additional layers of isolation.

current and future generations. Specifically, the act required DOE to study several locations around the country for possible repository sites and develop a contractual relationship with industry for disposal of the nuclear waste. The Congress amended NWPA in 1987 to restrict scientific study and characterization of a possible repository to only Yucca Mountain. (Fig. 2 shows the north crest of Yucca Mountain and a cut-out of the proposed mined repository.)

Figure 2: Aerial View and Cut-Out of the Yucca Mountain Repository

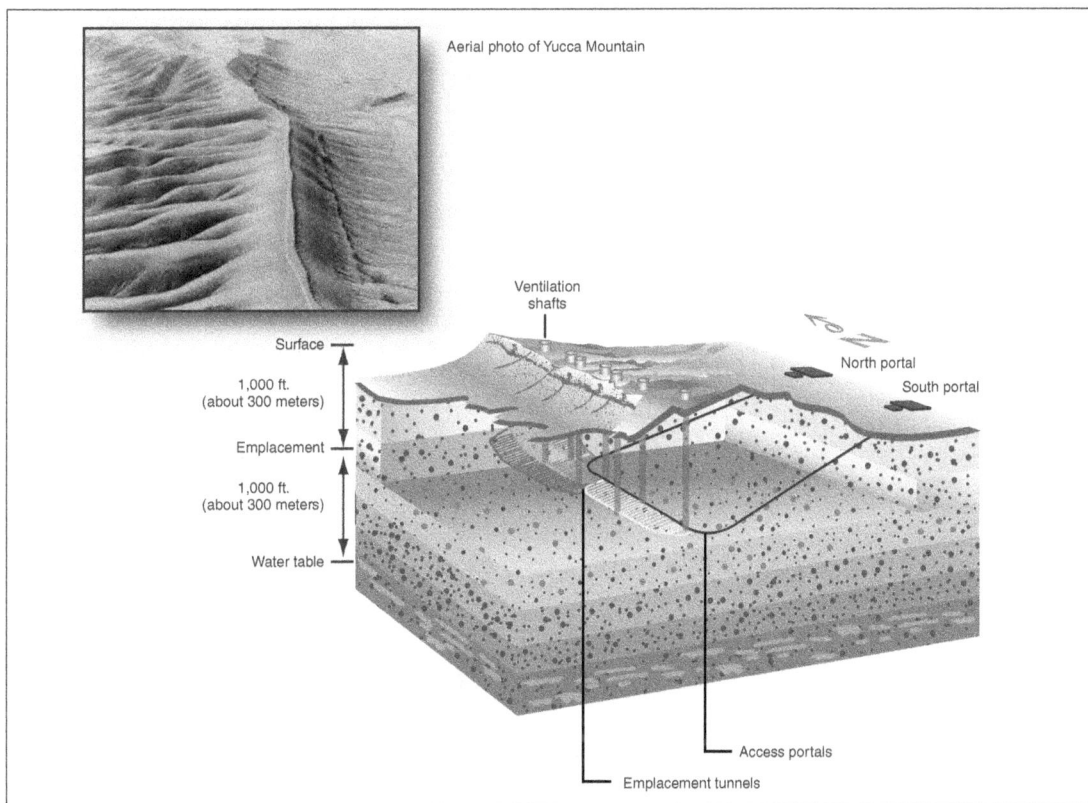

Aerial photo of Yucca Mountain

Ventilation shafts

Surface

1,000 ft. (about 300 meters)

Emplacement

1,000 ft. (about 300 meters)

Water table

North portal

South portal

Access portals

Emplacement tunnels

Source: DOE.

After the Congress approved Yucca Mountain as a suitable site for the development of a permanent nuclear waste repository in 2002, DOE began preparing a license application for submittal to NRC, which has regulatory authority over commercial nuclear waste management facilities. DOE submitted its license application to NRC in June 2008, and NRC accepted the license application for review in September 2008. NWPA requires NRC to complete its review of DOE's license application for the Yucca Mountain repository in 3 years, although a fourth year is allowed if NRC deems it necessary and complies with certain reporting requirements.

To pay the nuclear power industry's share of the cost for the Yucca Mountain repository, NWPA established the Nuclear Waste Fund, which is funded by a fee of one mill (one-tenth of a cent) per kilowatt-hour of nuclear-generated electricity that the federal government collects from electric power companies. DOE reported that, at the end of fiscal year 2008, the Nuclear Waste Fund contained $22 billion, with an additional $1.9 billion projected to be added in 2009. DOE receives money from the Nuclear Waste Fund through congressional appropriations. Additional funding for the repository comes from an appropriation which provides for the disposal cost of DOE-managed spent nuclear fuel and high-level waste.

NWPA caps nuclear waste that can be disposed of at the Yucca Mountain repository at 70,000 metric tons until a second repository is available. However, the nation has already accumulated about 70,000 metric tons of nuclear waste at current reactor sites and DOE facilities. Without a change in the law to raise the cap or to allow the construction of a second repository, DOE can dispose of only the current nuclear waste inventory. The nation will have to develop a strategy for an additional 83,000 metric tons of waste expected to be generated if NRC issues 20-year license extensions to all of the currently operating nuclear reactors.[5] This amount does not include any nuclear waste generated by new reactors or future defense activities, or greater than class C nuclear waste.[6] According to

---

[5]NRC has already issued license extensions for 54 reactors, enabling them to operate for a total of 60 years. Extension requests for 21 units are currently under review and requests for as many as 25 more are anticipated through 2017.

[6]As of October 2009, NRC has received 18 applications for 29 new reactors. In addition to spent nuclear fuel and DOE-managed high-level waste, the nation also generates so-called greater than class C nuclear waste from the maintenance and decommissioning of nuclear power plants, from radioactive materials that were once used for food irradiation or for medical purposes, and from miscellaneous radioactive waste, such as contaminated equipment from industrial research and development. DOE, which is required to dispose of this nuclear waste, has not issued an environmental impact statement describing potential options, which could include disposal of the waste at the Yucca Mountain repository.

Page 7   GAO-10-48 Nuclear Waste Management

DOE and industry studies, three to four times the 70,000 metric tons—and possibly more—could potentially be disposed safely in Yucca Mountain, which could address current and some future waste inventories, potentially delaying the need for a second repository for several generations.

Nuclear waste has continued to accumulate at the nation's commercial and DOE nuclear facilities over the past 60 years. Facility managers must actively manage the nuclear waste by continually isolating, confining, and monitoring it to keep humans and the environment safe. Most spent nuclear fuel is stored at reactor sites, immersed in pools of water designed to cool and isolate it from the environment. With nowhere to dispose of the spent nuclear fuel, the racks holding spent fuel in the pools have been rearranged to allow for more dense storage of assemblies. Even with this re-racking, spent nuclear fuel pools are reaching their capacities. Some critics have expressed concern about the remote possibility of an overcrowded spent nuclear fuel pool releasing large amounts of radiation if an accident or other event caused the pool to lose water, potentially leading to a fire that could disperse radioactive material. As reactor operators have run out of space in their spent nuclear fuel pools, they have turned in increasing number to dry cask storage systems that generally consist of stainless steel canisters placed inside larger stainless steel or concrete casks. (See fig. 3.) NRC requires protective shielding, routine inspections and monitoring, and security systems to isolate the nuclear waste to protect humans and the environment.

Figure 3: Dry Cask Storage System for Spent Nuclear Fuel

At some nuclear reactors across the country, spent fuel is kept on site, above ground, in systems basically similar to the one shown here.

1 Once the spent fuel has cooled, it is loaded into special canisters, each of which is designed to hold about two dozen assemblies. Water and air are removed. The canister is filled with inert gas, welded shut, and rigorously tested for leaks. It may then be placed in a "cask" for storage or transportation.

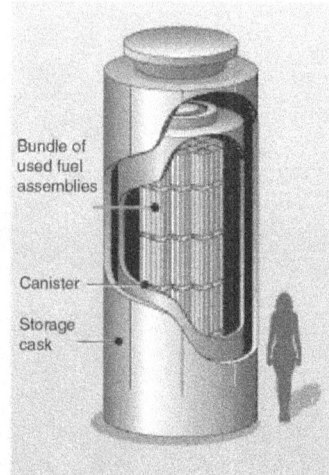

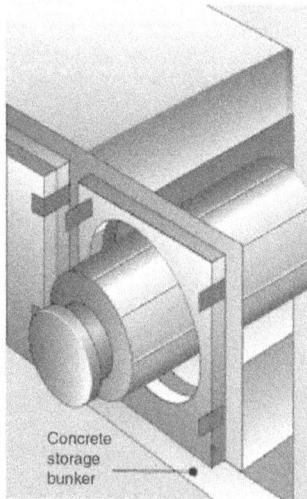

Bundle of used fuel assemblies

Canister

Storage cask

2 The canisters can also be stored in above ground concrete bunkers, each of which is about the size of a one-car garage. Eventually they may be transported elsewhere for storage.

Concrete storage bunker

Source: NRC.

Page 9   GAO-10-48 Nuclear Waste Management

NRC has determined that these dry cask storage systems can safely store nuclear waste, but NRC considers them to be interim measures. In 1990, NRC issued a revised waste confidence rule, stating that it had confidence that the waste generated by a reactor can be safely stored in either wet or dry storage for 30 years beyond a reactor's life, including license extensions. NRC further determined that it had reasonable assurance that safe geologic disposal was feasible and that a geologic repository would be operational by about 2025. More recently, NRC has published a notice of proposed rulemaking to revise that rule, proposing that waste generated by a reactor can be safely stored for 60 years beyond the life of a reactor and that geologic disposal would be available in 50 to 60 years beyond a reactor's life.[7] NRC is currently considering whether to republish its proposed rule to seek additional public input on certain issues. Forty-five reactor sites or former reactor sites in 30 states have dry storage facilities for their spent nuclear fuel as of June 2009, and the number of reactor sites storing spent nuclear fuel is likely to continue to grow until an alternative is implemented.

Implementing a permanent, safe, and secure disposal solution for the nuclear waste is of concern to the nation, particularly state governments and local communities, because many of the 80 sites where nuclear waste is currently stored are near large populations or major water sources or consist of shutdown reactor sites that tie up land that could be used for other purposes. In addition, states that have DOE facilities with nuclear waste storage are concerned because of possible contamination to aquifers, rivers, and other natural resources. DOE's Hanford Reservation, located near Richland, Washington, was a major component of the nation's nuclear weapons defense program from 1943 until 1989, when operations ceased. In the settlement of a lawsuit filed by the state of Washington in 2003, DOE agreed not to ship certain nuclear waste to Hanford until environmental reviews were complete. In August 2009, the U.S. government stated that the preferred alternative in DOE's environmental review would include limitations on certain nuclear waste shipments to Hanford until the process of immobilizing tank waste in glass begins,

[7] See 73 Fed. Reg. 59551-59570 (Oct. 9, 2008).

expected in 2019.[8] Moreover, some commercial and DOE sites where the nuclear waste is stored may not be able to accommodate much additional waste safely because of limited storage space or community objections. These sites will require a more immediate solution.

The nation has considered proposals to build centralized storage facilities where waste from reactor sites could be consolidated. The 1987 amendment to NWPA established the Office of the Nuclear Waste Negotiator to try to broker an agreement for a community to host a repository or interim storage facility. Two negotiators worked with local communities and Native American tribes for several years, but neither was able to conclude a proposed agreement with a willing community by January 1995, when the office's authority expired. Subsequently, in 2006 after a 9-year licensing process, a consortium of electric power companies called Private Fuel Storage obtained a NRC license for a private centralized storage facility on the reservation of the Skull Valley Band of the Goshute Indians in Utah. NRC's 20-year license—with an option for an additional 20 years—allows storage of up to 40,000 metric tons of commercial spent nuclear fuel. However, construction of the Private Fuel Storage facility has been delayed by Department of the Interior decisions not to approve the lease of tribal lands to Private Fuel Storage and declining to issue the necessary rights-of-way to transport nuclear waste to the facility through Bureau of Land Management land. Private Fuel Storage and the Skull Valley Band of Goshutes filed a federal lawsuit in 2007 to overturn Interior's decisions.

Reprocessing nuclear waste could potentially reduce, but not eliminate, the amount of waste for disposal. In reprocessing, usable uranium and plutonium are recovered from spent nuclear fuel and are used to make new fuel rods. However, current reprocessing technologies separate weapons usable plutonium and other fissionable materials from the spent nuclear fuel, raising concerns about nuclear proliferation by terrorists or

---

[8]The U.S. government made this statement in a letter related to a tentative settlement agreement in the lawsuit of *State of Washington v. Chu*, No. CV-08-5085-FVS (E.D. Washington, filed Nov. 26, 2008). In 2008, the state of Washington filed suit claiming DOE had violated the Tri-Party Agreement among DOE, the state, and the Environmental Protection Agency by failing to meet enforceable cleanup milestones in the agreement. On August 10, 2009, DOE and the state announced they had reached a tentative settlement, including new cleanup milestones and a 2047 completion date for certain key cleanup activities. We have questioned DOE's ability to meet this date. See GAO, *Nuclear Waste: Uncertainties and Questions about Costs and Risks Persist with DOE's Tank Waste Cleanup Strategy at Hanford*, GAO-09-913 (Washington, D.C.: Sept. 30, 2009).

Page 11   GAO-10-48 Nuclear Waste Management

enemy states. Although the United States pioneered the reprocessing technologies used by other countries, such as France and Russia, presidents Gerald Ford and Jimmy Carter ended government support for commercial reprocessing in the United States in 1976 and 1977, respectively, primarily due to proliferation concerns. Although President Ronald Reagan lifted the ban on government support in 1981, the nation has not embarked on any reprocessing program due to proliferation and cost concerns—the Congressional Budget Office recently reported that current reprocessing technologies are more expensive than direct disposal of the waste in a geologic repository.[9] DOE's Fuel Cycle Research and Development program is currently performing research in reprocessing technologies that would not separate out weapons usable plutonium, but it is not certain whether these technologies will become cost-effective.[10]

The general consensus of the international scientific community is that geologic disposal is the preferred long-term nuclear waste management alternative. Finland, Sweden, Canada, France, and Switzerland have decided to construct geologic disposal facilities, but none have yet completed any such facility, although DOE reports that Finland and Sweden have announced plans to begin emplacement operations in 2020 and 2023, respectively. Moreover, some countries employ a mix of complementary storage alternatives in their national waste management strategies, including on-site storage, consolidated interim storage, reprocessing, and geologic disposal. For example, Sweden plans to rely on on-site storage until the waste cools enough to move it to a centralized storage facility, where the waste will continue to cool and decay for an additional 30 years. This waste will then be placed in a geologic repository for disposal. France reprocesses the spent nuclear fuel, recycling usable portions as new fuel and storing the remainder for eventual disposal.

[9]Congressional Budget Office, *Costs of Reprocessing Versus Directly Disposing of Spent Nuclear Fuel; Testimony before the Committee on Energy and Natural Resources* (Washington, D.C.: Nov. 14, 2007).

[10]DOE changed the name of this program from the Advanced Fuel Cycle Initiative to the Fuel Cycle Research and Development program in its fiscal year 2010 budget submission.

GAO-10-48 Nuclear Waste Management

The Yucca Mountain Repository Would Provide a Permanent Solution for Nuclear Waste, but Its Implementation Faces Challenges and Significant Upfront Costs

The Yucca Mountain repository—mandated by NWPA, as amended—would provide a permanent nuclear waste management solution for the nation's current inventory of about 70,000 metric tons of waste. According to DOE and industry studies, the repository potentially could be a disposal site for three to four times that amount of waste. However, the repository lacks the support of the administration and the state of Nevada, and faces regulatory and other challenges. Our analysis of DOE's cost projections found that the Yucca Mountain repository would cost from $41 billion to $67 billion (in 2009 present value) for disposing of 153,000 metric tons of nuclear waste.[11] Most of these costs are up-front capital costs. However, once the Yucca Mountain repository is closed—in 2151 for our 153,000-metric-ton model—it is not expected to incur any significant additional costs, according to DOE.

As Designed, the Yucca Mountain Repository Would Be a Permanent Solution and Would Reduce the Uncertainty Associated with Future Nuclear Waste Safety

The Yucca Mountain repository is designed to isolate nuclear waste in a safe and secure environment long enough for the waste to degrade into a form that is less harmful to humans and the environment. As nuclear waste ages, it cools and decays, becoming less radiologically dangerous. In October 2008, after years of legal challenges, the Environmental Protection Agency (EPA) promulgated standards that require DOE to ensure that radioactive releases from the nuclear waste disposed of at Yucca Mountain do not harm the public for 1 million years.[12] This is because some waste components, such as plutonium 239, take hundreds of thousands of years to decay into less harmful materials. To meet EPA's standards and keep the waste safely isolated, DOE's license application proposes the use of both natural and engineered barriers. Key natural barriers of Yucca Mountain include its dry climate, the depth and isolation

---

[11]Our cost range for a permanent repository differs from DOE's most recent estimate of $96 billion for the following reasons: First, our cost range is in 2009 present value, while DOE uses 2007 constant dollars, which are not discounted. Our present value analysis reflects the time value of money—costs incurred in the future are worth less today—so that streams of future costs become smaller. Second, our cost range does not include about $14 billion in previously incurred costs. Third, our cost range is for 153,000 metric tons of nuclear waste while DOE's estimated cost is for 122,100 metric tons. Finally, we use a range while DOE provides a point estimate.

[12]The Energy Policy Act of 1992 directed EPA to base its health standards on a National Academy of Sciences study of the health issues related to radioactive releases. NRC has promulgated rules based on EPA's October 2008 standards that require the Yucca Mountain repository to limit the annual radiation dose of the public to at most 15 millirem for the first 10,000 years after disposal and at most 100 millirem from 10,001 years to 1 million years after disposal. In contrast, the average American is exposed to about 360 millirem of radiation annually, mainly from natural background sources.

GAO-10-48 Nuclear Waste Management

of the Death Valley aquifer in which the mountain resides, its natural physical shape, and the layers of thick rock above and below the repository that lie 1,000 feet below the surface of the mountain and 1,000 feet above the water table. Key engineered barriers include the solid nature of the nuclear waste; the double-shelled transportation, aging, and disposal canisters that encapsulate the waste and prevent radiation leakage; and drip shields that are composed of corrosion-resistant titanium to ward off any dripping water inside the repository for many thousands of years.

The construction of a geologic repository at Yucca Mountain would provide a permanent solution for nuclear waste that could allow the government to begin taking possession of the nuclear waste in the near term—about 10 to 30 years. The nuclear power industry sees this as an important consideration in obtaining the public support necessary to build new nuclear power reactors. The industry is interested in constructing new nuclear power reactors because, among other reasons, of the growing demand for electricity and pressure from federal and state governments to reduce reliance on fossil fuels and curtail carbon emissions. Some electric power companies see nuclear energy as an important option for noncarbon emitting power generation. According to NRC, 18 electric power companies have filed license applications to construct 29 new nuclear reactors.[13] Nuclear industry representatives, however, have expressed concerns that investors and the public will not support the construction of new nuclear power reactors without a final safe and secure disposition pathway for the nuclear waste, particularly if that waste is generated and stored near major waterways or urban centers. Moreover, having a permanent disposal option may allow reactor operators to thin-out spent nuclear fuel assemblies from densely packed spent fuel pools, potentially reducing the risk of harm to humans or the environment in the event of an accident, natural disaster, or terrorist event.

In addition, disposal is the only alternative for some DOE and commercial nuclear waste—even if the United States decided to reprocess the waste—because it contains nuclear waste residues that cannot be used as nuclear reactor fuel. This nuclear waste has no safe, long-term alternative other than disposal, and the Yucca Mountain repository would provide a near-term, permanent disposal pathway for it. Moreover, DOE has agreed to

---

[13]As of October 2, 2009, NRC had suspended or deferred five applications to build and operate six reactors at the request of the applicants.

remove spent nuclear fuel from at least two states by certain dates or face penalties. Specifically, DOE has an agreement with Colorado stating that if the spent nuclear fuel at Fort St. Vrain is not removed by January 1, 2035, the government will, subject to certain conditions, pay the state $15,000 per day until the waste is removed. In addition, the state of Idaho sued DOE to remove inventories of spent nuclear fuel stored at DOE's Idaho National Laboratory. Under the resulting settlement DOE agreed to (1) remove the spent nuclear fuel by January 1, 2035, or incur penalties of $60,000 per day and (2) curtail or suspend future shipments of spent nuclear fuel to Idaho.[14] Some of the spent nuclear fuel stored at the Idaho National Laboratory comes from refueling the U.S. Navy's submarines and aircraft carriers, all of which are nuclear powered. Special facilities are maintained at the Idaho National Laboratory to examine naval spent nuclear fuel to obtain information for improving future fuel performance and to package the spent nuclear fuel following examination to make it ready for rail shipment to its ultimate destination. According to Navy officials, refueling these warships, which necessitates shipment of naval spent nuclear fuel from the shipyards conducting the refuelings to the Idaho National Laboratory, is part of the Navy's national security mission. Consequently, curtailing or suspending shipments of spent nuclear fuel to Idaho raises national security concerns for the Navy.

The Yucca Mountain repository would help the government fulfill its obligation under NWPA to electric power companies and ratepayers to take custody of the commercial spent nuclear fuel and provide a permanent repository using the Nuclear Waste Fund. When DOE missed its 1998 deadline to begin taking custody of the waste, owners of spent fuel with contracts for disposal services filed lawsuits asking the courts to require DOE to fulfill its statutory and contractual obligations by taking custody of the waste. Though a court decided that it would not order DOE to begin taking custody of the waste, the courts have, in subsequent cases, ordered the government to compensate the utilities for the cost of storing the waste. DOE projected that, based on a 2020 date for beginning operations at Yucca Mountain, the government's liabilities from the 71 lawsuits filed by electric power companies could sum to about $12.3 billion, though the outcome of pending and future litigation could

---

[14]The penalties in the settlement agreement specifically apply to spent nuclear fuel and not to other high-level waste. However, the agreement specifies that DOE must have the other high-level waste treated and ready for shipment out of Idaho for disposal by 2035. DOE officials acknowledged that Idaho could take further court action if its milestones toward meeting these goals are not being met.

Page 15   GAO-10-48 Nuclear Waste Management

substantially affect the ultimate total liability.[15] DOE estimates that the federal government's future liabilities will average up to $500 million per year. Furthermore, continued delays in DOE's ability to take custody of the waste could result in additional liabilities. Some experts noted that without immediate plans for a permanent repository, reactor operators and ratepayers may demand that the Nuclear Waste Fund be refunded.[16]

Finally, disposing of the nuclear waste now in a repository facility would reduce the uncertainty about the willingness or the ability of future generations to monitor and maintain multiple surface waste storage facilities and would eliminate the need for any future handling of the waste. As a 2001 report of the National Academies noted, continued storage of nuclear waste is technically feasible only if those responsible for it are willing and able to devote adequate resources and attention to maintaining and expanding the storage facilities, as required to keep the waste safe and secure.[17] DOE officials noted that the waste packages at Yucca Mountain are designed to be retrievable for more than 100 years after emplacement, at which time DOE would begin to close the repository, allowing future generations to consider retrieving spent nuclear fuel for reprocessing or other uses. However, the risks and costs of retrieving the nuclear waste from Yucca Mountain are uncertain because planning efforts for retrieval are preliminary. Once closed, Yucca Mountain will require minimal monitoring and little or no maintenance, and all future controls will be passive.[18] Some experts stated that the current generation has a moral obligation to not pass on to future

---

[15]As of July 2009, of the 71 lawsuits filed by electric power companies, 51 cases were pending either in the Court of Federal Claims or in the Court of Appeals for the Federal Circuit, 10 had been settled, 6 were voluntarily withdrawn, and 4 had been litigated through final unappealable judgment.

[16]DOE estimated the Nuclear Waste Fund at about $23 billion in June 2009, some of which is interest that has accrued. DOE is required to invest the Nuclear Waste Fund in U.S. Treasury securities, resulting in the government paying about $11.2 billion interest to the fund. Both the principal and the interest might be returned, if the fund is returned to the electric power companies.

[17]National Research Council of the National Academies, *Disposition of High-Level Waste and Spent Nuclear Fuel: The Continuing Societal and Technical Challenges*, (Washington, D.C., 2001).

[18]Section 801 (c) of the Energy Policy Act of 1992 requires DOE to provide indefinite oversight to prevent any activity at the site that poses an unreasonable risk of (1) breaching the repository's engineered or geologic barriers or (2) increasing the exposure of the public to radiation beyond allowable limits. Pub. L. No. 102-486, 106 Stat. 2776, 2921-2922.

generations the extensive technical and financial responsibilities for managing nuclear waste in surface storage.

Yucca Mountain Faces Many Challenges, Including a Lack of Key Support and License Approval

There are many challenges to licensing and constructing the Yucca Mountain repository, some of which could delay or potentially terminate the program. First, in March 2009, the Secretary of Energy stated that the administration planned to terminate the Yucca Mountain repository and to form a panel of experts to review alternatives. During the testimony, the Secretary stated that Yucca Mountain would not be considered as one of the alternatives. The administration's fiscal year 2010 budget request for Yucca Mountain was $197 million, which is $296 million less than what DOE stated it needs to stay on its schedule and open Yucca Mountain by 2020.

In July 2009 letters to DOE, the Nuclear Energy Institute and the National Association of Regulatory Utility Commissioners raised concerns that, despite the announced termination of Yucca Mountain, DOE still intended on collecting fees for the Nuclear Waste Fund.[19] The letters requested that DOE suspend collection of payments to the Nuclear Waste Fund. Some states have raised similar concerns and legislators have introduced legislation that could hold payments to the Nuclear Waste Fund until DOE begins operating a federal repository.[20]

Nevertheless, NWPA still requires DOE to pursue geologic disposal at Yucca Mountain. If the administration continues the licensing process for Yucca Mountain, DOE would face a variety of other challenges in licensing and constructing the repository. Many of these challenges—though unique to Yucca Mountain—might also apply in similar form to other future repositories, should they be considered.

One of the most significant challenges facing DOE is to satisfy NRC that Yucca Mountain meets licensing requirements, including ensuring the repository meets EPA's radiation standards over the required 1 million year time frame, as implemented by NRC regulation. For example, NRC's

---

[19]The Nuclear Energy Institute represents the nuclear power industry and the National Association of Regulatory Utility Commissioners represents state public utility commissions that regulate the electric power industry.

[20]Minnesota House File No. 894, introduced February 16, 2009, and Michigan Senate Concurrent Resolution No. 8, introduced March 25, 2009.

GAO-10-48 Nuclear Waste Management

regulations require that DOE model its natural and engineered barriers in a performance assessment, including how the barriers will interact with each other over time and how the repository will meet the standards even if one or more barriers do not perform as expected. NRC has stated that there are uncertainties inherent in the understanding of the performance of the natural and engineered barriers and that demonstrating a reasonable expectation of compliance requires the use of complex predictive models supported by field data, laboratory tests, site-specific monitoring, and natural analog studies. The Nuclear Waste Technical Review Board has also stated that the performance assessment may be "the most complex and ambitious probabilistic risk assessment ever undertaken" and the Board, as well as other groups or individuals, have raised technical concerns about key aspects of the engineered or natural barriers in the repository design.

DOE and NRC officials also stated that budget constraints raise additional challenges. DOE officials told us that past budget shortfalls and projected future low budgets for the Yucca Mountain repository create significant challenges in DOE's ability to meet milestones for licensing and for responding to NRC's requests for additional information related to the license application. In addition, NRC officials told us budget shortfalls have constrained their resources. Staff members they originally hired to review DOE's license application have moved to other divisions within NRC or have left NRC entirely. NRC officials stated that the pace of the license review is commensurate with funding levels. Some experts have questioned whether NRC can meet the maximum 4-year time requirement stipulated in NWPA for license review and have pointed out that the longer the delays in licensing Yucca Mountain, the more costly and politically vulnerable the effort becomes.

In addition, the state of Nevada and other groups that oppose the Yucca Mountain repository have raised technical points, site-specific concerns, and equity issues and have taken steps to delay or terminate the repository. For example, Nevada's Agency for Nuclear Projects questioned DOE's reliance on engineered barriers in its performance assessment, indicating that too many uncertainties exist for DOE to claim human-made systems will perform as expected over the time frames required. In addition, the agency reported that Yucca Mountain's location near seismic and volcanic zones creates additional uncertainty about DOE's ability to predict a recurrence of seismic or volcanic events and to assess the performance of its waste isolation barriers should those events occur some time during the 1-million-year time frame. The agency also has questioned whether Yucca Mountain is the best site compared with other locations and has raised issues of equity, since Nevada is being asked to accept nuclear waste generated in other states. In addition to the Agency for Nuclear Projects' issues, Nevada has taken other steps to delay or terminate the project. For example, Nevada has denied the water rights DOE needs for construction of a rail spur and facility structures at Yucca Mountain. DOE officials told us that constructing the rail line or the facilities at Yucca Mountain without those water rights will be difficult.

Based on DOE's Cost Estimates, Yucca Mountain Will Likely Cost from $41 Billion to $67 Billion for 153,000 Metric Tons of Nuclear Waste, but Costs Could Increase

Our analysis of DOE's cost estimates found that (1) a 70,000 metric ton repository is projected to cost from $27 to $39 billion in 2009 present value over 108 years and (2) a 153,000 metric ton repository is projected to cost from $41 to $67 billion and take 35 more years to complete. These estimated costs include the licensing, construction, operation, and closure of Yucca Mountain for a period commensurate with the amount of waste. Table 1 shows each scenario with its estimated cost range over time.

Table 1: Estimated Cost of the Yucca Mountain Scenarios

Dollars in billions

Amount of nuclear waste disposed	Time period covered[a]	Present value estimate range[a]
70,000 metric tons	2009 to 2116 (108 years)	$27 to $39
153,000 metric tons	2009 to 2151 (143 years)	$41 to $67

Source: GAO analysis based on DOE data.

[a]These costs are in 2009 present value and thus different than the values presented by DOE which are in constant 2007 dollars. Also, these costs do not include more than $14 billion, in constant fiscal year 2009 dollars, that DOE spent from 1983 through 2008 for the Yucca Mountain repository. In addition, we did not include potential schedule delays and costs associated with licensing. DOE reported that each year of delay could cost DOE about $373 million in constant 2009 dollars.

As shown in figure 4, the Yucca Mountain repository costs are expected to be high during construction, followed by reduced, but consistent costs during operations, substantially reduced costs for monitoring, then a period of increased costs for installation of the drip shields, and finally costs tapering off for closure. Once the drip shields are installed, by design, the waste packages will no longer be retrievable. After closure, Yucca Mountain is not expected to incur any significant additional costs.

Figure 4: Cost Profile for the Yucca Mountain Repository, Assuming 70,000 Metric Tons

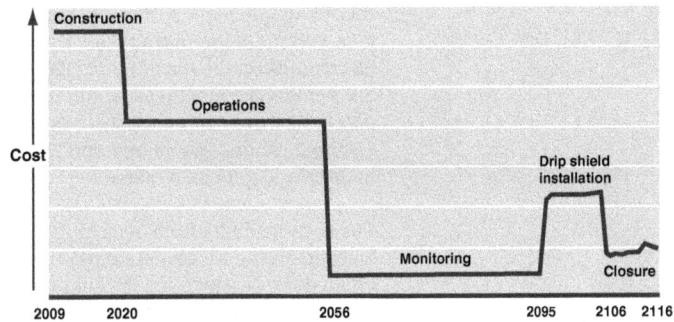

Source: GAO analysis of DOE data.

Costs for the construction of a repository, regardless of location, could increase based on a number of different scenarios, including delays in license application, funding shortfalls, and legal or technical issues that cause delays or changes in plans. For example, we asked DOE to assess the cost of a year's delay in license application approval from the current 3 years to 4 years, the maximum allowed by NWPA. DOE officials told us that each year of delay would cost DOE about $373 million in constant 2009 dollars. Although the experts with whom we consulted did not agree on how long the licensing process for Yucca Mountain might take, several experts told us that the 9 years it took Private Fuel Storage to obtain its license was not unreasonable. This licensing time frame may not directly apply to the Yucca Mountain repository because the repository has a significantly different licensing process and regulatory scheme, including extensive pre-licensing interactions, a federal funding stream, and an extended compliance period and, because of the uncertainties, could take shorter or longer than the Private Fuel Storage experience. A nine-year licensing process for construction authorization would add an estimated $2.2 billion to the cost of the repository, mostly in costs to maintain current systems, such as project support, safeguards and security, and its licensing support network. In addition to consideration of the issuance of a construction authorization, NRC's repository licensing process involves two additional licensing actions necessary to operate and close a repository, each of which allows for public input and could potentially adversely affect the schedule and cost of the repository. The second action is the consideration of an updated DOE application for a license to receive and possess high-level radioactive waste. The third action is the

GAO-10-48 Nuclear Waste Management

consideration of a DOE application for a license amendment to permanently close the repository. Costs could also increase if unforeseen technical issues developed. For example, some experts told us that the robotic emplacement of waste packages could be difficult because of the heat and radiation output from the nuclear waste, which could impact the electronics on the machinery. DOE officials acknowledged the challenges and told us the machines would have to be shielded for protection. They noted, however, that industry has experience with remote handling of shielded robotic machinery and DOE should be able to use that experience in developing its own machinery.

The responsibility for Yucca Mountain's costs would come from the Nuclear Waste Fund and taxpayers through annual appropriations. NWPA created the Nuclear Waste Fund as a mechanism for the nuclear power industry to pay for its share of the cost for building and operating a permanent repository to dispose of nuclear waste. NWPA also required the federal taxpayers to pay for the portion of permanent repository costs for DOE-managed spent nuclear fuel and high-level waste. DOE has responsibility for determining on an annual basis whether fees charged to industry to finance the Nuclear Waste Fund are sufficient to meet industry's share of costs. As part of that process, DOE developed a methodology in 1989 that uses the total system life cycle cost estimate as input for determining the shares of industry and the federal government by matching projected costs against projected assets. The most recent published assessment, published in July 2008, showed that 80.4 percent of the disposal costs would come from the Nuclear Waste Fund and 19.6 percent would come from appropriations for the DOE-managed spent nuclear fuel and high-level waste.

In addition, the Department of the Treasury's judgment fund will pay the government's liabilities for not taking custody of the nuclear waste in 1998, as required by DOE's contract with industry. Based on existing judgments and settlements, DOE has estimated these costs at $12.3 billion through 2020 and up to $500 million per year after that, though the outcome of pending litigation could substantially affect the government's ultimate liability. The Department of Justice has also spent about $150 million to defend DOE in the litigation.

GAO-10-48 Nuclear Waste Management

We Identified Two Nuclear Waste Management Alternatives and Developed Cost Models by Consulting with Experts

We used input from experts to identify two nuclear waste management alternatives that could be implemented if the nation does not pursue disposal at Yucca Mountain—centralized storage and continued on-site storage, both of which could be implemented with final disposal, according to experts. To understand the implications and likely assumptions of each alternative, as well as the associated costs for the component parts, we systematically solicited facts, advice, and opinions from experts in nuclear waste management. Finally, we used the data and assumptions that the experts provided to develop large-scale cost models that estimate ranges of likely total costs for each alternative.

We Consulted with Experts to Identify and Develop Assumptions for Two Generic Alternatives to Analysis

To identify waste management alternatives that could be implemented if the waste is not disposed of at Yucca Mountain, we solicited facts, advice, and opinions from nuclear waste management experts. Specifically, we interviewed dozens of experts from DOE, NRC, the Nuclear Energy Institute, the National Association of Regulatory Utility Commissioners, the National Conference of State Legislatures, and the State of Nevada Agency for Nuclear Projects. We also reviewed documents they provided or referred us to.

Based on this information, we chose to analyze (1) centralized interim dry storage and (2) on-site dry storage (both interim and long-term). Centralized storage has been attempted to varying degrees in the United States, and on-site storage has become the country's status quo. Consequently, the experts believe these two alternatives are currently among the most likely for this country in the near-term, in conjunction with final disposal in the long-term. The experts also told us that current nuclear waste reprocessing technologies raise proliferation concerns and are not considered commercially feasible, but they noted that reprocessing has future potential as a part of the nation's nuclear waste management strategy. Because nuclear waste is not reprocessed in this country, we found a lack of sufficient and reliable data to provide meaningful analysis for this alternative. Experts have largely dismissed other alternatives that have been identified, such as disposal of waste in deep boreholes, because of cost or technical constraints.

We developed a set of key assumptions to establish the scope of our alternatives by initially consulting with a small group of nuclear waste management experts. For example, we asked the experts about how many storage sites should be used and whether waste would have to be repackaged. These discussions occurred in an iterative manner—we followed up with experts with specific expertise to refine our assumptions as we learned more. Based on this input, we formulated several key assumptions and defined the alternatives in a generic manner by taking into account some, but not all, of the complexities involved with nuclear waste management (see table 2). We made this choice because experts advised us that trying to consider all of the variability among reactor sites would result in unmanageable models since each location where nuclear waste is currently stored has a unique set of environmental, management, and regulatory considerations that affect the logistics and costs of waste management. For example, reactor sites use different dry cask storage systems with varying costs that require different operating logistics to load the casks.

Table 2: Key Assumptions Used to Define Alternatives

Centralized storage	
Type of storage	Conventional dry cask storage (for commercial spent nuclear fuel).
Number of sites	Two centralized interim storage sites, located in different geographic regions of the country.
Reactor operations	All currently operating reactors receive a 20-year license extension and continue operating until the extensions expire. Reactors will be decommissioned when operations cease, and only spent nuclear fuel dry storage will remain on site.
Transportation	Transportation to the centralized site will be via rail using dedicated trains.
Repackaging	Waste will not be repackaged at the centralized facilities.
Final disposition[a]	After 100 years, the waste will be disposed of in a geologic repository.
On-site storage	
Type of storage	Conventional dry cask storage (for commercial spent nuclear fuel).
Number of sites	Commercial spent nuclear fuel will be stored on independent spent fuel storage installations at 75 reactor sites, which includes operating reactor sites, decommissioned reactor sites, and the Morris facility.[b] DOE high-level waste and spent nuclear fuel will remain at five current sites.[c] DOE spent nuclear fuel will be moved to dry storage. DOE high-level waste will be vitrified and stored in facilities like the Glass Waste Storage Building at the Savannah River Site.
Reactor operations	All currently operating reactors receive a 20-year license extension and continue operating until the extensions expire. Reactors will be decommissioned when operations cease, and only spent nuclear fuel dry storage will remain on site.
Transportation	Waste will not be transported between reactor sites.
Repackaging	Dry cask storage systems will need to be replaced after 100 years, requiring repackaging into new inner canisters and outer casks. Only our 500-year on-site storage model assumes repackaging.
Final disposition or long-term management[c]	We analyzed two final disposition scenarios: The waste will be disposed of in a geologic repository after 100 years or the waste will remain on site for 500 years and be repackaged every 100 years.

Source: GAO analysis based on expert-provided data.

[a]We analyzed some scenarios associated with these alternatives that did not include final disposition of the waste.

ᵇThe Morris facility is an independent spent nuclear fuel storage installation located in Illinois that is operated by General Electric Corporation, which originally intended to operate a fuel reprocessing plant at the site. The Morris facility is the only spent nuclear fuel pool licensed by NRC that is not at a reactor site.

ᶜHanford Reservation, Washington; Idaho National Laboratory, Idaho; Fort St. Vrain, Colorado; West Valley, New York; and Savannah River Site, South Carolina.

In addition, there were some instances in which we made assumptions that, while not entirely realistic, were necessary to keep our alternatives generic and distinct from one another. For example, some electric power companies would likely consolidate nuclear waste from different locations by transporting it between reactor sites, but to keep the on-site storage alternative generic and distinct from the centralized storage alternative, we assumed that there would be no consolidation of waste. These simplifying assumptions make our alternatives hypothetical and not entirely representative of their real-world implementation.

We also consulted with experts to formulate more specific assumptions about processes that reflect the sequence of activities that would occur within each alternative (see fig. 5). In addition, we identified the components of these processes that have associated costs. For example, one of the processes associated with both alternatives is packaging the nuclear waste in dry storage canisters from the pools of water where they are stored. The component costs associated with this process include the dry storage canisters and operations to load the spent nuclear fuel into the canisters.

Figure 5: Process Assumptions and Cost Components for Hypothetical Nuclear Waste Management Alternatives

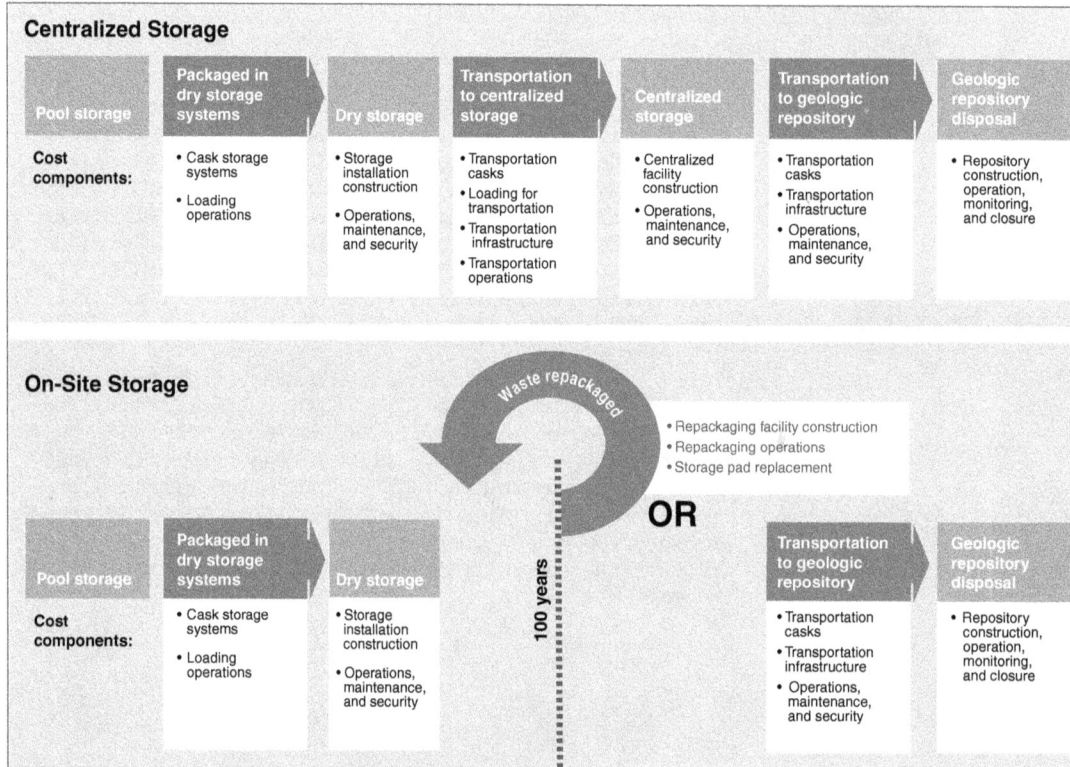

Centralized Storage

Pool storage	Packaged in dry storage systems	Dry storage	Transportation to centralized storage	Centralized storage	Transportation to geologic repository	Geologic repository disposal

Cost components:

- Cask storage systems
- Loading operations

- Storage installation construction
- Operations, maintenance, and security

- Transportation casks
- Loading for transportation
- Transportation infrastructure
- Transportation operations

- Centralized facility construction
- Operations, maintenance, and security

- Transportation casks
- Transportation infrastructure
- Operations, maintenance, and security

- Repository construction, operation, monitoring, and closure

On-Site Storage

Waste repackaged

- Repackaging facility construction
- Repackaging operations
- Storage pad replacement

OR

100 years

Pool storage	Packaged in dry storage systems	Dry storage		Transportation to geologic repository	Geologic repository disposal

Cost components:

- Cask storage systems
- Loading operations

- Storage installation construction
- Operations, maintenance, and security

- Transportation casks
- Transportation infrastructure
- Operations, maintenance, and security

- Repository construction, operation, monitoring, and closure

Source: GAO analysis based on expert-provided data.

We then began to gather data on specific processes and component costs, such as the kind of cask systems we would use in our model and their cost. We gathered initial data from a core group of experts with specialized knowledge in different aspects of nuclear waste management, such as cask systems, waste loading operations, and transportation. We then solicited comments on the initial data from a broader group of experts using a data collection instrument that asked specific questions about how reasonable the data were. We received almost 70 sets of comments and used them to refine or modify our assumptions and component costs and develop the input data that we would use to estimate

Page 25   GAO-10-48 Nuclear Waste Management

the overall costs of the alternatives. (See app. I for additional information about our scope and methodology, app. II for our methodology for soliciting comments from nuclear waste management experts, and app. III for these experts.)

We Developed Cost Ranges for Each Alternative Using Large-scale Cost Models that Addressed Uncertainties and Discounted Future Costs

To generate cost ranges for the centralized storage and on-site storage alternatives, we developed four large-scale cost models that analyzed the costs for each alternative of storing 70,000 metric tons and 153,000 metric tons of nuclear waste and created scenarios within these models to analyze different storage durations and final dispositions. (See table 3.) We generated cost ranges for each alternative for storing 153,000 metric tons of waste for 100 years followed by disposal in a geologic repository. We also generated cost ranges for each alternative of storing 70,000 metric tons and 153,000 metric tons of nuclear waste for 100 years, and for storing 153,000 metric tons of waste on site for 500 years without including the cost of subsequent disposal in a geologic repository. For each of the models, which rely upon data and assumptions provided by nuclear waste management experts, the cost range was based on the annual volume of commercial spent nuclear fuel that became ready to be packaged and stored in each year.[21] In general, each model started in 2009 by annually tracking costs of initial packaging and related costs for the first 100 years and for every 100 years thereafter if the waste was to remain on site and be repackaged. Since our models analyzed only the costs associated with storing commercial nuclear waste management, we augmented them with DOE's cost data for (1) managing its spent nuclear fuel and high-level waste and (2) constructing and operating a permanent repository. Specifically, we used DOE's estimated costs for the Yucca Mountain repository to represent cost for a hypothetical permanent repository.[22]

---

[21]NWPA caps the amount of nuclear waste that can be disposed of at Yucca Mountain at 70,000 metric tons. The estimated amount of current waste plus additional commercial spent nuclear fuel that would be generated if all currently operating commercial reactors received license extensions is 153,000 metric tons. Our analysis did not consider new reactors because of the uncertainty if or when new reactors would be built, how many would be built, and their impact on waste streams.

[22]We excluded historical costs for the Yucca Mountain repository because these costs represent challenges unique to Yucca Mountain and may not be applicable to a future repository. However, the bulk of future cost for construction, operation, and closure may be representative of a new repository.

Table 3: Models and Scenarios Used for Cost Ranges

Model		Scenario	
Nuclear waste management alternative	Waste volume (metric tons)	Storage duration (years)	Final disposition or long-term management
On-site storage	153,000	100	None
		100	Permanent repository
		500	Waste repackaged every 100 years
On-site storage	70,000	100	None
Centralized storage	153,000	100	None
		100	Permanent repository
Centralized storage	70,000	100	None

Source: GAO analysis.

One of the inherent difficulties of analyzing the cost of any nuclear waste management alternative is the large number of uncertainties that need to be addressed. In addition to general uncertainty about the future, there is uncertainty because of the lack of knowledge about the waste management technologies required, the type of waste and waste management systems that individual reactors will eventually employ, and cost components that are key inputs to the models and could occur over hundreds or thousands of years. Given these numerous uncertainties, it is not possible to precisely determine the total costs of each alternative. However, much of the uncertainty that we could not easily capture within our models can be addressed through the use of several alternative models and scenarios. As shown in table 3, we developed two models for each alternative to address the uncertainty regarding the total volume of waste for disposal. We then developed different scenarios within each model to address different time frames and disposal paths. Furthermore, we used a risk analysis modeling technique that recognized and addressed uncertainties in our data and assumptions. Given the different possible scenarios and uncertainties, we generated ranges, rather than point estimates, for analyzing the cost of each alternative.

One of the most important uncertainties in our analysis was uncertainty over component costs. To address this, we used a commercially available risk analysis software program that enabled us to model specific uncertainties associated with a large number of cost inputs and assumptions. Using a Monte Carlo simulation process,[23] the program explores a wide range of values, instead of one single value, for each cost input and estimates the total cost. By repeating the calculations thousands of times with a different set of randomly chosen input values, the process produces a range of total costs for each alternative and scenario. The process also specifies the likelihood associated with values in the estimated range.

Another inherent difficulty in estimating the cost of nuclear waste management alternatives is the fact that the costs are spread over hundreds or thousands of years. The economic concept of discounting is central to such long-term analysis because it allows us to convert costs that occur in the distant future to present value—equivalent values in today's dollars. Although the concept of discounting is an accepted and standard methodology in economics, the concept of discounting values over a very distant future—known as "intergenerational discounting"—is still subject to considerable debate. Furthermore, no consensus exists among economists regarding the exact value of the discount rate that should be used to discount values that are spread over many hundreds or thousands of years.

To develop an appropriate discounting methodology and to choose the discount rates for our analysis, we reviewed a number of economic studies published in peer-reviewed journals that addressed intergenerational discounting. Based on our review, we designed a discounting methodology for use in our models. Because our review did not find a consensus on discount rates, we used a range of values for discount rates that we developed based on the economic studies we reviewed, rather than using one single rate. Consequently, because we used ranges for the discount rate along with the Monte Carlo simulation process, the present value of estimated costs does not depend on one single discount rate, but rather reflect a range of discount rate values taken from peer-reviewed studies. (See app. IV for details of our modeling and discounting methodologies, assumptions, and results.)

---

[23]We used a commercially available risk analysis program called Crystal Ball for our Monte Carlo simulation. Crystal Ball is a commonly used spreadsheet-based software for predictive modeling and forecasting.

Centralized Storage Would Provide a Near-Term Alternative, Allowing Other Options to Be Studied, but Faces Implementation Challenges

Centralized storage would provide a near-term alternative for managing nuclear waste, allowing the government to begin taking possession of the waste within approximately the next 30 years, and giving additional time for the nation to consider long-term waste management options. However, centralized storage does not preclude the need for final disposal of the waste. In addition, centralized storage faces several implementation challenges including that DOE (1) lacks statutory authority to provide centralized storage under NWPA, (2) is expected to have difficulty finding a location willing to host a centralized storage facility, and (3) faces potential transportation risks. The estimated cost of implementing centralized storage for 100 years ranges from $15 billion to $29 billion for 153,000 metric tons of nuclear waste, and the total cost ranges from $23 billion to $81 billion if the nuclear waste is centrally stored and then disposed in a geologic repository.

Centralized Storage Would Provide a Near-Term Alternative to Managing Nuclear Waste but Does Not Eliminate the Need for Final Disposal

As the administration re-examines the Yucca Mountain repository and national nuclear waste policy, centralized dry cask storage could provide a near-term alternative for managing the waste that has accumulated and will continue to accumulate. This would provide additional time—NRC has stated that spent nuclear fuel storage is safe and environmentally acceptable for a period on the order of 100 years—to consider other long-term options that may involve alternative policies and new technologies and allow some flexibility for their implementation. For example, centralized storage would maintain nuclear waste in interim dry storage configurations so that it could be easily accessible for reprocessing in case the nation decided to pursue reprocessing as a waste management option and developed technologies that address current proliferation and cost concerns. In fact, reprocessing facilities could be built near or adjacent to centralized facilities to maximize efficiencies. However, even with reprocessing, some of the spent nuclear fuel and high-level waste in current inventories would require final disposal.

Centralized storage would consolidate the nation's nuclear waste after reactors are decommissioned, thereby decreasing the complexity of securing and overseeing the waste and increasing the efficiency of waste storage operations. This alternative would remove nuclear waste from all DOE sites and nine shutdown reactor sites that have no operations other than nuclear waste storage, allowing these sites to be closed. Some of these storage sites occupy land that potentially could be used for other purposes, imposing an opportunity cost on states and communities that no longer receive the benefits of electricity generation from the reactors. To compensate for this loss, industry officials noted that at least two states

GAO-10-48 Nuclear Waste Management

where decommissioned sites are located have tried to raise property taxes on the sites, and at one site, the state collects a per cask fee for storage. In addition, the continued storage of nuclear waste at decommissioned sites can cost the power companies between about $4 million and $8 million per year, according to several experts.

Centralized storage could allow reactor operators to thin-out spent nuclear fuel assemblies from densely packed spent fuel pools and may also prevent operating reactors from having to build the additional dry storage capacity they would need if the nuclear waste remained on site. According to an industry official, 28 reactor sites could have to add dry storage facilities over the next 10 years in order to maintain a desired capacity in their storage pools. These dry storage facilities could cost about $30 million each, but this cost would vary widely by site. In addition, some current reactor sites use older waste storage systems and are near large cities or large bodies of fresh water used for drinking or irrigation. Although NRC's licensing and inspection process is designed to ensure that these existing facilities appropriately protect public health and safety, new centralized facilities could use state-of-the-art design technology and be located in remote areas with fewer environmental hazards, in order to protect public health and enhance safety.

Finally, if DOE uses centralized facilities to store commercial spent nuclear fuel, this alternative could allow DOE to fulfill its obligation to take custody of the commercial spent nuclear fuel until a long-term strategy is implemented. As a result, DOE could curtail its liabilities to the electric power companies, potentially saving the government up to $500 million per year after 2020, as estimated by DOE. The actual impact of centralized storage on the amount of the liabilities would depend on several factors, including when centralized storage is available, whether reactor sites had already built on-site dry storage facilities for which the government may be liable for a portion of the costs, how soon waste could be transported to a centralized site, and the outcome of pending litigation that may affect the government's total liability. DOE estimates that if various complex statutory, regulatory, siting, construction, and financial issues were expeditiously resolved, a centralized facility to accept nuclear waste could begin operations as early as 6 years after its development began. However, a centralized storage expert estimated that the process from site selection until a centralized facility opens could take between 17 and 33 years.

Although centralized storage has a number of positive attributes, it provides only an interim alternative and does not eliminate the need for final disposal of the nuclear waste. To keep the waste safe and secure, a centralized storage facility relies on active institutional controls, such as monitoring, maintenance, and security. Over time, the storage systems may degrade and institutional controls may be disrupted, which could result in increased risk of radioactive exposure to humans or the environment. For example, according to several experts on dry cask systems, the vents on the casks—which allow for passive cooling—must be periodically inspected to ensure no debris clogs them, particularly during the first several decades when the spent nuclear fuel is thermally hot. If the vents become clogged, the temperature in the canister could rise, which could impact the life of the dry cask storage system. Over a longer time frame, concrete on the exterior casks could degrade, requiring more active maintenance. Although some experts stated that the risk of radiation being released into the environment may be low, such risks can be avoided by permanently isolating the waste in a manner that does not require indefinite, active institutional controls, such as disposal in a geologic repository.

Legal and Community Challenges Contribute to the Complexity of Implementing Centralized Storage

A key challenge confronting the centralized storage alternative is the lack of authority under NWPA for DOE to provide such storage. Provisions in NWPA that allow DOE to arrange for centralized storage have either expired or are unusable because they are tied to milestones in repository development that have not been met. For example, NWPA authorized DOE to provide temporary storage for a limited amount of spent nuclear fuel until a repository was available, but this authority expired in 1990. Some industry representatives have stated that DOE still has the authority to accept and store spent nuclear fuel under the Atomic Energy Act of 1954, as amended, but DOE asserts that NWPA limits its authority under the Atomic Energy Act.[24] In addition, NWPA provided authority for DOE to site, construct, and operate a centralized storage facility, but such a facility could not be constructed until NRC authorized construction of the Yucca Mountain repository, and the facility could only store up to 10,000 metric

---

[24]DOE acknowledged that the Atomic Energy Act of 1954, as amended, does provide the authority for DOE to accept and store spent nuclear fuel under certain circumstances, which DOE has used in the past to accept and store spent nuclear fuel. For example, pursuant to the Atomic Energy Act authority, DOE has accepted and stored U.S.-supplied spent nuclear fuel from foreign reactors, as well as damaged spent nuclear fuel from the Three Mile Island reactor site. However, DOE asserts that the NWPA's detailed statutory scheme limits its authority to accept spent nuclear fuel under Atomic Energy Act authority except in compelling circumstances, such as an emergency involving spent nuclear fuel threatening public health.

Page 31   GAO-10-48 Nuclear Waste Management

tons of nuclear waste until the repository started accepting spent nuclear fuel. Therefore, unless provisions in NWPA were amended, centralized storage would have to be funded, owned, and operated privately. A privately operated centralized storage facility alternative, such as the proposed Private Fuel Storage Facility in Utah, would not likely resolve DOE's liabilities with the nuclear power companies.[25]

A second, equally important, challenge to centralized storage is the likelihood of opposition during site selection for a facility. Experts noted that affected states and communities would raise concerns about safety, security, and the likelihood that an interim centralized storage facility could become a de facto permanent storage site if progress is not being made on a permanent repository. Even if a local community supports a centralized storage facility, the state may not. For example, the Private Fuel Storage facility was generally supported by the Skull Valley Band of the Goshute Indians, on whose reservation the facility was to be located, but the state of Utah and some tribal members opposed its licensing and construction. Other states have indicated their opposition to involuntarily hosting a centralized facility through means such as the Western Governors' Association, which issued a resolution stating that "no such facility, whether publicly or privately owned, shall be located within the geographic boundaries of a Western state without the written consent of the governor."[26] Some experts noted that a state or community may be willing to serve as a host if substantial economic incentives were offered and if the party building the site undertook a time-consuming and expensive process of site characterization and safety assessment. However, DOE officials stated that in their previous experience—such as with the Nuclear Waste Negotiator about 15 to 20 years ago—they have found no incentive package that has successfully encouraged a state to voluntarily host a site.

A third challenge to centralized storage is that nuclear waste would likely have to be transported twice—once to the centralized site and once to a permanent repository—if a centralized site were not colocated with a

---

[25]In addition, lawsuits filed against the government by nuclear reactor owners have included claims to recover the cost of the Private Fuel Storage facility. At least one utility has recovered these costs from the government, while a court did not allow another utility to recover these costs.

[26]Western Governors' Association Policy Resolution 09-5: Interim Storage and Transportation of Commercial Spent Nuclear Fuel.

Page 32   GAO-10-48 Nuclear Waste Management

repository.[27] Therefore, the total distance over which nuclear waste is transported is likely to be greater than with other alternatives, an important factor because, according to one expert, transportation risk is directly tied to this distance. However, according to DOE, nuclear waste has been safely transported in the United States since the 1960s and National Academy of Sciences, NRC, and DOE-sponsored reports have found that the associated risks are well understood and generally low. Yet, there are also perceived risks associated with nuclear waste transportation that can result in lower property values along transportation routes, reductions in tourism, and increased anxiety that create community opposition to nuclear waste transportation. According to experts, transportation risks could be mitigated through such means as shipping the least radioactive fuel first, using trains that only transport nuclear waste, and identifying routes that minimize possible impacts on highly populated areas. In addition, the hazards associated with transportation from a centralized facility to a repository would decline as the waste decayed and became less radioactive at the centralized facility.

Cost Ranges for Centralized Storage Will Vary Depending on Waste Volume and Final Disposition

As shown in table 4, our models generated cost ranges from $23 billion to $81 billion for the centralized storage of 153,000 metric tons of spent nuclear fuel and high-level waste for 100 years followed by geologic disposal. For centralized storage without disposal, costs would range from $12 billion to $20 billion for 70,000 metric tons of waste and from $15 billion to $29 billion for 153,000 metric tons of waste. These centralized model scenarios include the cost of on-site operations required to package and prepare the waste for transportation, such as storing the waste in dry-cask storage until it is transported off site, developing and operating a system to transport the waste to centralized storage, and constructing and operating two centralized storage facilities. (See app. IV for information about our modeling methodology, assumptions, and results.)

---

[27]NWPA prohibits development of a centralized storage facility in any state where a site is being characterized for development of a repository.

Table 4: Estimated Cost Range for Each Centralized Storage Scenario

Dollars in billions

Centralized storage scenario	Time period covered[a]	2009 present value estimate range
Storage of 70,000 metric tons	2009 to 2108 (100 years)	$12 to $20
Storage of 153,000 metric tons	2009 to 2108 (100 years)	$15 to $29
Storage of 153,000 metric tons, with disposal in a permanent repository after 100 years	2009 to 2240 (232 years[b])	$23 to $81

Source: GAO analysis of data provided by nuclear waste management experts and DOE.

[a]See appendix IV for an explanation of the periods covered by the scenarios.

[b]This period was chosen to capture costs of the hypothetical geologic repository through closure.

Actual centralized storage costs may be more or less than these cost ranges if a different centralized storage scenario is implemented. For example, our models assume that there would be two centralized facilities, but licensing, construction, and operations and maintenance costs would be greater if there were more than two facilities and lower if there was only one facility. Some experts told us that centralized storage would likely be implemented with only one facility because it would be too difficult to site two. But other experts noted that having more sites could reduce the number of miles traveled by the waste and provide a greater degree of geographic equity. The length of time the nuclear waste is stored could also impact the cost ranges, particularly if the nuclear waste were stored for less than or more than the time period assumed in our model. For periods longer than 100 years, experts told us that the dry storage cask systems may be subject to degradation and require repackaging, substantially raising the costs, as well as the level of uncertainty in those costs. Transportation is another area where costs could vary if, for example, transportation was not by rail or if the transportation system differed significantly from what is assumed in our models.

Furthermore, costs could be outside our ranges if the final disposition of the waste is different. Our scenario that includes geologic disposal is based on the current cost projections for Yucca Mountain, but these costs could be significantly different for another repository site or if much of the nuclear waste is reprocessed. A different geologic repository would have unique site characterization costs, may use an entirely different design than Yucca Mountain, and may be more or less difficult to build. Also, reprocessing could contribute significantly to the cost of an alternative.

For example, we previously reported that construction of a reprocessing plant with an annual production throughput of 3,000 metric tons of spent nuclear fuel could cost about $44 billion.[28] Studies analyzed by the Congressional Budget Office estimate that once a reprocessing plant is constructed, spent nuclear fuel could be reprocessed at between $610,000 and $1.4 million per-metric-ton, when adjusted to 2009 constant dollars.[29] This would result in an annual cost of about $2 billion to $4 billion, assuming a throughput of 3,000 metric tons per year.

Finally, the actual cost of implementing one of our centralized storage scenarios would likely be higher than our estimated ranges indicate because our models omit several location-specific costs. These costs could not be quantified in our generic models because we did not make an assumption about the specific location of the centralized facilities. For example, a few experts noted that incentives may be given a state or locality as a basis for allowing a centralized facility to be built, but the incentive amount may vary from location to location based on what agreement is reached. Also, several experts said that rail construction may be required for some locations, which could add significant cost depending on the distance of new rail line required at a specific location. Experts could not provide data for these location-dependent costs to any degree of certainty, so we did not use them in our models. Also, the funding source for government-run centralized storage is unclear. The Nuclear Waste Fund, which electric power companies pay into, was established by NWPA to fund a permanent repository and cannot be used to pay for centralized storage without amending the act. Without such a change, the cost for the federal government to implement this alternative would likely have to be borne by the taxpayers.

---

[28]GAO, *Global Nuclear Energy Partnership: DOE Should Reassess Its Approach to Designing and Building Spent Nuclear Fuel Recycling Facilities,* GAO-08-483 (Washington, D.C.: April 2008).

[29]The studies used in the Congressional Budget Office's analysis were: Boston Consulting Group, *Economic Assessment of Used Nuclear Fuel Management in the United States* (study prepared for AREVA Inc., July 2006); and Matthew Bunn and others, *The Economics of Reprocessing vs. Direct Disposal of Spent Nuclear Fuel,* Belfer Center for Science and International Affairs, John F. Kennedy School of Government, Harvard University, (Cambridge, Massachusetts, December 2003).

On-Site Storage Would Provide an Intermediate Option with Minimal Effort but Poses Challenges that Could Increase Over Time

On-site storage of nuclear waste provides an intermediate option to manage the waste until the government can take possession of it, requiring minimal effort to change from what the nation is currently doing to manage its waste. In the meantime, other longer term policies and strategies could be considered. Such strategies would eventually be required because the on-site storage alternative would not eliminate the need for final disposal of the waste. Some experts believe that legal, community, and technical challenges associated with on-site storage will intensify as the waste remains on site without plans for final disposition because, for example, communities are more likely to oppose recertification of on-site storage. The estimated cost to continue storing 153,000 metric tons of nuclear waste on site for 100 years range from $13 billion to $34 billion, and total costs would range from $20 billion to $97 billion if the nuclear waste is stored on site for 100 years and then disposed in a geologic repository.

On-Site Storage Would Require Minimal Near-Term Logistics and Provide Time to Decide on Long-Term Waste Management Strategies

Because of delays in the Yucca Mountain repository, on-site storage has continued as the nation's strategy for managing nuclear waste, thus its continuation would require minimal near-term effort and allow time for the nation to consider alternative long-term nuclear waste management options. This alternative maintains the waste in a configuration where it is readily retrievable for reprocessing or other disposition, according to an expert. However, like centralized storage, on-site storage is an interim strategy that relies on active institutional controls, such as monitoring, maintenance, and security. To permanently isolate the waste from humans and the environment without the need for active institutional controls some form of final disposal would be required, even if some of the waste were reprocessed.

The additional time in on-site storage may also make the waste safer to handle because older spent nuclear fuel and high-level waste has had a chance to cool and become less radioactive. As a result, on-site storage could reduce transportation risks, particularly in the near-term, since the nuclear waste would be cooler and less radioactive when it is finally transported to a repository. In addition, some experts state that older, cooler waste may provide more predictability in repository performance and be some degree safer than younger, hotter waste. However, NRC cautioned that the ability to handle the waste more safely in the future also depends on other factors, including how the waste or waste packages might degrade over time. In particular, NRC stated that there are many uncertainties with the behavior of spent nuclear fuel as it ages, such as potential fracturing of the structural assemblies, possibly increasing the risks of release. If the waste has to be repackaged, for example, the process may require additional safety measures. Some experts noted that continuing to store nuclear waste on site would be more equitable than consolidating it in one or a few areas. As a result, the waste, along with its associated risks, would be kept in the location where the electrical power was generated, leaving the responsibility and risks of the waste in the communities that benefited from its generation.

On-Site Storage Poses Legal, Community, and Technical Challenges that Are Likely to Intensify over Time

With on-site storage of DOE-managed spent nuclear fuel and high-level waste, DOE would have difficulty meeting enforceable agreements with states, which could result in significant costs being incurred the longer spent nuclear fuel remains on site. In addition to Idaho's agreement to impose a penalty of $60,000 per day if spent nuclear fuel is not removed from the state by 2035, DOE has an agreement with Colorado stating that if the spent fuel at Fort St. Vrain is not removed by January 1, 2035, the government will, subject to certain conditions, pay the state $15,000 per day until it is removed. Other states where DOE spent nuclear fuel and high-level waste are currently stored may seek similar penalties if the spent fuel and waste remain on-site with no progress toward a permanent repository or centralized storage facility.

A second challenge is the cost due to the government's possible legal liabilities to commercial reactor operators. Leaving waste on site under the responsibility of the electric power companies does not relieve the government of its obligation to take custody of the waste, thus the liability debt could continue to mount. For every year after 2020 that DOE fails to take custody of the waste in accordance with its contracts with the reactor operators, DOE estimates that the government will continue to accumulate up to $500 million per year beyond the estimated $12 billion in liabilities that will have accrued up to that point; however, the outcome of pending litigation could substantially affect the government's total liability.[30] The government will no longer incur these costs if DOE takes custody of the waste. Some representatives from industry have stated that it is not practical for DOE to take custody of the waste at commercial reactor sites. Moreover, some electric power company executives have stated that their ratepayers are paying for DOE to provide a geologic repository through

---

[30]Legislative action by the Congress could also affect the amount of compensation the government ultimately pays to the reactor operators. For example, the Congress could amend NWPA to change contract provisions that would be applicable to newly constructed reactors.

Page 37   GAO-10-48 Nuclear Waste Management

their contributions to the Nuclear Waste Fund, and the executives believe that simply taking custody of the waste is not sufficient. A DOE official stated that if DOE were to take custody of the waste on site, it would be a complex undertaking due to considerations such as liability for accidents.

Third, continued use of on-site storage would likely also face community opposition. Some experts noted that without progress on a centralized storage facility or repository site to which waste will be moved, some state and local opposition to reactor storage site recertification will increase, and so will challenges to nuclear power companies' applications for reactor license extensions and combined licenses to construct and operate new reactors. Also, experts noted that many commercial reactor sites are not suitable for long-term storage, and none has had an environmental review to assess the impacts of storing nuclear waste at the site beyond the period for which it is currently licensed. One expert noted that if on-site storage were to become a waste management policy, the long-term health, safety, and environmental risks at each site would have to be evaluated. Because waste storage would extend beyond the life of nuclear power reactors, decommissioned reactor sites would not be available for other purposes, and the former reactor operators may have to stay in business for the sole purpose of storing nuclear waste.

Finally, although dry cask storage is considered reliable in the short term, the longer-term costs, maintenance requirements, and security requirements are not well understood. Many experts said waste packages will likely retain their integrity for at least 100 years, but eventually dry storage systems may begin to degrade and the waste in those systems would have to be repackaged. However, commercial dry storage systems have only been in existence since 1986, so nuclear utilities have little experience with long-term system degradation and requirements for repackaging. Some experts suggested that only the outer protective cask would require replacement, but the inner canister would not have to be replaced. Yet, other experts said that, over time, the inner canister would also be exposed to environmental conditions by vents in the outer cask, which could cause corrosion and require a total system replacement. In addition, experts disagreed on the relative safety risks and costs associated with using spent fuel pools to transfer the waste during repackaging compared to using a dry transfer system, which industry representatives said had not been used on a commercial scale. Finally, future security requirements for extended storage are uncertain because as spent nuclear waste ages and becomes cooler and less radioactive, it becomes less lethal to anyone attempting to handle it without protective shielding. For example, a spent nuclear fuel assembly can lose nearly 80 percent of its heat 5 years after it has been removed from a reactor, thereby reducing one of the inherent deterrents to thieves and terrorists attempting to steal or sabotage the spent nuclear fuel and potentially creating a need for costly new security measures.

Cost Ranges for On-Site Storage Will Vary Depending on Waste Volume, Final Disposition, and Duration of Storage

As shown in table 5, our models generated cost ranges from $20 billion to $97 billion for the on-site storage of 153,000 metric tons of spent nuclear fuel and high-level waste for 100 years followed by geologic disposal. For only on-site storage for 100 years without disposal, costs would range from $10 billion to $26 billion for 70,000 metric tons of waste and from $13 billion to $34 billion for 153,000 metric tons of waste. On-site storage costs would increase significantly if the waste were stored for longer periods—storing 153,000 metric tons on site for 500 years would cost from $34 billion to $225 billion—because it would have to be repackaged every 100 years for safety. The on-site storage model scenarios include the costs of on-site operations required to package the waste into dry canister storage, build additional dry storage at the reactor sites, prepare the waste for transportation, and operate and maintain the on-site storage facilities. Most of the costs for the first 100 years would result from the initial loading of materials into dry storage systems. (See app. IV for information on our modeling methodology, assumptions, and results.)

Table 5: Estimated Cost Range for Each On-site Storage Scenario

Dollars in billions

On-site storage scenario	Period covered[a]	2009 present value estimate range
Storage of 70,000 metric tons	2009 to 2108 (100 years)	$10 to $26
Storage of 153,000 metric tons	2009 to 2108 (100 years)	$13 to $34
Storage of 153,000 metric tons, with disposal in a permanent repository after 100 years	2009 to 2240 (232 years[b])	$20 to $97
Storage of 153,000 metric tons with repackaging every 100 years	2009 to 2508 (500 years)	$34 to $225

Source: GAO analysis of data provided by nuclear waste management experts and DOE.

[a]See appendix IV for an explanation of the periods covered by the scenarios.

[b]This period was chosen to capture costs of the hypothetical geologic repository through closure.

Actual on-site storage costs may be more or less than these cost ranges if a different on-site storage scenario is implemented. For example, to keep it distinct from the centralized storage models, our on-site storage models assume that there would be no transportation or consolidation of waste between the reactor sites. However, several experts noted that in an actual on-site storage scenario, reactor operators would likely consolidate their waste to make operations more efficient and reduce costs. Also, as with the centralized storage alternative, costs for the on-site storage scenario that includes geologic disposal could differ for a repository site other than Yucca Mountain or for additional waste management technologies.

Finally, our models did not include certain costs that were either location-specific or could not be predicted sufficiently to be quantified for our purposes, which would make the actual costs of on-site storage higher than our cost ranges. For example, the taxes and fees associated with on-site storage could vary significantly by state and over time. Also, repackaging operations in our 500-year on-site storage scenario would generate low-level waste that would require disposal. However, the amount of waste generated and the associated disposal costs could vary depending on the techniques used for repackaging. Finally, the total amount of the government's liability for failure to begin taking spent nuclear fuel for disposal in 1998 will depend on the outcome of pending and future litigation.

Like the centralized storage alternative, the funding source for the on-site storage alternative is uncertain. Currently, the reactor operators have been paying for the cost to store the waste, but have filed lawsuits to be compensated for storage costs of waste that the federal government was required to take title to under standard contracts. Payments resulting from these lawsuits have come from the Department of the Treasury's judgment fund, which is funded by the taxpayer, because a court determined that the Nuclear Waste Fund could not be used to compensate electric power companies for their storage costs. Without legislative or contractual changes—such as allowing the Nuclear Waste Fund to be used for on-site storage—taxpayers would likely bear the ultimate costs for on-site storage.

Concluding Observations

Developing a long-term national strategy for safely and securely managing the nation's high-level nuclear waste is a complex undertaking that must balance health, social, environmental, security, and financial factors. In addition, virtually any strategy considered will face many political, legal, and regulatory challenges in its implementation. Any strategy selected will need to have geologic disposal as a final disposition pathway. In the case of the Yucca Mountain repository, these challenges have left the nation with nearly three decades of experience. In moving forward, whether the nation commits to the same or a different waste management strategy, federal agencies, industry, and policy makers at all levels of government can benefit from the lessons of Yucca Mountain. In particular, stakeholders can better understand the need for a sustainable national focus and community commitment. Federal agencies, industry, and policymakers may also want to consider a strategy of complementary and parallel interim and long-term disposal options—similar to those being pursued by some other nations—which might provide the federal government with maximum flexibility, since it would allow time to work with local communities and to pursue research and development efforts in key areas, such as reprocessing.

Agency Comments

We provided DOE and NRC with a draft of this report for their review and comment. In their written comments, DOE and NRC generally agreed with the report. (See apps. V and VI.) In addition, both DOE and NRC provided comments to improve the draft report's technical accuracy, which we have incorporated as appropriate.

We also discussed the draft report with representatives of the Nuclear Waste Technical Review Board, the Nuclear Energy Institute, and the State of Nevada Agency for Nuclear Projects. These representatives provided comments to clarify information in the draft report, which we have incorporated as appropriate.

As agreed with your offices, unless you publicly announce the contents of this report earlier, we plan no further distribution until 30 days from the report date. At that time, we will send copies of this report to other appropriate congressional committees, the Secretary of Energy, the Chairman of NRC, the Director of the Office of Management and Budget, and other interested parties. The report also will be available at no charge on the GAO Web site at http://www.gao.gov.

If you or your staffs have any questions about this report, please contact me at (202) 512-3841 or gaffiganm@gao.gov. Contact points for our Offices of Congressional Relations and Public Affairs can be found on the last page of this report. GAO staff who made major contributions to this report are listed in appendix VII.

Mark E. Gaffigan
Director, Natural Resources
and Environment

Appendix I: Scope and Methodology

For this report we examined (1) the key attributes, challenges, and costs of the Yucca Mountain repository; (2) alternative nuclear waste management approaches; (3) the key attributes, challenges, and costs of storing the nuclear waste at two centralized sites; and (4) the key attributes, challenges, and costs of continuing to store the nuclear waste at its current locations.

Developing Information on Key Attributes, Challenges, and Costs of Yucca Mountain

To provide information on the key attributes and challenges of the Yucca Mountain repository, we reviewed documents and interviewed officials from the Department of Energy's (DOE) Office of Civilian Radioactive Waste Management and Office of Environmental Management; the Nuclear Regulatory Commission's (NRC) Division of Spent Fuel Storage and Transportation and Division of High Level Waste Repository Safety, both within the Office of Nuclear Material Safety and Safeguards; and the Department of Justice's Civil Division. We also reviewed documents and interviewed representatives from the National Academy of Sciences, the Nuclear Waste Technical Review Board, and other concerned groups. Once we developed our preliminary analysis of Yucca Mountain's key attributes and challenges, we solicited input from nuclear waste management experts. (See app. II for our methodology for soliciting comments from nuclear waste management experts and app. III for a list of these experts.)

To analyze the costs for the Yucca Mountain repository through to closure, we started with the cost information in DOE's Yucca Mountain Total System Lifecycle Cost report, which used 122,100 metric tons of nuclear waste in its analysis.[1] We asked DOE officials to provide a breakdown of the component costs on a per-metric-ton basis that DOE used in the Total System Lifecycle Cost report. We used this information to calculate the costs of a repository at Yucca Mountain for 70,000 metric tons and 153,000 metric tons, changing certain component costs based on the ratio between 70,000 and 122,100 or 153,000 and 122,100. For example, we modified the cost of constructing the tunnels for emplacing the waste for the 70,000-metric-ton scenario by 0.57, the ratio of 70,000 metric tons to 122,100 metric tons. We applied this approach to component costs that would be

---

[1]DOE, *Analysis of the Total System Lifecycle Cost of the Civilian Radioactive Waste Management Program, Fiscal Year 2007*, DOE/RW-0591 (Washington, D.C., July 2008). The 122,100 metric tons of nuclear waste included the spent nuclear fuel expected to be generated from all commercial nuclear reactors that had received NRC license extensions through January 2007.

Page 43   GAO-10-48 Nuclear Waste Management

Appendix I: Scope and Methodology

impacted by the ratio difference, particularly for transporting and emplacing the waste and installing drip shields. We also incorporated DOE's cost estimates for potential delays to licensing the Yucca Mountain repository into our analysis and made modifications to the analysis based on comments by cognizant DOE officials. Finally, we discounted DOE's costs, which were in 2008 constant dollars, to 2009 present value using the methodology described in appendix IV.

Examining and Identifying Nuclear Waste Management Alternatives

To examine and identify alternatives, we started with a series of interviews among federal and state officials and industry representatives. We also gathered and reviewed numerous studies and reports on managing nuclear waste— along with interviewing the authors of many of these studies— from federal agencies, the National Academy of Sciences, the Nuclear Waste Technical Review Board, the Massachusetts Institute of Technology, the American Physical Society, Harvard University, the Boston Consulting Group, and the Electric Power Research Institute. To better understand how commercial spent nuclear fuel is stored, we visited the Dresden Nuclear Power Plant in Illinois and the Hope Creek Nuclear Power Plant in New Jersey, which both store spent nuclear fuel in pools and in dry cask storage. We also visited DOE's Savannah River Site in South Carolina and Fort St. Vrain site in Colorado to observe how DOE-managed spent nuclear fuel and high-level waste are processed and stored.

As we began to identify potential alternatives to analyze, we shared our initial approach and methodology with nuclear waste management experts—including members of the National Academy of Sciences and the Nuclear Waste Technical Review Board to obtain their feedback—and revised our approach accordingly. Many of these experts advised us to develop generic, hypothetical alternatives with clearly defined assumptions about technology and environmental conditions. Industry representatives and other experts advised us that trying to account for the thousands of variables relating to geography, the environment, regional regulatory differences, or differences in business models would result in infeasible and unmanageable models. They also advised us against trying to predict changes in the future for technologies or environmental conditions because they would purely conjectural and fall beyond the scope of this analysis.

Based on this information, we identified two generic, hypothetical alternatives to use as the basis of our analysis: centralized storage and on-site storage. Within each of these alternatives, we identified different scenarios that examined the costs associated with the management of

Appendix I: Scope and Methodology

70,000 metric tons and 153,000 metric tons of nuclear waste and whether or not the waste is shipped to a repository for disposal after 100 years.

Once we identified the alternatives, we again consulted with experts to establish assumptions regarding commercial spent nuclear fuel management and its associated components to define the scope and specific processes that would be included in each alternative. To identify a more complete, qualified list of nuclear waste management experts with relevant experience who could provide and critique this information, we used a technique known as snowballing. We started with experts in the field who were known to us, primarily from DOE, NRC, National Council of State Legislators, the State of Nevada Agency for Nuclear Projects, the Nuclear Energy Institute, and the National Association of Regulatory Utility Commissioners and asked them to refer us to other experts, focusing on U.S.-based experts. We then contacted these individuals and asked for additional referrals. We continued this iterative process until additional interviews did not lead us to any new names or we determined that the qualified experts in a given technical area had been exhausted.

We conducted an initial interview with each of these experts by asking them questions about the nature and extent of their expertise and their views on the Yucca Mountain repository. Specifically, we asked each expert:

- What is the nature of your expertise? How many years have you been doing work in this area? Does your expertise allow you to comment on planning assumptions and costs of waste management related to storage, disposal, or transport?

- If you were to classify yourself in relation to the Yucca Mountain repository, would you classify yourself as a proponent, an opponent, an independent, an undecided or uncommitted, or some combination of these?

We then narrowed our list down to those individuals who identified themselves or whom others identified as having current, nationally recognized expertise in areas of nuclear waste management that were relevant to our analysis. For balance, we ensured that we included experts who reflected (1) key technical areas of waste management; (2) a range of industry, government, academia, and concerned groups; and (3) a variety of viewpoints on the Yucca Mountain repository. (See app. III for 147 experts we contacted.)

GAO-10-48 Nuclear Waste Management

Once we developed our list of experts, we classified them into three groups:

- Those whose expertise would allow them to provide us with specific information and advice on the processes that should be included in each alternative and the best estimates of expected cost ranges for the components of each alternative, such as a typical or reasonable price for a dry cask storage.

- Those who could weigh in on these estimates, as well as give us insight and comments on assumptions that we planned to use to define our alternatives.

- Those whose expertise was not in areas of component costs, but who could nonetheless give us valuable information on other assumptions, such as transportation logistics.

To define our alternatives and develop the assumptions and cost components we needed for our analysis, we started with the experts from the first group who had the most direct and reliable knowledge of the processes and costs associated with the alternatives we identified. This group consisted of seven experts and included federal government officials and representatives from industry. We worked closely with these experts to identify the key assumptions that would establish the scope of our alternatives, the more specific assumptions to identify the processes associated with each alternative, the components of these processes that we could quantify in terms of cost, and the level of uncertainty associated with each component cost. For example, two of the experts in this first group told us that for the on-site alternative, commercial reactor sites that did not already have independent spent nuclear fuel storage installations would have to build them during the next 10 years and that the cost for licensing, design, and construction of each installation would range from $24 million to $36 million. Once we had gathered our initial assumptions and cost components, we used a data collection instrument to solicit comments on them from all of our experts. We then used the experts' comments to refine our assumptions and component costs. (See app. II for our methodology for consulting with this larger group of nuclear waste management experts.)

DOE officials provided assumptions and cost data for managing DOE spent nuclear fuel and high-level waste, which we incorporated into our analysis of the centralized storage and on-site storage alternatives. These assumptions and cost information covered management of spent nuclear fuel and high-level waste at DOE's Idaho National Laboratory, Hanford Reservation, Savannah River Site, and West Valley site.

Developing Information on Key Attributes, Challenges, and Costs of the Centralized Storage and On-Site Storage Alternatives

To gather information on the key attributes and challenges of our alternatives, we interviewed agency officials and nuclear waste management experts from industry, academic institutions, and concerned groups. We also reviewed the reports and studies and visited the locations that were mentioned in the previous section. To ensure that the attributes and challenges we developed were accurate, comprehensive, and balanced, we asked our snowballed list of experts to provide their comments on our work, using the data collection instrument that is described in appendix II. We used the comments that we received to expand the attributes or challenges on our list or, where necessary, to modify our characterization of individual attributes or challenges.

To generate cost ranges for the centralized storage and on-site storage alternatives, we developed four large-scale cost models that analyzed the costs for each alternative of storing 70,000 metric tons and 153,000 metric tons of nuclear waste for 100 years followed by disposal in a geologic repository. (See app. IV.) We also generated cost ranges for each alternative of storing the waste for 100 years without including the cost of subsequent disposal in a geologic repository for storing 153,000 metric tons of waste on site for 500 years. For each model, which rely upon data and assumptions provided by nuclear waste management experts, the cost range was based on the annual volume of commercial spent nuclear fuel that became ready to be packaged and stored in each year. In general, each model started in 2009 by annually tracking costs of initial packaging and related costs for the first 100 years and for every 100 years thereafter if the waste was to remain on site and be repackaged. Since our models analyzed only the costs associated with storing commercial nuclear waste management, we augmented them with DOE's cost data for (1) managing its spent nuclear fuel and high-level waste and (2) constructing and operating a permanent repository. Specifically, we used DOE's estimated costs for the Yucca Mountain repository to represent cost for a hypothetical permanent repository.[2]

---

[2]We excluded historical costs for the Yucca Mountain repository because these costs represent challenges unique to Yucca Mountain and may not be applicable to a future repository. However, the bulk of future cost for construction, operation, and closure may be representative of a new repository.

Appendix I: Scope and Methodology

We conducted this performance audit from April 2008 to October 2009 in accordance with generally accepted government auditing standards. These standards require that we plan and perform the audit to obtain sufficient, appropriate evidence to provide a reasonable basis for our findings and conclusions based on our audit objectives. We believe that the evidence obtained provides a reasonable basis for our findings and conclusions based on our audit objectives.

Appendix II: Our Methodology for Obtaining Comments from Nuclear Waste Management Experts

As discussed in appendix I, we gathered the assumptions and associated component costs used to define our nuclear waste management alternatives by consulting with experts in an iterative process of identifying initial assumptions and component costs and revising them based on expert comments. This appendix (1) describes the data collection instrument we used to obtain comments on the initial assumptions and component costs, (2) describes how we analyzed the comments and revised our assumptions, and (3) provides a list of the assumptions and cost data that we derived through this process and used in our cost models.

To obtain comments from a broad group of nuclear waste management experts, we compiled the initial assumptions and component costs that we gathered from a small group of experts into a data collection instrument that included

- a description of the Yucca Mountain repository and our proposed nuclear waste management alternatives—on-site storage and centralized storage— and attributes and challenges associated with them;

- our initial assumptions that would identify and define the processes, time frames, and major components used to bound our hypothetical centralized and on-site storage alternatives;

- the major component costs of each alternative, including definitions and initial cost data; and

- components associated with each alternative with a high degree of uncertainty that we did not attempt to quantify in terms of costs.

The data collection instrument asked the experts to answer specific questions about each piece of information that we provided (see table 6).

Table 6: Our Data Collection Instrument for Nuclear Waste Management Experts

Section of the data collection instrument	Questions asked of the experts
Description of each alternative and its attributes and challenges	What additional issues do you suggest we consider, or is there one listed that you would modify?
List of initial assumptions for each alternative	To what extent to you think this assumption is reasonable or unreasonable?[a]
	If this assumption does not seem reasonable, please describe.[a]
	Are there additional assumptions defining our scenario not mentioned above that you would recommend GAO consider? Please describe.
List of component costs and initial cost data	Is this estimate reasonable or unreasonable?[a]
	If this estimate is not reasonable, please describe why (estimate too high, estimate too low, range too broad, range too narrow) and, if possible, provide specific alternative cost estimates.[a]
	Please tell us anything about this cost item that might make it difficult (or not difficult) to estimate accurately?[a]
	Are there additional cost categories not mentioned above that you would recommend GAO consider? Please provide a generic cost estimate or potential source of such an estimate, if possible.
List of uncertain components	In your opinion, do you think any of these items can be quantified? If so, please provide suggestions for how to quantify them, along with supporting data, if available.

Source: GAO.

[a]This question was asked after each assumption or component.

We pretested our instrument with several individual experts to ensure that our questions were clear and would provide us with the information that we needed, and then refined the instrument accordingly. Next, we sent the instrument to 114 experts who were identified through our snowballing methodology (see apps. I and III). Each expert received the sections of our data collection instrument that included the attributes and challenges of the alternatives and the initial assumptions, but only those experts with the type and level of expertise to comment on costs received the cost component sections.

We received 67 sets of comments from independent experts and experts representing industry, federal government, state governments, and other concerned groups.[1] These experts also represented a range of viewpoints on the Yucca Mountain repository. Each of their responses was compiled

---

[1]The 67 sets of comments do not reflect the total number of experts who responded because some groups of affiliated experts compiled their comments into a single response. For example, DOE's Office of Civilian Radioactive Waste Management provided a consolidated set of comments for its nine experts.

into a database organized by each individual assumption or cost element for the on-site storage and centralized interim storage alternatives.

To arrive at the final assumptions and cost component data for our models, we qualitatively analyzed the experts' comments. The comments we received on the assumptions differed in nature from those we received on the component costs, so our analysis and disposition of comments differed slightly. For the assumptions, we took the comments on each assumption that were made when an expert did not believe it was entirely reasonable and grouped comments that were similar. We determined the relevance of a comment to our assumption based on whether the comment provided a basis upon which we could modify the assumption or was within the scope or capability of our models. For example, we received several comments about how an assumption may be affected by nuclear waste from new reactors, including potential liabilities if the Department of Energy (DOE) does not take custody of that waste, but in the key assumptions defining our alternatives, we explicitly excluded new reactors because we could not predict how many new reactors would be built, when they would operate, and the amount of waste that they would generate. For those comments that were relevant, we weighed the expertise of those making the comments and determined whether the balance of the comments warranted a modification to our preliminary assumption. In some instances, we conducted followup interviews with selected experts to clarify issues that the broad group of experts raised.

For the component costs, we organized the comments on a particular component based on whether an expert thought the cost and uncertainty range was reasonable, too high, too low, the range was too broad, or the range was too narrow. We developed a ranking system to identify which experts had the greatest degree of direct experience or knowledge with the cost and weighed their comments accordingly to determine whether our preliminary cost should be modified. Also, we took into account the incidence of expert agreement or disagreement when deciding how much uncertainty to apply to a particular cost.

Through this analysis, we determined that the preponderance of our preliminary assumptions and cost data were reasonable for use in our models either because the experts generally agreed it was reasonable, or the experts who thought it was reasonable had a greater degree of relevant expertise or knowledge than those who commented otherwise. However, some of the experts' responses indicated that a modification to our model was needed. Table 7 presents a summary of the modifications we made to

Appendix II: Our Methodology for Obtaining
Comments from Nuclear Waste Management
Experts

our model assumptions and cost data based on the expert comments received.

Table 7: Initial Assumptions and Component Cost Estimates for Our Centralized Storage and On-site Storage Alternatives and Modifications Made Based on Experts' Responses to Our Data Collection Instrument

Centralized storage

Key aspect of the alternative	Initial key assumption	Modification based on expert comments
Number of sites	Two sites located in different geographic regions of the country.	None
Reactor operations	Current reactors will receive, if they have not already, a 20-year license extension and will operate until the end of their licensed life.	None
	When reactors cease operations, they will be decommissioned and only spent nuclear fuel dry storage will remain on site.	None
Transportation	Transportation will be the similar to what is assumed for the Yucca Mountain repository—via rail, using dedicated trains.	None
Repackaging	Waste will not be repackaged at the centralized facilities.[a]	None
Final disposition	Waste will be stored at the centralized sites until 100 years from now and then be disposed of in a geologic repository.[b]	None

Process	Initial process assumption	Modification based on expert comments
Waste packaged into dry storage casks	Reactor operators will only move the amount of waste from pools into dry storage that is necessary to preserve full-core offload capability—the capacity in their spent nuclear fuel pools to store all of the fuel in the reactor core.	None
	The overall amount of fuel moved from the pools to dry storage will be equal to estimated annual rates at which fuel is discharged from the reactors.	None
	Dual-purpose canister systems will be used until Transportation, Aging and Disposal systems become widely available.	Only dual-purpose systems will be used.
	Transportation, Aging and Disposal systems will have a capacity of 8.5 metric tons plus or minus 5 percent.	None (although this assumption became obsolete when we no longer assumed transportation, aging, and disposal systems would be used).

Appendix II: Our Methodology for Obtaining
Comments from Nuclear Waste Management
Experts

Centralized storage

Reactor site dry storage	All reactor sites without dry storage facilities will construct them at the time they lose full-core offload capability—the capacity in their spent nuclear fuel pools to store all of the fuel in the reactor core.	None
	Dry storage operations and maintenance costs vary by nature of the site, such as operating versus decommissioned.	None
	On average, 1.5 decommissioned reactor sites will be cleared of their waste each year.	None
Transportation to centralized storage	Once running at full capacity, transportation rates will be approximately 3,000 metric tons per year (what is assumed for Yucca Mountain).	None
	Waste from decommissioned sites and GE Morris will be transported before waste from operating sites. This waste would not be converted to dry storage prior to transportation.	None
	133 transportation casks will be required (what is assumed for Yucca Mountain) and will be acquired over a 7-year period.	None
	No new rail construction will be required.	None
	Transportation system infrastructure, system support, and operations will be analogous to what DOE assumes for Yucca Mountain.	None
Centralized storage	The two centralized facilities will begin accepting waste in 2028.	None
	The sites will be built at existing federal facilities and be owned and operated by DOE.	None
Geologic disposal	Waste will not be repackaged before being disposed of in a permanent repository.	None
	Any spent nuclear fuel not originally packaged into a Transportation, Aging and Disposal canister will be repackaged at the geologic repository.	This assumption became obsolete when we no longer assumed transportation, aging, and disposal canisters would be used.

Process component	Initial component cost estimate	Modification based on expert comments
Dry cask storage systems: • transportation, aging, and disposal • dual-purpose	• $1.1 million plus or minus 10 percent • $900,000 plus or minus 5 percent	• Obsolete • $900,000 plus or minus 25 percent
Loading operations: • cost per cask to load fuel into dry storage canisters • loading campaign consisting, on average, of five casks (including set-up, clean up, training, and labor)	• $150,000 plus or minus 5 percent • $750,000 plus or minus 5 percent	• $275,000 plus or minus 45 percent • None

GAO-10-48 Nuclear Waste Management

Appendix II: Our Methodology for Obtaining
Comments from Nuclear Waste Management
Experts

Centralized storage

Design, licensing, and construction of dry storage installations at reactor sites	$30 million plus or minus 20 percent	$30 million plus or minus 40 percent
Annual operations and maintenance: • operating reactor site dry storage • decommissioned reactor site dry storage • decommissioned reactor site wet storage	• $100,000 plus or minus 20 percent • $3 million plus or minus 20 percent • $10 million plus or minus 20 percent	• $100,000 plus or minus 50 percent • $4.5 million plus or minus 40 percent • None
Transportation casks	$4.5 million plus or minus 10 percent	None
Loading for transportation cost per canister	$250,000 plus or minus 5 percent	$150,000 plus or minus 40 percent
Transportation infrastructure: • rolling stock and facilities • transportation system support	• $400 million plus or minus 10 percent • $2.5 billion plus or minus 10 percent	• None • None
Transportation operations per-metric-ton	$26,000 plus or minus 10 percent	None
Centralized facility licensing and construction: • 70,000 metric ton facility • 153,000 metric ton facility	• $168 million plus or minus 10 percent • $232 million plus or minus 10 percent	• $218 million plus or minus 20 percent • $302 million plus or minus 20 percent
Centralized facility annual operations and maintenance	$8.8 million plus or minus 10 percent	None

On-site storage

Key aspect of the alternative	Initial key assumption	Modification based on expert comments
Number of commercial sites	Commercial spent nuclear fuel spent nuclear fuel will be stored at 75 reactor sites.	None
Number of DOE sites	DOE high-level waste and spent nuclear fuel will remain at five current sites.	None
Reactor operations	Current reactors will receive, if they have not already, a 20-year license extension and will operate until the end of their licensed life.	None
	When reactors cease operations, they will be decommissioned and only spent nuclear fuel dry storage will remain on site.	None
Transportation	There will be no transportation of waste between sites.	None
Repackaging	Dry cask storage systems would require repackaging every 100 years.	None

Appendix II: Our Methodology for Obtaining Comments from Nuclear Waste Management Experts

On-site storage

Process	Initial process assumption	Modification based on expert comments
Waste packaged into dry storage casks	Reactor operators will use generic dual-purpose canisters for dry storage with a capacity of 13 metric tons plus or minus 5 percent.	Range increased to plus or minus 15 percent.
	Reactor operators will only move the amount of waste from pools into dry storage that is necessary to preserve full-core offload capability.	None
	The overall amount of fuel moved from the pools to dry storage will be equal to estimated annual rates at which fuel is discharged from the reactors.	None
Reactor site dry storage	All reactor sites without dry storage facilities will construct them at the time they lose full-core offload capability.	None
	Dry storage operations and maintenance costs vary by nature of the site, such as operating versus decommissioned.	None
Repackaging	Wet transfer facilities will need to be built at each site for every packaging interval (i.e. every 100 years).	We will assume a generic transfer system that could be either wet or dry.
	All sites will need to replace their dry storage pad and infrastructure every 100 years when they repackage.	None

Process component	Initial component cost estimate	Modification based on expert comments
Dry cask storage system	$900,000 plus or minus 5 percent	$900,000 plus or minus 25 percent
Loading operations: • cost per cask to load fuel into dry storage canisters • loading campaign consisting, on average, of five casks (including set-up, clean up, training, and labor)	• $150,000 plus or minus 5 percent • $750,000 plus or minus 5 percent	• $275,000 plus or minus 45 percent • None
Design, licensing, and construction of dry storage installations at reactor sites	$30 million plus or minus 20 percent	$30 million plus or minus 40 percent
Annual operations and maintenance: • operating reactor site dry storage • decommissioned reactor site dry storage • decommissioned reactor site wet storage	• $100,000 plus or minus 20 percent • $3 million plus or minus 20 percent • $10 million plus or minus 20 percent	• $200,000 plus or minus 50 percent • $4.5 million plus or minus 40 percent • None
Construction of a transfer facility for repackaging	$300 million plus or minus 50 percent (for a wet transfer facility)	$300 million plus or minus 50 percent (for either a wet or a dry transfer facility)

GAO-10-48 Nuclear Waste Management

On-site Storage

Repackaging operations:

• repackaging costs per cask	• $1.2 million plus or minus 10 percent	• $1.6 million plus or minus 10 percent
• repackaging campaign consisting, on average, of 5 casks (including set-up, clean up, training, and labor)	• $750,000 plus or minus 10 percent	• None
Storage pad replacement	$30 million plus or minus 20 percent	$30 million plus or minus 40 percent

Source: GAO analysis based on expert-provided data.

Note: Unless specifically noted, all assumptions and costs apply specifically to commercial nuclear power sites. We used information provided by DOE for the assumptions and costs related to DOE-managed spent nuclear fuel and high-level waste.

[a]We did not explicitly solicit comment on this assumption in the data collection instrument for the centralized storage alternative because we solicited comments on the repackaging requirements in the on-site alternative.

[b]This assumption applies only to the version of our centralized storage alternative that includes final disposal.

Appendix III: Nuclear Waste Management Experts We Interviewed

	Name	Affiliation
1	Mark D. Abkowitz	U.S. Nuclear Waste Technical Review Board (member)
2	John Ahearne	Sigma Xi
3	Joonhong Ahn	National Academy of Sciences/Nuclear and Radiation Studies Board
4	David Applegate	U.S. Geological Survey
5	Wm. Howard Arnold	U.S. Nuclear Waste Technical Review Board (member)
6	Tom Baillieul	The Chamberlain Group
7	James David Ballard	California State University, Northridge
8	William D. Barnard	U.S. Nuclear Waste Technical Review Board (retired) (staff)
9	Lake Barrett	DOE/Office of Civilian Radioactive Waste Management (retired)
10	Barbara Beller	DOE/Office of Environmental Management
11	David W. Bland	TriVis Incorporated
12	Ted Borst	CH2M-WG Idaho, LLC
13	David C. Boyd	Minnesota Public Utilities Commission
14	Michele Boyd	Physicians for Social Responsibility
15	William Boyle	DOE/Office of Civilian Radioactive Waste Management
16	E. William Brach	Nuclear Regulatory Commission (NRC)/Division of Spent Fuel Storage and Transportation
17	Bruce Breslow	State of Nevada Agency for Nuclear Projects
18	Philip Brochman	NRC/Office of Nuclear Security and Incident Response
19	Tom Brookmire	Dominion Resources, Inc.
20	Robert J. Budnitz	Lawrence Berkeley National Laboratory
21	Susan Burke	Idaho Department of Environmental Quality
22	Barbara Byron	Western Interstate Energy Board
23	Robert Capstick	The Yankee Nuclear Power Companies
24	Thure E. Cerling	U.S. Nuclear Waste Technical Review Board (member)
25	Margaret Chu	M.S. Chu & Associates
26	Tom Clements	Friends of the Earth
27	Jean Cline	University of Nevada Las Vegas
28	Thomas Cochran	Natural Resources Defense Council
29	Marshall Cohen	Nuclear Energy Institute
30	Kevin Crowley	Nuclear and Radiation Studies Board, National Research Council of the National Academies
31	Jeanne Davidson	U.S. Department of Justice/Civil Division
32	Bradley Davis	DOE/Office of Nuclear Energy
33	Jack Davis	NRC/Division of High Level Waste Repository Safety
34	Jay C. Davis	Lawrence Livermore National Laboratory (retired)
		Nuclear and Radiation Studies Board, National Research Council of the National Academies

	Name	Affiliation
35	Scott DeClue	DOE/Office of Environmental Management
36	Edgardo DeLeon	DOE/Office of Environmental Management
37	Fred Dilger	Black Mountain Research
38	David J. Duquette	U.S. Nuclear Waste Technical Review Board (member)
39	Doug Easterling	Wake Forest University
40	Steven Edwards	Progress Energy
41	Randy Elwood	CH2M-WG Idaho, LLC
42	Rod Ewing	University of Michigan
43	Steve Fetter	University of Maryland
44	James Flynn	Pacific World History Institute
45	Charles Forsberg	Massachusetts Institute of Technology
46	Derrick Freeman	Nuclear Energy Institute
47	Steve Frishman	State of Nevada Nuclear Waste Project Office
48	Robert Fronczak	Association of American Railroads
49	B. John Garrick	U.S. Nuclear Waste Technical Review Board (chairman)
50	Ron Gecan	U.S. Congressional Budget Office
51	Lynn Gelhar	Massachusetts Institute of Technology
52	Christine Gelles	DOE/Office of Environmental Management
53	Robert Gisch	Department of Defense/Department of the Navy
54	Aubrey Godwin	Arizona Radiation Regulatory Agency
55	Charles R. Goergen	Washington Savannah River Company[a]
56	Stephen Goldberg	Argonne National Laboratory
57	Steven Grant	Bechtel SAIC Company, LLC[b]
58	Paul Gunter	Beyond Nuclear
59	Brian Gustems	PSEG Nuclear, LLC
60	Brian Gutherman	ACI Nuclear Energy Solutions
61	Roger L. Hagengruber	University of New Mexico Nuclear and Radiation Studies Board, National Research Council of the National Academies
62	R. Scott Hajner	Bechtel SAIC Company, LLC[b]
63	Robert Halstead	Transportation Advisor, State of Nevada Agency for Nuclear Projects
64	Paul Harrington	DOE/Office of Civilian Radioactive Waste Management
65	Ronald Helms	Bechtel SAIC Company, LLC[b]
66	Damon Hindle	Bechtel SAIC Company, LLC[b]
67	James Hollrith	DOE/Office of Civilian Radioactive Waste Management
68	Greg Holden	Department of Defense/Department of the Navy
69	Mark Holt	U.S. Congressional Research Service
70	George M. Hornberger	U.S. Nuclear Waste Technical Review Board (member)

Appendix III: Nuclear Waste Management
Experts We Interviewed

	Name	Affiliation
71	William Hurt	Idaho National Laboratory
72	Thomas H. Isaacs	Stanford University Lawrence Livermore National Laboratory Nuclear and Radiation Studies Board, National Research Council of the National Academies
73	Lisa R. Janairo	Council of State Governments, Midwestern Office
74	Andrew C. Kadak	U.S. Nuclear Waste Technical Review Board (member)
75	Kevin Kamps	Beyond Nuclear
76	Anthony Kluk	DOE/Office of Environmental Management
77	Lawrence Kokajko	NRC/Division of High Level Waste Repository Safety
78	Leonard Konikow	U.S. Geological Survey
79	Christopher Kouts	DOE/Office of Civilian Radioactive Waste Management
80	Steven Kraft	Nuclear Energy Institute
81	Darrell Lacy	Nye County, State of Nevada
82	Gary Lanthrum	DOE/Office of Civilian Radioactive Waste Management
83	Doug Larson	Western Interstate Energy Board
84	Ned Larson	DOE/Office of Civilian Radioactive Waste Management
85	Ronald M. Latanision	U.S. Nuclear Waste Technical Review Board (member)
86	Thomas Leschine	University of Washington
87	Adam H. Levin	Exelon Corporation
88	David Little	Washington Savannah River Company[c]
89	David Lochbaum	Union of Concerned Scientists
90	Bob Loux	Consultant
91	Edwin Lyman	Union of Concerned Scientists
92	Allison Macfarlane	George Mason University
93	Arjun Makhijani	Institute for Energy and Environmental Research
94	Zita Martin	Tennessee Valley Authority
95	Rodney McCullum	Nuclear Energy Institute
96	John McKenzie	Department of Defense/Department of the Navy
97	Richard A. Meserve	Carnegie Institution for Science Nuclear and Radiation Studies Board, National Research Council of the National Academies
98	Barry Miles	Department of Defense/Department of the Navy
99	Thomas Minvielle	Department of Defense/Department of the Navy
100	Bob Mitchell	Yankee Rowe
101	Ali Mosleh	U.S. Nuclear Waste Technical Review Board (member)
102	William M. Murphy	U.S. Nuclear Waste Technical Review Board (member)
103	Connie Nakahara	Utah Department of Environmental Quality
104	Irene Navis	Clark County, Nevada
105	Tara Neider	Transnuclear, Inc.

	Name	Affiliation
106	Brian O'Connell	National Association of Regulatory Utility Commissioners
107	Mary Olson	Nuclear Information and Resource Service
108	Pierre Oneid	Holtec International
109	Ronald S. Osteen	DOE/Office of Environmental Management
110	Jean Ridley	DOE/Office of Environmental Management
111	John Parkyn	Private Fuel Storage
112	Stan Pedersen	Bechtel SAIC Company, LLC[b]
113	Charles W. Pennington	NAC International
114	Mark Peters	Argonne National Laboratory
115	Per Peterson	University of California at Berkeley
116	Henry Petroski	U.S. Nuclear Waste Technical Review Board (member)
117	Max Power	Oregon Hanford Cleanup Board
118	Kenneth Powers	DOE/Office of Civilian Radioactive Waste Management
119	Jay Ray	DOE/Office of Environmental Management
120	Jeffrey Ray	Washington Savannah River Company[c]
121	Everett Redmond II	Nuclear Energy Institute
122	James Robert	Tennessee Valley Authority
123	Gene Rowe	U.S. Nuclear Waste Technical Review Board (staff)
124	Karyn Severson	U.S. Nuclear Waste Technical Review Board (staff)
125	David Shoesmith	University of Western Ontario
126	Linda Sikkema	National Conference of State Legislators
127	Kris Singh	Holtec International
128	Brian M. Smith	Department of Defense/Department of the Navy
129	Susan Smith	DOE/Office of Civilian Radioactive Waste Management
130	Joseph D. Sukaskas	Maine Public Utilities Commission
131	Jane Summerson	DOE/Office of Civilian Radioactive Waste Management
132	Eileen Supko	Energy Resources International, Inc.
133	Bill Swift	Washington Savannah River Company[c]
134	Peter Swift	Sandia National Laboratories
135	Raymond Termini	Exelon Corporation
136	Mike Thorne	Mike Thorne and Associates Limited
137	John Till	Risk Assessment Corporation
138	Richard Tosetti	Bechtel SAIC Company, LLC[b]
139	Brian Wakeman	Dominion Resources, Inc.
140	John Weiss, Jr.	Entergy Corporation
141	Christopher U. Wells	Southern States Energy Board
142	Chris Whipple	ENVIRON International Corporation

**Appendix III: Nuclear Waste Management
Experts We Interviewed**

	Name	Affiliation
143	James Williams	Western Interstate Energy Board
144	Wayne Worthington	Progress Energy
145	David Zabransky	DOE/Civilian Radioactive Waste Management Board
146	Paul L. Ziemer	Purdue University (retired) Nuclear and Radiation Studies Board, National Research Council of the National Academies
147	Louis Zeller	Blue Ridge Environmental Defense League

Source: GAO.

[a]On August 1, 2008, Savannah River Nuclear Solutions, LLC replaced Washington Savannah River Company as the primary contractor for DOE's Savannah River site. Expert affiliation was with Washington Savannah River Company at the time of our interviews.

[b]On April 1, 2009, USA Repository Services, LLC, replaced Bechtel SAIC Company, LLC, as the primary contractor for the Yucca Mountain repository. Expert affiliation was with Bechtel SAIC Company, LLC at the time of our interviews.

[c]On July 1, 2009, Savannah River Remediation, LLC replaced Washington Savannah River Company as the liquid waste program contractor. Expert affiliation was with Washington Savannah River Company at the time of our interviews.

Appendix IV: Modeling Methodology, Assumptions, and Results

The methodology and results of the models we developed to analyze the total costs of two alternatives for managing nuclear waste are based on cost data and assumptions we gathered from experts. Specifically, this appendix contains information on the following:

- The modeling methodology we developed to generate a range of total costs for the two nuclear waste management alternatives with two different volumes of waste.

- The Monte Carlo simulation process we used to address uncertainties in input data.

- The discounting methodology we developed to derive the present value of total costs in 2009 dollars.

- The individual models and scenarios within each model.

- The results of our cost estimations for each scenario.

- Caveats to our modeling work.

Appendixes I and II describe our methodology for collecting cost data and assumptions and how we ensured their reliability.

Modeling Methodology

The general framework for our models was an Excel spreadsheet that annually tracked all costs associated with packaging, transportation, construction, operation, and maintenance of nuclear waste facilities as well as repackaging of nuclear waste every 100 years when applicable. The starting time period for all models was the year 2009, but the end dates vary depending on the specifics of the scenario. The cost inputs were collected in constant 2008 dollars, but the range of total costs for each scenario was converted to and reported in 2009 present value dollars. Our analysis began with an estimate of existing and future annual volume of nuclear waste ready to be packaged and stored. We chose to model two amounts of waste: 70,000 metric tons and 153,000 metric tons.[1] For ease of

---

[1] The 70,000 metric tons is the statutory limit placed on the amount of waste that can be disposed of at Yucca Mountain. The 153,000 metric tons is the estimated amount of current waste plus additional commercial spent nuclear fuel that would be generated by 2055 if all currently operating commercial reactors received license extensions.

calculation, we converted all input costs to cost per-metric-ton of waste, when applicable.

The total cost range for each scenario was developed in four steps. First, we developed the total costs for commercial spent nuclear fuel volumes of about 63,000 metric tons and 140,000 metric tons, respectively. Second, we added DOE cost data for its managed waste.[2] Third, we discounted all annual costs to 2009 present value by a discounting methodology discussed later in this appendix. Finally, for scenarios where we assumed that the waste would be moved to a permanent repository after 100 years, we added DOE's cost estimate for the Yucca Mountain repository to represent cost for a permanent repository.[3] To ensure compatibility of cost data that DOE provided with cost ranges generated by our models, we converted DOE cost data to 2009 present value.

Monte Carlo Simulation Process

To address the uncertainties inherent in our analysis, we used a commercially available risk analysis software program called Crystal Ball to incorporate uncertainties associated with the data. This program allowed us to explore a wide range of possible values for all the input costs and assumptions we used to build our models. The Crystal Ball program uses a Monte Carlo simulation process, which repeatedly and randomly selects values for each input to the model from a distribution specified by the user. Using the selected values for cells in the spreadsheet, Crystal Ball then calculates the total cost of the scenario. By repeating the process in thousands of trials, Crystal Ball produces a range of estimated total costs for each scenario as well as the likelihood associated with any specific value in the range.

[2]DOE management costs include spent nuclear fuel managed at the Hanford Reservation, Idaho National Laboratory, and Fort St. Vrain, in Colorado, and high-level waste at the Hanford Reservation, Savannah River Site, Idaho National Laboratory, and West Valley.

[3]We used DOE estimates for Yucca Mountain to represent the cost of a permanent repository. We, however, did not include historical costs for Yucca Mountain as we felt that these historical costs represent challenges unique to Yucca Mountain and may not be applicable to a future repository whereas the bulk of future cost for construction, operation, and closure would be replicated for a new repository.

Appendix IV: Modeling Methodology,
Assumptions, and Results

Discount Rates and Present Value Analysis

One of the inherent difficulties in developing the cost for a nuclear waste disposal option is that costs are spread over thousands of years. The economic concept of discounting is central to such analyses as it allows costs incurred in the distant future to be converted to present equivalent worth. We selected discount rates primarily based on results of studies published in peer reviewed journals. That is, rather than subjectively selecting a single discount rate, we developed our discounting approach based on a methodology and values for discount rates that were recommended by a number of published studies.

We selected studies that addressed issues related to discounting activities whose costs and effects spread across the distant future or many generations, also known as "intergenerational discounting." In general, we found that these studies were in near consensus on two points: (1) discounting is an appropriate methodology when analyzing projects and policies that span many generations and (2) rates for discounting the distant future should be lower than near term discount rates and/or should decline over time. However, we found no consensus among the studies as to any specific discount rate that should be used. Consequently, we developed a discounting methodology using the following steps:

- We divided the entire time frame of our analysis into five different discounting intervals: immediate, near future, medium future, far future, and far-far future.

- We assumed that within each interval the discount rates were distributed with a triangular distribution.

- Based on all published rates, we developed the maximum, minimum, and mode values for each of the five specified intervals.

- We discounted all costs, using Crystal Ball to randomly and repeatedly select a rate from the appropriate interval and discount cost values using a different rate for each trial.

- Using these steps, we discounted all annual costs to 2009 present value.

Our methodology builds on a wide range of published rates from a number of different sources in concert with the Crystal Ball program. This enabled us, to the extent possible, to address the general lack of consensus on any specific discount rate and, at the same time, address the uncertainties that were inherent in intergenerational discounting and long-term analyses of nuclear waste management alternatives.

GAO-10-48 Nuclear Waste Management

Individual Models

We developed the following four models to estimate the cost of several hypothetical nuclear waste disposal alternatives, and we incorporated a number of scenarios within each model to address all uncertainties that we could not easily capture with Crystal Ball:

- **Model I:** Centralized storage for 153,000 metric tons, which included the following scenarios:

  - *Scenario 1:* Centralized storage for 100 years.

  - *Scenario 2:* Centralized storage for 100 years plus a permanent repository after 100 years.

- **Model II:** Centralized storage for 70,000 metric tons, which included one scenario:

  - *Scenario 1:* Centralized storage for 100 years.

- **Model III:** On-site storage using total waste volume of 153,000 metric tons which included the following scenarios:

  - *Scenario 1:* On-site storage for 100 years.

  - *Scenario 2:* On-site storage for 100 years plus a permanent repository after 100 years.

  - *Scenario 3:* On-site storage for 500 years.

- **Model IV:** On-site storage using total waste volume of 70,000 metric tons, which included one scenario:

  - *Scenario 1:* On-site storage for 100 years.

Model I: Centralized Storage (153,000 metric tons)

For this model we assumed that nuclear waste would remain on site until interim facilities are constructed and ready to receive the waste. Two centralized storage facilities would be constructed over 3 years—from 2025 through 2027—and then start accepting waste. The first scenario for this model includes the costs to store waste at the centralized facilities through 2108. In the second scenario, these facilities would stay in

GAO-10-48 Nuclear Waste Management

operation through 2155, or 47 years after a permanent repository for the waste would become available. The total analysis period for the cost of this alternative plus permanent repository continues until 2240, when a permanent repository would be expected to close. In general, the costs include the following:

- Initial costs, which include costs of casks, costs for loading of casks, cost of loading campaigns, and operating and maintenance costs by three types of nuclear sites, i.e., operating sites with dry storage, decommissioned sites with dry storage, and decommissioned sites with wet storage. The uncertainty ranges for these costs were from plus or minus 5 percent to plus or minus 50 percent, depending on specific cost variable.

- Costs associated with centralized facilities, including construction costs for centralized facilities, transportation cost for transfer of nuclear waste to centralized facilities, capital and operation and maintenance costs for transportation of waste to centralized facilities and operation and maintenance of centralized facilities. The uncertainty ranges for these costs are from plus or minus 10 percent to plus or minus 40 percent, depending on the cost category.

Figure 6: Scenario and Cost Time Frames for the Centralized 153,000 Metric Ton Models

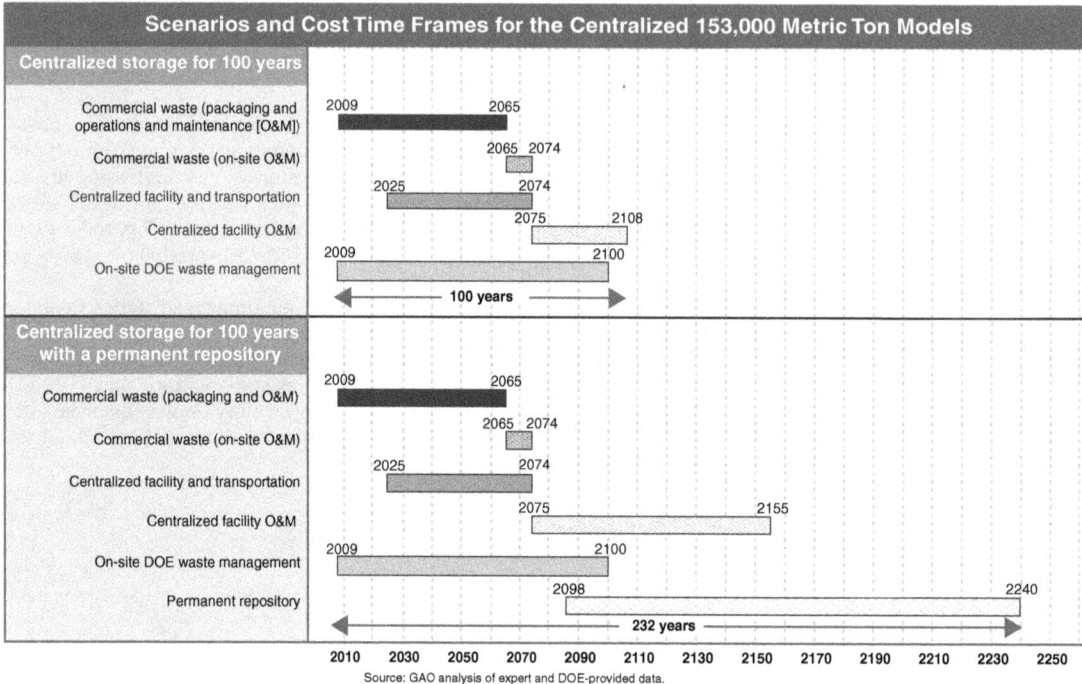

Source: GAO analysis of expert and DOE-provided data.

Model II: Centralized Storage (70,000 metric tons)

This model was developed under the assumption that total existing and newly generated waste from the private sector and DOE will be 70,000 metric tons. The stream of new annual waste ready to be moved to dry storage will continue through 2030. The cost categories and uncertainty ranges assumed for this storage alternative are the same as those assumed in the centralized storage model for 153,000 metric tons.

Appendix IV: Modeling Methodology,
Assumptions, and Results

Figure 7: Scenario and Cost Time Frames for the Centralized 70,000 Metric Ton Model

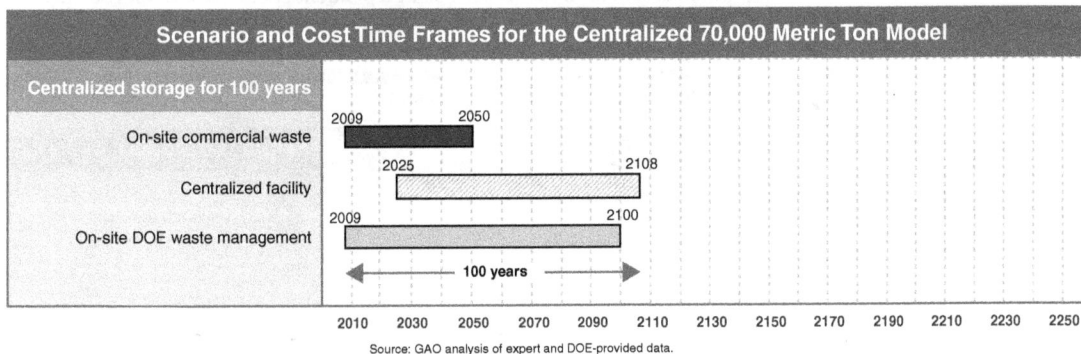

Scenario and Cost Time Frames for the Centralized 70,000 Metric Ton Model

Centralized storage for 100 years

On-site commercial waste	2009 — 2050	
Centralized facility	2025 — 2108	
On-site DOE waste management	2009 — 2100	

100 years

2010 2030 2050 2070 2090 2110 2130 2150 2170 2190 2210 2230 2250

Source: GAO analysis of expert and DOE-provided data.

Model III: On-Site Storage (153,000 metric tons)

We developed this model under the assumption that total existing and newly generated nuclear waste by the private sector and DOE would be 153,000 metric tons. The stream of new waste ready to be moved to dry storage would continue through 2065. In general, the costs include the following:

- Initial costs, which include costs of casks, costs for loading of casks, cost of loading campaigns, and operating and maintenance costs by three types of nuclear sites, i.e., operating sites with dry storage, decommissioned sites with dry storage, and decommissioned sites with wet storage. The uncertainty ranges for these costs were from plus or minus 5 percent to plus or minus 50 percent, depending on specific cost variable.

- Repackaging costs, which include the costs for casks; construction of transfer facilities, site pools, and other needed infrastructure; and repackaging campaigns. Because these costs are first incurred after 100 years and then every 100 years thereafter, they are included only in the model scenarios covering more than 100 years. The uncertainty for these costs range from plus or minus 10 percent to plus or minus 50 percent, depending on the specific cost variable.

- Dry storage pad costs, including initial costs when dry storage is first established, as well as replacement costs. Because the replacement costs are first incurred after 100 years and then every 100 years thereafter, they are included only in the model scenarios covering more than 100 years. The cost of these pads, collectively referred to as independent spent fuel

Page 68   GAO-10-48 Nuclear Waste Management

storage installations, include costs related to licensing, design, and construction of dry storage. The independent spent nuclear fuel storage installation costs have an uncertainty range of plus or minus 40 percent.

Figure 8: Scenarios and Cost Time Frames for the On-Site 153,000 Metric Ton Models

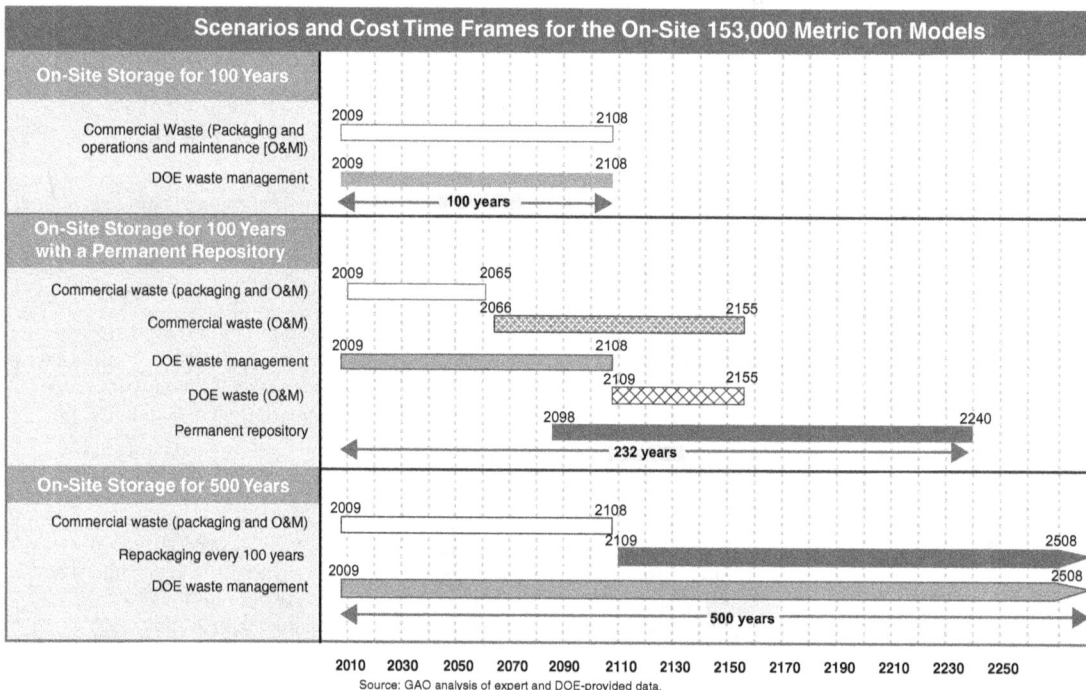

Scenarios and Cost Time Frames for the On-Site 153,000 Metric Ton Models

On-Site Storage for 100 Years
- Commercial Waste (Packaging and operations and maintenance [O&M]): 2009–2108
- DOE waste management: 2009–2108
- 100 years

On-Site Storage for 100 Years with a Permanent Repository
- Commercial waste (packaging and O&M): 2009–2065
- Commercial waste (O&M): 2066–2155
- DOE waste management: 2009–2108
- DOE waste (O&M): 2109–2155
- Permanent repository: 2098–2240
- 232 years

On-Site Storage for 500 Years
- Commercial waste (packaging and O&M): 2009–2108
- Repackaging every 100 years: 2109–2508
- DOE waste management: 2009–2508
- 500 years

2010 2030 2050 2070 2090 2110 2130 2150 2170 2190 2210 2230 2250

Source: GAO analysis of expert and DOE-provided data.

Model IV: On-Site Storage (70,000 metric tons)

We developed this model under the assumption that total existing and newly generated nuclear waste by the private sector and DOE will be 70,000 metric tons. The stream of new annual waste ready to be moved to dry storage will continue through 2030. The cost categories and uncertainty ranges assumed for this storage alternative are the same as those for the on-site model for storing 153,000 metric tons for 100 years.

Page 69   GAO-10-48 Nuclear Waste Management

Figure 9: Scenario and Cost Time Frames for the On-Site 70,000 Metric Ton Model

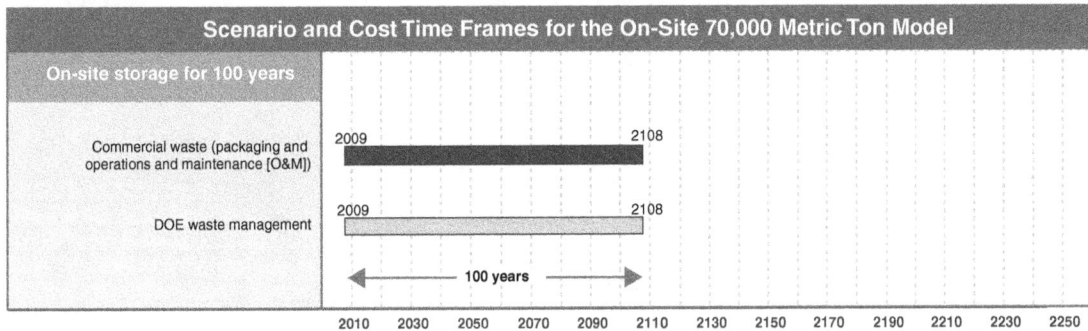

Source: GAO analysis of expert and DOE-provided data.

Costs for a Permanent Repository

For two scenarios, we assumed that at the end of 100 years the nuclear waste would be transferred to a permanent repository for disposal. To estimate the cost for a repository, we used DOE's cost data for the Yucca Mountain repository and made three adjustments to ensure compatibility with costs generated by our models. First, we included only DOE's future cost estimates for the Yucca Mountain repository. Second, because DOE provided costs in 2008 constant dollars, we converted all costs for the permanent repository to costs to 2009 present value using corresponding ranges of interest rates as previously described in this appendix. Finally, we assumed that repository construction and operating costs would be incurred from 2098 to 2240 when we added these cost ranges to our alternatives after 100 years.

Modeling Results

Table 8 shows the results of our analysis for all scenarios.

GAO-10-48 Nuclear Waste Management

Appendix IV: Modeling Methodology,
Assumptions, and Results

Table 8: Model Results for All Scenarios

Dollars in billions

Models and scenarios	Range of total costs[a]	Mean[a]
Permanent repository (153,000 metric tons)		
Permanent repository[b]	$41 to $67	$53
Permanent repository (70,000 metric tons)		
Permanent repository[b]	$27 to $39	$32
Model I: centralized storage (153,000 metric tons)		
Centralized 100 years	$15 to $29	$21
Centralized 100 years plus permanent repository	$23 to $81	$47
Model II: centralized storage (70,000 metric tons)		
Centralized 100 years	$12 to $20	$15
Model III: on-site storage (153,000 metric tons)		
On-site 100 years	$13 to $34	$22
On-site 100 years plus permanent repository	$20 to $97	$51
On-site for 500 years	$34 to $225	$89
Model IV: on-site storage (70,000 metric tons)		
On-site 100 years	$10 to $26	$18

Source: GAO.

Note: All costs are in 2009 present value and represent costs regardless of who will pay or is legally responsible to pay for them and as such do not address the issue of liabilities. Furthermore, these costs do not include other potential costs, such as decommissioning and environmental costs and the government's penalties for delays in moving waste from the Idaho National Laboratory under the settlement agreement with Idaho.

[a]The cost estimates do not present exact values rather order-of-magnitude estimates as both the maximum and minimum as well as mean values will be somewhat different each time the simulation is repeated. This is because the Monte Carlo methodology will randomly select a different set of input data from one simulation run to the next.

[b]While our cost ranges for a permanent repository are based on DOE's estimate for the Yucca Mountain repository, our cost ranges differ from DOE's of $96 billion estimate for the following reasons: First, our cost ranges are in 2009 present value, while DOE uses 2007 constant dollars, which are not discounted. Our present value analysis reflects the time value of money—costs incurred in the future are worth less today—so that streams of future costs become smaller. Second, our cost ranges do not include about $14 billion in previously incurred costs. Third, our cost ranges are for 153,000 metric tons and 70, 000 metric tons of nuclear waste, while DOE's estimated cost is for 122,100 metric tons. Finally, we use ranges while DOE provides a point estimate.

Figures 10 and 11 show ranges of total costs, as well as the probabilities for two selected scenarios. In the figures, each bar indicates a range of values for total cost and the height of the each bar indicates the probability associated with those values.

Figure 10: Total Cost Ranges for Centralized Storage for 100 Years with Final Disposition

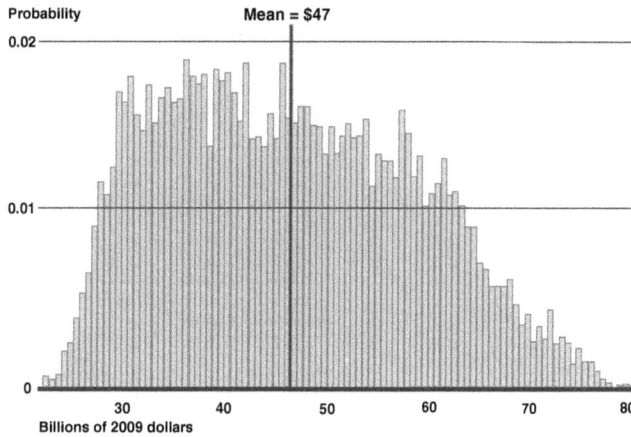

Source: GAO analysis of expert and DOE provided data.

Note: The values on the horizontal axis of the figure are to provide a scale and do not correspond exactly to the ranges for total costs which are provided in table 8.

Figure 11: Total Cost Ranges for On-site Storage for 100 years with Final Disposition

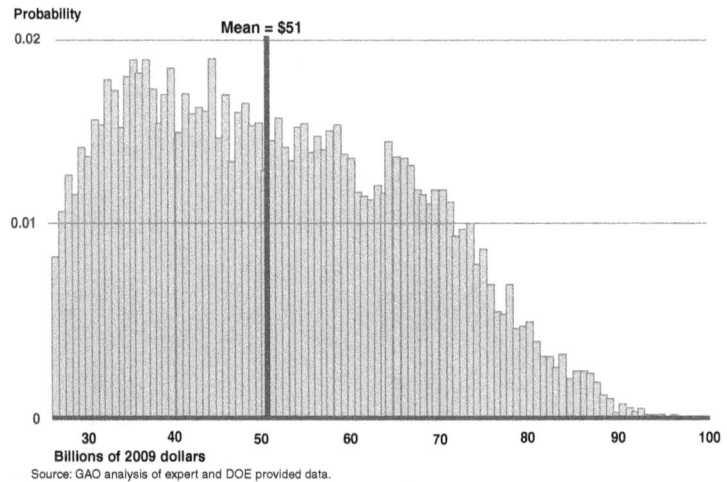

Source: GAO analysis of expert and DOE provided data.

Note: The values on the horizontal axis of the figure are to provide a scale and do not correspond exactly to the ranges for total costs which are provided in table 8.

Figure 12 shows the present value of the total cost ranges of storing the nuclear waste on site over 2,000 years. The shaded areas indicate the probability that the values fall within the indicated ranges and are the result of combinations of uncertainties from a large number of input data. Specifically, we estimate that these costs could range from $34 billion to $225 billion over 500 years, from $41 billion to $548 billion over 1,000 years, and from $41 billion to $954 billion over 2,000 years, indicating and substantial level of uncertainty in making long-term cost projections.

GAO-10-48 Nuclear Waste Management

Figure 12: Total Cost Ranges of On-Site Storage over 2,000 Years

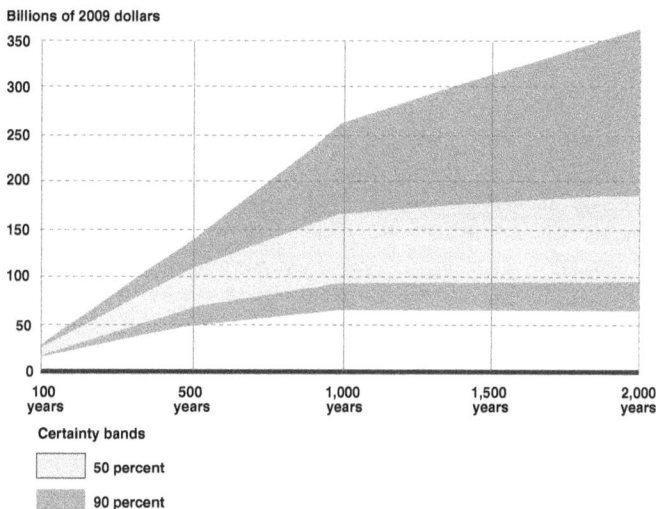

Billions of 2009 dollars

Certainty bands

☐ 50 percent

▨ 90 percent

Source: GAO analysis of expert and DOE-provided data.

Note: The values on the vertical axis of the figure are to provide a scale and do not correspond exactly to the total cost ranges presented in table 8.

Modeling Caveats

Our models are based on ranges of average costs for each major cost category that is applicable to the alternative under analysis. As a result, the costs do not reflect storage costs for any specific site. Since we did not attempt to capture specific characteristics of each site, our values for any cost factor, if applied to any specific site, are likely incorrect. Nevertheless, since we used ranges rather than single values for a wide range of cost inputs to the models, we expect that our cost range for each variable includes the true cost for any specific site. Moreover, we expect the total cost point estimate for any scenario is within the range of total costs we developed.

Our models are designed to develop total cost ranges for each scenario within each alternative, regardless of who will pay or is legally responsible for the costs. Issues related to assignment of the costs and potentially responsible entities are discussed elsewhere in this report but are not incorporated into our ranges. Also, our cost ranges focus on actual expenditures that would be incurred over the period of analysis and do not assume a particular funding source and do not necessarily represent costs to the federal government. Finally, because a number of cost categories are not included in our final estimated ranges, we cannot predict their impact on our final costs ranges. For example, we did not include (1) decontamination and decommissioning costs for existing facilities or facilities yet to be built within each scenario and (2) estimates for local and state taxes or fees, which would be required to establish new sites or for continued operation of on-site storage facilities after nuclear reactors are decommissioned.

Table 8 and figures 10 and 11 present the results of our analysis by individual scenario. Because the purpose of our analysis was primarily to provide cost ranges for various nuclear waste management alternatives, we did not attempt to provide a comparison of results across scenarios. For a number of reasons, we believe such a comparison would have been misleading. The alternatives we have considered are inherently different in a large number of characteristics that could not be captured in our modeling work or they were not within the scope of our analysis. For example, differences in safety, health, and environmental effects, and ease of implementation characteristics of these alternatives should have an integral role in the policy debate on waste management decisions. However, because these effects cannot be readily quantified, they were outside the scope of our modeling work and are not reflected in the total cost ranges we generated.

Appendix V: Comments from the Department of Energy

Department of Energy
Washington, DC 20585

October 28, 2009

Mr. Mark E. Gaffigan
Director, Natural Resources and Environment
U.S. Government Accountability Office
441 G Street, NW
Washington, D.C. 20548

Dear Mr. Gaffigan:

Thank you for the opportunity to review and submit comments on the draft report, "NUCLEAR WASTE MANAGEMENT: Key Attributes, Challenges and Costs for the Yucca Mountain Repository and Two Potential Alternatives" (GAO-10-48). The U.S. Department of Energy appreciates the amount of time and effort that you and your staff have taken to review this important topic.

Specific comments from Naval Reactors, the Office of General Counsel, and the Office of Environmental Management on the draft report are enclosed. If you have any questions, please feel free to call me on 202-586-6850.

Sincerely,

Christopher A. Kouts
Acting Director
Office of Civilian Radioactive
   Waste Management

Enclosure

Printed with soy ink on recycled paper

Page 76   GAO-10-48 Nuclear Waste Management

Appendix VI: Comments from the Nuclear Regulatory Commission

UNITED STATES
NUCLEAR REGULATORY COMMISSION
WASHINGTON, D.C. 20555-0001

October 26, 2009

Mr. Richard Cheston
Assistant Director
U.S. Government Accountability Office
441 G Street, N.W.
Washington, DC 20548

Dear Mr. Cheston:

Thank you for providing the U.S. Nuclear Regulatory Commission (NRC) the opportunity to review and comment on the U.S. Government Accountability Office's (GAO) draft report GAO-10-48, "NUCLEAR WASTE MANAGEMENT – Key Attributes, Challenges, and Costs for the Yucca Mountain Repository and Two Potential Alternatives." The NRC staff has reviewed the draft report. Although we did not identify any significant issues regarding accuracy, completeness, or sensitivity of information, we have separately transmitted several technical and editorial comments to your staff.

If you have any questions regarding this response, please contact Mr. Jesse Arildsen of my staff, at (301) 415-1785.

Sincerely,

R. W. Borchardt
Executive Director
for Operations

Enclosure:
NRC Staff Comments on Draft
Report GAO-10-48

Page 77

GAO-10-48 Nuclear Waste Management

Appendix VII: GAO Contact and Staff Acknowledgments

GAO Contact	Mark Gaffigan, (202) 512-3841 or gaffiganm@gao.gov
Staff Acknowledgments	In addition to the individual named above, Richard Cheston, Assistant Director; Robert Sánchez; Ryan Gottschall; Carol Henn; Anne Hobson; Anne Rhodes-Kline; Mehrzad Nadji; Omari Norman; and Benjamin Shouse made key contributions to this report. Also contributing to this report were Nancy Kingsbury, Karen Keegan, and Timothy Persons.

GAO's Mission	The Government Accountability Office, the audit, evaluation, and investigative arm of Congress, exists to support Congress in meeting its constitutional responsibilities and to help improve the performance and accountability of the federal government for the American people. GAO examines the use of public funds; evaluates federal programs and policies; and provides analyses, recommendations, and other assistance to help Congress make informed oversight, policy, and funding decisions. GAO's commitment to good government is reflected in its core values of accountability, integrity, and reliability.
Obtaining Copies of GAO Reports and Testimony	The fastest and easiest way to obtain copies of GAO documents at no cost is through GAO's Web site (www.gao.gov). Each weekday afternoon, GAO posts on its Web site newly released reports, testimony, and correspondence. To have GAO e-mail you a list of newly posted products, go to www.gao.gov and select "E-mail Updates."
Order by Phone	The price of each GAO publication reflects GAO's actual cost of production and distribution and depends on the number of pages in the publication and whether the publication is printed in color or black and white. Pricing and ordering information is posted on GAO's Web site, http://www.gao.gov/ordering.htm.

Place orders by calling (202) 512-6000, toll free (866) 801-7077, or TDD (202) 512-2537.

Orders may be paid for using American Express, Discover Card, MasterCard, Visa, check, or money order. Call for additional information. |
| To Report Fraud, Waste, and Abuse in Federal Programs | Contact:

Web site: www.gao.gov/fraudnet/fraudnet.htm
E-mail: fraudnet@gao.gov
Automated answering system: (800) 424-5454 or (202) 512-7470 |
| Congressional Relations | Ralph Dawn, Managing Director, dawnr@gao.gov, (202) 512-4400
U.S. Government Accountability Office, 441 G Street NW, Room 7125
Washington, DC 20548 |
| Public Affairs | Chuck Young, Managing Director, youngc1@gao.gov, (202) 512-4800
U.S. Government Accountability Office, 441 G Street NW, Room 7149
Washington, DC 20548 |

Please Print on Recycled Paper

552

DOE/NE-0070

NUCLEAR POWERPLANT SAFETY:

Operations

NUCLEAR ENERGY

U.S. Department of Energy
Office of Nuclear Energy, Science,
and Technology

703-739-3790 TCNNuclear.com **553**

Nuclear Powerplant Safety: Operations

Table of Contents

On the cover:
Palo Verde Nuclear Reactor,
Arizona

U.S. Department of Energy
Office of Nuclear Energy,
Science, and Technology
Washington, D.C. 20585

1

Nuclear Powerplant Safety: Operations

Nuclear powerplants are a major source of electricity in the United States and throughout most of the world. They generate about 20 percent of our nation's total electricity. Careful planning, good engineering and design, strict licensing and regulation, intensive training of operators, and thorough environmental monitoring help to ensure that nuclear powerplants operate safely. In fact, no powerplant accident has harmed a member of the public in the entire history of U.S. commercial nuclear power.

Hope Creek Generating Station, located along the Delaware River in Lower Alloways Creek Township, N.J. The reactor containment building is in the foreground. The cooling tower in the background emits steam from the cooling of water from the turbine.

3

How Nuclear Reactors Work

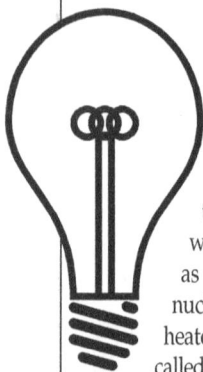

Both nuclear and non-nuclear powerplants produce electricity in much the same way. Water is heated and converted to steam, which in turn drives a turbine attached to a generator. In non-nuclear plants, the water is heated by fossil fuels such as coal, natural gas, or oil. In nuclear powerplants, the water is heated by the energy from a process called nuclear fission.

In the 1930s, scientists discovered that if they bombarded atoms of uranium with high-energy neutrons the uranium atoms would split apart, or fission. When this happens, they release excess energy as well as more neutrons, which can then strike other atoms and keep the fission process going. This chain reaction is the basis for nuclear powerplant operations.

For use as fuel in a reactor, uranium is formed into small pellets, which are stacked inside protective metal rods. Fuel rods are bundled together into fuel assemblies, which make up the reactor core. In addition to the fuel rods, there are other rods in the fuel assemblies which are made of a material which can absorb the neutrons. They are called the control rods, because they are used to control the rate at which the fission reaction takes place, or to stop it altogether.

4

Operational Safety

When uranium atoms split, they give off radiation as well as heat. Radiation is a natural form of energy that has always existed on earth and throughout the universe. Although the term can include such forms of energy as light and radio waves, it usually refers to ionizing radiation. Ionizing radiation can change the chemical composition of many things, including living tissue. Therefore, we must limit our exposure to it.

The following measures help ensure that nuclear powerplants in the United States are operated to protect the health and safety of workers, the public, and the environment.

- State and Federal regulations limit the amount of radiation that nuclear powerplants are allowed to release.

- Multiple barriers protect against the release of radioactive materials, with the final barrier being a robust steel or concrete containment structure that surrounds the major portion of the nuclear system.

- All U.S. nuclear powerplants must be licensed by the Nuclear Regulatory Commission (NRC). The NRC is also responsible for inspection and enforcement, standards development, and research into what kind of regulations will best ensure safety.

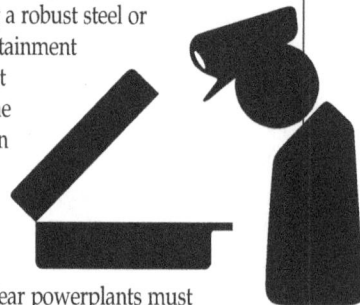

5

556

- Powerplant personnel must establish programs and conduct audits, inspections, and environmental monitoring to ensure that the plants are operating safely and are in compliance with the limits on radioactive material releases.

- Reactor operators are licensed by the NRC. They are thoroughly trained and periodically retrained on operating procedures.

- Powerplants must always maintain strict security to prevent the loss of nuclear materials and acts of sabotage.

- The transport and storage of spent nuclear fuel are strictly regulated.

The Role of the Nuclear Regulatory Commission

The NRC (formerly the Atomic Energy Commission) was created in 1974 by the Federal Energy Reorganization Act to regulate the basic functions of the nuclear power industry. The NRC establishes safety requirements, provides inspection and oversight, and sponsors ongoing safety research programs. These programs are both domestic and international in scope.

The NRC's Office of Nuclear Reactor Regulation is responsible for licensing all nuclear powerplants. In order to obtain a license, a powerplant must provide detailed information to show that its design, construction, and operation will ensure safe conditions for workers and the public.

After licensing, the NRC's plant inspection program ensures continual attention to public health and safety, environmental protection, and security of nuclear materials and facilities. The NRC conducts routine inspections and investigates any accidents, unusual incidents, or even claims of unusual events that may occur during powerplant operation. Resident inspectors at each nuclear powerplant observe and monitor licensee activities and respond to operational events at the plant.

Violations of NRC requirements call for corrective actions and may result in heavy fines, plant closure, or suspending or revoking of the utility's operating license.

Control of Radioactive Material Releases

Releases of radioactive materials from nuclear powerplants could have harmful effects on

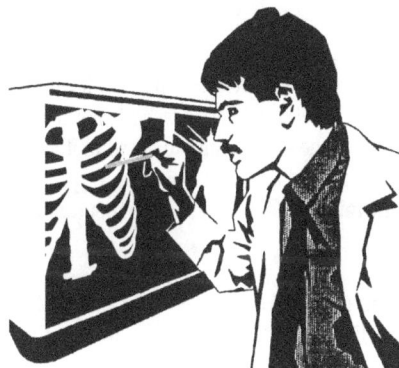

Sources of Radiation o the General Public

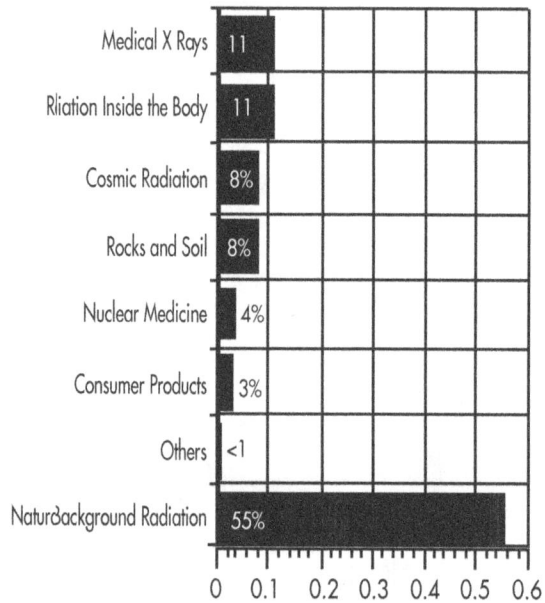

	0	0.1	0.2	0.3	0.4	0.5	0.6
Medical X Rays	11						
Rliation Inside the Body	11						
Cosmic Radiation	8%						
Rocks and Soil	8%						
Nuclear Medicine		4%					
Consumer Products		3%					
Others		<1					
Natur ackground Radiation						55%	

people and the environment. Utilities are required to monitor releases to the environment and ensure that they do not exceed limits established by the NRC.

Scientists measure radiation's effect on humans in units called rems. However, measuring the radiation emitted from powerplants in rems is like measuring the dimensions of a desk top in miles. It can be done, but it is not very practical. For that reason, the most commonly used unit to measure the radiation from a powerplant is the millirem (a millirem is one thousandth of a rem).

The average American receives about 360 millirem of radiation each year from both natural and man-made sources.

About 55 percent of this radiation comes from radon — a radioactive gas from the decay of naturally occurring uranium and thorium. The remaining portion comes from the earth's soil, food, and water, and from medical applications. A very small fraction comes from various other sources, including nuclear powerplants. Less than one-tenth of one percent of the average American's exposure to radiation comes from the nuclear power industry.

NRC regulations state that people who live near a nuclear powerplant cannot be exposed to more than 100 millirem of radiation from that facility annually. Operating data show that powerplants expose their neighbors to far less than 1 millirem per year.

8

9

Personnel Training Programs

To train their operators, the utilities use exact mockups of an actual powerplant control room. These simulators are computer controlled like the ones pilots use in flight training. They allow the operators to gain practical experience in managing all types of normal and unusual situations without affecting the plant. The operators react to simulated plant functions, using operating procedures. Instructors discuss the best actions to take in given situations.

The Institute of Nuclear Power Operations (INPO), an organization supported by the utility industry, conducts detailed evaluations of operating practices at all nuclear powerplants. The U.S. Department of Energy (DOE), INPO, and the NRC work together to upgrade the training of reactor operators. The recommendations from DOE and INPO supplement NRC's requirements and help utilities achieve very high standards for selecting and training the people who operate the plants.

Plant Operators

The powerplant staff is highly qualified, skilled, and trained in reactor operations and safety. Operators and senior operators are required to be licensed by the NRC for that particular facility. The operator works the controls of the facility. The senior operator has the additional responsibility of directing the activities of the other licensed operators.

Hope Creek station control room

At least one Licensed Senior Reactor Operator, two Licensed Reactor Operators, and two Unlicensed Plant Operators must be present during each working shift at a nuclear powerplant. In addition, a technical adviser, who is usually a graduate engineer, provides advanced technical help and information for both normal and unexpected conditions.

Operator licenses are granted only to qualified personnel who must demonstrate a high degree of proficiency and pass in-depth examinations (written and oral) on plant operations and safety. All licensed operators must complete periodic retraining courses to keep their skills at the highest levels. This retraining consists of classroom lectures, drills, and exercises with a control room simulator.

10

11

Nuclear Powerplant Emergency Plans

Every nuclear powerplant has to have an emergency plan for dealing with any unexpected event that could affect public health and safety. Utilities also conduct periodic exercises and drills to ensure that all personnel know what to do in case of an onsite or offsite emergency. These activities are coordinated with other Federal, state, and local agencies.

Nuclear Powerplant Security

Nuclear powerplants always maintain strict security to help ensure their operating safety and protect against the loss of nuclear materials or acts of sabotage that could cause an unplanned release of radioactive material. Commercial nuclear powerplants employ well trained security forces, set up physical barriers and elaborate electronic surveillance systems, and screen visitors to keep unauthorized persons from entering the site.

12

Spent Fuel — Handle With Care

For a nuclear powerplant to operate efficiently, part of its nuclear fuel must be periodically replaced with new fuel. Upon removal from the reactor, this highly radioactive spent fuel, or high-level waste, must be handled with care. Plant personnel transfer the spent fuel to a 30-foot-deep pool of water within the powerplant. The water removes heat from the spent fuel and serves as a radiation shield. Although the present method of storing spent fuel has been proven safe, available storage space is filling up.

The Department of Energy (DOE) is working closely with industry to find the best solution to this problem. Using existing technology, it is possible to design, construct, and safely operate a high-level waste repository. The DOE is currently studying Yucca Mountain in Nevada to determine whether it is a suitable location for permanent storage of this high-level waste.

13

560

Shipping Spent Fuel

Disposing of spent fuel requires shipping it from one point to another. There are certain risks involved in shipping radioactive waste, such as theft or release of radioactive material due to a transportation accident.

Heavy-duty shipping containers and highly trained security personnel help prevent theft and accidental releases. Shipping containers have been tested in collisions up to 80 miles per hour, immersed in water, placed in fires, and dropped on hard surfaces and steel bars. In all of the tests, the damage that the containers received would not have permitted the release of any radioactive material. In the long history of transporting spent fuel, there has never been an accident that released radioactive material.

14

Operational Safety — A Continuing Commitment

Each aspect of a nuclear powerplant's design, construction, operation, and security reinforces its overall operational safety. Every nuclear powerplant has to be licensed by the NRC. Ongoing NRC oversight ensures that the plant continues to be operated safely and in compliance with the regulations and the requirements set forth in the facility license.

In recent years, the government has further upgraded requirements for safety as a result of new information from ongoing research programs and operating experience. As a result of this careful approach and ongoing commitment, nuclear power production in the U.S. has an excellent safety record.

15

The Department of Energy produces
publications to fulfill a statutory
mandate to disseminate information to
the public on all energy sources and
energy conservation technologies. These
materials are for public use and do not
purport to present an exhaustive
treatment of the subject matter.

This is one in a series of publications on
nuclear energy.

U.S. Department of Energy

Printed on recycled paper

U.S.NRC Security Spotlight

UNITED STATES NUCLEAR REGULATORY COMMISSION
Protecting People and the Environment

Overview

The NRC requires nuclear power plants to protect against threats. These plants are some of the most fortified civilian facilities in the country. After 9/11, the NRC used its independent regulatory authority to order the nuclear industry to implement new defensive capabilities, more rigorous guard training and many other security enhancements. In response, the industry has met the increased requirements regardless of cost. The process of upgrading security continues.

■ **Safety and Security**

The NRC requires that nuclear power plants be both safe and secure. Safety refers to operating the plant in a manner that protects the public and the environment. Security refers to protecting the plant—using people, equipment and fortifications—from intruders who wish to damage or destroy it in order to harm people and the environment.

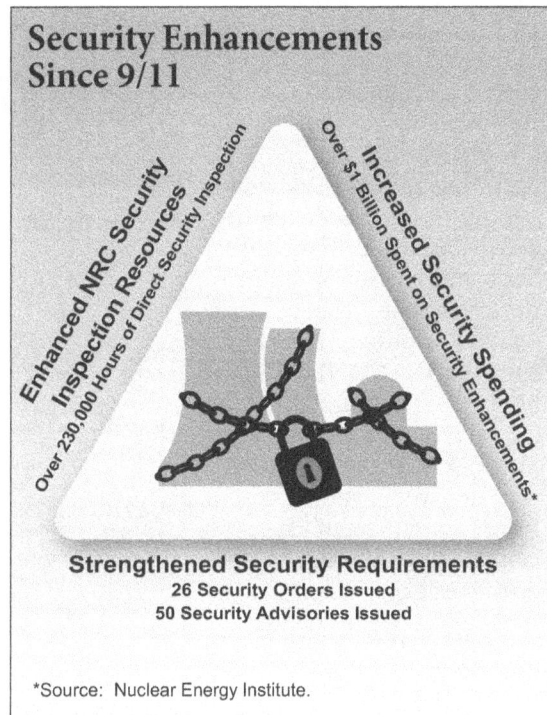

Security Enhancements Since 9/11

Enhanced NRC Security Inspection Resources
Over 230,000 Hours of Direct Security Inspection

Increased Security Spending
Over $1 Billion Spent on Security Enhancements*

Strengthened Security Requirements
26 Security Orders Issued
50 Security Advisories Issued

*Source: Nuclear Energy Institute.

> *"The NRC requires that nuclear power plants be both safe and secure."*
>
> **www.nrc.gov**

Layers of Defense

Protecting Against Aircraft

■ A combination of factors protect nuclear power plants from air attacks, including the fact they are robust structures of steel and concrete, and relatively small targets. Cooperation with other federal agencies also reduces the risk of an aircraft attack.

Securing Materials

■ Thousands of industrial and medical devices safely use small amounts of radioactive material to improve our quality of life. Some of these materials must be licensed and tracked to prevent them from being misused.

Defending Against Adversaries

■ There are many layers protecting a nuclear power plant from a ground or water attack, including well-trained and armed security officers, and defensive barriers. The NRC routinely tests the security of the plants through realistic exercises.

Strengthening Regulations

■ Enforcing regulations—also called rules—is how the NRC ensures the safety of the public and the environment. Three new or revised rules will further enhance the security of nuclear power plants.

May 2007

U.S.NRC
UNITED STATES NUCLEAR REGULATORY COMMISSION
Protecting People and the Environment

Security Spotlight

Protecting Against Aircraft

Since 9/11, the issue of an airborne attack on this nation's infrastructure, including both operating and potential new nuclear power plants, has been widely discussed. The NRC has comprehensively studied the effect of an airborne attack on nuclear power plants. Shortly after 9/11, the NRC began a security and engineering review of operating nuclear power plants. Assisting the NRC were national experts from Department of Energy laboratories, who used state-of-the-art experiments, and structural and fire analyses.

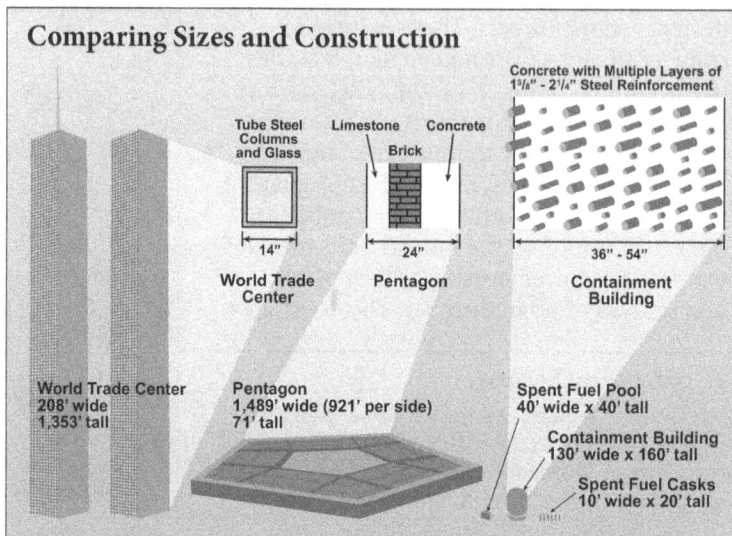

Comparing Sizes and Construction

Concrete with Multiple Layers of 1⅝" - 2¼" Steel Reinforcement

Tube Steel Columns and Glass

Limestone   Concrete

Brick

14"   24"   36" - 54"

World Trade Center   Pentagon   Containment Building

World Trade Center
208' wide
1,353' tall

Pentagon
1,489' wide (921' per side)
71' tall

Spent Fuel Pool
40' wide x 40' tall

Containment Building
130' wide x 160' tall

Spent Fuel Casks
10' wide x 20' tall

These classified studies confirm that there is a low likelihood that an airplane attack on a nuclear power plant would affect public health and safety, thanks in part to the inherent robustness of the structures. A second study identified new methods plants could use to minimize damage and risk to the public in the event of any kind of large fire or explosion. Nuclear power plants subsequently implemented many of these methods.

The NRC is now considering new regulations for future reactors' security. The goal is to include inherent safety and security features to minimize potential damage from an airborne attack.

Integrated Federal Response

It is the federal government and military's responsibility to protect the nation against an aircraft attack. To that end, the NRC works closely with its federal partners to identify and implement enhanced security programs, including:

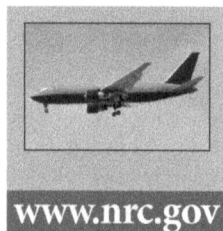

- Military and Department of Homeland Security program to identify and protect critical infrastructure
- Criminal history checks on flight crews
- Reinforced cockpit doors
- Checking of passenger lists against "no-fly" lists
- Increased control of cargo
- Random inspections
- Increased Federal Air Marshal presence
- Improved screening of passengers and baggage
- Controls on foreign passenger carriers
- Improved coordination and communication between civilian and military authorities.

www.nrc.gov

May 2007

U.S.NRC Security Spotlight
UNITED STATES NUCLEAR REGULATORY COMMISSION
Protecting People and the Environment

Defending Against Adversaries

Commercial nuclear power plants are heavily fortified with well-trained and armed guards. They also have layered physical security measures, such as access controls, water barriers, intrusion detection and strategically placed guard towers. Together, these make up the plants' response to the Design Basis Threat – usually called the DBT. The DBT is developed from real-world intelligence information and describes the adversary force – coming from both ground and water – the plants must defend against. DBT specifics are not public in order to protect sensitive information that could aid terrorists. The NRC regularly reviews the DBT and adds new requirements when necessary.

■ Category I Fuel-Cycle Facilities

There are two NRC-licensed Category I Fuel-Cycle Facilities in the U.S. that make reactor fuel for nuclear plants. Since these plants handle nuclear material that could be targeted by adversaries, they also must defend against a DBT similar to that for nuclear power plants.

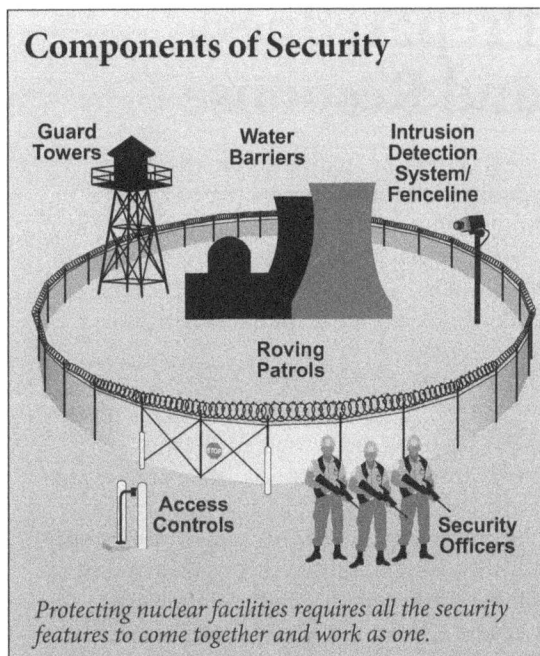

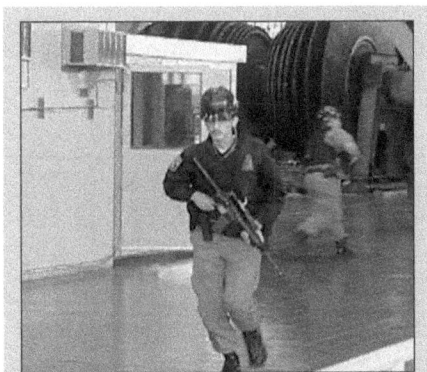

Components of Security

Guard Towers

Water Barriers

Intrusion Detection System/Fenceline

Roving Patrols

Access Controls

Security Officers

Protecting nuclear facilities requires all the security features to come together and work as one.

www.nrc.gov

Force-on-Force Exercises

■ The NRC routinely tests the security at nuclear facilities with realistic exercises using a well-trained mock adversary force. These force-on-force exercises are designed to test a security force's ability to defend against the DBT. The NRC oversees every aspect of these exercises and evaluates them using rigorous standards. These exercises typically span several days. During the attack, the mock adversary force tries to reach and damage key safety systems. Any significant security problems are promptly identified, reviewed, and fixed prior to NRC's inspection team leaving the facility. The NRC tests every plant with a force-on-force exercise a minimum of every three years. The plants also must conduct their own yearly exercises.

May 2007

U.S.NRC Security Spotlight

UNITED STATES NUCLEAR REGULATORY COMMISSION
Protecting People and the Environment

Preparedness and Response

No matter how small the risk, the NRC requires all nuclear power plants to have and periodically test emergency plans that are coordinated with federal, state and local responders. The goal of preparedness is to reduce the risk to the public during an emergency.

In an emergency, the NRC and the licensee would activate their Incident Response Programs. Licensee specialists would evaluate the situation and identify ways to end the emergency, while the NRC would monitor the event closely, keeping government offices informed. If a radiation release occurred, the plant would make protective action recommendations to state and local officials, such as evacuating areas around the plant.

The Team Approach

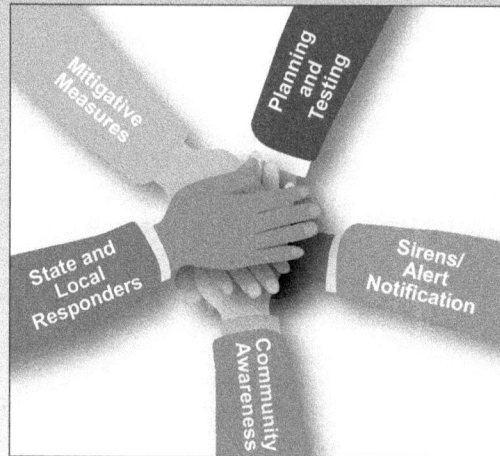

Mitigative Measures

Planning and Testing

State and Local Responders

Sirens/ Alert Notification

Community Awareness

Effective preparedness and response requires cooperation among the federal government, state and local officials, the public, and the nuclear plants.

■ **Emergency Planning Zones (EPZs)**

Each nuclear power plant has two EPZs. Each EPZ considers the specific conditions and geography at the site, and the community. The first is the Plume Exposure Pathway EPZ, which has a radius of about 10 miles from the reactor. People living there may be asked to evacuate or "shelter in place" during an emergency, to avoid or reduce their radiation dose. The second is the Ingestion Exposure Pathway EPZ. This has a radius of about 50 miles from the reactor. Protective action plans for this area aim to avoid or reduce the radiation dose from consuming contaminated food and water.

"The goal of preparedness is to reduce the risk to the public during an emergency."

www.nrc.gov

Response Modes

The NRC uses these modes for responding to events:

■ Monitoring - A heightened state of readiness for getting and accessing incident information.

■ Activation - A team of Reactor and Preparedness specialists begin staffing the Headquarters Operations Center and Regional Incident Response Centers to respond to the event. Another team of specialists travels to the site, if needed.

May 2007

U.S.NRC Security Spotlight
UNITED STATES NUCLEAR REGULATORY COMMISSION
Protecting People and the Environment

Securing Materials

Radioactive materials are used in many beneficial ways, including medical, academic and industrial uses. Cancer treatment is just one way that radioactive materials benefit the public. Despite these benefits, some materials can potentially harm people and the environment if misused. For these reasons, their security, including use and handling, is strictly regulated in the United States by the NRC.

■ "Dirty bombs"

A "dirty bomb," also called a "radiological dispersal device" (RDD), combines explosives, such as dynamite, with radioactive material. A dirty bomb is NOT a nuclear weapon. Most dirty bombs would not be highly destructive and would not release enough radiation to kill people or cause severe illness. Instead, a dirty bomb is a "Weapon of Mass <u>Disruption</u>" that could cause panic and fear, and require costly cleanup. Some materials licensed by the NRC could possibly be used in a dirty bomb, which is why they are strictly regulated.

Securing Radioactive Materials

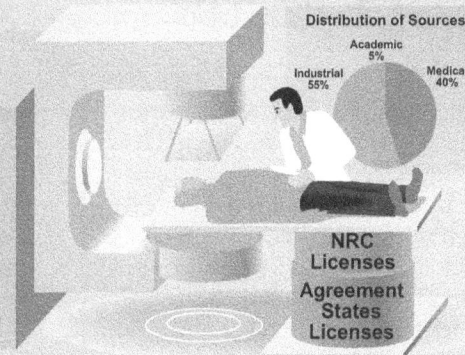

Distribution of Sources
Academic 5%
Industrial 55%
Medical 40%

NRC Licenses
Agreement States Licenses

The NRC and Agreement States have issued about 22,000 licenses for radioactive material. Through the Agreement State Program, the NRC shares its regulatory authority to license and oversee the use of certain types of radioactive material. The NRC regularly reviews the programs set up by the states to verify that they can effectively protect public health and safety.

National Source Tracking System (NSTS)

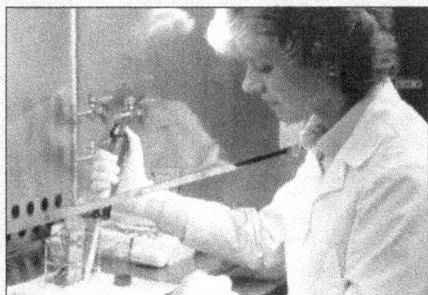

www.nrc.gov

■ The NRC will implement the NSTS in 2008 to enhance controls for certain radioactive materials considered to be of the greatest concern from a safety and security standpoint. Until the NSTS is deployed, the NRC and Agreement States perform an annual inventory of these sources. The tracking system is being developed with other federal and state agencies, and international partners.

The NSTS will require licensees to report the manufacture, transfer, receipt, disassembly, and disposal of nationally tracked sources. The NSTS is an important component of the NRC's effort to enhance the control of radioactive material and prevent its use by the nation's adversaries. There are approximately 54,000 of these sources in use in the United States.

May 2007

U.S.NRC Security Spotlight
UNITED STATES NUCLEAR REGULATORY COMMISSION
Protecting People and the Environment

Research and Test Reactors

Research and Test Reactors – also called RTRs or "non-power" reactors – are low-power nuclear reactors that are primarily used for research, training and development. There are 32 operating NRC-licensed RTRs around the country that are used to study almost every field of science. Regulating the safety and security of RTRs is one of NRC's jobs.

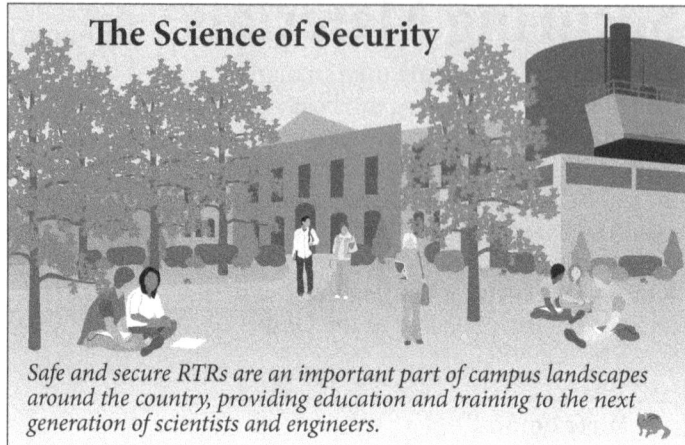

The Science of Security

Safe and secure RTRs are an important part of campus landscapes around the country, providing education and training to the next generation of scientists and engineers.

RTRs are designed and operated so that material is not easily handled or dispersed. This protects the public and environment against potential radiological exposure or theft of the material. RTRs are licensed to have only small amounts of radioactive material on site. The NRC evaluates and inspects each RTR's security plans, procedures and systems to verify that effective security measures are in place to protect the reactors.

■ Size Matters

NRC-licensed RTRs range in size from 20 Megawatts (MW) to 5 Watts (about the size of a child's nightlight). In comparison, the typical operating nuclear power plant is 3,000MW and can power over 1 million homes.

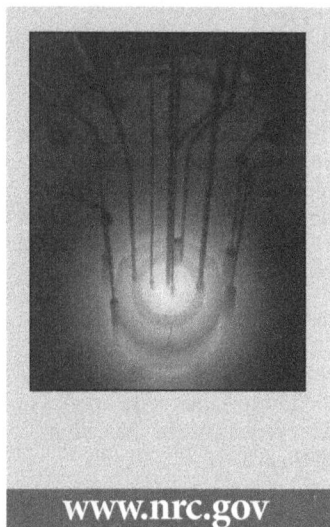

Rules of Regulation

Because NRC-licensed RTRs operate at significantly lower power levels than their power plant cousins and have a limited amount of radioactive material on site, the standard for regulating these reactors is different. In fact, the NRC is federally mandated to apply the minimum regulation needed to protect the public health and safety at RTRs so they can effectively conduct education and research.

After 9/11, the NRC established additional security measures and inspected RTRs to ensure the measures were followed. The NRC identified several potential enhancements and RTRs around the country voluntarily implemented many of the improvements. With these security measures in place, the NRC has determined that these reactors pose minimal risk to public health and safety.

Today, the NRC continues to monitor RTR security through our regulatory processes. If threat conditions change, such that they could potentially affect public health and safety, the NRC will act promptly to further enhance security at RTRs.

September 2007

www.nrc.gov

U.S.NRC Security Spotlight

UNITED STATES NUCLEAR REGULATORY COMMISSION
Protecting People and the Environment

Rulemaking Overview

Immediately after the 9/11 terrorist attacks, the NRC advised nuclear facilities to go to the highest level of security. After that, the NRC issued a series of mandates – called Orders – to further strengthen security. The NRC is taking a multifaceted approach to security enhancements in the post-9/11 threat environment. The NRC has raised the security of existing nuclear power plants while also requiring new security features in the design of new reactors that may be built in coming years.

Most recently, three new rulemakings provide additional security enhancements.

- One rule, issued by the NRC in March 2007 after extensive public comment, modifies and enhances the Design Basis Threat.

- A second rule, which was issued for public comment in 2006, proposes enhancements to the physical security at nuclear power plants. Among other things, the proposed rule addresses

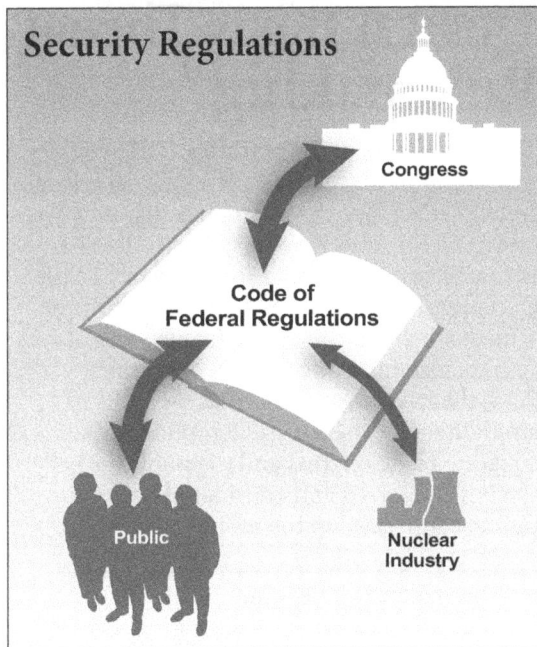

Security Regulations

Congress

Code of
Federal Regulations

Public

Nuclear
Industry

access controls, event reporting, security personnel training, safety and security activity coordination, contingency planning, cyber and radiological sabotage protection.

- A third rule, still in the early stages, will propose additional aircraft impact assessments for new power reactor designs.

How Rulemaking Works

www.nrc.gov

- Rules – or regulations – and their enforcement are how the NRC protects people and the environment. Nuclear power plants must adhere to the rules or risk serious repercussions – up to closing a plant down. A new rule may be proposed by the NRC's five-member Commission, because of a petition from the public or as suggested by the NRC staff based on research or actual events. Once developed, a proposed rule is published in the *Federal Register* for a public comment period, usually 75 to 90 days. Once the comment period has closed, the NRC staff analyzes the comments, makes any needed changes, and forwards the final rule to the NRC Commissioners for approval. If approved, the final rule is published in the *Federal Register* and usually becomes effective in 30 days.

May 2007

U.S.NRC
UNITED STATES NUCLEAR REGULATORY COMMISSION
Protecting People and the Environment

Security Spotlight

Design Basis Threat Rulemaking

The revised Design Basis Threat (DBT) rule was issued in March 2007. The rule describes general adversary characteristics that nuclear power plants must defend against. All existing nuclear power plants and Category I Fuel Cycle Facilities – and any built in the future – must adhere to this rule. The new rule also reflects insights gained by the NRC since 9/11, the latest threat information and a strengthened cyber threat component, as suggested by Congress and the public. In all, the NRC received and considered over 900 public comments on the rule.

Strengthening Security Regulations

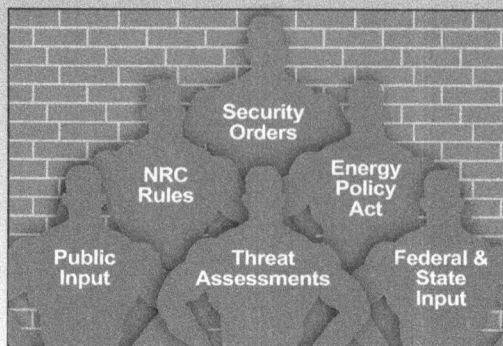

Security Orders

NRC Rules

Energy Policy Act

Public Input

Threat Assessments

Federal & State Input

By incorporating new threat information and Congressional and public input, the NRC has strengthened the DBT against which all nuclear power plants must defend.

■ **Chairman Dale Klein:**

"This rule is an important piece, but only one piece, of a broader effort to enhance nuclear power plant security. Overall we are taking a multifaceted approach to security enhancements in this post 9/11 threat environment, and looking at how best to secure existing nuclear power plants and how to incorporate security enhancements into design features of new reactors that may be built in coming years."

"Overall we are taking a multifaceted approach to security enhancements..."

www.nrc.gov

Energy Policy Act of 2005

In this legislation, Congress outlined 12 factors that the NRC considered when developing the new DBT rule. Among those factors were:

■ An assessment of physical, cyber, biochemical, and other terrorist threats;

■ The potential for attack on facilities by multiple coordinated teams of a large number of individuals and several insiders;

■ The potential for suicide attacks;

■ The potential for water-based and air-based threats;

■ The potential use of explosive devices of considerable size and other modern weaponry;

■ The potential for attacks by persons with a sophisticated knowledge of facility operations;

■ The potential for possibly long-lived fires.

May 2007

U.S.NRC Security Spotlight
UNITED STATES NUCLEAR REGULATORY COMMISSION
Protecting People and the Environment

Physical Protection Rulemaking

A significant rulemaking on physical protection of nuclear power plants is currently underway. Originally published for public comment in the *Federal Register* on October 26, 2006, the proposed rule enhances requirements for access controls, event reporting, security personnel training, safety and security activity coordination, contingency planning and radiological sabotage protection. It would also add requirements related to background checks for firearms users and authorization for enhanced weapons to fulfill certain provisions in the Energy Policy Act of 2005.

- ### Safety/Security Interface Requirements
  The proposed rule's safety/security requirements define how nuclear power plants are to minimize potential adverse interactions between security activities and other plant activities. The goal is to ensure that neither plant security nor plant safety is compromised.

The Chain of Security

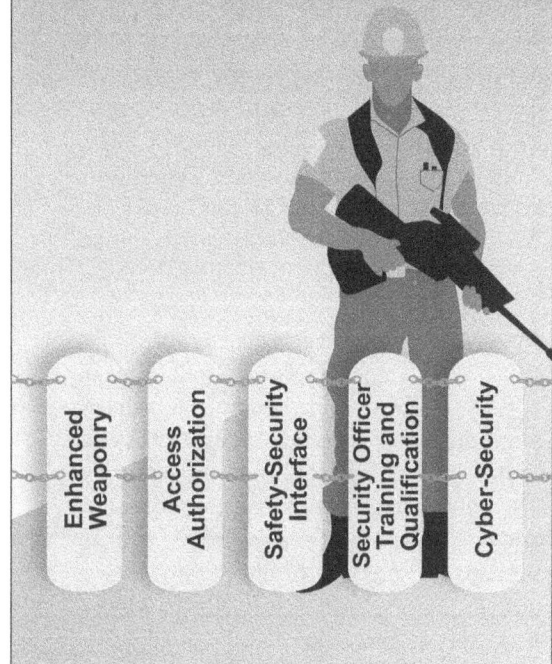

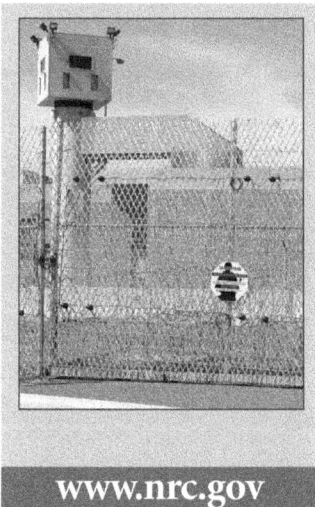

Enhanced Weaponry

Access Authorization

Safety-Security Interface

Security Officer Training and Qualification

Cyber-Security

www.nrc.gov

Amended Regulations - 10 CFR Part 73 "Physical Protection"

- 73.55 "Requirements for physical protection of licensed activities in nuclear power reactors against radiological sabotage"
- 73.56 "Personnel access authorization requirements for nuclear power plants"
- 73.71 "Reporting of safeguards events"
- Part 73 Appendix B "General criteria for security personnel"
- Part 73 Appendix C "Licensee safeguards contingency plans"
- Part 73 Appendix G "Reportable safeguards events"

New Regulations

- 73.18 "Firearms background checks for armed security personnel"
- 73.19 "Authorization for preemption of firearms laws and use of enhanced weapons"
- 73.58 "Safety/security interface requirements for nuclear power reactors"

May 2007

U.S.NRC
UNITED STATES NUCLEAR REGULATORY COMMISSION
Protecting People and the Environment

Security Spotlight

New Reactor Rulemaking

Although the NRC has not received an application to construct a new reactor for nearly 30 years, there is growing interest in nuclear power in the U.S. – some have called it a nuclear renaissance. Based on conversations with energy companies, the NRC expects to receive at least 19 applications for new reactors in the coming years.

Many of the new reactors would be built based on designs the NRC has already approved. These "next generation" nuclear plant designs have benefited from the current plants' decades of operating experience. The new designs are inherently more safe and secure, using many "passive" systems that ensure safety without operator action.

■ **Current Activities**

In 2005, the NRC Commission directed its staff to review and propose changes to the regulation of new nuclear power plants. The purpose is to integrate the expectations for security and

Building Security Into the Design

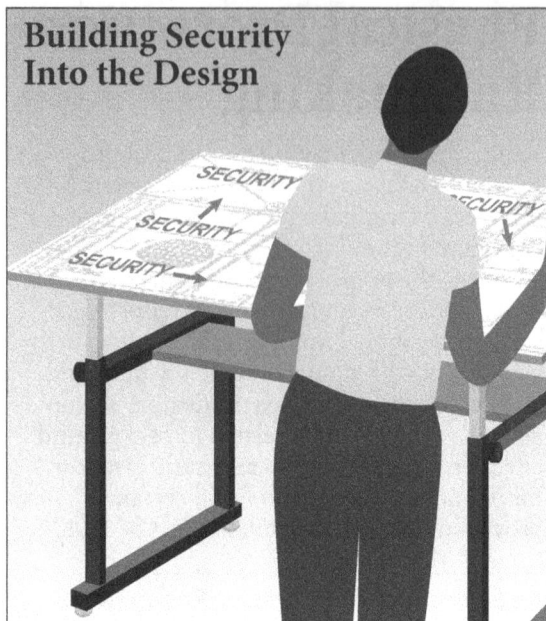

preparedness with the current expectations for safety. The Commission has also directed the staff to develop a proposed rule that would require the assessment of a commercial aircraft impact on new reactor designs. New reactor designs will be required to include an evaluation of their specific features, capabilities, and strategies that can prevent or lessen the effect of an impact. The NRC will publish this rule for public comment.

Certified Reactor Designs

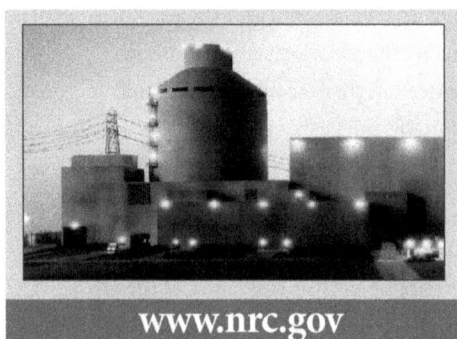

www.nrc.gov

There are currently four certified reactor designs that can be referenced in an application for a combined license. A design certification is good for 15 years. The certified designs and their date of approval are:

■ Advanced Boiling Water Reactor design by GE Nuclear Energy (May 1997);

■ System 80+ design by Westinghouse (May 1997);

■ AP600 design by Westinghouse (December 1999); and

■ AP1000 design (pictured) by Westinghouse (February 2006).

May 2007

U.S.NRC Security Spotlight
UNITED STATES NUCLEAR REGULATORY COMMISSION
Protecting People and the Environment

Conclusion

While security of the nation's nuclear power plants has always been a top priority, the NRC has responded to today's threat environment with heightened scrutiny and increasingly stringent requirements. NRC-regulated nuclear facilities are, in fact, considered among the most secure of the nation's critical infrastructure.

The key is layers of defense. As a first layer, nuclear power plants are inherently robust structures, built to withstand hurricanes, tornadoes and earthquakes. Additional security measures as previously explained are then layered on top. A final layer of protection is NRC's close coordination with the Department of Homeland Security (DHS), intelligence agencies, the Department of Defense and local law enforcement. This coordination is focused on building an integrated federal, state and local response to protect the public. The NRC Operations Center, at NRC headquarters in Rockville, Md., provides an around-the-clock conduit for information and coordinated response.

The Layers of Robust Security

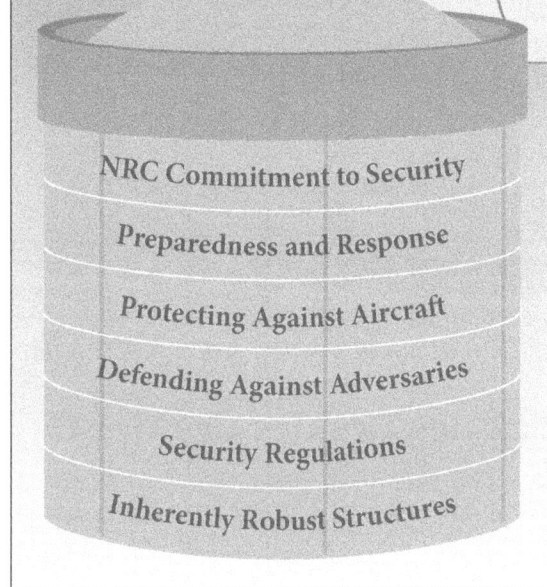

- NRC Commitment to Security
- Preparedness and Response
- Protecting Against Aircraft
- Defending Against Adversaries
- Security Regulations
- Inherently Robust Structures

Together, these layers make a formidable defense – they provide a level of security second to none in the commercial power sector.

"NRC-regulated nuclear facilities are considered among the most secure private facilities."

www.nrc.gov

NRC's Security Highlights

- The NRC's budget for nuclear security has increased more than ten-fold since 9/11;

- The defenses of nuclear plants are being tested through the force-on-force program nearly three times as often as before and in a much more realistic fashion;

- A NRC-DHS review of the nuclear sector has yielded additional improvements in plant security; and

- The nation has substantially enhanced its system to secure risk-significant radioactive material.

– NRC Commissioners' letter to DHS Secretary Michael Chertoff, August 28, 2006

May 2007

Nuclear Power Plant Security and Vulnerabilities

Mark Holt
Specialist in Energy Policy

Anthony Andrews
Specialist in Energy and Energy Infrastructure Policy

March 18, 2009

Congressional Research Service

7-5700

www.crs.gov

RL34331

CRS Report for Congress
Prepared for Members and Committees of Congress

Summary

The physical security of nuclear power plants and their vulnerability to deliberate acts of terrorism was elevated to a national security concern following the September 11, 2001 attacks. Since then, Congress has repeatedly focused oversight and legislative attention on nuclear power plant security requirements established and enforced by the Nuclear Regulatory Commission (NRC).

The Energy Policy Act of 2005 (P.L. 109-58) imposed specific criteria for NRC to consider in revising the "Design Basis Threat" (DBT), which specifies the maximum severity of potential attacks that a nuclear plant's security force must be capable of repelling. In response to the legislative mandate, NRC revised the DBT (10 C.F.R. Part 73.1) on April 18, 2007. Among other changes, the revisions expanded the assumed capabilities of adversaries to operate as one or more teams and attack from multiple entry points.

To strengthen nuclear plant security inspections, the Energy Policy Act of 2005 act required NRC to conduct "force-on-force" security exercises at nuclear power plants at least once every three years. In these exercises, a mock adversary force from outside a nuclear plant attempts to penetrate the plant's vital area and simulate damage to key safety components. The first three-year cycle of force-on-force exercises was completed for all U.S. nuclear plants at the end of calendar year 2007. During that period, 172 force-on-force exercises were conducted (an average of three per site), and 10 security inadequacies were cited. Two of the exercises, both in 2007, resulted in simulated damage or destruction of a vital target by the adversary team. In both cases, NRC ordered corrective actions and conducted follow-up exercises to confirm the improvements.

The Energy Policy Act also included provisions for fingerprinting and criminal background checks of security personnel, their use of firearms, and the unauthorized introduction of dangerous weapons. The designation of facilities subject to enforcement of penalties for sabotage was expanded to include treatment and disposal facilities.

Nuclear power plant vulnerability to deliberate aircraft crashes has been a continuing issue. After much consideration, NRC voted February 17, 2009, to require all new nuclear power plants to incorporate design features that would ensure that, in the event of a crash by a large commercial aircraft, the reactor core would remain cooled or the reactor containment would remain intact, and radioactive releases would not occur from spent fuel storage pools.

NRC rejected proposals that existing reactors also be required to protect against aircraft crashes, such as by adding large external steel barriers. However, NRC did impose some additional requirements related to aircraft crashes on all reactors, both new and existing, after the 9/11 attacks. In 2002, NRC ordered all nuclear power plants to develop strategies to mitigate the effects of large fires and explosions that could result from aircraft crashes or other causes. An NRC regulation on fire mitigation strategies, along with requirements that reactors establish procedures for responding to specific aircraft threats, was approved December 17, 2008.

Other ongoing nuclear plant security issues include the vulnerability of spent fuel pools, which hold highly radioactive nuclear fuel after its removal from the reactor, standards for nuclear plant security personnel, and nuclear plant emergency planning. NRC's December 2008 security regulations addressed some of those concerns and included a number of other security enhancements.

Nuclear Power Plant Security and Vulnerabilities

Contents

Contacts

Congressional Research Service

Overview of Reactor Security

Physical security at nuclear power plants involves the threat of radiological sabotage—a deliberate act against a plant that could directly or indirectly endanger public health and safety through exposure to radiation. The Nuclear Regulatory Commission (NRC) establishes security requirements at U.S. commercial nuclear power plants based on its assessment of plant vulnerabilities to, and the consequences of, potential attacks. The stringency of NRC's security requirements and its enforcement program have been a significant congressional issue, especially since the September 11, 2001, terrorist attacks on the United States.

Nuclear plant security measures are designed to protect three primary areas of vulnerability: controls on the nuclear chain reaction, cooling systems that prevent hot nuclear fuel from melting even after the chain reaction has stopped, and storage facilities for highly radioactive spent nuclear fuel. U.S. plants are designed and built to prevent dispersal of radioactivity, in the event of an accident, by surrounding the reactor in a steel-reinforced concrete containment structure.

NRC requires commercial nuclear power plants to have a series of physical barriers and a trained security force, under regulations already in place prior to the 9/11 attacks (10 C.F.R. 73—Physical Protection of Plants and Materials). The plant sites are divided into three zones: an "owner-controlled" buffer region, a "protected area," and a "vital area." Access to the protected area is restricted to a portion of plant employees and monitored visitors, with stringent access barriers. The vital area is further restricted, with additional barriers and access requirements. The security force must comply with NRC requirements on pre-hiring investigations and training.[1]

A fundamental concept in NRC's physical security requirements is the design basis threat (DBT), which establishes the severity of the potential attacks that a nuclear plant's security force must be capable of repelling. The DBT includes such characteristics as the number of attackers, their training, and the weapons and tactics they could use. Specific details are classified. Critics of nuclear plant security have contended that the DBT should be strengthened to account for potentially larger and more sophisticated terrorist attacks.

Reactor vulnerability to deliberate aircraft crashes has also been a major concern since 9/11. Most existing nuclear power plants were not specifically designed to withstand crashes from large jetliners, although analyses differ as to the damage that could result. NRC has determined that commercial aircraft crashes are beyond the DBT but voted in February 2009 to require that new reactor designs be able to withstand such crashes without releasing radioactivity. Nuclear power critics have called for retrofits of existing reactors as well.

Since the 9/11 attacks, NRC and Congress have taken action to increase nuclear power plant security. NRC issued a series of security measures beginning in 2002, including a strengthening of the DBT and establishing the Office of Nuclear Security and Incident Response (NSIR). The office centralizes security oversight of all NRC-regulated facilities, coordinates with law enforcement and intelligence agencies, and handles emergency planning activities. In 2004, NRC implemented a program to conduct "force on force" security exercises overseen by NSIR at each nuclear power plant at least every three years. The Energy Policy Act of 2005 (P.L. 109-58) required NRC to further strengthen the DBT, codified the force-on-force program, and established

---

[1] General NRC requirements for nuclear power plant security can be found in 10 C.F.R. 73.55.

a variety of additional nuclear plant security measures. In December 2008, NRC approved a series of security regulations that require power plants to prepare cyber security plans, develop strategies for dealing with the effects of aircraft crashes, strengthen access controls, improve training for security personnel, and take other new security measures.

Design Basis Threat

The design basis threat describes general characteristics of adversaries that nuclear plants and nuclear fuel cycle facilities must defend against to prevent radiological sabotage and theft of strategic special nuclear material. NRC licensees use the DBT as the basis for implementing defensive strategies of a specific nuclear plant site through security plans, safeguards contingency plans, and guard training and qualification plans.

General requirements for the DBT are prescribed in NRC regulations,[2] while specific attributes of potential attackers, such as their weapons and ammunition, are contained in classified adversary characteristics documents (ACDs).

Fundamental policies on nuclear plant security threats date back to the Cold War. In 1967, the Atomic Energy Commission (AEC) instituted a rule that nuclear plants are not required to protect against an attack directed by an "enemy of the United States."[3] That so-called "Enemy of the State Rule" specifies that nuclear power plants are

> not required to provide for design features or other measures for the specific purpose of protection against the effects of (a) attacks and destructive acts, including sabotage, directed against the facility by an enemy of the United States, whether a foreign government or other person, or (b) use or deployment of weapons incident to U.S. defense activities.[4]

The Nuclear Regulatory Commission (NRC), the AEC's successor regulatory agency, says that the rule "was primarily intended to make clear that privately-owned nuclear facilities were not responsible for defending against attacks that typically could only be carried out by foreign military organizations."[5] NRC's initial DBT, established in the late 1970s, was intended to be consistent with the enemy of the state rule, which remains in effect.

However, the 9/11 attacks drew greater attention to the potential severity of credible terrorist threats. Following the attacks, NRC evaluated the extent to which nuclear plant security forces should be able to defend against such threats, and ordered a strengthening of the DBT, along with other security measures, on April 29, 2003. That order changed the DBT to "represent the largest reasonable threat against which a regulated private guard force should be expected to defend under existing law," according to the NRC announcement.[6]

---

[2] 10 C.F.R. § 73.1.

[3] It was feared that Cuba might launch an attack on Florida reactors. Government Accountability Office, *Nuclear Power Plants—Efforts Made to Upgrade Security, but the Nuclear Regulatory Commission's Design Basis Threat Process Should Be Improved* (GAO-06-388), March 2006, p. 2. Regulations at 10 CFR 50.13.

[4] 10 C.F.R. § 50.13. Attacks and destructive acts by enemies of the United States; and defense activities.

[5] Nuclear Regulatory Commission, "Design Basis Threat," 72 *Federal Register* 12714, March 19, 2007.

[6] *Federal Register*, May 7, 2003 (vol. 68, no. 88). NRC, All Operating Power Reactor Licensees; Order Modifying Licenses.

In the Energy Policy Act of 2005 (EPACT), Congress imposed a statutory requirement on the NRC to initiate rulemaking for revising the design basis threat.[7] EPACT required NRC to consider 12 factors in revising the DBT, such as an assessment of various terrorist threats, sizable explosive devices and modern weapons, attacks by persons with sophisticated knowledge of facility operations, and attacks on spent fuel shipments.

NRC approved its final rule amending the DBT (10 C.F.R. Part 73.1) on January 29, 2007, effective April 18, 2007.[8] Although specific details of the revised DBT were not released to the public, in general the final rule

- clarifies that physical protection systems are required to protect against diversion and theft of fissile material;

- expands the assumed capabilities of adversaries to operate as one or more teams and attack from multiple entry points;

- assumes that adversaries are willing to kill or be killed and are knowledgeable about specific target selection;

- expands the scope of vehicles that licensees must defend against to include water vehicles and land vehicles beyond four-wheel-drive type;

- revises the threat posed by an insider to be more flexible in scope; and

- adds a new mode of attack from adversaries coordinating a vehicle bomb assault with another external assault.

The DBT final rule excluded aircraft attacks, a decision that raised considerable controversy. In approving the rule, NRC rejected a petition from the Union of Concerned Scientists to require that nuclear plants be surrounded by aircraft barriers made of steel beams and cables (the so-called "beamhenge" concept). Critics of the rule charged that deliberate aircraft crashes were a highly plausible mode of attack, given the events of 9/11. However, NRC contended that power plants were already required to mitigate the effects of aircraft crashes and that "active protection against airborne threats is addressed by other federal organizations, including the military."[9] Additional NRC action on aircraft threats is discussed below.

NRC Commissioners in January 2009 rejected a proposal by the NRC staff to strengthen the classified portion of the DBT to include additional capabilities by potential attackers, according to news reports. The staff proposal lost in a 2-2 vote, with one commissioner position currently vacant. In an interview afterward, NRC Chairman Dale Klein said the vote could be reconsidered after completion of an ongoing interagency study.[10]

Critics of NRC's security regulations have pointed out that licensees are required to employ only a minimum of five security personnel on duty per plant, which they argue is not enough for the job.[11] Nuclear spokespersons responded that the actual security force for the nation's 65 nuclear

[7] P.L. 109-58, Title VI, Subtitle D—Nuclear Security (Secs. 651-657). Sec. 651 adds Atomic Energy Act Sec. 170E. Design Basis Threat Rulemaking.

[8] *Federal Register*, March 19, 2007 (vol. 72, no. 52), NRC, Design Basis Threat, Final Rule, pp. 12705-12727.

[9] NRC, "NRC Approves Final Rule Amending Security Requirements," News Release No. 07-012, January 29, 2007.

[10] Jeff Beattie, "NRC Chairman Questions Case for Tougher DBT," *Energy Daily*, February 17, 2009, p. 1.

[11] 10 C.F.R. 73.55 (h)(3) states: "The total number of guards, and armed, trained personnel immediately available at the (continued...)

plant sites numbers more than 5,000, an average of about 75 per site (covering multiple shifts). Nuclear plant security forces are also supposed to be aided by local law enforcement officers if an attack occurs.

Large Aircraft Crashes

Nuclear power plants were designed to withstand hurricanes, earthquakes, and other extreme events. But deliberate attacks by large airliners loaded with fuel, such as those that crashed into the World Trade Center and Pentagon, were not analyzed when design requirements for today's reactors were determined.[12] Concern about aircraft crashes was intensified by a taped interview shown September 10, 2002, on the Arab TV station al-Jazeera, which contained a statement that Al Qaeda initially planned to include a nuclear plant in its list of 2001 attack sites.

In light of the possibility that an air attack might penetrate the containment structure of a nuclear plant or a spent fuel storage facility, some interest groups have suggested that such an event could be followed by a meltdown or spent fuel fire and widespread radiation exposure. Nuclear industry spokespersons have countered by pointing out that relatively small, low-lying nuclear power plants are difficult targets for attack, and have argued that penetration of the containment is unlikely, and that even if such penetration occurred it probably would not reach the reactor vessel. They suggest that a sustained fire, such as that which melted the steel support structures in the World Trade Center buildings, would be impossible unless an attacking plane penetrated the containment completely, including its fuel-bearing wings. According to former NRC Chairman Nils Diaz, NRC studies "confirm that the likelihood of both damaging the reactor core and releasing radioactivity that could affect public health and safety is low."[13]

NRC proposed in October 2007 to amend its regulations to require newly designed power reactors to take into account the potential effects of the impact of a large commercial aircraft.[14] As discussed in the previous section, NRC considers an aircraft attack to be beyond the design basis threat that plants must be able to withstand, so the requirements of the proposed rule were intended to provide an additional margin of safety. The proposed rule would affect only new reactor designs not previously certified by NRC, because the previous designs were still considered adequately safe. Nevertheless, Westinghouse submitted changes in the certified design of its AP1000 reactor to NRC on May 29, 2007, proposing to line the inside and outside of the reactor's concrete containment structure with steel plates to increase resistance to aircraft penetration.[15]

(...continued)

facility to fulfill these response requirements shall nominally be ten (10), unless specifically required otherwise on a case by case basis by the Commission; however, this number may not be reduced to less than five (5) guards."

[12] Meserve, Richard A., NRC Chairman, "Research: Strengthening the Foundation of the Nuclear Industry," Speech to Nuclear Safety Research Conference, October 29, 2002.

[13] Letter from NRC Chairman Nils J. Diaz to Secretary of Homeland Security Tom Ridge, September 8, 2004.

[14] *Federal Register*, October 3, 2007 (vol. 72, no. 191), Consideration of Aircraft Impacts for New Nuclear Power Reactor Designs.

[15] MacLachlan, Ann, "Westinghouse Changes AP1000 Design to Improve Plane Crash Resistance," Nucleonics Week, June 21, 2007.

Under NRC's 2007 proposed rule, applicants for new certified designs or for new reactor licenses using uncertified designs would have been required to assess the effects that a large aircraft crash would have on the proposed facilities. Each applicant would then describe how the plant's design features, capabilities, and operations would avoid or mitigate the effects of such a crash, particularly on core cooling, containment integrity, and spent fuel storage pools.

In response to comments, the NRC staff proposed in October 2008 that the aircraft impact assessments be conducted by all new reactors, including those using previously certified designs.[16] The NRC Commissioners, in a 3-1 vote, approved the change February 17, 2009, and added specific design requirements that all new reactors would have to meet:[17]

> Each applicant subject to this section shall perform a design-specific assessment of the effects on the facility of the impact of a large, commercial aircraft. Using realistic analyses, the applicant shall identify and incorporate into the design those design features and functional capabilities to show that, with reduced use of operator actions:
>
> (A) the reactor core remains cooled, or the containment remains intact; and
>
> (B) spent fuel cooling or spent fuel pool integrity is maintained.

As noted above, NRC rejected proposals that existing reactors—in addition to new reactors—be required to protect against aircraft crashes, such as by adding "beamhenge" barriers. However, NRC did impose some additional requirements related to aircraft crashes on all reactors after the 9/11 attacks. In 2002, NRC ordered all nuclear power plants to develop strategies to mitigate the effects of large fires and explosions that could result from aircraft crashes or other causes.[18] As part of a broad security rulemaking effort, NRC proposed in October 2006 to incorporate the 2002 order on fire and explosion strategies into its security regulations (10 CFR Part 73).[19] In response to comments, NRC published a supplemental proposed rule in April 2008 to move the fire and explosion requirements into its reactor licensing regulations at 10 CFR Part 50, along with requirements that reactors establish procedures for responding to specific aircraft threat notifications.[20] Those regulations received final approval by the NRC Commissioners December 17, 2008.[21]

Force-On-Force Exercises

EPACT codified an NRC requirement that each nuclear power plant conduct security exercises every three years to test its ability to defend against the design basis threat. In these "force-on-

[16] Nuclear Regulatory Commission, *Final Rule—Consideration of Aircraft Impacts for New Nuclear Power Reactors*, Rulemaking Issue Affirmation, SECY-08-0152, October 15, 2008.

[17] Nuclear Regulatory Commission, *Final Rule—Consideration of Aircraft Impacts for New Nuclear Power Reactors*, Commission Voting Record, SECY-08-0152, February 17, 2009.

[18] Nuclear Regulatory Commission, *Final Rule—Consideration of Aircraft Impacts for New Nuclear Power Reactors*, Rulemaking Issue Affirmation, SECY-08-0152, October 15, 2008, p. 2.

[19] Nuclear Regulatory Commission, "Power Reactor Security Requirments, Proposed Rule," 71 *Federal Register* 62664, October 26, 2006.

[20] Nuclear Regulatory Commission, "Power Reactor Security Requirements, Supplemental Proposed Rule," 73 *Federal Register* 19443, April 10, 2008.

[21] Nuclear Regulatory Commission, "NRC Approves Final Rule Expanding Security Requirements for Nuclear Power Plants," press release, December 17, 2008, http://www.nrc.gov/reading-rm/doc-collections/news/2008/08-227.html.

force" exercises, monitored by NRC, a mock adversary force from outside the plant attempts to penetrate the plant's vital area and simulate damage to key safety components. Participants in the tightly controlled exercises carry weapons modified to fire only blanks and laser bursts to simulate bullets, and they wear laser sensors to indicate hits. Other weapons and explosives, as well as destruction or breaching of physical security barriers, may also be simulated. While one squad of the plant's guard force is participating in a force-on-force exercise, another squad is also on duty to maintain normal plant security. Plant defenders know that a mock attack will take place sometime during a specific period of several hours, but they do not know what the attack scenario will be. Multiple attack scenarios are conducted over several days of exercises.

Full implementation of the force-on-force program began in late 2004. Standard procedures and other requirements have been developed for using the force-on-force exercises to evaluate plant security and as a basis for taking regulatory enforcement action. Many tradeoffs are necessary to make the exercises as realistic and consistent as possible without endangering participants or regular plant operations and security.

NRC required the nuclear industry to develop and train a "composite adversary force" comprising security officers from many plants to simulate terrorist attacks in the force-on-force exercises. However, in September 2004 testimony, GAO criticized the industry's selection of Wackenhut, a security company that guards about half of U.S. nuclear plants, to also provide the adversary force. In addition to raising "questions about the force's independence," GAO noted that Wackenhut had been accused of cheating on previous force-on-force exercises by the Department of Energy.[22] Exelon terminated its security contracts with Wackenhut in late 2007 after guards at the Peach Bottom reactor in York County, Pennsylvania, were discovered sleeping while on duty.[23] EPACT requires NRC to "mitigate any potential conflict of interest that could influence the results of a force-on-force exercise, as the Commission determines to be necessary and appropriate." NRC's 2007 annual security report to Congress found that the industry adversary teams "continued to meet expectations for a credible, well-trained, and consistent mock adversary force."[24]

The first three-year cycle of force-on-force exercises was completed for all 64 U.S. nuclear plant sites at the end of calendar year 2007.[25] During that period, 172 force-on-force exercises were conducted (an average of three per site), and 10 security inadequacies were cited. Two of the exercises, both in 2007, resulted in simulated damage or destruction of a vital target by the adversary team. If such an attack had been real, the plant could have released unacceptable levels of radioactivity. Both cases resulted from "the failure of licensee armed security personnel to interpose themselves between the mock adversary and the vital areas and target set components,"

[22] GAO. "Nuclear Regulatory Commission: Preliminary Observations on Efforts to Improve Security at Nuclear Power Plants." Statement of Jim Wells, Director, Natural Resources and Environment to the Subcommittee on National Security, Emerging Threats, and International Relations, House Committee on Government Reform. September 14, 2004. p. 14.

[23] *Washington Post*, "Executive Resigns in Storm Over Sleeping Guards," January 10, 2008.

[24] Nuclear Regulatory Commission, Office of Nuclear Security and Incident Response, Report to Congress on the Security Inspection Program for Commercial Power Reactor and Category 1 Fuel Cycle Facilities: Results and Status Update; Annual Report for Calendar Year 2007, NUREG-1885, July 2008, p. 7, http://www.nrc.gov/reading-rm/doc-collections/congress-docs/correspondence/2008/boxer-07-01-2008.pdf.

[25] NRC generally lists 65 U.S. plant sites, but the adjacent Hope Creek and Salem sites in New Jersey are considered to be a single site for security exercises. E-mail message from David Decker, NRC Office of Congressional Affairs, March 13, 2009.

according to NRC's 2007 security report to Congress. In response to the failures, NRC imposed "immediate compensatory measures followed by long-term corrective actions." Follow-up force-on-force exercises were conducted to verify that the necessary security improvements had been made.[26]

Emergency Response

After the 1979 accident at the Three Mile Island nuclear plant near Harrisburg, PA, Congress required that all nuclear power plants be covered by emergency plans. NRC requires that within an approximately 10-mile Emergency Planning Zone (EPZ) around each plant, the operator must maintain warning sirens and regularly conduct evacuation exercises monitored by NRC and the Federal Emergency Management Agency (FEMA). In light of the increased possibility of terrorist attacks that, if successful, could result in release of radioactive material, proposals have been made to expand the EPZ to include larger population centers.

The release of radioactive iodine during a nuclear incident is a particular concern, because iodine tends to concentrate in the thyroid gland of persons exposed to it. Emergency plans in many states include distribution of iodine pills to the population within the EPZ. Taking non-radioactive iodine before exposure would prevent absorption of radioactive iodine but would afford no protection against other radioactive elements. In 2002, NRC began providing iodine pills to states requesting them for populations within the 10-mile EPZ.

Spent Fuel Storage

When no longer capable of sustaining a nuclear chain reaction, highly radioactive "spent" nuclear fuel is removed from the reactor and stored in a pool of water in the reactor building and at some sites later transferred to dry casks on the plant grounds. Because both types of storage are located outside the reactor containment structure, particular concern has been raised about the vulnerability of spent fuel to attack by aircraft or other means. If terrorists could breach a spent fuel pool's concrete walls and drain the cooling water, the spent fuel's zirconium cladding could overheat and catch fire.

The National Academy of Sciences (NAS) released a report in April 2005 that found that "successful terrorist attacks on spent fuel pools, though difficult, are possible," and that "if an attack leads to a propagating zirconium cladding fire, it could result in the release of large amounts of radioactive material." NAS recommended that the hottest spent fuel be interspersed with cooler spent fuel to reduce the likelihood of fire, and that water-spray systems be installed to cool spent fuel if pool water were lost. The report also called for NRC to conduct more analysis of the issue and consider earlier movement of spent fuel from pools into dry storage.[27] The

---

[26] Nuclear Regulatory Commission, Office of Nuclear Security and Incident Response, Report to Congress on the Security Inspection Program for Commercial Power Reactor and Category 1 Fuel Cycle Facilities: Results and Status Update; Annual Report for Calendar Year 2007, NUREG-1885, July 2008, p. 8, http://www.nrc.gov/reading-rm/doc-collections/congress-docs/correspondence/2008/boxer-07-01-2008.pdf.

[27] National Academy of Sciences, Board on Radioactive Waste Management, Safety and Security of Commercial Spent Nuclear Fuel Storage, Public Report (online version), released April 6, 2005.

FY2006 Energy and Water Development Appropriations Act (P.L. 109-103, H.Rept. 109-275) provided $21 million for NRC to carry out the site-specific analyses recommended by NAS.

NRC has long contended that the potential effects of terrorist attacks are too speculative to include in environmental studies for proposed spent fuel storage and other nuclear facilities. However, the U.S. Court of Appeals for the 9th Circuit ruled in June 2006 that terrorist attacks must be included in the environmental study of a dry storage facility at California's Diablo Canyon nuclear plant. NRC reissued the Diablo Canyon study May 29, 2007, to comply with the court ruling, but it did not include terrorism in other recent environmental studies.[28]

Long-term management of spent nuclear fuel is currently undergoing review, but spent fuel stored at reactor sites is expected to be moved eventually to central storage, permanent disposal, or reprocessing facilities. Large-scale transportation campaigns would increase public attention to NRC transportation security requirements and related security issues.

Security Personnel and Other Issues

After video recordings of inattentive security officers at the Peach Bottom (PA) nuclear power plant were aired on local television, an NRC inspection in late September 2007 confirmed that there had been multiple occasions on which multiple security officers were inattentive.[29] However, after a follow-up inspection into security issues at the Peach Bottom plant, run by Exelon Nuclear, the NRC concluded that the plant's security program had not been significantly degraded as a result of the guards' inattentiveness. NRC issued a bulletin December 12, 2007, requiring all nuclear power plants to provide written descriptions of their "managerial controls to deter and address inattentiveness and complicity among licensee security personnel."[30]

The incident drew harsh criticism from the House Committee on Energy and Commerce. "The NRC's stunning failure to act on credible allegations of sleeping security guards, coupled with its unwillingness to protect the whistleblower who uncovered the problem, raises troubling questions," said Representative John D. Dingell, then-Chairman of the Committee.[31] NRC proposed a $65,000 fine on Exelon Nuclear on January 6, 2009.[32]

Following the 9/11 terrorist attacks, NRC conducted a "top-to-bottom" review of its nuclear power plant security requirements. On February 25, 2002, the agency issued "interim compensatory security measures" to deal with the "generalized high-level threat environment" that continued to exist, and on January 7, 2003, it issued regulatory orders that tightened nuclear plant access. On April 29, 2003, NRC issued orders to restrict security officer work hours,

---

[28] Beattie, Jeff, "NRC Takes Two Roads on Terror Review Issue," Energy Daily, February 27, 2007.

[29] NRC, *NRC Commences Follow-up Security Inspection at Peach Bottom*, November 5, 2007 http://www.nrc.gov/reading-rm/doc-collections/news/2007/07-057.i.html.

[30] Nuclear Regulatory Commission, *Security Officer Attentiveness*, NRC Bulletin 2007-1, Washington, DC, December 12, 2007.

[31] Committee on Energy and Commerce, *Energy and Commerce Committee to Probe Breakdowns in NRC Oversight*, January 7, 2008 http://energycommerce.house.gov/Press_110/110nr149.shtml.

[32] Nuclear Regulatory Commission, "NRC Proposes $65,000 Fine for Violations Associated with Inattentive Security Guards at Peach Bottom Nuclear Plant," press release, January 6, 2009, http://www.nrc.gov/reading-rm/doc-collections/news/2009/09-001.i.html.

establish new security force training and qualification requirements, and increase the DBT that nuclear security forces must be able to defend against, as discussed previously.

In October 2006, NRC proposed to amend the security regulations and add new security requirements that would codify the series of orders issued after 9/11 and respond to requirements in the Energy Policy Act of 2005.[33] The new security regulations were approved by the NRC Commissioners on December 17, 2008, with the following provisions[34]:

- *Safety and Security Interface.* Explicit requirements are established for nuclear plants to ensure that necessary security measures do not compromise plant safety.

- *Mixed-Oxide Fuel.* Enhanced physical security requirements are established to prevent theft or diversion of plutonium-bearing mixed-oxide (MOX) fuel.

- *Cyber Security.* Nuclear plants must submit security plans to prevent cyber attacks on digital computer and communications systems and networks. The cyber security plan will become a license condition for each plant.

- *Aircraft Attack Mitigative Strategies and Response.* As discussed in the earlier section on vulnerability to aircraft crashes, nuclear plants must prepare strategies for responding to warnings of an aircraft attack and for mitigating the effects of large explosions and fires.

- *Plant Access Authorization.* Nuclear plants must implement more rigorous programs for authorizing access, including enhanced psychological assessments and behavioral observation.

- *Security Personnel Training and Qualification.* Modifications to security personnel requirements include additional physical fitness standards, increased minimum qualification scores for mandatory personnel tests, and requirements for on-the-job training.

- *Physical Security Enhancements.* New requirements are intended to ensure the availability of backup security command centers, uninterruptible power supplies to detection systems, enhanced video capability, and protection from waterborne vehicles.

A proposal by NRC staff to release more details about the results of nuclear plant security inspections was defeated by the NRC Commissioners in a 2-2 vote on January 21, 2009. Under current policy, NRC announces after a security inspection whether any violations that were found were of low safety significance or moderate-or-higher safety significance. Critics of the current policy contend that the public needs more detail to be assured of plant security. The policy's supporters counter that greater information about security inspection findings could inadvertently provide useful information to terrorists.[35]

---

[33] *Federal Register*, October 26, 2006 (vol. 71, no. 207), NRC, Power Reactor Security Requirements, Proposed Rule.

[34] E-mail message from David Decker, NRC Office of Congressional Affairs, February 27, 2009.

[35] Jenny Weil, "Commissioners Reach Stalemeat on Security-Related Amendment," *Inside NRC*, February 2, 2009.

Author Contact Information

Mark Holt
Specialist in Energy Policy
mholt@crs.loc.gov, 7-1704

Anthony Andrews
Specialist in Energy and Energy Infrastructure
Policy
aandrews@crs.loc.gov, 7-6843

United States Government Accountability Office

GAO

Report to Congressional Requesters

June 2008

NUCLEAR SAFETY

NRC's Oversight of Fire Protection at U.S. Commercial Nuclear Reactor Units Could Be Strengthened

G A O
Accountability ★ Integrity ★ Reliability

GAO-08-747

June 2008

GAO
Accountability· Integrity· Reliability

Highlights

Highlights of GAO-08-747, a report to congressional requesters

NUCLEAR SAFETY

NRC's Oversight of Fire Protection at U.S. Commercial Nuclear Reactor Units Could Be Strengthened

Why GAO Did This Study

After a 1975 fire at the Browns Ferry nuclear plant in Alabama threatened the unit's ability to shut down safely, the Nuclear Regulatory Commission (NRC) issued prescriptive fire safety rules for commercial nuclear units. However, nuclear units with different designs and different ages have had difficulty meeting these rules and have sought exemptions to them. In 2004, NRC began to encourage the nation's 104 nuclear units to transition to a less prescriptive, risk-informed approach that will analyze the fire risks of individual nuclear units. GAO was asked to examine (1) the number and causes of fire incidents at nuclear units since 1995, (2) compliance with NRC fire safety regulations, and (3) the transition to the new approach.

GAO visited 10 of the 65 nuclear sites nationwide, reviewed NRC reports and related documentation about fire events at nuclear units, and interviewed NRC and industry officials to examine compliance with existing fire protection rules and the transition to the new approach.

What GAO Recommends

GAO recommends that NRC obtain and monitor data on the status of compliance with its fire safety regulations, and address long-standing fire safety issues concerning interim compensatory measures, fire wrap effectiveness, and multiple spurious actuations. NRC commented the report was accurate and complete but did not address the recommendations.

To view the full product, including the scope and methodology, click on GAO-08-747. For more information, contact Mark Gaffigan at (202) 512-3841 or gaffiganm@gao.gov.

What GAO Found

According to NRC, all 125 fires at 54 of the nation's 65 nuclear sites from January 1995 through December 2007 were classified as being of limited safety significance. According to NRC, many of these fires were in areas that do not affect shutdown operations or occurred during refueling outages, when nuclear units are already shut down. NRC's characterization of the location, significance, and circumstances of those fire events was consistent with records GAO reviewed and statements of utility and industry officials GAO contacted.

NRC has not resolved several long-standing issues that affect the nuclear industry's compliance with existing NRC fire regulations, and NRC lacks a comprehensive database on the status of compliance. These long-standing issues include (1) nuclear units' reliance on manual actions by unit workers to ensure fire safety (for example, a unit worker manually turns a valve to operate a water pump) rather than "passive" measures, such as fire barriers and automatic fire detection and suppression; (2) workers' use of "interim compensatory measures" (primarily fire watches) to ensure fire safety for extended periods of time, rather than making repairs; (3) uncertainty regarding the effectiveness of fire wraps used to protect electrical cables necessary for the safe shutdown of a nuclear unit; and (4) mitigating the impacts of short circuits that can cause simultaneous, or near-simultaneous, malfunctions of safety-related equipment (called "multiple spurious actuations") and hence complicate the safe shutdown of nuclear units. Compounding these issues is that NRC has no centralized database on the use of exemptions from regulations, manual actions, or compensatory measures used for long periods of time that would facilitate the study of compliance trends or help NRC's field inspectors in examining unit compliance.

Primarily to simplify units' complex licensing, NRC is encouraging nuclear units to transition to a risk-informed approach. As of April 2008, some 46 units had stated they would adopt the new approach. However, the transition effort faces significant human capital, cost, and methodological challenges. According to NRC, as well as academics and the nuclear industry, a lack of people with fire modeling, risk assessment, and plant-specific expertise could slow the transition process. They also expressed concern about the potentially high costs of the new approach relative to uncertain benefits. For example, according to nuclear unit officials, the costs to perform the necessary fire analyses and risk assessments could be millions of dollars per unit. Units, they said, may also need to make costly new modifications as a result of these analyses.

_____ United States Government Accountability Office

Contents

Abbreviations

NFPA   National Fire Protection Association
NRC   Nuclear Regulatory Commission

G A O
Accountability * Integrity * Reliability

United States Government Accountability Office
Washington, DC 20548

June 30, 2008

The Honorable Peter Visclosky
Chairman
Subcommittee on Energy and Water Development
Committee on Appropriations
House of Representatives

The Honorable David Price
House of Representatives

On March 22, 1975, a fire involving electrical cables at unit 1 of the three-unit Browns Ferry nuclear power plant in Alabama damaged numerous safety systems and reduced unit operators' ability to monitor the nuclear unit. The fire raised awareness of the potential danger that fires pose to the ability of the nation's commercial nuclear units to safely shutdown. The Nuclear Regulatory Commission (NRC), which approves nuclear units' licenses to operate, responded by issuing numerous guidance documents and in 1980 promulgating new fire safety regulations for nuclear units. These regulations, commonly called Appendix R, are intended to (1) prevent fires from starting; (2) rapidly detect, control, and extinguish fires that do occur; and (3) protect a nuclear unit's structures, systems, and components important to safety so that a fire that is not promptly extinguished will not prevent its safe shutdown.[1]

NRC's fire safety regulations for the nation's commercial nuclear units establish the design requirements in commercial nuclear reactor units for mitigating the effects of a fire on the unit's ability to shut down safely. As of May 2008, 104 commercial nuclear units operated at 65 sites in 31 states, with between one and three units located at each site. Among other things, these prescriptive (or deterministic) regulations call for nuclear units to have at least one redundant system of electric cables and equipment available to safely shut down the unit free from fire damage. When two such systems are in the same area of a nuclear unit, the regulations require that they be separated (1) horizontally by at least 20 feet with automatic

---

[1] *10 CFR part 50*, Appendix R applies to commercial nuclear units that were operating prior to January 1, 1979. Units that began operation on or after that date are required to meet specific requirements in their licensing conditions that are similar to Appendix R.

GAO-08-747 Fire Safety and Nuclear Reactor Units

fire suppression and detections systems and without intervening combustibles or (2) by a fire barrier, such as a fire-proof wall or floor, or by a material (fire wrap) that protects important cables.[2] The fire barriers must be able to withstand fire for at least 1 hour in areas with automatic fire detection and suppression equipment, such as smoke detectors and sprinklers, or at least 3 hours where such features are not present. NRC required nuclear units that were operating prior to January 1, 1979, to make necessary modifications, if possible, to meet NRC's fire regulations or request exemptions from the requirements. Units that NRC licensed after that date incorporate the principles of NRC's fire regulations as conditions to their operating licenses.

Over the years, NRC approved exemptions or deviations[3] from the fire regulations for units that could not meet the regulations if these units could otherwise demonstrate the ability to safely shut down. According to NRC's records, by 2001 NRC had granted over 900 exemptions for the nation's nuclear units. Many of these exemptions take the form of operator manual actions, whereby nuclear unit staff manually activate or control unit functions by hand outside of the unit's control room, such as stopping a pump that malfunctions during a fire and could impair a unit's ability to safely shut down. In addition, NRC allows nuclear units, in accordance with their NRC-approved fire protection program, to institute interim compensatory measures, which are temporary measures that units can take without prior approval to compensate for equipment that needs to be repaired or replaced. These interim compensatory measures often consist of roving or continuously manned fire watches[4] that occur while nuclear units take corrective actions. Under NRC rules, the repairs or replacements should take place as soon as practicable, thereby limiting the time an interim compensatory measure is in effect. Many operator manual actions or interim compensatory measures were instituted because some fire wraps did not meet the requirements to withstand a fire for 1 hour or 3 hours. In lieu of reliance on such a fire wrap, a unit might

---

[2]NRC's technical term for such a wrap is "Electrical Raceway Fire Barrier System." However, in this report we use the term "fire wrap" because this term is widely used in practice by industry.

[3]Nuclear units licensed prior to January 1, 1979, pursuant to Appendix R are issued "exemptions" to the regulations NRC, while those licensed after 1979 are issued "deviations" from conditions in their licenses. For purposes of clarity, hereafter, our report will use the generic term "exemptions."

[4]Fire watches are teams of nuclear unit employees who can be posted continuously in a single location or can rove throughout the unit site to detect signs of fire.

Page 2   GAO-08-747 Fire Safety and Nuclear Reactor Units

594

opt to use a fire watch as an interim compensatory measure while repairs are made.

In 2004, NRC issued a regulation that allowed the transition of nuclear units from its existing, prescriptive fire safety regulations to a less prescriptive, risk-informed, performance-based approach that complies with the National Fire Protection Association (NFPA) standard 805.[5] Under this approach, nuclear units can use tools, such as fire modeling and risk analysis, to determine which areas of the unit are most at risk from fire. According to NRC officials, these analyses could enable units to focus their resources on addressing these higher-risk areas and reduce the number of future exemptions in areas that are no longer considered to be at high risk from fire. Reductions in exemptions would, thus, simplify the units' licenses.

Resolving any issues about the fire safety of nuclear units will be important for assuring the public that nuclear power is safe. Providing such assurances is especially significant given the scope of the nuclear power industry's plans for expanding the nation's capacity to generate electricity using nuclear reactors. According to the Nuclear Energy Institute, which represents the nuclear power industry, as of April 2008, electric utilities planned to build 29 new nuclear power units at 23 sites nationwide. Currently, 104 nuclear units are operating in the nation, so the planned expansion will be significant.

In this context, we were asked to examine (1) the number, reported safety significance, and causes of fire incidents at U.S. nuclear units since 1995, (2) commercial nuclear reactor units' compliance with NRC's fire protection regulations, and (3) the status of the nuclear industry's implementation of the risk-informed approach to fire safety advocated by NRC.

In conducting our work, we met with officials from NRC, industry, public interest groups, and experts on fire safety and risk analysis in academia

---

[5]NRC, through 60 *Fed. Reg.* 33536 (June 16, 2004)(codified at *10 C.F.R. 50.48(c)*), endorsed the use of key aspects of National Fire Protection Association, *NFPA-805, Performance-Based Standard for Fire Protection for Light Water Reactors Electric Generating Plants*, 2001 Edition (Quincy, Massachusetts, 2001). NRC differentiates between "risk-informed" and "risk-based" regulation, noting that the former uses risk analysis to augment other information used to support management decisions, while the latter approach relies solely on the numerical results of risk assessments. NRC does not endorse a risk-based approach for fire protection.

and government. We also selected and visited 10 nuclear unit sites, constituting a sample that is not generalizable to all nuclear units at all nuclear unit sites. We selected sites based on covering each of NRC's four regional offices, varying levels of unit performance, different unit licensing characteristics, and reactor types. At each site visit, we reviewed documentation on fire events, use of operator manual actions and interim compensatory measures, and analysis justifying decisions about whether to transition to the risk-informed approach. In addition, we reviewed fire event data from NRC and the industry for all fires in calendar years 1995 through 2007 to provide us with a reasonable time frame of data. Finally, we reviewed relevant fire protection regulations and guidance from NRC and industry.[6]

We conducted this performance audit from September 2007 to June 2008 in accordance with generally accepted government auditing standards. Those standards require that we plan and perform the audit to obtain sufficient, appropriate evidence to provide a reasonable basis for our findings and conclusions based on our audit objectives. We believe that the evidence obtained provides a reasonable basis for our findings and conclusions based on our audit objectives.

Results in Brief

According to NRC, nuclear unit operators reported 125 fires at 54 sites from January 1995 through December 2007; all were classified as having limited safety significance, and no fire since the 1975 Browns Ferry fire has threatened a nuclear unit's ability to safely shut down.[7] The most commonly reported cause of fires was electrical followed by maintenance-related causes and the ignition of oil-based lubricants or coolant. Although 13 fires were classified as significant alerts, and some of these fires damaged or destroyed unit equipment, NRC officials stated that none of these fires degraded units' safe shutdown capabilities or resulted in damage to nuclear units' core or containment buildings. These officials noted that most of these fires occurred in areas that do not affect

---

[6]The scope of our work focuses on fire safety as it pertains to a nuclear unit's ability to achieve safe shutdown. NRC is also overseeing plans and actions undertaken by unit operators to safeguard against fires resulting from a catastrophic event in which containment structures surrounding a unit's core and spent fuel pool are damaged or destroyed. We did not analyze this issue because it falls outside the scope of our audit.

[7]NRC only collects data on events that meet certain reporting thresholds including (1) whether a fire lasts longer than 10 or 15 minutes and (2) whether the fire affects plant equipment necessary for safe shutdown.

shutdown operations or happened during refueling outages, when nuclear units are already shut down.

NRC has not fully resolved the long-standing issues that complicate the commercial nuclear industry's compliance with NRC's fire regulations; moreover, NRC lacks a comprehensive database on the use of exemptions, manual actions, and compensatory measures for long periods of time that would facilitate the study of compliance trends or help NRC's field inspectors in examining unit compliance. Specifically, these issues include:

- *The use of operator manual actions.* After regular triennial fire inspections began in 2000, NRC fire safety inspectors found that nuclear units were using unapproved or undocumented operator manual actions. Nuclear unit operators told us that, in some cases, NRC officials approved these actions verbally but did not document their approval in writing; however, in other cases, unit officials said they applied operator manual actions that were not explicitly approved by NRC but that NRC had approved for similar situations. NRC has directed nuclear units to resolve these issues by March 2009, either by applying for licensing exemptions for these operator manual actions or by modifying the units' designs. Compounding this issue is a lack of a centralized database of approved manual actions (exemptions), as well as those that are unapproved or undocumented.

- *The long-term use of interim compensatory measures.* Some nuclear units have used compensatory measures for extended periods of time—for years, in some cases—rather than repairing or replacing the damaged equipment. For example, at one nuclear unit we visited, unit staff used fire watches for more than 5 years instead of replacing faulty seals to cover openings in structural barriers. Although NRC guidance tells units to repair fire protection features as quickly as possible, it does not specify how long units can rely on interim compensatory measures. NRC has no immediate plans for resolving this issue. Compounding this issue is a lack of a centralized database of compensatory measures that can be used for long periods of time.

- *Concerns about the effectiveness of fire wraps.* NRC has not resolved the uncertainty regarding the effectiveness of some types of fire wraps used to protect cables that are important for safely shutting down the nuclear units. Until this issue is resolved, nuclear unit operators are continuing to rely on operator manual actions and interim compensatory measures. During testimony before Congress in 1993, a then-NRC chairman committed to assess the effectiveness of fire wraps, and NRC officials maintain that the agency has satisfied this commitment. According to NRC officials, licensees are responsible for conducting endurance tests on fire wraps used at nuclear units. However, in January 2008 the NRC Office of Inspector General reported that no fire endurance tests have been conducted to qualify a key fire wrap as an NRC-approved 1- or 3-hour fire barrier.

- *Mitigating the effects of short circuits on safety-related equipment.* Nuclear units must plan for short circuits that could cause safety-related equipment to start or malfunction spuriously (instances called spurious actuations). To date, units typically account only for spurious actuations that occur one at a time or in isolation. In 2001, industry tests demonstrated that spurious actuations could occur simultaneously or in rapid succession and that units' current fire protection plans do not account for this possibility. NRC has not endorsed guidance or developed a timeline for industry to resolve this issue, but NRC staff stated they expect to recommend a plan of action by June 2008.

As of May 2008, 46 nuclear units had announced they would adopt the new risk-informed approach to fire safety that NRC is endorsing. Four nuclear units are piloting the new approach, and NRC plans to evaluate the results for the pilot program units by March 2009. According to NRC officials, 22 additional units will begin submitting their license amendment requests for the risk-informed approach by March 2009. Operators at the units that plan to adopt the new approach told us that identifying and focusing their resources on the areas most at risk from fire and areas that are significant to safely shutting down the unit would help them better focus their resources and reduce the need for some operator manual actions to meet regulations. However, experts we contacted noted that while the risk-informed approach may have some safety benefits, the small number of fires at nuclear units has resulted in limited real-world data for use in the probabilistic risk assessments that units will conduct under the new approach. NRC and nuclear unit operators also face possible shortages of personnel with expertise in developing and evaluating probabilistic risk assessments and related analyses, which could delay the transition process. Operators of some of the 58 nuclear units that have not indicated their intention to adopt the new approach also said the costs and outcomes of the new approach are uncertain. For example, the operators believe that NRC's guidance for conducting the fire models that are used in the probabilistic risk assessments assumes worst-case fire scenarios, and thus the resulting analyses would not provide a realistic assessment of risk. According to these officials, following those fire models could require them to spend millions of dollars to install modifications that likely would not provide a substantial increase in safety. These officials also questioned NRC's encouragement of units to adopt the new risk-informed approach before the two pilot programs are complete.

We are recommending that the Commissioners direct NRC staff to (1) develop a central database for tracking the status of exemptions, manual actions, and compensatory measures used for long periods of time both nationwide and at individual commercial nuclear units; (2) address safety concerns related to the extended use of interim compensatory measures; (3) analyze the effectiveness of existing fire wraps and undertake efforts to ensure that the fire endurance tests have been conducted to qualify fire wraps as NRC-approved 1- or 3-hour fire barriers; and (4) ensure that nuclear units are able to safeguard against multiple spurious actuations by committing to a specific date for developing guidelines to prevent multiple spurious actuations.

In commenting on a draft of this report, NRC found that it was accurate, complete, and handled sensitive information appropriately and stated that it intends to give GAO's findings and conclusions serious consideration. However, in its response, NRC did not provide comments on our recommendations. NRC's comments are reprinted in appendix II.

Background

In 1971, the Atomic Energy Commission,[8] NRC's predecessor, promulgated the first regulations for fire protection at commercial nuclear power units in the United States. These regulations—referred to as General Design Criterion 3—provided basic design requirements and broad performance objectives for fire protection,[9] but lacked implementation guidance or

---

[8]In 1974, Congress abolished the Atomic Energy Commission and created two new agencies in its place—NRC and the Energy Research and Development Administration (now the Department of Energy). NRC continued to function with the same regulations and guidance developed under the Atomic Energy Commission and currently codified in Parts 1–199 of Title 10 of the U.S. Code of Federal Regulations.

[9]Appendix A to 10 C.F.R. 50, "General Design Criteria for Nuclear Power Plants," Criterion 3 – Fire protection: Structures, systems, and components important to safety shall be designed and located to minimize, consistent with other safety requirements, the probability and effect of fires and explosions. Noncombustible and heat resistant materials shall be used wherever practical throughout the unit, particularly in locations such as the containment and control room. Fire detection and fighting systems of appropriate capacity and capability shall be provided and designed to minimize the adverse effects of fires on structures, systems, and components important to safety. Firefighting systems shall be designed to assure that their rupture or inadvertent operation does not significantly impair the safety capability of these structures, systems, and components.

GAO-08-747 Fire Safety and Nuclear Reactor Units

assessment criteria. As such, NRC generally deemed a unit's fire protection program to be adequate if it complied with standards set by the National Fire Protection Association (NFPA)—an international organization that promotes fire prevention and safety—and received an acceptable rating from a major fire insurance company.[10] However, at that time the fire safety requirements for commercial nuclear power units were similar to those for conventional, fossil-fueled power units.

NRC and nuclear industry officials did not fully perceive that fires could threaten a nuclear unit's ability to safely shut down until 1975, when a candle that a worker at Browns Ferry nuclear unit 1 was using to test for air leaks in the reactor building ignited electrical cables. The resulting fire burned for 7 hours and damaged more than 1,600 electrical cables, more than 600 of which were important to unit safety. Nuclear unit workers eventually used water to extinguish the fire, contrary to the existing understanding of how to put out an electrical fire. The fire damaged electrical power, control systems, and instrumentation cables and impaired cooling systems for the reactor. During the fire, operators could not monitor the unit normally.

NRC's investigation of the Browns Ferry fire revealed deficiencies in the design of fire protection features at nuclear units and in procedures for responding to a fire, particularly regarding safety concerns that were unique to nuclear units, such as the ability to protect redundant electrical cables and equipment important for the safe shutdown of a reactor.[11] In response, NRC developed new guidance in 1976 that required units to take steps to isolate and protect at least one system of electrical cables and equipment to ensure a nuclear unit could be safely shut down in the event of a fire. NRC worked with licensees throughout the late 1970s to help them meet this guidance.

In November 1980, NRC published two new sets of regulations to formalize the regulatory approach to fire safety. First, NRC required all nuclear units to have a fire protection plan that satisfies General Design Criteria 3 and that describes an overall fire protection program.[12] Second,

---

[10]NRC typically documents its acceptance of a fire protection program by issuing safety evaluation reports.

[11]See NUREG 0050, "Recommendations Related to Browns Ferry Fire" (February 1976).

[12]45 *Fed. Reg.* 76610 (Nov. 19, 1980) codified as amended at 10 CFR 50.48.

GAO-08-747 Fire Safety and Nuclear Reactor Units

NRC published Appendix R,[13] which requires nuclear units operating prior to January 1, 1979 (called "pre-1979 units"), to implement design features—such as fire walls, fire wraps, and automatic fire detection and suppression systems—to protect a redundant system of electrical cables and equipment necessary to safely shut down a nuclear unit during a fire. Among other things, Appendix R requires units operating prior to 1979 to protect one set of cables and equipment necessary for safe shutdown through one of the following means:[14]

1. Separating the electrical cables and equipment necessary for safe shutdown by a horizontal distance of more than 20 feet from other systems, with no combustibles or fire hazards between them. In addition, fire detectors and an automatic fire suppression system (for example, a sprinkler system) must be installed in the fire area.

2. Protecting the electrical cables and equipment necessary for safe shutdown by using a fire barrier able to withstand a 3-hour fire, as conducted in a laboratory test (thereby receiving a 3-hour rating).

3. Enclosing the cable and equipment necessary for safe shutdown by using a fire barrier with a 1-hour rating and combining that with automatic fire detectors and an automatic fire suppression system.

If a nuclear unit's fire protection systems do not satisfy those requirements or if redundant systems required for safe shutdown could be damaged by fire suppression activities, Appendix R requires the nuclear unit to maintain an alternative or dedicated shutdown capability and its associated circuits. Moreover, Appendix R requires all units to provide emergency lighting in all areas needed for operating safe shutdown equipment.[15]

Nuclear units that began operating on or after January 1, 1979 (called "post-1979 units") must satisfy the broad requirements of General Design Criteria 3[16] but are not subject to the requirements of Appendix R.

[13]*45 Fed. Reg.* 76611 (*Nov. 19, 1980*).

[14]Appendix R also includes other requirements for fire safety, such as requirements governing fire brigades at nuclear units.

[15]These requirements are contained in paragraphs G.3 and J of Section III of Appendix R.

[16]See 10 CFR 50.48(a).

However, NRC has imposed or attached conditions similar to the requirements of Appendix R to these units' operating licenses.

When promulgating these regulations, NRC recognizes that strict compliance for some older units would not significantly enhance the level of fire safety. In those cases, NRC allows nuclear units licensed before 1979 to apply for an exemption to Appendix R. The exemption depends on if the nuclear unit can demonstrate to NRC that existing or alternative fire protection features provided safety equivalent to those imposed by the regulations.[17] Since 1981, NRC has issued approximately 900 unit-specific exemptions to Appendix R. Nuclear units licensed after 1979 can apply for "deviations" against their licensing conditions.[18]

Many exemptions take the form of NRC-approved operator manual actions, whereby nuclear unit staff manually activate or control unit operations from outside the unit's control room, such as manually stopping a pump that malfunctions during a fire and could affect a unit's ability to safely shut down. NRC also allows nuclear units to institute, in accordance with their NRC-approved fire protection program, "interim compensatory measures"—temporary measures that units can take without prior approval to compensate for equipment that needs to be repaired or replaced. Interim compensatory measures often consist of roving or continuously staffed fire watches that occur while nuclear units take corrective actions.

In part to simplify the licensing of nuclear units that have many exemptions, NRC recently began encouraging units to transition to a more risk-informed approach to nuclear safety in general. In 2004, NRC promulgated 10 C.F.R. 50.48(c), which allows—but does not require— nuclear units to adopt a risk-informed approach to fire protection. The risk-informed approach considers the probability of fires in conjunction with a unit's engineering analysis and operating experience. The NRC rule allows licensees to voluntarily adopt and maintain a fire protection program that meets criteria set forth by the NFPA's fire protection

---

[17]Licensees request exemptions from fire protection requirements in accordance with 10 CFR 50.12.

[18]As previously noted, post-1979 units documented their differences in licensing "deviations" against the criteria with which NRC approved their fire protection programs. For clarity purposes, we use the term "exemptions" to refer to both exemptions and deviations.

standard 805[19]— which describes the risk-informed approach endorsed by NRC—as an alternative to meeting the requirements or unit-specific fire-protection license conditions represented by Appendix R and related rules and guidance. Nuclear units that choose to adopt the risk-informed approach must submit a license amendment request to NRC asking NRC to approve the unit's adoption of the new risk-informed, regulatory approach.[20] NRC is overseeing a pilot program at two nuclear unit locations and expects to release its evaluation report on these programs by March 2009.

According to NRC, Recent Fires at U.S. Commercial Nuclear Units Have Had Limited Safety Significance

NRC officials told us that none of the 125 fires at 54 sites[21] that nuclear unit operators reported from January 1995 to December 2007 has posed significant risk to a commercial unit's ability to safely shut down. No fires since the 1975 Browns Ferry fire have threatened a nuclear unit's ability to safely shut down.[22] Most of the 125 fires occurred outside areas that are considered important for safe shutdown of the unit or happened during refueling outages when nuclear units were already shut down.

Nuclear units categorized 13 of the 125 reported fires as "alerts" under NRC's Emergency Action Level rating system, meaning that the reported situation involved an actual or potential substantial degradation of unit safety, but none of the fires actually threatened the safe shutdown of the unit. NRC further characterizes alerts as providing early and prompt notification of minor events that could lead to more serious consequences. As shown in the table 1, the primary reported causes of these fires were electrical fires.

---

[19]National Fire Protection Association *NFPA 805: Performance-Based Standard for Fire Protection for Light Water Reactor Electric Generating Plants*, 2001 ed. (Quincy, Massachusetts, 2001).

[20]10 CFR. 50.90 provides the requirements for making license amendment applications. 10 C.F.R. 50.48(c)(3) describes the required content of the application for adopting the risk-informed, performance-based approach to fire safety.

[21]The nation's 104 nuclear units operate at 65 sites in 31 states.

[22]NRC directs nuclear units to report fires to the agency in accordance with their approved fire protection programs. Typically, this includes fires that meet certain criteria, such as (1) whether a fire lasts longer than 10 or 15 minutes and (2) whether the fire affects plant equipment necessary for safe shutdown.

Table 1: Characteristics of Fires Rising to "Alert" Status at U.S. Commercial Nuclear Units, 1995-2007

Year	Unit	State	Location within unit	Cause
2007	Arkansas Nuclear One, Unit 2	Arkansas	Auxiliary building	Electrical
2007	Columbia Generating Station	Washington	Equipment room	Electrical
2007	Callaway Nuclear Plant	Missouri	Control building	Electrical
2006	Arkansas Nuclear One, Unit 2	Arkansas	Breaker compartment	Electrical
2006	Perry Nuclear Power Plant	Ohio	Ventilation fan	Bearing
2003	Palisades Power Plant	Michigan	Cable spreading room	Electrical
2002	D.C. Cook Nuclear Plant	Michigan	Switchyard	Electrical
2001	Cooper Nuclear Station	Nebraska	Startup transformer	Unreported
2001	Fermi Unit 2	Michigan	Emergency diesel generator	Bearing
2000	Farley Unit 2	Alabama	Service water pump motor	Unreported
1998	Fermi Unit 2	Michigan	Emergency diesel generator	Unreported
1997	Limerick Generating Station Unit 2	Pennsylvania	Emergency diesel generator exhaust	Unreported
1996	Clinton Power Station	Illinois	Pump turbine insulation	Oil-Soaked Insulation

Source: GAO analysis of NRC data.

Nuclear units classified the remaining 112 reported fires in categories that do not imply a threat to safe shutdown. Specifically, 73 were characterized as being "unusual events"—a category that is less safety-significant than "alerts"—and 39 fires as being "non-emergencies." No reported fire event rose to the level of "site area emergency" or "general emergency"—the two most severe ratings in the Emergency Action Level system.[23]

As shown in table 2 below, about 41 percent of the 125 reported fires were electrical fires, 14 percent were maintenance related, 7 percent were caused by oil-based lubricants or insulation, and the remaining 38 percent

---

[23]NRC requires units to categorize events according to the following four classes of Emergency Action Levels in increasing order of seriousness: Notification of Unusual Event, Alert, Site Area Emergency, and General Emergency. The first two levels are to provide early and prompt notification of minor events that could lead to more serious consequences. In particular, an Alert describes a situation that involves an actual or potential substantial degradation of the level of safety of the plant, with any resulting radiological releases expected to be limited to small fractions based on guidance from the Environmental Protection Agency. A Site Area Emergency reflects conditions where some significant radiological releases are likely but where a core melt situation is not indicated, and a General Emergency involves actual or imminent substantial core degradation or melting with the potential for loss of containment.

either had no reported causes or the causes were listed as "other," including brush fires, cafeteria grease fires, and lightning.

Table 2: Information on Reported Causes of Fires at Nuclear Units from January 1995 through December 2007

Cause of fire	Number of reported fire events	Percentage of total reported fire events
Electrical-related	51	41
Maintenance-related	17	14
Oil-based lubricants or insulation	9	7
Other causes[a] or cause not reported	48	38
Totals	**125**	**100**

Source: GAO analysis of NRC data.

[a]Includes brush fires, cafeteria grease fires, and lightning.

We also gathered information on fire events that had occurred at nuclear unit sites we visited. NRC's data on the location and circumstances surrounding fire events was consistent with the statements of unit officials whom we contacted at selected nuclear units. Although unit officials told us that some recent fires necessitated the response of off-site fire departments to supplement the units' on-site firefighting capabilities, they confirmed that none of the fires adversely affected the units' ability to safely shut down. Additionally, officials at two units told us that, although fires affected the units' auxiliary power supply, the events caused both units to "trip"—an automatic power down as a precaution in emergencies.

NRC Has Not Resolved Long-standing Issues Affecting Industry's Compliance with NRC's Fire Regulations

NRC has not fully resolved several long-standing issues that affect the commercial nuclear industry's compliance with existing NRC fire regulations. These issues include (1) nuclear units' use of operator manual actions; (2) nuclear units' long-term use of interim compensatory measures; (3) uncertainties regarding the effectiveness of fire wraps for protecting electrical cables necessary for the safe shutdown of a nuclear unit; and (4) the regulatory treatment of fire-induced multiple spurious actuations of equipment that could prevent the safe shutdown of a nuclear unit. Moreover, NRC lacks a central system of records that would enhance its ability to oversee and address the use of operator manual actions and extended interim compensatory measures, among other related issues. According to an NRC Commissioner, the current "patchwork of requirements" is characterized by too many exemptions, as well as by unapproved or undocumented operator manual actions. He said the current regulatory situation was not the ideal, transparent, or safest way to deal with the issue of fire safety.

Many Nuclear Units Are Using Operator Manual Actions That May Not Comply with NRC's Fire Regulations

NRC's oversight of fire safety is complicated by nuclear units' use of operator manual actions that NRC has not explicitly approved. NRC's initial Appendix R regulations required that nuclear units protect at least one redundant system—or "train"—of equipment and electrical cables required for a unit's safe shutdown through the use of fire protection measures, such as 1-hour or 3-hour fire barriers, 20 feet of separation between redundant systems, and automatic fire detection and suppression systems.[24] The regulations do not list operator manual actions as a means of protecting a redundant system from fire. However, according to NRC officials and NRC's published guidance, units licensed before January 1979 can receive approval for a specific operator manual action by applying for a formal exemption to the regulations. For example, unit officials at one site told us they rely on 584 operator manual actions that are approved by 15 NRC exemptions for safe shutdown. (NRC allows units to submit multiple operator manual actions under one exemption.) Units licensed after January 1979 may use operator manual actions for fire protection if these actions are permitted by the unit's license and if the unit can demonstrate that the actions will not adversely affect safe shutdown. NRC and nuclear unit officials told us that units have been using operator manual actions since Appendix R became effective in 1981. These officials added that a majority of nuclear units that use operator manual actions started using them beginning in the mid-1990s in response to the failure of Thermo-Lag—a widely used fire wrap—to meet fire endurance testing.

A lack of clear understanding between NRC and industry over the permissible use of operator manual actions in lieu of passive measures emerged over the years. For example, officials at several of the sites we visited produced documentation—some dating from the 1980s—showing NRC's documented approval of some, but not all, operator manual actions. In some other cases, unit operators told us that NRC officials verbally

---

[24]See NRC, *NRC Regulatory Issues Summary 2006-10, Regulatory Expectations with Appendix R Paragraph III.G.2 Operator Manual Actions* (Washington, D.C., June 30, 2006). These regulations also require a trained fire brigade with adequate capability to fight fires in all areas of the unit containing structures, systems, and components important to safety.

approved certain operator manual actions but did not document their approval in writing. In some other instances, without explicit NRC approval, unit officials applied operator manual actions that NRC had previously approved for similar situations. NRC officials explained that NRC inspectors may not have cited units for violations for these operator manual actions because they believed the actions were safe; however, NRC's position is that these actions do not comply with NRC's fire regulations. Moreover, in fire inspections initiated in 2000 of nuclear units' safe shutdown capabilities, NRC found that units were continuing to use operator manual actions without exemptions in lieu of protecting safe shutdown capabilities through the required passive measures. For example, management officials for some nuclear units authorized staff to manually turn a valve to operate a pump if it failed due to fire damage rather than protecting the cables that operate the valve automatically. Unit officials at one site stated that they rely on more than 20 operator manual actions that must be implemented within 25 minutes for safe shutdown in the event of a fire.

In March 2005 NRC published a proposal to revise Appendix R to allow feasible and reliable operator manual actions if units maintained or installed automatic fire detection and suppression systems. The agency stated that this would reduce the regulatory burden by decreasing the need for licensees to prepare exemption requests and the need for NRC to review and approve them.[25] However, industry officials stated, among other things, that the requirement for suppression would be costly without a clear safety enhancement and, therefore, would likely not reduce the number of exemption requests. Officials at one unit told us that this requirement, in conjunction with other NRC proposed rules, could cost as much as $12 million at one unit, and they believe that the rule would have caused the industry to submit a substantial number of exemption requests to NRC. Due in part to these concerns, NRC withdrew the proposed rule in March 2006.[26]

NRC officials reaffirmed the agency's position that nuclear units using unapproved or undocumented operator manual actions are not in compliance with regulations. In published guidance sent to all operating nuclear units in 2006, NRC stated that this has been its position since

---

[25] 70 *Fed. Reg.* 10901 (Mar. 7, 2005)

[26] 71 *Fed. Reg.* 11169 (Mar. 6, 2006)

GAO-08-747 Fire Safety and Nuclear Reactor Units

607

Appendix R became effective in 1981.[27] The guidance further stated that NRC has continued to communicate this position to licensees via various public presentations, proposed rulemaking, and industry wide communications.

In June 2006, NRC directed nuclear units to complete corrective actions for these operator manual actions by March 2009, either by applying for licensing exemptions for undocumented or unapproved operator manual actions or by making design modifications to the unit to eliminate the need for operator manual actions.[28] Staff at most nuclear units we visited said they would resolve this issue either by transitioning to the new risk-informed approach, or by applying to NRC for licensing exemptions because making modifications would be resource-intensive. In March 2006, NRC also stated in the *Federal Register* that the regulations allow licensees to use the risk-informed approach in lieu of seeking an exemption or license amendment.[29]

NRC officials told us that, at least for the short-term, they have no plans to examine unapproved or undocumented operator manual actions for units that have sought exemptions to determine if these units are compliant with regulations. They said that NRC has already received exemption requests for operator manual actions, and it expects about 25 units—mostly units licensed before 1979 that do not intend to adopt the new risk-informed approach—to submit additional exemption requests by March 2009.[30] They estimated that about half of the 58 units that have not decided to transition to the risk-informed approach do not have compliance issues regarding operator manual actions and, therefore, will not need to submit related requests for exemptions. These officials anticipate that the remaining units that are not transitioning to the risk-informed approach will submit exemptions in the following two broad groups: (1) license amendment requests that should be short and easy to process because the technical review has already been completed, showing that the operator manual actions in place do not degrade unit safety; and (2) exemption

---

[27]See NRC, *NRC Regulatory Issue Summary 2006-10.*

[28]See NRC, *NRC Regulatory Issue Summary 2006-10.*

[29]71 *Fed. Reg.* 11169 (Mar. 6, 2006).

[30]NRC officials told us that the actual number of exemptions will be less than 25 because units will submit them by site, not per nuclear unit.

Page 16   GAO-08-747 Fire Safety and Nuclear Reactor Units

requests that require more detailed review because the units have been using unapproved operator manual actions.

NRC Has Not Yet Acted to Address Extended Use of Interim Compensatory Measures

Some nuclear units have used interim compensatory measures for extended periods of time—in some cases, for years—rather than perform the necessary repairs or procure the necessary replacements. As of April, 2008, NRC has no firm plans for resolving this problem. For example, at one nuclear unit we visited, unit officials chose to use fire watches for over 5 years instead of replacing faulty penetration seals covering openings in structural fire barriers. Officials at several units told us that they typically use fire watches with dedicated unit personnel as interim compensatory measures whenever they have deficiencies in fire protection features. NRC regional officials confirmed that most interim compensatory measures are currently fire watches and that many of these were implemented at nuclear units after tests during the 1980s and 1990s determined that Thermo-Lag and, later, Hemyc fire wraps, used to protect safe shutdown cables from fire damage, were deficient. According to a statement released by an NRC commissioner in October 2007, interim compensatory measures are not the most transparent or safest way to deal with this issue. Moreover, NRC inspectors have reported weaknesses in certain interim compensatory measures used at some units, including an over reliance on 1-hour roving fire watches rather than making the necessary repairs.

Although NRC regulations state that all deficiencies in fire protection features must be promptly identified and corrected,[31] they do not limit how long units can rely on interim compensatory measures—such as hourly fire watches—before taking corrective actions or include a provision to compel licensees to take corrective actions. In the early 1990s, NRC issued guidance addressing the timeliness of corrective actions, stating that the agency expected units to promptly complete all corrective actions in a timely manner commensurate with safety and thus eliminate reliance on the interim compensatory measures. In 1997, NRC issued additional guidance, stating that if a nuclear unit does not resolve a corrective action at the first available opportunity or does not appropriately justify a longer completion schedule, the agency would conclude that corrective action has not been timely and would consider taking enforcement action. NRC's current guidance for its inspectors states that a unit may implement interim compensatory measures until final corrective action is completed

---

[31]See Appendix B to 10 C.F.R. 50.

and reliance on an interim compensatory measure for operability should be an important consideration in establishing the time frame for completing the corrective action.[32] This guidance further states that conditions calling for interim compensatory measures to restore operability should be resolved quickly because such conditions indicate a greater degree of degradation or nonconformance than conditions that do not rely on interim compensatory measures. For example, the guidance states that NRC expects interim compensatory measures that substitute an operator manual action for automatic safety-related functions to be resolved expeditiously. Officials from several different units that we visited confirmed that NRC has not implemented a standard timeframe for when corrective actions must be made regarding safe shutdown deficiencies.

NRC officials further state that interim compensatory measures could remain in place at some units until they fully transition to the risk-informed approach to fire protection. They stated that this was because many of the interim compensatory measures are in place for Appendix R issues that are not risk significant, and nuclear units will be able to eliminate them after they implement the risk-informed approach.

NRC Has Not Resolved Uncertainty Regarding the Effectiveness of Fire Wraps

NRC has not resolved uncertainty regarding fire wraps used at some nuclear units for protecting cables critical for safe shutdown. NRC's regulations state that fire wraps protecting shutdown-related systems must have a fire rating of either 1 or 3 hours. NRC guidance further states that licensees should evaluate fire wrap testing results and related data to ensure it applies to the conditions under which they intend to install the fire wraps. If all possible configurations cannot be tested, an engineering analysis must be performed to demonstrate that cables would be protected adequately during and after exposure to fire. NRC officials told us that the agency prefers passive fire protection, such as fire barriers—including fire

---

[32]This inspection guidance states the following: In determining whether the licensee is making reasonable efforts to complete corrective actions promptly, the NRC will consider safety significance, the effects on operability, the significance of the degradation, and what is necessary to implement the corrective action. The NRC may also consider the time needed for design, review, approval, or procurement of the repair or modification; the availability of specialized equipment to perform the repair or modification; and the need for the unit to be in hot or cold shutdown to implement the actions. If the licensee does not resolve the degraded or nonconforming condition at the first available opportunity or does not appropriately justify a longer completion schedule, the staff would conclude that corrective action has not been timely and would consider taking enforcement action.

wraps—because such protection is more reliable than other forms of fire protection, for example, human actions for fire protection.

Following the 1975 fire at Browns Ferry, manufacturers of fire wraps performed or sponsored fire endurance tests to establish that their fire wraps met either the 1-hour or 3-hour rating period required by NRC regulations. However, NRC became concerned about fire wraps in the late 1980s when Thermo-Lag—a fire wrap material commonly used in units at the time—failed performance tests to meet its intended 1-hour and 3-hour ratings, even though it had originally passed the manufacturer's fire qualification testing. In 1992, NRC's Inspector General found that NRC and nuclear licensees had accepted qualification test results for Thermo-Lag that were later determined to be falsified. From 1991 to 1995, NRC issued a series of information notices on performance test failures and installation deficiencies related to Thermo-Lag fire wrap systems. As a result, in the early 1990s, NRC issued several generic communications informing industry of the test results and requested that licensees implement appropriate interim compensatory measures and develop plans to resolve any noncompliance. One such communication included the expectation that licensees would review other fire wrap materials and systems and consider actions to avoid problems similar to those identified with Thermo-Lag.

Deficiencies emerged in other fire wrap materials starting in the early 1990s, and NRC suggested that industry conduct additional testing. It took NRC over 10 years to initiate and complete its program of large-scale testing of Hemyc—another commonly used fire wrap—and then direct units to take corrective actions after small-scale test results first indicated that Hemyc might not be suitable as a 1-hour fire wrap. In 1993, NRC conducted pilot-scale fire tests on several fire wrap materials, but because the tests were simplified and small-scale models were used, NRC applied test results for screening purposes only. These tests involved various fire wraps assembled in different configurations. The test results indicated unacceptable performance in approximately one-third of the assemblies tested, and NRC reported that the results for Hemyc were inconclusive, although NRC's Inspector General recently reported that Hemyc had failed this testing. In 1999 and 2000, several NRC inspection findings raised concerns about the performance of Hemyc and MT—another fire wrap—including: (1) whether test acceptance criteria for insurance purposes is valid for fire barrier endurance tests and (2) the performance of fire wraps when those wraps are used in untested configurations. In 2001, NRC initiated testing for typical Hemyc and MT installations used in units in the United States, and the test results indicated that the Hemyc configuration

Page 19   GAO-08-747 Fire Safety and Nuclear Reactor Units

did not pass the 1-hour criteria and that the MT configuration did not pass the 3-hour criteria. In 2005, NRC held a public meeting with licensees to discuss these test results and how to achieve compliance.

In 2006, NRC published guidance stating that fire wraps installed in configurations that are not capable of providing the designed level of protection are considered nonconforming installations and that licensees that use Hemyc and MT—previously accepted fire wraps—may not be conforming with their licenses. This guidance further stated that if licensees identify nonconforming conditions, they may take the following corrective actions: (1) replace the failed fire wraps with an appropriately rated fire wrap material, (2) upgrade the failed fire barrier to a rated barrier, (3) reroute cables or instrumentation lines through another fire area, or (4) voluntarily transition to the risk-informed approach to fire protection.

According to NRC's Inspector General, during testimony before Congress in 1993 on the deficiencies of Thermo-Lag, the then-NRC Chairman committed NRC to assess all fire wraps to determine what would be needed in order to meet NRC requirements. The testimony also contained an attachment of an NRC task force that made the following two recommendations: (1) NRC should sponsor new tests to evaluate the fire endurance characteristics of other fire wraps and (2) NRC should review the original fire qualification test reports from fire wrap manufacturers.[33]

Although NRC maintains that it has satisfied this commitment, the NRC Inspector General reported in January 2008 that the agency had yet to complete these assessments. NRC officials told us that licensees are required to conduct endurance tests on fire wraps used at nuclear units; however, the NRC Inspector General noted that, to date, no test has been conducted certifying Hemyc as a 1- or 3- hour fire wrap. Licensees' proposed resolutions for this problem ranged from making replacements with another fire wrap material to requesting license exemptions. In addition, although NRC advised licensees that corrective actions associated with Hemyc and MT are subject to future inspection, the Inspector General noted that NRC has not yet scheduled or budgeted for inspections of licensees' proposed resolutions. The Inspector General's report indicated that several different fire wraps failing endurance tests

---

[33]Nuclear Regulatory Commission, Office of Inspector General, *NRC's Oversight of Hemyc Fire Barriers*, Case 05-46 (Washington, D.C., Jan. 22, 2008).

are still installed at units across the country, but NRC does not maintain current records of these installations. Until issues regarding the effectiveness of fire wraps are resolved, utilities may not be able to use the wraps to their potential and instead rely on other measures, including operator manual actions.

NRC Has Not Yet Acted to Resolve How to Protect against Multiple Spurious Actuations That Could Affect a Nuclear Unit's Ability to Safely Shut Down

NRC has not finalized guidance on how nuclear units should protect against short-circuits that could cause safety-related equipment to start or malfunction spuriously (instances called spurious actuations). In the early 1980s, NRC issued guidance clarifying the requirements in its regulations for safeguarding against spurious actuations that could adversely affect a nuclear unit's ability to safely shut down.[34] However, NRC approved planning for spurious actuations occurring only one at a time or in isolation. In the late 1990s, nuclear units identified problems related to multiple spurious actuations occurring simultaneously. Due to uncertainty over this issue, in 1998 NRC exempted units from enforcement actions related to spurious actuations, and in 2000 the agency temporarily suspended the electrical circuit analysis portion of its fire inspections at nuclear units. Cable fire testing performed by industry in 2001 demonstrated that multiple spurious actuations occurring simultaneously or in rapid succession without sufficient time to mitigate the consequences may have a relatively high probability of occurring under certain circumstances, including fire damage.[35]

Following the 2001 testing, NRC notified units that it expects them to plan for protecting electrical systems against failures due to fire damage, including multiple spurious actuations in both safety-related systems and

---

[34]Specifically, Appendix R requires plants to protect cables or equipment necessary for safe shutdown from fire damage, including (1) electrical systems used directly to perform a safe-shutdown function and (2) associated nonsafety circuits—electrical systems not directly related to performing safe-shutdown functions but for which a spurious actuation might prevent safe shutdown. For example, an associated nonsafety system might control a valve necessary for keeping a storage tank full of water used to cool a reactor, whereas a safety-related system might control a pump responsible for transporting the water to the reactor.

[35]See Electric Power Research Institute, *Spurious Actuation of Electrical Circuits Due to Cable Fires: Results of an Expert Elicitation*, Report No. 1006961 (Palo Alto, California, May 2002); and NRC, *Cable Insulation Resistance Measurements Made During Cable Fire Tests*, NUREG/CR-6776 (Washington, D.C., June 2002).

associated nonsafety systems.[36] NRC resumed electrical inspections in 2005 and proposed that licensees review their fire protection programs to confirm compliance with NRC's stated regulatory position on this issue and report their findings in writing. The proposal suggested that noncompliant units could come into compliance by (1) reperforming their circuit analyses and making necessary design modifications, (2) performing a risk-informed evaluation, or (3) adopting the overall risk-informed approach to fire protection advocated by NRC. In 2006, however, NRC decided not to issue the proposal, stating that further thought and care can be taken to ensure the resolution of this issue has a technically sound and traceable regulatory footprint that would provide permanent closure.

The nuclear industry has issued statements disagreeing with NRC's proposed regulatory approach for multiple spurious actuations. Industry officials noted that NRC approved licenses for many units that require operators to plan for spurious actuations from a fire event that occur one at a time or in isolation and that NRC's current approach amounts to a new regulatory position on this issue. Furthermore, the industry asserts that units only need to plan for protecting against spurious actuations occurring one at a time or in isolation because, in industry's view, multiple spurious actuations occurring are highly improbable and should not be considered in safety analyses. Industry officials told us that the 2001 test results were generated under worst-case scenarios, which operating experience has shown may not represent actual conditions at nuclear units. These officials further told us that NRC's requirements are impossible to achieve.

In December 2007, the nuclear industry proposed an approach for evaluating the effects on circuits from two or more spurious actuations occurring simultaneously, but NRC had not officially commented on the proposal as of May 2008. NRC has stated that draft versions of the proposal it has reviewed do not achieve regulatory compliance. As of May 2008, despite numerous meetings and communications with industry, NRC has not endorsed guidance or developed a timeline for resolving

---

[36]NRC has also stated that plants cannot use operator manual actions to mitigate multiple spurious actuations because Appendix R does not mention operator manual actions as an acceptable method of fire protection. As discussed previously, many plants believe that operator manual actions are allowed without explicit approval from NRC. However, industry testing in 2001 indicates that some operator manual actions may not be able to mitigate multiple spurious actuations due to insufficient time to act.

disagreements with industry about how to plan for multiple spurious actuations of safety-related equipment due to fire damage. However, NRC officials told us they have recently developed a closure plan for this issue that they intend to propose to NRC's Commissioners for approval in June 2008. NRC officials told us that after this plan is approved, their planned next steps are to determine (1) the analysis tools, such as probabilistic risk assessments or fire models, that units can use to analyze multiple spurious actuations; and (2) a time frame for ending its ongoing exemption of units from enforcement actions related to spurious actuations.

NRC Lacks a Comprehensive Database to Track Nuclear Units' Use of Operator Manual Actions, Interim Compensatory Measures, and Exemptions

NRC has no comprehensive database of the operator manual actions or interim compensatory measures implemented at nuclear units since its regulations were first promulgated in 1981, in addition to the hundreds of related licensing exemptions. NRC does not require units to report operator manual actions upon which they rely for safe shutdown. Although NRC reports operator manual actions in the inspection reports it generates through its triennial fire inspections, it does not track these operator manual actions industrywide nor does it compile them on a unit by unit basis. NRC does not maintain a central database of interim compensatory measures being used in place of permanent fire protection features at units for any duration of time. In addition, NRC regional officials told us that triennial fire inspectors do not typically track the status of interim compensatory measures used for fire protection or which units are using them. However, units record maintenance-related issues in their corrective action programs, including those issues requiring the implementation of interim compensatory measures. As a result, data are available to track interim compensatory measures that last for any period of time as well as to analyze their safety significance. NRC resident inspectors told us that they review these corrective action programs on a daily basis and that they are always aware of the interim compensatory measures in place at their units. They reported that this information is sometimes reviewed by NRC regional offices but rarely by headquarters officials.

NRC officials explained that the agency tracked the use of exemptions—including some operator manual actions—through 2001 but then stopped because the number of exemptions requested by units decreased. This information is available, in part, electronically through its public documents system and partly in microfiche format. These officials explained that part of the agency's inspection process is to test if licensees have copies of their license exemptions and, thus, are familiar with their own licensing basis. Inspectors have the ability to confirm an exemption, but once the inspectors are in the field, they often rely on the licensee's documentation. According to these officials, NRC has no central repository for all the exemptions for a unit, but agency inspectors can easily validate a licensee's exemption documentation by looking it up in their public documents system. They said that they conduct the triennial inspections over 2 weeks at the unit because they realize licensees may not be able to locate documentation immediately. They notify licensees what documents they need during the first week onsite so the licensees can have time to prepare them for NRC's return trip. NRC regional officials told us that it is difficult to inspect fire safety due to the complicated licensing basis and inability to track documents.

An NRC commissioner told us that nuclear power units have adopted many different fire safety practices with undocumented approval status. The commissioner further stated that NRC does not have good documentation of which units are using interim compensatory measures or operator manual actions for fire protection and that it needs a centralized database to track these issues. The commissioner stated the lack of a centralized database does not necessarily indicate that safety has been compromised.

However, without a database that contains information about the existence, length, nature, and safety significance of interim compensatory measures, operator manual actions, and exemptions in general, NRC may not have a way to easily track which units have had significant numbers of extended interim compensatory measures and possibly unapproved operator manual actions. Moreover, the database could help NRC make informed decisions about how to resolve these long-standing issues. Also, the database could help NRC inspectors more easily determine whether specific operator manual actions or extended interim compensatory measures have, in fact, been approved through exemptions.

To Date, 46 Nuclear Unit Operators Have Announced They Will Adopt a New Risk-Informed Approach to Fire Safety, but the Transition Effort Faces Challenges

Officials at 46 nuclear units have announced their intention to adopt the risk-informed approach to fire safety. Officials from NRC, industry, and units we visited that plan to adopt the risk-informed approach stated that they expect the new approach will make units safer by reducing reliance on unreliable operator manual actions and help identify areas of the unit where multiple spurious actuations could occur. Academic and industry experts believe that the risk-informed approach could provide safety benefits, but they stated that NRC must address inherent complexities and unknowns related to the development of probabilistic risk assessments used in the risk-informed approach. Furthermore, the shortage of skilled personnel and concerns about the potential cost of conducting risk analyses could slow the transition process and limit the number of units that ultimately make the transition to the new approach.

GAO-08-747 Fire Safety and Nuclear Reactor Units

616 703-739-3790 TCNNaturalGas.com

Nuclear Units Adopting the Risk-Informed Approach Expect It to Improve Safety

As of May 2008, 46 nuclear units at 29 sites have announced that they will transition to the risk-informed approach endorsed by NRC (see fig. 1). To facilitate the transition process for the large number of units that will change to the new approach within the next 5 years, NRC is overseeing a pilot program involving three nuclear units at the Oconee Nuclear Power Plant in South Carolina and one unit at the Shearon Harris Nuclear Power Plant in North Carolina, and NRC expects to release its evaluation of these units' license amendment requests supporting their transition to the risk-informed approach by March 2009. At that point, 22 nuclear units will have submitted their license amendment requests for NRC's review, followed by other units in a staggered fashion.

Figure 1: The 46 Commercial Nuclear Reactors in the United States That Are Transitioning to the Risk-Informed Approach, as of May 2008

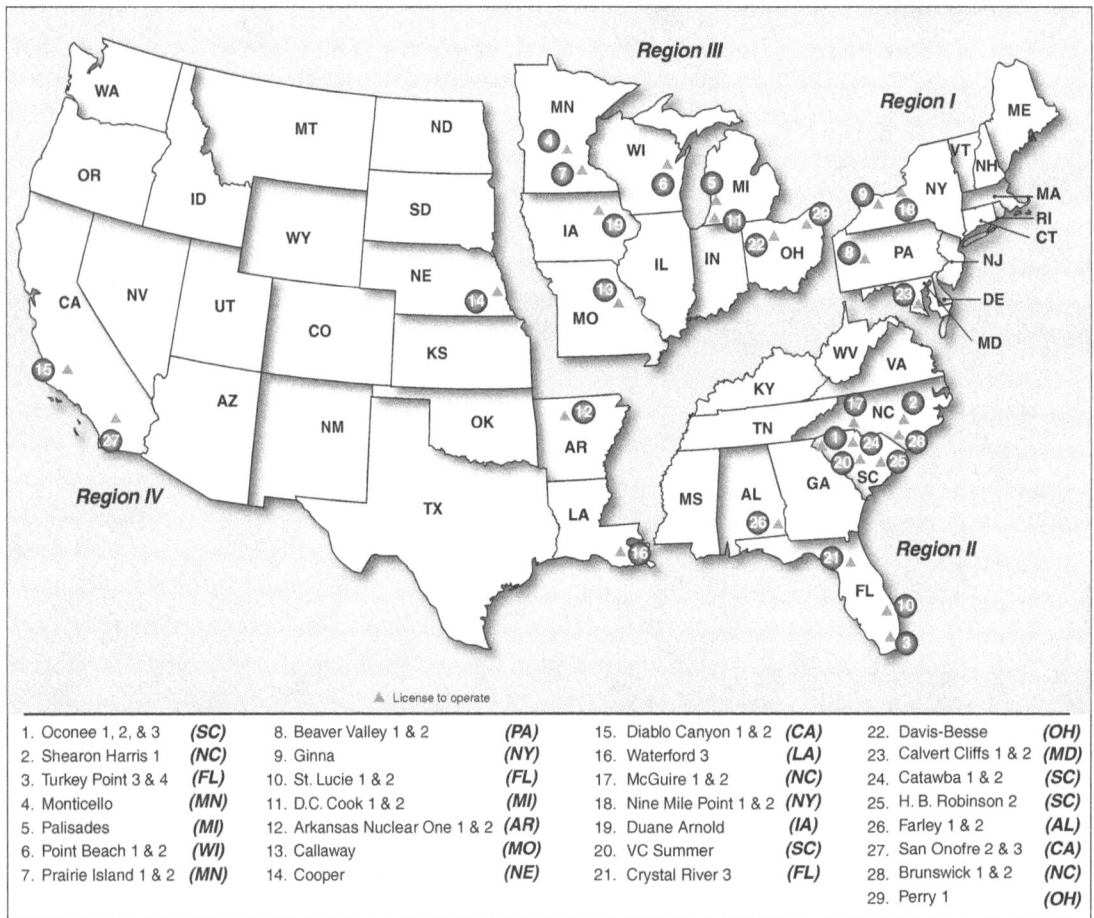

▲ License to operate

1. Oconee 1, 2, & 3 **(SC)**	8. Beaver Valley 1 & 2 **(PA)**	15. Diablo Canyon 1 & 2 **(CA)**	22. Davis-Besse **(OH)**
2. Shearon Harris 1 **(NC)**	9. Ginna **(NY)**	16. Waterford 3 **(LA)**	23. Calvert Cliffs 1 & 2 **(MD)**
3. Turkey Point 3 & 4 **(FL)**	10. St. Lucie 1 & 2 **(FL)**	17. McGuire 1 & 2 **(NC)**	24. Catawba 1 & 2 **(SC)**
4. Monticello **(MN)**	11. D.C. Cook 1 & 2 **(MI)**	18. Nine Mile Point 1 & 2 **(NY)**	25. H. B. Robinson 2 **(SC)**
5. Palisades **(MI)**	12. Arkansas Nuclear One 1 & 2 **(AR)**	19. Duane Arnold **(IA)**	26. Farley 1 & 2 **(AL)**
6. Point Beach 1 & 2 **(WI)**	13. Callaway **(MO)**	20. VC Summer **(SC)**	27. San Onofre 2 & 3 **(CA)**
7. Prairie Island 1 & 2 **(MN)**	14. Cooper **(NE)**	21. Crystal River 3 **(FL)**	28. Brunswick 1 & 2 **(NC)**
			29. Perry 1 **(OH)**

Sources: NRC (data); Map Resources (map).

NRC and transitioning unit officials we spoke with expected that transitioning to the new approach could simplify nuclear units' licensing

Page 26   GAO-08-747 Fire Safety and Nuclear Reactor Units

bases by reducing the number of future exemptions significantly at each unit.[37] Furthermore, officials from each of the 12 units we contacted that plan to adopt the approach said that one of the main reasons for their transition is to reduce the number of exemptions, including those involving operator manual actions, that are required to ensure safe shutdown capability under NRC's existing regulations. Specifically, these officials told us that they expected that conducting fire modeling and probabilistic risk assessments—aspects of the risk-informed approach—would allow the nuclear units to demonstrate that fire protection features in an area with shutdown-related systems would be acceptable based on the expected fire risk in that area. According to some of these officials, under these circumstances units would no longer need to use exemptions—including those involving operator manual actions—to demonstrate compliance with the regulations. Officials at 10 of the units we visited stated that, as a result, the approach could eliminate the need for some operator manual actions. For example, officials at one site that contained two nuclear units expected that by transitioning to the new risk-informed approach, the units could eliminate the need for over 1,200 operator manual actions currently in place. Other unit officials conceded that the outcomes of probabilistic risk assessments may demonstrate the need for new operator manual actions that are currently not required under the current regulations. These officials added that any new actions or other safety features could be applied only to those areas subject to fire risk, rather than to the entire facility, thereby allowing units to maximize resources.

According to nuclear unit officials, adopting the risk-informed approach could also help resolve concerns about multiple spurious actuations that could occur as a result of fire events. Officials from six units we visited told us that conducting the probabilistic risk assessments would allow them to identify where multiple spurious actuations are most likely to occur and which circuit systems would be most likely affected. These officials told us that limiting circuit analyses to the most critical areas would make such analyses feasible. NRC has repeatedly promoted the transition to the new risk informed approach as a way for nuclear units to address the multiple spurious actuation issue.

---

[37]NRC has stated that it also expects that the risk-informed approach to fire protection will (1) focus licensee and regulatory attention on design and operational issues commensurate with their importance to public health and safety, (2) identify areas with insufficient safety margin, and (3) provide the bases for additional requirements or regulatory actions.

GAO-08-747 Fire Safety and Nuclear Reactor Units

Industry and Academic Experts Expressed Concern about Probabilistic Risk Assessments That Would Be Used under the Risk-Informed Approach

According to industry officials and academic experts we consulted, the results of a probabilistic risk assessment used in the risk-informed approach could help units direct safety resources to areas where risk from accidents could be minimized or where the risk of damage to the core or a unit's safe shutdown capability is highest; however, officials also noted that the absence of significant fire events since the 1975 Browns Ferry fire limits the relevant data on fire events at nuclear units. Specifically, these experts noted the following:

- Probabilistic risk assessments require large amounts of data; therefore the small number of fires since the Browns Ferry fire and the subsequent lack of real-world data may increase the amount of uncertainty in the analysis.

- Probabilistic risk assessments are limited by the range of scenarios that practitioners include in the analysis. If a scenario is not examined, its risks cannot be considered and mitigated.

- The role of human performance and error in a fire scenario—especially those scenarios involving operator manual actions—is difficult to model.

Finally, these parties stated that probabilistic risk assessments in general are difficult for a regulator to review and are not as enforceable as a prescriptive approach, in which compliance with specific requirements can be inspected and enforced.

NRC and Industry Face a Possible Shortage in Personnel with Skills Relevant to the Risk-Informed Approach

Numerous NRC, industry, and academic officials we spoke with expressed concern that the transition to the new risk-informed approach could be delayed by a limited number of personnel with the necessary skills and training to design, review, and inspect against probabilistic risk assessments. Several nuclear unit officials told us that the pool of fire protection engineers with expertise in these areas is already heavily burdened with developing probabilistic risk assessments for the pilot program units and other units, including the 38 units that had already begun transitioning as of October 2007.

Academic experts, consultants, and industry officials told us that the current shortage of skilled personnel is due to (1) an increased demand for individuals with critical skills under the risk-informed approach and (2) a shortage of academic programs specializing in fire protection engineering. According to these experts and officials, the current number of individuals skilled in conducting probabilistic risk assessments is insufficient to handle the increased work expected to be generated by the transition to a risk-informed approach. NRC officials we spoke with expressed concern that the nuclear industry has not trained or developed sufficient personnel with needed fire protection skills. These officials also told us that they expect that, as demand for work increases, more engineering students will choose to go into the fire protection field. However, to date, only one university has undergraduate and graduate programs in the fire protection engineering field, and the ability to produce graduates is limited. Other officials we spoke with noted that engineers in other fields can be trained in fire protection but that this training takes a significant amount of time.

Academic experts and industry officials stated that without additional skilled personnel, units would not be able to perform all of the necessary activities, especially probabilistic risk assessments, within the 3-year enforcement discretion "window" that NRC has granted each transition unit as an incentive to adopt the new approach. Most nuclear units that responded to an industry survey on this issue indicated that they expected that they will need NRC to extend the discretion deadline for each unit. Delays in individual units' transition processes could create a significant backlog in the entire transition process.

NRC also faces an aging workforce and the likelihood that it will be competing with industry for engineers with skills in the fire protection area. As we reported in January 2007, the agency as a whole faces significant human capital challenges, in part because approximately 33 percent of its workforce will be eligible to retire in 2010.[38] To address this issue, we reported that NRC identified several critical skill gaps that it must address, such as civil engineering and operator licensing. In relation to needed skill areas, the agency has taken steps, including supporting key university programs, to attract greater numbers of students into mission-critical skills areas and to offer scholarships to those studying in these fields. In relation to fire protection, and probabilistic risk assessments in particular, NRC officials told us that they expect to address future resource needs through the use of a multiyear budget and by contracting with the Department of Energy's National Laboratories to help manage the process. Further, these officials stated that part of the purpose of the pilot program is to help them determine future resource needs for the transition to the risk-informed approach, and, as a result, they do not intend to finalize resource planning until the pilot programs are complete. A number

---

[38]GAO, *Human Capital: Retirements and Anticipated New Reactor Applications Will Challenge NRC's Workforce*, GAO-07-105 (Washington, D.C.: Jan. 17, 2007).

GAO-08-747 Fire Safety and Nuclear Reactor Units

of experts in the engineering field, including academics and fire engineers, stated that it will be difficult for NRC to compete with industry over the projected numbers of graduates in this field over the next few years. Also, NRC's total workload, in addition to fire protection, is expected to increase as nuclear unit operators submit license applications to build new units, extend the lives of existing units, or increase the generating capacity of existing units. For example, NRC staff are currently reviewing license applications for units at six sites and have recently announced that operators have submitted licenses for two additional units at a seventh site. The agency expects to review or receive 12 more applications during 2008.

Operators of 58 Nuclear Units Have Not Announced Whether They Will Transition to the New Approach, in Part Due to Concerns about NRC's Risk-Assessment Guidance and Pilot Program Timetable

To date 58 of the nation's 104 nuclear units have not announced whether they will adopt the risk-informed approach. NRC and industry officials stated that they expected that newer units and units with relatively few exemptions from existing regulations would be less likely to transition to the new approach, while those with older licenses and extensive exemptions would make the transition. However, to date, 25 units licensed prior to 1979 have yet to announce whether they will make the transition. Officials from nontransitioning units we visited told us that concerns over NRC's guidance and time table have been key reasons why they have not yet announced their intent to transition.

According to industry and nuclear unit officials we spoke with, the costs associated with conducting fire probabilistic risk assessments for the units may be too high to justify transitioning to the new approach. For example, some officials told us that performing the necessary analysis of circuits and fire area features in support of the probabilistic risk assessment could cost millions of dollars without substantially improving fire safety. These officials noted that both pilot sites currently expect to spend approximately $5 million to $10 million each in transition costs, including circuit analysis. Some of these officials also noted that updating probabilistic risk assessments—which units are required to do every 3 years or whenever any significant changes are made to a unit—would require units to dedicate staff to this effort on a long term or permanent basis.

Officials at transition and nontransition units stated that NRC's guidance for developing fire models that support probabilistic risk assessments is overly conservative. In effect, these models require engineers to assume that fires will result in massive damage, burn for significant periods of time, and require greater response and mitigation efforts than less conservative models. As such, these officials stated that the fire models

Page 30   GAO-08-747 Fire Safety and Nuclear Reactor Units

provided by NRC guidance would not provide an accurate assessment of risk at a given unit. Furthermore, these officials stated that unit modifications required by the risk analysis could cost more than seeking exemptions from NRC. Some of these officials stated that they expect NRC to revise the probabilistic risk assessment guidance to facilitate the transition process in the future. NRC officials told us that nuclear units have the option to develop and conduct their own fire models rather than follow NRC's guidance. Furthermore, in its initial review of one of the pilot unit's probabilistic risk assessments, NRC agreed with industry that models used in the development of the probabilistic risk assessment contained some overly conservative aspects and recommended that the unit conduct additional analysis to address this. However, nuclear unit officials expressed concern that the costs of developing site-specific fire models, a process that includes numerous iterations, could be prohibitive.

Nuclear industry officials identified another area of concern in the current transition schedule, in which 22 units are expected to submit their license amendment requests for the risk-informed approach before NRC finishes assessing the license amendment requests for the pilot program units in March 2009. Although NRC has established a steering committee and a frequently asked question process to disseminate information learned in the ongoing pilot programs to other transition units, a number of nuclear unit officials expressed concern about beginning the transition process before the transition pilot programs are complete and lessons learned from the pilot programs are available. For example, an official at one of the pilot sites noted that the success of the pilot program probably will not be known until after the first triennial safety inspection conducted by NRC, which will occur after March 2009. The transition project manager for two nonpilot transition units expressed his opinion that, due to uncertainties regarding the work units must perform in order to comply with the risk-informed standard, no unit should commit itself to transitioning to the new approach until 2 years after the completion of the pilot programs.

Conclusions

NRC's ability to regulate fire safety at nuclear power units has been adversely affected by several long-standing issues. To its credit, NRC has required that nuclear units come into compliance with requirements related to the use of unapproved operator manual actions by March 2009. However, NRC has not effectively resolved the long-term use of interim compensatory measures or the possibility of multiple spurious actuations. Especially critical, in our opinion, is the need for NRC to test and resolve the effectiveness of fire wraps at nuclear units, because units have instituted many manual actions and compensatory measures in response to fire wraps that were found lacking in

Page 31   GAO-08-747 Fire Safety and Nuclear Reactor Units

effectiveness in various tests. Compounding these issues, NRC has no central database of exemptions, operator manual actions, and extended interim compensatory measures. Such a system would allow it to track trends in compliance, devise solutions to compliance issues, and help provide important information to NRC's inspection activities.

Unless NRC deals effectively with these issues, units will likely continue to postpone making necessary repairs and replacements, choosing instead to rely on unapproved or undocumented manual actions as well as compensatory measures that, in some cases, continue for years. According to NRC, nuclear fire safety can be considered to be degraded when reliance on passive measures is supplanted by manual actions or compensatory measures. By taking prompt action to address the unapproved use of operator manual actions, long-term use of interim compensatory measures, the effectiveness of fire wraps, and multiple spurious actuations, NRC would provide greater assurance to the public that nuclear units are operated in a way that promotes fire safety. Despite the transition of 46 units to a new risk-informed approach, for which the implementation timeframes are uncertain, the majority of the nation's nuclear units will remain under the existing regulatory approach, and the long-standing issues will continue to apply directly to them.

Recommendations for Executive Action

To address long-standing issues that have affected NRC's regulation of fire safety at the nation's commercial nuclear power units, we recommend that the NRC Commissioners direct NRC staff to take the following four actions:

- Develop a central database for tracking the status of exemptions, compensatory measures, and manual actions in place nationwide and at individual commercial nuclear units.

- Address safety concerns related to extended use of interim compensatory measures by

  - defining how long an interim compensatory measure can be used and identifying the interim compensatory measures in place at nuclear units that exceed that threshold,

  - assessing the safety significance of such extended compensatory measures and defining how long a safety-significant interim compensatory measure can be used before NRC requires the unit operator to make the necessary repairs or replacements or request an exemption or deviation from its fire safety requirements, and,

- developing a plan and deadlines for units to resolve those compensatory measures.

- Address long-standing concerns about the effectiveness of fire wraps at commercial nuclear units by analyzing the effectiveness of existing fire wraps and undertaking efforts to ensure that the fire endurance tests have been conducted to qualify fire wraps as NRC-approved 1- or 3-hour fire barriers.

- Address long-standing concerns by ensuring that nuclear units are able to safeguard against multiple spurious actuations by committing to a specific date for developing guidelines that units should meet to prevent multiple spurious actuations.

Agency Comments and Our Evaluation

We provided a draft of this report to the Commissioners of the Nuclear Regulatory Commission for their review and comment. In commenting on a draft of this report, NRC found that it was accurate, complete, and handled sensitive information appropriately and stated that it intends to give GAO's findings and conclusions serious consideration. However, in its response, NRC did not provide comments on our recommendations. NRC's comments are reprinted in appendix II.

We are sending copies of this report to the Commissioners of the Nuclear Regulatory Commission, the Nuclear Regulatory Commission's Office of the Inspector General, and interested congressional committees. We will also make copies available to others on request. In addition, this report will be available at no charge on the GAO Web site at http://www.gao.gov.

If you or your staff have any questions about this report, please contact me at (202) 512-3841 or gaffiganm@gao.gov. Contact points for our Offices of Congressional Relations and Public Affairs may be found on the last page of this report. GAO staff who made major contributions to this report are listed in appendix III.

Mark Gaffigan
Director, Natural Resources and Environment

Appendix I: Scope and Methodology

To examine the number, causes, and reported safety significance of fire incidents at nuclear reactor units since 1995, we analyzed Nuclear Regulatory Commission (NRC) data on fires occurring at operating commercial nuclear reactor units from January, 1995, to December, 2007.[1] NRC requires units to report fire events meeting certain criteria, including fires lasting longer than 15 minutes or those threatening safety.[2] To assess the reliability of the data, we (1) interviewed NRC officials about the steps they take to ensure the accuracy of the data; (2) confirmed details about selected fire events, NRC inspection findings, and local emergency responders with unit management officials and NRC resident inspectors during site visits to nuclear power units; (3) reviewed NRC inspection reports related to fire protection; and (4) checked the data for obvious errors. We determined that the data were sufficiently reliable for the purposes of this report.

To examine what is known about nuclear reactor units' compliance with NRC's deterministic fire protection regulations, we reviewed the relevant fire protection regulations and guidance from NRC and industry. We also met with and reviewed documents provided by officials from NRC, industry, academia, and public interest groups. In particular, we interviewed officials from NRC's Fire Protection Branch, Office of Enforcement, four regional offices, Office of the Inspector General, and Advisory Committee on Reactor Safeguards. In addition, we interviewed officials from the Nuclear Energy Institute, National Fire Protection Association, nuclear industry consultants, and nuclear insurance companies. We conducted site visits to nuclear power units, where we met with unit management officials and NRC resident inspectors. During these site visits, we discussed and received documentation on the use of

---

[1]The scope of our work focuses on fire safety as it pertains to a nuclear unit's ability to achieve safe shutdown. NRC is also overseeing plans and actions undertaken by unit operators to safeguard against fire resulting from a catastrophic event in which containment structures surrounding a unit's core and spent fuel pool are damaged or destroyed. We did not analyze this issue because it falls outside the scope of our audit.

[2]In most cases, however, fires only result in notification because there is a declaration of an emergency class, which is reportable under 10 C.F.R. 50.72. According to NRC guidance, a fire lasting longer than 10 or 15 minutes or which affects plant equipment important to safe operation would result in declaration of an emergency class. If there is an actual threat or significant hampering, a Licensee Event Report is also required. According to 10 C.F.R. 50.73, a plant must submit a Licensee Event Report for any event, including a fire, that posed an actual threat to the safety of the nuclear power plant or significantly hampered site personnel in the performance of duties necessary for the safe operation of the nuclear power plant. NRC guidance states that it generally considers a control room fire to constitute an actual threat and significant hampering.

operator manual actions, interim compensatory measures, and fire wraps, and we obtained views on multiple spurious actuations and their impact on safe shutdown. We also reviewed and discussed each unit's corrective action plan. Finally, we observed multiple NRC public meetings and various collaborations with industry concerning issues related to compliance with NRC's deterministic fire protection regulations.

To examine the status of the nuclear industry's implementation of the risk-informed approach to fire safety advocated by NRC, we met with and reviewed documents provided by officials from NRC, industry, and public interest groups, as well as academic officials with research experience in fire safety and risk analysis. In particular, we interviewed officials from NRC's Fire Protection Branch, Office of Enforcement, four regional offices, Office of the Inspector General, and Advisory Committee on Reactor Safeguards. We also interviewed officials from the Nuclear Energy Institute, National Fire Protection Association, nuclear industry consultants, and nuclear insurance companies. We conducted site visits to nuclear power units, where we met with unit management officials and NRC resident inspectors. During these site visits, we discussed and received documentation on the risk-informed approach to fire safety, including resource planning and analysis justifying decisions on whether or not to transition to NFPA-805. We also observed multiple NRC public meetings and collaborations with industry concerning issues related to the risk-informed approach to fire safety. Finally, we reviewed relevant fire protection regulations and guidance from NRC and industry.

In addressing each of our three objectives, we conducted visits to sites containing one or more commercial nuclear reactor units. These visits allowed us to obtain in-depth knowledge about fire protection at each site. We selected a nonprobability sample of sites to visit because certain factors—including custom designs that differ according to each nuclear unit, hundreds of licensing exemptions and deviations in place at units nationwide, and the geographic dispersal of units units across 31 states—complicate collecting data and reporting generalizations about the entire population of units.[3] We chose 10 sites (totaling 20 operating nuclear reactor units out of a national total of 104 operating nuclear units) that provided coverage of each of NRC's four regional offices and that represented varying levels of unit fire safety performance, unit licensing characteristics, reactor types, and NRC oversight. At the time of our visits, 5 of the 10 sites we visited (totaling 10 of the 20 nuclear reactor units we visited) had notified NRC that they intend to transition to the new risk-informed approach to fire safety. Over the course of our work, we visited the following sites: (1) D.C. Cook (2 units), located near Benton Harbor, Michigan; (2) Diablo Canyon (2 units), located near San Luis Obispo, California; (3) Dresden (2 units), located near Morris, Illinois; (4) Indian Point (2 units), located near New York, New York; (5) La Salle (2 units), located near Ottawa, Illinois; (6) Nine Mile Point (2 units), located near Oswego, New York; (7) Oconee (3 units), located near Greenville, South Carolina; (8) San Onofre (2 units), located near San Clemente, California; (9) Shearon Harris (1 unit), located near Raleigh, North Carolina; and (10) Vogtle (2 units), located near Augusta, Georgia.

We selected the nonprobability sample from the entire population of commercial nuclear power units currently operating in the United States.[4] In order to capture variations that could play a role in how these units address fire safety, we designed our site visit selection criteria to represent the following: (1) geographic diversity; (2) units licensed to operate before and after 1979; (3) sites choosing to remain under the deterministic regulations and those transitioning to the risk-informed approach; (4) pressurized and boiling water reactor types; (5) a variety of safety problems in which inspection findings or performance indicators of higher

---

[3]The information gathered on these site visits cannot be used to generalize findings to, or make inferences about, the entire population of plants, or the nuclear power industry as a whole. Although the sample provides some variety, it is unlikely to capture the full variability of conditions under which fire protection takes place at the plants, and it cannot provide comprehensive insight into the effects of any one set of conditions. This is because, in a nonprobability sample, some elements of the population being studied have no chance, or an unknown chance, of being selected. However, the information gathered during these site visits allows us to make qualified comparisons between different groups of plants and to discuss issues faced by each group. It also helps us interpret the quantitative data, documentation, guidance, and testimonial evidence we have collected. In addition, it provides anecdotal and illustrative evidence about fire protection at plants under various conditions, as well as providing important context overall.

[4]As of May 2008, the commercial nuclear power industry in the United States was composed of 104 operating nuclear reactor units at 65 sites in 31 states. Each site had one to three units often operated and licensed by the same utility, and therefore combined for NRC oversight purposes.

risk significance (white, yellow, or red) were issued; (6) units that have been subjected to at least some level of increased oversight since regular fire inspections were initiated in 2000; and (7) sites with various numbers of fires reportable to NRC since 1995. We received feedback on our selection criteria from nuclear insurance company officials, nuclear industry consultants, NRC officials, and academic officials with research experience in fire safety and risk analysis. We interviewed NRC resident inspectors and unit management officials at each site to learn about the fire protection program at the site. We also observed fire protection features at each site, including safe-shutdown equipment and areas of the units where operator manual actions, interim compensatory measures, and fire wraps are used for fire safety. Finally, we observed part of an NRC triennial fire inspection at one site.

We conducted this performance audit from September 2007 to June 2008 in accordance with generally accepted government auditing standards. Those standards require that we plan and perform the audit to obtain sufficient, appropriate evidence to provide a reasonable basis for our findings and conclusions based on our audit objectives. We believe that the evidence obtained provides a reasonable basis for our findings and conclusions based on our audit objectives.

Appendix II: Comments from the Nuclear Regulatory Commission

UNITED STATES
NUCLEAR REGULATORY COMMISSION
WASHINGTON, D.C. 20555-0001

June 17, 2008

Mr. Mark Gaffigan, Director
Natural Resources and Environment
U.S. Government Accountability Office
441 G Street, NW
Washington, D.C. 20548

Dear Mr. Gaffigan:

Thank you for providing the U.S. Nuclear Regulatory Commission (NRC) the opportunity to review and comment on the U.S. Government Accountability Office's (GAO's) draft report GAO-08-747, "Nuclear Safety: NRC's Oversight of Fire Protection at U.S. Commercial Nuclear Reactor Units Could Be Strengthened." The NRC staff has reviewed the draft report and found that it was accurate, complete, and handled sensitive information appropriately. We intend to give GAO's findings and conclusions serious consideration.

If you have any questions regarding this response, please contact Jesse Arildsen. Mr. Arildsen can be reached by telephone at (301) 415-1785.

Sincerely,

R. W. Borchardt
Executive Director
for Operations

Page 38   GAO-08-747 Fire Safety and Nuclear Reactor Units

Appendix III: GAO Contact and Staff Acknowledgments

GAO Contact	Mark Gaffigan, (202) 512-3841 or gaffiganm@gao.gov
Staff Acknowledgments	In addition to the contact named above, Ernie Hazera (Assistant Director), Cindy Gilbert, Chad M. Gorman, Mehrzad Nadji, Omari Norman, Alison O'Neill, Steve Rossman, and Jena Sinkfield made key contributions to this report.

Related GAO Products

Nuclear Energy: NRC Has Made Progress in Implementing Its Reactor Oversight and Licensing Processes but Continues to Face Challenges. GAO-08-114T. Washington, D.C.: October 3, 2007.

Nuclear Energy: NRC's Workforce and Processes for New Reactor Licensing are Generally in Place, but Uncertainties Remain as Industry Begins to Submit Applications. GAO-07-1129. Washington, D.C.: September 21, 2007.

Human Capital: Retirements and Anticipated New Reactor Applications Will Challenge NRC's Workforce. GAO-07-105. Washington, D.C.: January 17, 2007.

Nuclear Regulatory Commission: Oversight of Nuclear Power Plant Safety Has Improved, but Refinements Are Needed. GAO-06-1029. Washington, D.C.: September 27, 2006.

Nuclear Regulatory Commission: Preliminary Observations on Its Process to Oversee the Safe Operation of Nuclear Power Plants. GAO-06-888T. Washington, D.C.: June 19, 2006.

Nuclear Regulatory Commission: Preliminary Observations on Its Oversight to Ensure the Safe Operation of Nuclear Power Plants. GAO-06-886T. Washington, D.C.: June 15, 2006.

Nuclear Regulatory Commission: Challenges Facing NRC in Effectively Carrying Out Its Mission. GAO-05-754T. Washington, D.C.: May 26, 2005.

Nuclear Regulation: Challenges Confronting NRC in a Changing Regulatory Environment. GAO-01-707T. Washington, D.C.: May 8, 2001.

Major Management Challenges and Performance Risks: Nuclear Regulatory Commission. GAO-01-259. Washington, D.C.: January 2001.

Fire Protection: Barriers to Effective Implementation of NRC's Safety Oversight Process. GAO/RCED-00-39. Washington, D.C.: April 19, 2000.

Nuclear Regulation: Regulatory and Cultural Changes Challenge NRC. GAO/T-RCED-00-115. Washington, D.C.: March 9, 2000.

Nuclear Regulatory Commission: Strategy Needed to Develop a Risk-Informed Safety Approach. GAO/T-RCED-99-071. Washington, D.C.: February 4, 1999.

GAO's Mission

The Government Accountability Office, the audit, evaluation, and investigative arm of Congress, exists to support Congress in meeting its constitutional responsibilities and to help improve the performance and accountability of the federal government for the American people. GAO examines the use of public funds; evaluates federal programs and policies; and provides analyses, recommendations, and other assistance to help Congress make informed oversight, policy, and funding decisions. GAO's commitment to good government is reflected in its core values of accountability, integrity, and reliability.

Obtaining Copies of GAO Reports and Testimony

The fastest and easiest way to obtain copies of GAO documents at no cost is through GAO's Web site (www.gao.gov). Each weekday, GAO posts newly released reports, testimony, and correspondence on its Web site. To have GAO e-mail you a list of newly posted products every afternoon, go to www.gao.gov and select "E-mail Updates."

Order by Mail or Phone

The first copy of each printed report is free. Additional copies are $2 each. A check or money order should be made out to the Superintendent of Documents. GAO also accepts VISA and Mastercard. Orders for 100 or more copies mailed to a single address are discounted 25 percent. Orders should be sent to:

U.S. Government Accountability Office
441 G Street NW, Room LM
Washington, DC 20548

To order by Phone: Voice: (202) 512-6000
TDD: (202) 512-2537
Fax: (202) 512-6061

To Report Fraud, Waste, and Abuse in Federal Programs

Contact:

Web site: www.gao.gov/fraudnet/fraudnet.htm
E-mail: fraudnet@gao.gov
Automated answering system: (800) 424-5454 or (202) 512-7470

Congressional Relations

Ralph Dawn, Managing Director, dawnr@gao.gov, (202) 512-4400
U.S. Government Accountability Office, 441 G Street NW, Room 7125
Washington, DC 20548

Public Affairs

Chuck Young, Managing Director, youngc1@gao.gov, (202) 512-4800
U.S. Government Accountability Office, 441 G Street NW, Room 7149
Washington, DC 20548

PRINTED ON RECYCLED PAPER

633

Resources from TheCapitol.Net

Live Training
<www.CapitolHillTraining.com>

- Capitol Hill Workshop
  <www.CapitolHillWorkshop.com>

- Understanding Congressional Budgeting and Appropriations
  <www.CongressionalBudgeting.com>

- Advanced Federal Budget Process
  <www.BudgetProcess.com>

- The President's Budget
  <www.PresidentsBudget.com>

- Understanding the Regulatory Process: Working with Federal Regulatory Agencies
  <www.RegulatoryProcess.com>

- Drafting Effective Federal Legislation and Amendments
  <www.DraftingLegislation.com>

Capitol Learning Audio Courses™
<www.CapitolLearning.com>

- Congress and Its Role in Policymaking
  ISBN: 158733061X

- Understanding the Regulatory Process Series
  ISBN: 1587331398

- Authorizations and Appropriations in a Nutshell
  ISBN: 1587330296

Other Resources

Internet Resources

Government

- The Office of Nuclear Energy DOE <*http://www.ne.doe.gov/*>

- Nuclear Energy Research Initiative <*http://nuclear.energy.gov/neri/neNERIresearch.html*>

- Gen IV Nuclear Energy Systems <*http://www.ne.doe.gov/genIV/neGenIV1.html*>

- Idaho National Laboratory <*http://www.ne.doe.gov/np2010/overview.html*>

- International Nuclear Energy Research Initiative <*http://sites.energetics.com/ineri_client/index.aspx*>

- Fuel Cycle Research and Development Program <*http://nuclear.gov/FuelCycle/neFuelCycle.html*>

- Nuclear Reactors <*http://www.nrc.gov/reactors.html*>

- Nuclear Materials <*http://www.nrc.gov/materials.html*>

- Radioactive Waste <*http://www.nrc.gov/waste.html*>

- Nuclear Security <*http://www.nrc.gov/security.html*>

- U.S. Energy Information Administration <*http://eia.doe.gov/*>

- State of Nevada Agency for Nuclear Projects <*http://www.state.nv.us/nucwaste/index.htm*>

- Energy Policy Act of 1992 <*http://thomas.loc.gov/cgi-bin/query/z?c102:H.R.776.ENR:*>

- Energy Policy Act of 2005 <*http://frwebgate.access.gpo.gov/cgi-bin/getdoc.cgi?dbname=109_cong_bills&docid=f:h6enr.txt.pdf*>

Associations, Coalitions, News

- Nuclear Energy Institute <*http://www.nei.org*>

- American Nuclear Society <*http://www.new.ans.org/*>

- World Nuclear Association <*http://www.world-nuclear.org/*>

- The Nuclear Industry Association (UK) <*http://www.niauk.org/*>

- International Atomic Energy Agency (IAEA) <*http://www.iaea.org/*>

- World Nuclear News <*http://www.world-nuclear-news.com/*>

- World Association of Nuclear Operators (WANO) <*http://www.wano.org.uk*>

- Beyond Nuclear <*http://www.beyondnuclear.org/*>

- Natural Resource Defense Council *<http://www.nrdc.org/nuclear/plants/plants.pdf>*

- Nuclear Suppliers Association *<http://www.nuclearsuppliers.org>*

- NuclearFiles.org Project of the Nuclear Age Peace Foundation
  <http://www.nuclearfiles.org/menu/key-issues/nuclear-energy/basics/introduction.htm>

- European Atomic Forum (FORATOM) *<http://www.foratom.org>*

- European Nuclear Society *<http://www.euronuclear.org/>*

- World Information Service on Energy *<http://www10.antenna.nl/wise/>*

- Environmentalists for Nuclear Energy (EFN) *<http://www.ecolo.org/>*

- Freedom for Fission *<http://www.freedomforfission.org.uk/>*

- Union of Concerned Scientists *<http://www.ucsusa.org/nuclear_power/>*

- You Tube Nuclear Power *<http://www.youtube.com/watch?v=MRC5KDqpZ_I>*

- NuclearStreet *<http://nuclearstreet.com/>*

- Scientific American *<http://www.scientificamerican.com/report.cfm?id=nuclear-future>*

- RadTown USA Nuclear Power Plants *<http://www.epa.gov/radtown/nuclear-plant.html>*

- Public Citizen Energy Program *<http://www.citizen.org/cmep/>*

Articles

- "Conditions and Policy Reforms Must Accompany Nuclear Loan Guarantee
  Boost" by Jack Spencer, The Heritage Foundation February 2, 2010
  <http://www.heritage.org/Research/EnergyandEnvironment/wm2789.cfm>

- "In Brief: What Pending Climate Legislation Does for Nuclear Energy,"
  The Pew Center on Global Climate Change (October 2009)
  <http://www.pewclimate.org/docUploads/brief-climate-legislation-nuclear-power-oct2009.pdf>

- "Nuclear Power is True 'Green' Energy," by Stuart M.Butler, Ph.D., The Heritage Foundation,
  January 30, 2009 *<http://www.heritage.org/Press/Commentary/ed012909a.cfm>*

- "Three Mile Island and Chernobyl: What Went Wrong and Why Today's Reactors Are
  Safe," by Jack Spencer and Nicholas Loris, The Heritage Foundation, March 27, 2009
  <http://www.heritage.org/Research/EnergyandEnvironment/wm2367.cfm>

- "Security Implications for the Expansion of Nuclear Energy" by Charles
  K. Ebinger and Kevin Massy, The Brookings Institution, October 2009
  <http://www.brookings.edu/papers/2009/10_nuclear_energy_ebinger.aspx>

- "The Real Risk of Nuclear Power" by Nathan Hultman and
  Jonathan G. Koomey, The Brookings Institution, December 2, 2009
  <http://www.brookings.edu/opinions/2009/1202_nuclear_power_hultman.aspx>

638 703-739-3790 TCNNuclear.com

Books

- *Power to Save the World: The Truth About Nuclear Energy*, by Gwyneth Cravens, Vintage Books Copyright 2007 ISBN: 0307385876

- *Atomic Awaking: A New Look at the History and Future of Nuclear Power*, by James Mahaffey Pegasus Books Copyright 2009 ISBN: 1605980404

- *Terrestrial Energy: How Nuclear Energy Will Lead the Green Revolution and End America's Energy Odyssey*, by William Tucker Pegasus Books Copyright June 2009 ISBN-10: 1605980404

- *Nuclear Power Is Not the Answer*, by Helen Caldicott, New Press Copyright September 1, 2007 ISBN-10: 1595582134

- *Nuclear Energy, Principles, Practices, and Prospects*, by David Bodansky Springer Science+Business Media, Inc. Copyright 2004 ISBN: 0387207783

- *Nuclear Energy Sixth Edition: An Introduction to the Concepts, Systems, and Applications of Nuclear Processes*, by Raymond L. Murray Elsevier, Inc. Copyright 2009 ISBN: 0123705479

- *Nuclear Power (Energy for Today)* by Tea Benduhn, Gareth Stevens Publishing Copyright 2008 ISBN: 0836893611

- *Nuclear Engineering: Theory and Technology of Commercial Nuclear Power*, by Ronald Allen Knief, ANS 2nd Edition Copyright 2008 ISBN: 0894484583

- *Nuclear Energy in the 21st Century: World Nuclear University Press*, by Ian Hore-Lacy (Academic Press 2006), ISBN-10: 0123736226

- *Commercial Nuclear Power: Assuring Safety for the Future*, by Charles Ramsey (BookSurge 2006), ISBN-10: 1419649795

- Building History—A Nuclear Power Plant, by Marcia Lusted and Greg Lusted (Lucent 2004), ISBN-10: 1590183924

- *Nuclear Renaissance: Technologies and Policies for the Future of Nuclear Power*, by W.J. Nuttall (Taylor & Francis 2004), ISBN-10: 0750309369

- *Principles of Fusion Energy: An Introduction to Fusion Energy for Students of Science and Engineering*, by A. A. Harms, K. F. Schoepf, G. H. Miley, and D. R. Kingdon (World Scientific Publishing 2000), ISBN-10: 9812380337

- *Nuclear Reactor Analysis*, by James J. Duderstadt and Louis J. Hamilton (Wiley 1976), ISBN-10: 0471223638

- *Introduction to Nuclear Power (Series in Chemical and Mechanical Engineering)*, by Geoffrey Hewitt (Taylor & Francis 2000), ISBN-10: 1560324546

- *Nuclear Safety*, by Gianni Petrangeli (Butterworth-Heinemann 2006), ISBN-10: 0750667230

- *Power Plant Engineering*, by Larry Drbal, Kayla Westra and Pat Bostony (Springer 1995), ISBN-10: 0412064014

- *Power Generation Handbook: Selection, Applications, Operation, Maintenance*, by Philip Kiameh (McGraw-Hill 2002), ISBN-10: 0071396047

- *Introductory Nuclear Physics*, by Kenneth S. Krane (Wiley 1987), ISBN-10: 047180553X

Toys

- Simpsons Nuclear Power Plant Playset with Radioactive Homer ASIN: B000IZG628

- O Scale Operating Nuclear Reactor, AEC, ASIN: B002WGDFKW

640 703-739-3790 TCNNuclear.com

About TheCapitol.Net

We help you understand Washington and Congress.™

For over 30 years, TheCapitol.Net and its predecessor, Congressional Quarterly Executive Conferences, have been training professionals from government, military, business, and NGOs on the dynamics and operations of the legislative and executive branches and how to work with them.

Instruction includes topics on the legislative and budget process, congressional operations, public and foreign policy development, advocacy and media training, business etiquette and writing. All training includes course materials.

TheCapitol.Net encompasses a dynamic team of more than 150 faculty members and authors, all of whom are independent subject matter experts and veterans in their fields. Faculty and authors include senior government executives, former Members of Congress, Hill and agency staff, editors and journalists, lobbyists, lawyers, nonprofit executives and scholars.

We've worked with hundreds of clients across the country to develop and produce a wide variety of custom, on-site training. All courses, seminars and workshops can be tailored to align with your organization's educational objectives and presented on-site at your location.

Our practitioner books and publications are written by leading subject matter experts.

TheCapitol.Net has more than 2,000 clients representing congressional offices, federal and state agencies, military branches, corporations, associations, news media and NGOs nationwide.

Our blog: Hobnob Blog—hit or miss ... give or take ... this or that ...

TheCapitol.Net is on Yelp.
Our recommended provider of government training in Brazil is
PATRI/EDUCARE <www.patri.com>

T.C. Williams
Debate Society

**TheCapitol.Net supports the T.C. Williams Debate Society,
Scholarship Fund of Alexandria, and Sunlight Foundation**

TheCapitol.Net

Non-partisan training and publications that show how Washington works.™

PO Box 25706, Alexandria, VA 22313-5706   703-739-3790   www.TheCapitol.Net

www.ingramcontent.com/pod-product-compliance
Lightning Source LLC
Chambersburg PA
CBHW080849300326

41935CB00040B/1557